χ^2	chi-squared random variable
$\chi^2_{df,\alpha}$	value of a χ^2 statistic with df degrees of freedom associated with a tail area of α
F	F-statistic
$F_{df_1,df_2,\alpha}$	F-statistic with df_1 numerator degrees of freedom and df_2 denominator degrees of freedom associated with a tail area of α
$\ln$	natural logarithm
η	population median

Chapter 5

y_i	observed response at time i
y_{ij}	j^{th} observed response at time i
$\mu(t)$	true population mean at time t
ε_i	random error at time i
μ	true population mean
σ^2	true population variance
σ	true population standard deviation
n	subgroup sample size
$\bar{y}$	sample mean
$\bar{y}_i$	sample or subgroup mean at time i
θ	generic parameter
$\hat{\theta}$	estimator of the parameter θ
$\sigma_{\hat{\theta}}$	standard error for $\hat{\theta}$
R_i	sample range at time i
$\overline{R}$	average range over the base period
σ_R	standard error for the sample range
ρ	expected value for the sample range
d_n^*	a constant that relates ρ to σ_R
D_3	control chart constant for the lower control limit for an R-chart
D_4	control chart constant for the upper control limit for an R-chart
d_2	constant that relates the expected value of the sample range to σ
A_2	control chart constant for the control limits of an $\overline{X}$-chart based on the sample range
$\overline{\overline{y}}$	overall mean response for the base period
s_i^2	sample variance at time i
$\bar{s}^2$	average sample variance over the base period
MR_i	moving range at time i
$\overline{MR}$	average moving range over the base period
$\bar{p}$	average sample proportion over the base period
c_i	observed count at time i

$\bar{c}$	average observed count over the base period
λ	expected value for the count
S_i	cumulative sum statistic at time i
Z_i	EWMA statistic at time i
ϕ	EWMA weighting constant
m	number of base periods

Chapter 6

y_i	response for the ith observed value
x_i	i^{th} value for the regressor in simple linear regression
β_0	y-intercept
β_j	regression coefficient for the jth regressor
b_0	estimated y-intercept
b_j	estimated regression coefficient for the jth regressor
$\hat{y}_i$	predicted value for the ith response
e_i	residual associated with the ith response
ε_i	random error associated with the ith response
SS_{res}	sum of squared residuals
n	total number of residuals
$\bar{x}$	average value of the regressor (simple linear regression)
$\bar{y}$	average response
SS_{total}	sum of squares, total
SS_{reg}	sum of squares, regression
MS_{res}	mean squared residual
MS_{reg}	mean square for regression
σ^2	variance of the random errors
$\hat{\sigma}_{b_j}$	estimated standard error for the estimated regression coefficient for the jth regressor
df_{reg}	degrees of freedom associated with MS_{reg}
df_{res}	degrees of freedom associated with MS_{res}
R^2	coefficient of determination
R^2_{adj}	adjusted coefficient of determination
$\hat{y}(x_0)$	predicted value of the response at x_0
x_{ij}	i^{th} value for the jth regressor
$\mathbf{X}$	model matrix
k	number of regressors
$\mathbf{b}$	vector of estimated regression coefficients
$\mathbf{y}$	vector of responses
$\boldsymbol{\beta}$	vector of regression coefficients
$\hat{\mathbf{y}}$	vector of predicted responses

e	vector of residuals
H	hat matrix
h_{ii}	i^{th} diagonal element of the hat matrix
r_i	i^{th} value of R-student
$S_{(i)}$	estimated standard deviation without the ith data point
α	Type I error rate

Chapter 7

y_i	i^{th} observed value for the response
x_{ij}	i^{th} value for the jth design variable
β_0	y-intercept
β_j	regression coefficient associated with the jth design variable
b_0	estimated y-intercept
b_j	estimated regression coefficient associated with the jth design variable
ε_i	random error associated with the ith response
(0)	average response for the center runs
(1)	average response when all factors are at their low levels
(a)	average response with a high and all other factors low
(b)	average response with b high and all other factors low
(ab)	average response with a and b high and all other factors low

k	number of factors
$\hat{y}_i$	predicted response for the ith setting of the design variables

Chapter 8

y_i	i^{th} observed value for the response
x_{ij}	i^{th} value for the jth design variable
β_0	y-intercept
β_j	regression coefficient associated with the jth design variable
b_0	estimated y-intercept
b_j	estimated regression coefficient associated with the jth design variable
ε_i	random error associated with the ith response
k	number of factors
$\hat{y}_i$	predicted response for the ith setting of the design variables
α	distance from the design center for the axial runs
n_f	number of factorial runs
D	overall desirability
Y_L	lower bound desirable value
Y_U	upper bound desirable value
Y_T	fully desirable target value
z_j	j^{th} design variable associated with a noise factor

Statistical Methods for Engineers

DUXBURY

Statistical Methods for Engineers

Second Edition

Geoffrey Vining

Virginia Polytechnic Institute and State University

Scott Kowalski

Minitab, Inc.

THOMSON

™

BROOKS/COLE

Australia • Brazil • Canada • Mexico • Singapore • Spain
United Kingdom • United States

Statistical Methods for Engineers, Second Edition
Geoffrey Vining and Scott Kowalski

Senior Acquisitions Editor: Carolyn Crockett
Assistant Editor: Ann Day
Editorial Assistant: Daniel Geller
Technology Project Manager: Burke Taft
Marketing Manager: Stacy Best
Marketing Assistant: Jessica Bothwell
Project Manager, Editorial Production: Belinda Krohmer
Creative Director: Rob Hugel
Art Director: Lee Friedman
Print Buyer: Judy Inouye

Permissions Editor: Kiely Sisk
Production Service: Matrix Productions
Text Designer: Mike Stratton
Copy Editor: Chuck Cox
Cover Designer: Hiroko Chastain
Cover Image: Corbis
Cover Printer: Phoenix Color
Compositor: International Typesetting and Composition
Printer: RR Donnelley

Printed in the United States of America

3 4 5 6 7 09 08 07

For more information about our products, contact us at:
Thomson Learning Academic Resource Center
1-800-423-0563

For permission to use material from this text or product, submit a request online at
http://www.thomsonrights.com.

Any additional questions about permissions can be submitted by e-mail to
thomsonrights@thomson.com.

Library of Congress Control Number: 2005922090

Student
ISBN-13: 978-0-534-38473-9
ISBN-10: 0-534-38473-0

International Student Edition
ISBN-13: 978-0-495-01940-4
ISBN-10: 0-495-01940-2

Thomson Higher Education
10 Davis Drive
Belmont, CA 94002-3098
USA

To Christopher, David, and Catherine

To Kim and Regan

Contents

Preface

Purpose

Creating a text book for a one semester course in engineering statistics is a daunting task. It requires making intelligent decisions about presentation that balance depth and breadth. Scott and I both believe that the first edition was a very good basic presentation for a first course in engineering statistics. Our intent in the second edition is to build on that foundation. We have not changed the basic philosophy of presentation from the first edition. The second edition continues to emphasize the appropriate uses of graphics and the computer. It continues to focus on the proper application of statistics within the engineering method. We continue to insist on the use of real engineering data for the examples and the exercises. The essential structure and the theme of the book remains the same.

Innovative Aspects of This Text

Streamlining all of the material that should appear in a single semester of engineering statistics requires a delicate balancing act and many compromises. Applications in industry clamor for emphasis on control charts, regression analysis, and response surface methodology. Providing enough time to cover these topics in any detail requires an efficient presentation of the more traditional topics.

In this spirit, we do not present a detailed, separate discussion of histograms. Instead, we introduce the histogram when illustrating how to generate graphical displays using software. We present normal probability plots in Chapter 3 immediately after introducing the normal distribution, when students can better understand what these plots are. We introduce scatter plots in Chapter 6 when we begin our discussion of simple linear regression, where we can make the most efficient use of these plots.

Instead of introducing descriptive statistics such as the sample mean and the sample variance in either Chapter 1 or Chapter 2 and then allowing these statistics to lie dormant until the text discusses sampling distributions and formal estimation, we introduce these statistics when we are really ready to use them. As a result, they first appear in Chapter 3.

Rather than devoting separate chapters to probability, discrete distributions, continuous distributions, sampling distributions, confidence intervals, and hypothesis tests, the text combines all of this material in two chapters by concentrating on the essentials and their appropriate application.

This text introduces control charts immediately after the chapter on estimation and testing, and it shows how we can view a control chart as a sequence of hypothesis tests. In the process, we achieve two goals: First, we establish an important link between two statistical methods, and second, we lay the appropriate foundation for discussing average run lengths.

Practicing engineers need and use the 2^k factorial design, its fractions, and response surface methodology (RSM) far more than the more traditional analysis of variance (ANOVA) models. A one-semester course really does not provide sufficient time to devote to both the ANOVA and the RSM approaches. Given that fact, this text unabashedly chooses to pursue a modern RSM approach, because we truly believe that it is much more valuable to the practicing engineer.

New to the Second Edition

The most extensive changes occur in Chapters 1 and 7. Chapter 1 now includes all of the basic discussion on how to plan an experiment. As a result, students have enough background from the first week of the course to start a project involving a factorial experiment. This change required moving some of the discussion on experimental design from Chapter 7 to Chapter 1. Fifteen exercises were added to Chapter 1 to help students understand the concepts of experimental design. Also in Chapter 1, we have added a discussion on statistical thinking and structured problem solving methodologies. More and more companies are requiring their engineers to use sound, structured problem solving methodologies such as Six Sigma. We intend our discussion to introduce students to this need and to make them aware of the importance of engineering statistics.

We concluded that the original Chapter 7 was too long. Often, instructors could only cover approximately half the chapter, through fractional factorial designs, in a single semester. The new Chapter 7 focuses on the two-level factorial designs. The new Chapter 8 discusses response surface methodology. Chapter 8 continues to discuss the use of experimental design for quality improvement; however, we did cut back on the details of the Taguchi methods. It is important for engineers to appreciate that better methods exist for analyzing such experiments.

An important new feature is our running **Voice of Experience**, which appears in the margins of the text throughout the book. The Voice of Experience provides snippets from our experience about the proper application of statistics within engineering. The Voice of Experience highlights important concepts and reinforces basic principles.

The second edition has nearly 500 exercises, including new exercises. These exercises continue to emphasize real engineering data. Chapter 2 has a new section illustrating how we can use boxplots to analyze ANOVA type experiments. We moved the discussion of the distribution of sample variances to an expanded section 4.7, and we included the two-variance case. This section emphasizes that one should not do the two-sample test on the variances in conjunction with the two-independent sample test for the means. We have integrated the discussion of p-values throughout Chapter 4 rather than as a separate section. This change allows easier incorporation of statistical software, which tends to emphasize p-values in analysis. Also in Chapter 4, the discussion on transformations and nonparametrics has been expanded to form a new section, including exercises. Chapter 5 discusses Phase I and Phase II control charts. In Phase I control charts are closer to an exploratory data analysis tool. In Phase II control charts are better viewed as extensions of statistical inference. Also, Chapter 5 now introduces the exponentially weighted moving average (EWMA) control chart. Chapters 6, 7, and 8 now all conclude with case studies like the other chapters in the book.

Scope and Organization

In developing this text, we were guided by the statistical tools really needed by a practicing engineer. The first four chapters lay the essential foundations for understanding the more important material presented in Chapters 5 through 7.

Chapter 1 introduces the engineering method and the proper role of statistics within it. It exposes the student to probabilistic models, data collection, and sampling, and it introduces experimental design.

Chapter 2 outlines simple graphical tools for data analysis. It emphasizes the stem-and-leaf display and the boxplot because they are easy to generate by hand. The book develops these techniques in sufficient detail that the instructor can spend less time lecturing on these displays and more time on interpreting them. This chapter also shows how to generate these plots, as well as histograms and timeplots, using appropriate software. Engineering students find stem-and-leaf displays much easier to construct by hand than histograms. However, appropriate use of the computer makes either tool equally easy to use, which is why we discuss the histogram in the section dedicated to the use of software. Throughout this chapter graphical tools are use to analyze real engineering data.

Chapter 3 develops modeling of random behavior through probability distributions. We use as little formal probability as possible, moving quickly to the major distributions used in probability and reliability. This chapter introduces the normal probability plot as a means for checking the usual normality assumptions required by classical statistics.

Chapter 4 covers basic estimation and testing. It honestly strives to let the instructor emphasize either confidence intervals or hypothesis tests, depending on his or her philosophical slant. The chapter makes extensive use of stem-and-leaf displays, boxplots, and normal probability plots to check underlying assumptions.

With Chapter 5, the true meat of the book begins. Here we show how good statistical thinking can convert an essentially enumerative tool—hypothesis testing—into a powerful analytic tool: control charts. This chapter follows fairly faithfully the spirit and the tenets of the peer review literature, which often views control charts as a sequence of hypothesis tests. This chapter presents such concepts as average run length, runs rules, and the CUSUM chart.

Chapter 6 introduces the student to linear models via regression analysis and emphasizes the use of appropriate software. It discusses in detail proper residual analysis, detection of leverage and influential points, multicollinearity, and transformations.

Chapter 7 presents the most important material in response surface methodology. It begins with a basic overview of experimental design. It then develops in detail the two-level factorial designs most commonly used in industry. The chapter moves on to process optimization through sequential experimentation. Students learn how to generate a path of steepest ascent and to construct central composite designs. This chapter teaches how to analyze multiple responses using commonly available software. It concludes with a presentation of robust parameter design and its proper analysis within RSM.

Suggested Coverage

Originally, we had hoped to produce a true one-semester course in engineering statistics, but "modern engineering statistics" means different things to different people. Instructors who teach a one-semester course have many options. Those who wish to emphasize industrial statistics will cover Chapter 1 and then carefully selected topics from Chapters 2 through 4. They then will cover in detail the vast majority of Chapters 5 through 7. Other instructors may want to cover Chapters 1 through 4 in more detail and then pursue carefully selected topics from Chapters 5 through 7. Experience indicates that the text works well with both approaches.

Instructors can comfortably cover the entire book in two quarters. We recommend Chapters 1 through 4 plus the first half of Chapter 5 for the first quarter and the remainder of the book in the second.

Case Studies and Student Projects

The case studies illustrate how statistical techniques can be applied to improve production. These case studies provide more examples of how engineers use statistics to analyze real data.

The chapters also include ideas for small student projects that complement homework assignments. We strongly believe in term-long projects, over and beyond these smaller activities. Typically, engineers are asked to plan, carry out, and analyze a factorial experiment of their own choosing. Chapter 1 provides enough details for students to begin thinking about their projects. By the end of the term, when they have conducted their experiments and collected their data, we develop in lecture the material required for them to analyze the results.

Examples and the Exercises

We aim to present each statistical method within the context of a real engineering problem. The student sees how statistics fits within engineering problem solving.

In the process, we try to teach engineering as much as statistics through the examples. The overwhelming majority of examples and exercises involve real engineering data that either appeared in appropriate journals or are based on consulting experience. To the extent possible, we try to give the full engineering context of both the examples and the exercises. The exercises emphasize good data analysis within specific engineering settings.

Use of Graphics

Modern statistical analysis requires extensive and thoughtful use of graphics. This text makes extensive use of graphical analysis in every chapter after the introduction. Chapter 2 illustrates how the appropriate use of graphics can provide significant insights into the analysis of engineering data. Chapter 3 introduces the normal probability plot and shows how we can use it to check the common normality assumption. Chapter 4 illustrates how we can use the stem-and-leaf display, the boxplot, and the normal probability plot to check the assumptions required for formal statistical inference. Chapter 5 not only introduces the control chart, a graphical tool in its own right, but it also illustrates how we can use the stem-and-leaf display to determine if the subgroup size seems appropriate for the process being monitored. Chapters 6 and 7 make extensive use of residual plots to confirm if the underlying assumptions for our analyses are reasonable.

Use of the Computer

Modern engineering statistics must focus on sound data analysis, which of necessity must incorporate extensive use of the computer. This text assumes that students have ready access to at least one major statistical software package, such as Minitab, Splus, or SAS, used extensively in developing this text, but the students can use any good package to work the exercises. We encourage students to use the computer to do all but very simple analyses and insist that students use the software to generate stem-and-leaf displays, boxplots, and normal probability plots to check their assumptions even if they perform the test or calculate the interval by hand. Students use the computer extensively in Chapter 5 (control charts) to process large data sets and to produce quality plots, Chapter 6 (regression analysis), and Chapter 7 (response surface methodology) to perform the necessary calculations and to generate appropriate residual plots.

Acknowledgments

We express my sincere and humble appreciation to all those who have contributed in one way or another to this book, especially Dr. Raymond H. Myers and Dr. Douglas C. Montgomery and the Department of Statistics at the University of Florida, particularly from Dr. Ronald H. Randles. We greatly appreciate support, guidance, and advice of Carolyn Crockett, whose constant involvement in this project greatly enhanced the final product. We thank all of the people at Duxbury who have contributed, especially Belinda Krohmer. We thank Minitab Inc. for their continued support. We also thank all of the reviewers, who went above and beyond the call of duty in suggesting improvements. Thanks are due to:

John E. Boyer, Kansas State University
Rick Cleary, Bentley College
Richard Deveaux, Williams College
Conrad A. Fung, University of Wisconsin
James Halarin, Rochester Institute of Technology
Larry Haugh, University of Vermont
Peter R. Nelson, Clemson University
John Ramberg, University of Arizona
David Ruppert, Cornell University
Carl Sorenson, Brigham Young University
John Spurrier, University of South Carolina
Roy Welsch, Massachusetts Institute of Technology

Last, but certainly not least, Geoff thanks his children, Christopher, David, and Catherine. Too often, they paid a price when he had to meet important deadlines. He greatly appreciates their love and patience, without which he could never have written this book. Scott thanks his wife Kim and daughter Regan for providing the inspiration to write this book.

G. Geoffrey Vining
Scott M. Kowalski

Overture: Engineering Method and Data Collection

1.1 Need for Statistical Methods in Engineering

The Role of Statistics

To appreciate the role of statistics in engineering, we first must understand that engineers contribute to society through their ability to apply basic scientific principles to real problems in an efficient and effective manner. One of our fathers, a mechanical engineer, describes an engineer as "someone who can build something for one dollar that any fool could build for two." We can achieve this goal only by studying and understanding the physical world around us. Engineers use as their basic approach the *scientific* or *engineering method* whereby models are developed to explain real phenomena. Engineers then must collect data to test these basic models. *Model building, data collection, data analysis*, and *data interpretation* form the very core of sound engineering practice.

Statistical methodologies are now considered vital components in engineering curricula, yet even more important to the engineer is the ability to think "statistically." Engineers must learn to dig deeper into data, understanding fundamental concepts such as *variability, correlation, uncertainty*, and *risk in the face of uncertainty*.

Global competition is encouraging this focus on statistics. General Motors must be as concerned about Toyota and Nissan as it has been about Ford. In a highly competitive environment, continuous improvement is essential for survival. Many successful companies have shown that the first step to continuous improvement is to integrate the widespread use of statistics and basic data analysis into business operations. Engineers are learning that the next step is *well-planned experimentation*, which can efficiently and effectively improve products and processes.

Example 1.1	**Taking a New Polymer from the Lab to Production**

Figure 1.1 illustrates the role of statistics in engineering. Consider a new polymer that has been developed in a research laboratory. The next step toward full production of this polymer is to set up a pilot plant operation. In the laboratory, the chemist developed the polymer under relatively pristine conditions. The pilot plant represents the first attempt to produce this polymer under realistic manufacturing conditions with impure raw materials. The engineer must determine the appropriate operating conditions for this process—that is, what temperatures, pressures, flow rates, and other factors produce "good" yields of this polymer. The chemist has given the engineer, at best, some rough idea of where "good" operating conditions should lie. The engineer now needs to plan an efficient experiment that will suggest appropriate settings for the process that produce good yields (an important statistical technique called *experimental design*). The determination of these settings requires the use of a model that explains how the "factors"—temperature, pressure, and flow rate—affect the yields. The basic purpose of the experiment is to estimate this model, which is accomplished by another statistical technique, *regression analysis*. The engineer also needs to decide which of these factors are important to the yield, which is done by *hypothesis testing*. The engineer next must determine the most appropriate settings for the important factors, *process optimization*. After the appropriate operating conditions are specified, the engineer needs to take measures to ensure that the process produces a consistent yield, which is often accomplished with *control charts*.

Figure 1.1	Taking a New Polymer from the Laboratory to Production

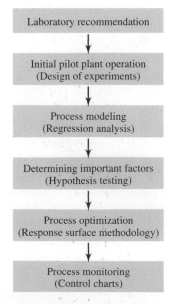

Engineering Statistics: A Symphony

Modern engineering statistics may be viewed as a symphony. Techniques such as those we have mentioned are the individual parts. Viewed in isolation, these parts may appear strange and abstract. Integrated, however, they form a single harmonic whole. Some of the methodologies we shall develop are powerful tools in their own right, soloists if you will. Some are necessary precursors to other valuable techniques. Although all parts are important in their own way, of even greater importance is how the parts are woven together. In order for the orchestra to reach its full potential, the skills of the individual players must be guided by a vision. The members of the orchestra must understand not only their own specific instruments but how their instruments blend together to create the piece. Thus, it is not enough to learn basic statistical methodologies in isolation from each other. Of even greater importance is *statistical thinking,* which provides the fundamental vision for data analysis and an understanding of how all these techniques relate to one another. The end product is a powerful approach for analyzing data to solve real engineering problems.

1.2 Engineering Method and Statistical Thinking

The Engineering Method

The scientific or engineering method is the fundamental approach for solving engineering problems and consists of these basic steps:

1. Clearly define a concrete problem.
2. Postulate the important factors that influence this problem.
3. Formulate a working model for the underlying mechanism.
4. Collect data concerning the problem (conduct an appropriate experiment).
5. Estimate the working model.
6. Determine the important factors and test the adequacy of the model.
7. Revise the working model as appropriate.
8. Collect additional data (conduct a confirmatory experiment).
9. If the model does not lead to a solution, return to step 2.

Engineering problems always start out as concrete problems in the physical universe. Their solutions, however, require abstraction. Thus, the engineering method involves a constant interplay between the concrete and the abstract. Example 1.2 and Figure 1.2 illustrate how the engineering method works. In the process, we can see the constant interplay between the engineer and the "statistician." By statistician, we mean someone well trained in data collection, data analysis, and data interpretation. The statistician can be the initial engineer, another engineer, or a degreed statistician.

Figure 1.2 | The Engineering Method

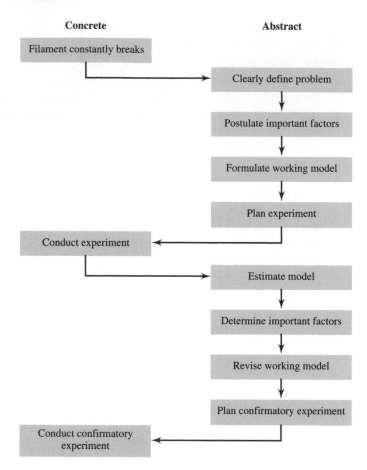

Example 1.2 | **Strength of a Polymer Filament**

Consider a production process of a new polymer filament that is later spun into yarn. At the startup of this new process, the filament constantly breaks when spun (*a concrete problem*). In this case, the problem is fairly well defined: The filament's elastic strength (*S*) is inadequate. A basic knowledge of polymer chemistry suggests that the filament's strength depends on the amount of catalyst (*C*) used, the polymerization temperature (*T*), and the polymerization pressure (*P*). The engineer thus is able to *postulate important factors*. The engineer then *formulates a working model* of the form

$$S = f(C, T, P)$$

where f is typically an unknown function. The statistician suggests using a first-order Taylor series approximation to produce this model:

$$S = \beta_0 + \beta_1 C + \beta_2 T + \beta_3 P$$

where β_0, β_1, β_2, and β_3 are all constants. This model represents an abstraction and serves as a reasonable basis for approximating the behavior of the "response" (strength) in terms of the "factors" (catalyst, temperature, and pressure). With a good model, the engineer can determine appropriate settings for the three factors that optimize the strength. The model is useful only to the extent that it actually explains the behavior of the filament's strength. The next step is to determine the adequacy of this approximation. The statistician must *plan the experiment* that should determine the exact impact of the three factors on the filament's strength. At this stage, the statistician develops a formal strategy that systematically changes the amount of catalyst, the polymerization temperature, and the polymerization pressure. The engineer then *conducts the experiment*. For each combination of these factors, the engineer operates the process, allows it to reach equilibrium, and takes a sample of filaments and measures their strengths. The statistician takes the experimental results and *estimates the model*, which means estimating the coefficients, the β's. The estimated model gives the statistician an appropriate basis for *determining the important factors*. In particular, the statistician will determine which of the estimated coefficients is really different from zero. Why is this important? Suppose the coefficient associated with the amount of catalyst is zero. In this situation, no matter what we do with the amount of the catalyst, it has no effect on the filament's strength. The results of this statistical analysis may suggest that the engineer *revise the working model*. The statistician then *plans a confirmatory experiment* to determine the appropriateness of the revised model. The engineer next *conducts the confirmatory experiment,* which the statistician analyzes. The engineering method continues in this manner until the engineer finds an adequate solution—that is, the settings for the amount of catalyst, temperature, and pressure that produce the optimum strength.

The Problem of Variability

Data lie at the very heart of the engineering method. Unfortunately, real data exhibit *variability,* which obscures our ability to make sound decisions.

Example 1.3 | **Outside Diameters of Pen Barrels**

In the manufacture of pens, each pen barrel is designed with a specific critical outside diameter, which determines the quality of the seal between the cap and the barrel. If the outside diameter of the barrel is too small, then either or both of these results occur:

- Air leaks may develop, which ultimately will dry out the pen and make it unusable.
- The cap may fall off, which can cause severe problems in the high-speed packaging equipment.

On the other hand, if the outside diameter is too large, then either or both of these results occur:

- The cap can be extremely difficult to remove, which upsets the final customer.
- The pen assembly machine may jam as it places the cap on the barrel.

For this particular product, the target for the critical outside diameter is 0.190 inch. Can we realistically expect each barrel produced by this process to have exactly the same critical outside diameter? Of course not! These barrels are made in "shots" of 64; that is, the die tool consists of 64 individual molds. Although these individual molds are precisely drilled, some differences exist among them. The melted plastic does not flow perfectly uniformly into the individual molds, which creates additional differences. Also the barrels shrink as they cool. Thus, temperature gradients across the die tool as the barrels cool contribute to the variability. The following list gives the critical outside diameters, in inches, from a production sample:

0.194	0.187	0.198	0.188
0.191	0.190	0.185	0.205
0.180	0.191	0.182	0.191

Several important questions arise: Is the "typical" outside diameter acceptable? What is the proportion of unacceptable outside diameters? How can we know if these outside diameters are becoming too large or too small?

Traditionally, engineers have responded to variability by creating *specification limits*. In this case, the specification limits on the critical outside diameter are 0.180–0.200 inch. In this light, does our sample indicate acceptable production? One value exceeds the specifications. Should a single value dictate whether an hour's worth of production is acceptable or not? Several other diameters are quite near the limits, both to the high and to the low side. Even if we ignore the single value outside the specifications, is there evidence to suggest that the critical diameter is drifting too large or too small? Specification limits provide some guidance, but they do not solve the fundamental problem of how to make a decision in the face of variability.

Thinking Statistically

Only by "thinking statistically" can engineers truly address the problems inherent in the variability in real data. When we think statistically, we come to know that all decisions based on real data involve risk and uncertainty. Good decisions require us to quantify this risk. As we become more mature in our thinking, we understand that there are sources or causes of variability. Discovering these sources and removing them are often the keys to engineering success.

Example 1.4 | **Operators and an Injection Molding Process**

Consider an injection molding process for pen barrels. The goal is to produce pen barrels with as uniform an outside diameter as possible. Many operators, those who have yet to learn how to think statistically, always set the machine to their favorite settings when they start the shift. An operator claims, "The machine runs best on these settings." Of course, each operator has a different idea of which settings run best. Because the equipment operates three shifts a day, the resulting outside diameters tend to vary quite a bit from shift to shift. Each time an operator shifts the settings, he or she is introducing additional variability; the operator is a source of unwanted variability.

In a better approach, which reflects statistical thinking, an operator looks at a sample of the production at the start of the shift. If this sample indicates a problem with the outside diameters, then the equipment settings are changed according to a well-estimated model. Otherwise, the settings are left alone. Operators "tamper" with the process only when there is a documented need, thus keeping variability to a minimum.

VOICE OF EXPERIENCE

Statistically grounded, structured problem-solving methodologies are essential to the practicing engineer.

⠶ **1.3** Statistical Thinking and Structured Problem Solving

According to the Quality and Productivity Section of the American Statistical Association, "*statistical thinking* is a philosophy of learning and making decisions based on three fundamental principles: All work occurs in a system of interconnected processes, variation exists in all processes, and understanding and reducing variation are keys to success." Clearly, statistical thinking goes far beyond the use of basic statistical analysis. Instead, it is a different way to view the world. Statistical thinking is a big-picture philosophy that recognizes that real-life processes involve systems and variability. Understanding any real-life process involves seeing the various stages in the system and the resulting sources of variation. Process improvement generally comes from simplifying the system and removing unwanted sources of variation.

Statisticians, by the nature of their profession, are generalists. Often, the biggest contribution a statistician makes on any team is getting other people to see clearly the nature of the system that forms the process of interest and to understand the causes of variation. Usually, the steps then needed to make improvements become very obvious. Statisticians, by themselves, rarely can improve a process or product. It takes a solid appreciation of the basic science to make fundamental improvements. Statisticians, however, are often the catalysts to improvement, without whom the improvements would never be made.

An important key to constant product and process improvement is the use of structured problem-solving methodologies. A very popular contemporary approach is "Six Sigma," which was originally developed at Motorola in the 1980s and was made popular by Jack Welch during the 1990s when he was CEO of General Electric. Six Sigma is a heavily statistical methodology within a team context. The core of the Six Sigma approach is the DMAIC (Define, Measure, Analyze, Improve, Control) cycle, which is an extension of Shewhart's Plan, Do, Check, Act cycle. The Shewhart cycle is nothing more than a formal, structured implementation of the basic engineering method.

In the define stage, upper management selects an appropriate cross-functional team and defines the team's mission. The team's first step is to clearly define the problem. Often, the team's success depends heavily upon how well it can clarify and define the problem.

In the measure stage, the team selects the important characteristics to study. It then defines the appropriate performance standards for these characteristics. An important aspect of this stage is to validate the measurement systems. It is difficult to improve a characteristic when it is poorly measured. The measurement systems need to be accurate, precise, and consistent. Statistics, particularly gauge repeatability and reproducibility studies, are essential at this stage.

The team begins the analyze stage once it completes the measure phase. The team establishes the product's and/or process's capability. It defines clear and specific performance objectives. The team also identifies potential sources of variation. Statistical tools are less important in this stage than sound statistical thinking.

Next, the team enters the improve phase. The team now screens potential causes through well-planned experiments. Through these experiments, the team discovers the relationships among the various variables and the characteristics of interest. Once these relationships are established, the team establishes operating tolerances. Central to the improve phase are regression analysis and design of experiments. We discuss the details of regression analysis in Chapter 6 and how to plan engineering experiments in this chapter and in more detail in Chapters 7 and 8.

The final phase of the DMAIC cycle is the control stage. The team institutes procedures to maintain the improvements. Control charts, which we discuss in Chapter 5, are essential for ensuring the maintenance of the improvements.

Six Sigma is an upper-management-driven, structured problem-solving methodology that has saved many corporations billions of dollars. These corporations have spent millions of dollars training virtually everyone in the company in basic statistical methods and in statistical thinking. The proper application of statistics by everyone in the company is essential to the success of this program. Even more important, though, is statistical thinking, which is at the core of the Six Sigma culture. In the future, companies will continue to use structured process-improvement strategies, well founded in statistics, in order to compete and to survive in the global market.

⋗ 1.4 Models

Deterministic Models

Traditionally, engineers have been taught to think in terms of *deterministic* models, which make no attempt to explain variability. In many engineers' minds, variability is the result either of factors that remain unaccounted for or of basic limitations in our ability to measure. Two classical examples are the ideal gas law and Ohm's law. Let P be a gas's pressure, let V be its volume, let T be its temperature, and let n be the "amount" of the gas present (technically, the number of moles). The ideal gas law then states

$$PV = nRT$$

where R is a constant. If this law is in fact true and if we know the volume and the temperature, then the pressure is already *exactly determined*. There is no real need to measure the pressure because knowledge of the volume and the temperature is the appropriate basis for determining it. Ohm's law states

$$V = IR$$

where V is now the voltage, I is the current, and R is the resistance. These laws are classical cases of abstraction and of deterministic models.

Deterministic models are almost always oversimplifications of reality. The ideal gas law has proven very useful over the years, but we must recognize that no truly "ideal" gases exist in reality. A common electrical engineering laboratory experiment involves setting up a simple circuit with known resistance and, for a given current, measuring the voltage. If 20 students conduct exactly the same experiment, using the same resistance and current, we should expect to see 20 different voltages, all close to the value predicted by Ohm's law. Does such a result invalidate the basic principle that voltage, current, and resistance are all closely related? Not at all; rather, it implies that we need a better model to account for the variation we should inherently expect in nature.

Probabilistic or Statistical Models

A better approach uses a *probabilistic* or *statistical model*, which formally takes into account the random variations from the values given by the deterministic model. We can improve Ohm's law by using

$$V = IR + \epsilon$$

where ϵ is a *random error*. In this specific case, it is probably best to view ϵ as a measurement error. In other cases, ϵ represents more fundamental and generally unpredictable deviations from the deterministic model.

Models provide an appropriate basis for predicting the physical universe. From a statistician's perspective, models describe *populations*.

Definition 1.1	Population

A *population* is the set of all possible observations of interest to the problem at hand.

Most engineering data come from processes that change over time. Within this context, the population corresponds to the state of the process at a specific interval in time. Examples of populations include:

- The measured outside diameters of a specific pen barrel from an injection molding process
- The elastic strengths of polymer yarns spun on two different machines
- The strengths of a polymer filament from a new chemical process

Example 1.5	**Model for Outside Diameters**

Consider the outside diameters of the pen barrels produced by an injection molding process. Let y_i be the observed outside diameter for the ith pen barrel inspected. Let μ represent the true average or *mean* outside diameter for this process. The simplest deterministic model would claim that each pen barrel has exactly this outside diameter. A more realistic model adds a term to account for the deviations of the individual outside diameters from the mean. Let ϵ_i represent the random error associated with the ith pen barrel. The resulting model is then

$$y_i = \mu + \epsilon_i.$$

In this case, ϵ_i represents much more than some error in measurement because each pen barrel's true outside diameter is different. We shall use this model extensively when we discuss single-sample inference in Chapter 4 and when we discuss basic control charts in Chapter 5.

Example 1.6	**Model for Yarn Strength**

Consider the strengths of the polymer yarns produced by two different spinning machines. Strengths tend to be fairly variable. A relatively few test specimens will contain small imperfections that cause premature failures. Slight differences in how the yarn was spun also contribute to variability in the strengths. The differences in these strengths do not follow any perceptible pattern; thus, we may treat them as random. Let y_{ij} be the strength of the jth sample yarn spun on the ith machine, where $i = 1$ or 2. An appropriate model must take into account that the inherent variability in yarn strength and the mean strength may be different for each machine. Thus, consider as our model

$$y_{ij} = \mu_i + \epsilon_{ij}$$

where μ_i represents the true mean strength of the yarn spun on the ith machine and ϵ_{ij} represents the random error associated with the jth sample yarn from the ith machine. We shall use this model when we discuss two-sample inference in Chapter 4 and when we begin to discuss the formal analysis of designed experiments in Chapter 7.

Example 1.7	**Model for Filament Strength**

Consider the strength of a polymer filament produced by the process outlined in Example 1.2 where we proposed

$$S = \beta_0 + \beta_1 C + \beta_2 T + \beta_3 P$$

to describe the relationship of the filament's strength (S) and the amount of catalyst (C), the polymerization temperature (T), and the polymerization pressure (P). We now should realize that this equation is a deterministic model, which allows neither for random variations from the expected value nor for the fact that this model is only an approximation subject to additional errors. Let y_i be the strength of the test filament from the ith batch produced of the polymer. A better model adds a term to account for both sources of error. Thus, consider the model

$$y_i = \beta_0 + \beta_1 C + \beta_2 T + \beta_3 P + \epsilon_i$$

where β_0, β_1, β_2, and β_3 are all constants and ϵ_i is the appropriate error term. In Chapter 6, we shall see that this is a specific example of a regression model.

▷ 1.5 Obtaining Data

The engineering method requires data to solve real problems. Too often, engineers assume that they should use statistics only after they have collected their data. Instead, statistics should play a pivotal role in the engineering method from the very beginning of the data-collection effort. Any statistical analysis is only as good as the data upon which it is based. Statistical thinking and statistical methodology can make the data-collection process as efficient and effective as possible.

Proper data collection requires a significant dialogue between the engineer and the statistician. This dialogue should result in a well-defined problem and at least a reasonable sketch for its solution. Without proper prior planning, the data may provide no useful information for the problem at hand. Even worse, improperly collected data can provide deep insights into the wrong problem! The ability of the engineer and the statistician to communicate often determines the ultimate success of the data-collection effort.

Issues in Data Collection

Important issues in the data-collection process include:

- The fundamental purpose for collecting the data
- The characteristic or characteristics of interest
- The presumed engineering model
- The parameters of interest to the study
- The physical constraints, if any, on the actual data collection

We always collect data for a specific purpose. The better we understand this purpose, the better we can plan the data-collection process. After we clearly understand why we are collecting data, we then need to define clearly what are the characteristics of interest and how we can measure these characteristics. These are important engineering issues. Next, we need to propose an engineering model to adequately describe the problem at hand. The model defines for us the parameters of interest, which in turn provide the keys for solving the problem. Finally, we need to consider what physical constraints could impede our ability to collect data. In many cases, these physical constraints force us to modify the data-collection procedure, which may require some modification to the engineering model.

Example 1.8	**Injection Molding of Pen Barrels**

An important quality characteristic for a pen is the fit between the barrel and the cap. Since separate injection molding processes produce the barrels and the caps, we control the fit by making each outside diameter of the barrel and each inside diameter of the cap as close to specified target values as possible. In this situation, a well-defined engineering problem is to keep the outside diameter of the pen barrel as close to its target value as possible. The characteristic of interest is the outside diameter of the pen barrel at the spot where it should seal with the pen cap.

Let y_i be this critical outside diameter for the ith pen barrel produced by this process. The simplest statistical model that approximates an engineering model is

$$y_i = \mu + \epsilon_i$$

where μ is the true average critical outside diameter for this process at the given time and ϵ_i is a random error associated with this particular pen barrel. Since we wish to make each barrel with a critical outside diameter as close to its target as possible, we must ensure that the typical outside diameter, μ, stays at the target value and that the random variability around this typical value is as small as possible. Thus, for this model, the two parameters of interest are μ and some measure of the random variation.

Our particular injection molding process produces "shots" of 64 barrels approximately every 90 seconds. The shot of 64 barrels drops directly into a box, which the operator then transfers to a larger container. As a result, we cannot determine directly which specific mold produced a specific pen barrel, and we face a physical constraint on how we can collect the data. If we have no reason to believe that the individual molds differ significantly from one another, then we probably should take samples of four or five barrels on a periodic basis. On the other hand, if we suspect problems with one or two of the specific molds, then we should periodically select entire shots and inspect each barrel.

Importance of Pairing Data

Often, the sampling units available for a study differ widely. The inherent variability among these units obscures our ability to study the engineering process. *Pairing* allows us to remove the sampling unit to sampling unit variability and to focus on the real issues at hand.

Example 1.9	**Testing Octane Blends—Paired Data**

Snee (1981) conducted an experiment to compare two different methods for measuring the octane rating of gasoline blends. In practice, petroleum engineers need to measure precisely the octane ratings for blends over a wide range of values.

Engineers often encounter situations where the units available for testing differ greatly. In this case, we know that gasoline blends differ dramatically among themselves and that these differences are not due to the testing method. This inherent blend-to-blend variability can obscure our ability to see whether the two test methods really agree with each other unless we can take this variability into account. For example, Snee had 32 different gasoline blends available for his study. One way he could have run his experiment was to randomly allocate 16 blends to test method I and the other 16 blends to test method II. He then could compare the average octane rating for the blends tested by method I to the average octane rating for the blends tested by method II.

A problem with this approach is that the differences in the true octane ratings among the blends increase the variability in the observed octane ratings for each test method. For example, if we compare a specific blend tested by method I to one tested by method II, we do not know if the difference in the octane rating is due to the difference in the blends or the difference in the test methods. By randomly allocating the blends to the test methods, we hope to balance the differences among the blends so that they "average out." We pay a price, however, with an increase in the variability among the observed octane ratings for each test method purely because of the inherent gasoline blend-to-blend variability.

A better approach to this situation *splits* each gasoline blend into two test samples. Snee then could randomly allocate one test sample to test method I and the other test sample to test method II. He then could look at the observed *difference* in the two observed octane ratings. Let y_{i1} be the observed octane rating for the ith gasoline blend when tested by method I, and let y_{i2} be the observed octane rating for the same blend when tested by method II. Snee could

VOICE OF EXPERIENCE

Blocking often improves statistical analyses.

Figure 1.3 | Pairing of Gasoline Blends

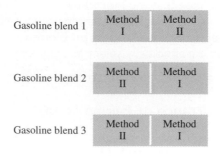

---

define the difference, d_i, for the ith blend by $d_i = y_{i1} - y_{i2}$. If $d_i > 0$, then test method I gives a higher octane rating than test method II for the ith blend. Figure 1.3 illustrates the design. By focusing on the blend-to-blend differences, Snee removed the effect of the true octane ratings from the analysis. If the ith gasoline blend has a higher octane rating, then both test methods should yield high observed octane ratings. If one test method indicates a high octane rating but the other does not, then Snee had evidence that the two methods do not agree.

Splitting each gasoline blend into two test samples is an example of *pairing*. Whenever possible, experimenters should use this concept of pairing to remove unwanted or extraneous sources of variability. Pairing data in this manner is the simplest example of the statistical principle called *blocking*.

Methods for Collecting Data

Three basic methods for collecting data are:

1. A retrospective study based on historical data
2. An observational study
3. A designed experiment

The next example illustrates these three methods.

Example 1.10 | **Strength of a Polymer Filament—Revisited**

In Example 1.2 we discussed a problem with the strength of a polymer filament that depended on the amount of catalyst (C), the polymerization temperature (T), and the polymerization pressure (P). Figure 1.4 illustrates this situation. The strengths of the polymer filaments produced by this process form the population of interest. For this process, production maintains and archives these records:

*   The average strength of a sample of filaments taken from every batch
*   The temperature controller log, which is a plot of the reactor temperature

Figure 1.4 | The Polymer Filament Process

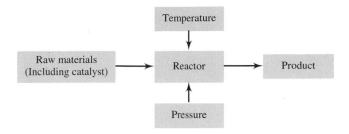

- The pressure controller log
- The approximate start and stop times for each batch

Supervision maintains strict controls on the weigh-out of the raw materials and therefore does not require a separate log of the specific weight of the catalyst used in each batch. Supervision *assumes* that each batch has the same amount of catalyst.

Retrospective Study   We could pursue a *retrospective study,* which would use either all or a sample of the historical process data over some period of time to determine the impact among *C, T,* and *P* on the strength of the filament. In so doing, we take advantage of previously collected data and minimize the cost of the study. We must note several problems, however:

1. We really cannot see the effect of *C* on the strength because we must assume that *C* did not vary over the historical period.
2. The data relating *T* and *P* with the filament strengths do not correspond directly. Constructing an approximate correspondence usually requires a great deal of effort.
3. Production maintains both *T* and *P* as close as possible to specific target values through the use of automatic controllers. Since *T* and *P* vary so little over time, we have a great deal of difficulty seeing their real impact on the strength.
4. Within the narrow confines that they do vary, *T* tends to increase with *P.* As a result, we have a great deal of difficulty separating out the effects of *T* and *P.*

Retrospective studies often offer limited amounts of useful information. In general, their primary disadvantages are:

- The nature of the data often does not allow us to address the problem at hand.
- Some of the relevant data often are missing.
- Logs, notebooks, and memories may not explain interesting phenomena identified by the data analysis.

Observational Study   We could use an *observational study* to collect data for this problem. As the name implies, an observational study simply observes the process or population. We interact or disturb the process only as much as is required to obtain relevant data. With proper planning, these studies can ensure accurate, complete, and reliable data. On the other hand, these studies often provide very limited information about specific *relationships* among the data.

In this example, we would set up a data-collection form that allows production personnel to record the actual amount of catalyst, the average polymerization temperature, and the average polymerization pressure. Such a procedure ensures accurate data collection and takes care of problems 1 and 2 described for retrospective studies. Unfortunately, an observational study cannot address problems 3 and 4.

Designed Experiment   The best data-collection strategy for this problem uses a *designed experiment* where we manipulate C, T, and P, which we call the *factors*, according to a well-defined strategy, called the *experimental design*. This strategy must ensure that we can separate out the *effects* of each factor on the filament strength. The specified values of the factors used in the experiment are called the *levels*. Typically, we use a small number of levels for each factor, such as two or three. For the polymer example, suppose we use a "high" or $+1$ and a "low" or -1 level for each of the factors C, T, and P. We thus use two levels for each of the three factors. A *treatment combination* is a specific combination of the levels of each factor. Each time we carry out a treatment combination is an experimental *run* or *setting*. The experimental *design* or *plan* consists of a series of runs.

In general, factors are either *categorical* or *continuous*. Examples of categorical factors are suppliers, operators, and types of polymer. With categorical factors, the levels represent distinct units. For example, we can construct an experiment using supplier 1, supplier 2, and supplier 3 where we arbitrarily designate the actual suppliers to these specific levels. Since we arbitrarily designate the available suppliers to these levels, supplier $1^1/_2$ really makes no sense.

Examples of continuous factors are pressure, temperature, and amount of carbon. With continuous factors, we choose specific levels from a continuum. For example, we can construct an experiment using pressures, with 2 atm as the low or -1 level and 4 atm as the high or $+1$ level. In this case, a level of 0 makes perfect sense because it corresponds to a pressure of 3 atm, which falls halfway between 2 and 4 atm.

The choice of the levels for a factor is very important. Suppose the response is linear between the two levels of a factor. If the slope of this line is steep, then it is likely that even choosing the levels of the factor fairly close together will result in identifying this factor as significant. However, if the slope is gradual, then the levels must be far apart to allow a sizable change in the response.

For the polymer example, a very reasonable experimental strategy uses every possible treatment combination to form a base experiment with eight different settings for the process. The following table outlines these combinations of high and low levels:

C	T	P
−1	−1	−1
+1	−1	−1
−1	+1	−1
+1	+1	−1
−1	−1	+1
+1	−1	+1
−1	+1	+1
+1	+1	+1

Figure 1.5 illustrates that this design forms a cube with these high and low levels. With each setting of the process conditions, we allow the process to reach equilibrium, produce a "batch" of the polymer, take a sample of filaments, and determine their strengths. In this case, we might call the resulting average of the strengths the *response*. We then can draw specific inferences about the effect of these factors. Such an approach allows us to proactively study a population or process.

Properly planned experiments provide a wealth of information for a minimal number of settings. Well-planned experiments also tend to produce unusable product. For example, a properly planned experiment should use some combinations of the high and low levels of C, T, and P that produce filaments that fail to meet the product's specifications. A well-planned experiment should produce some batches with extremely strong filaments, perhaps even too strong for the specific product. The experiment should also produce batches with extremely weak filaments. In either case, production cannot use the product and thus rejects these batches. However, these rejected batches often provide the

---

Figure 1.5 | The Experimental Design for the Polymer Study

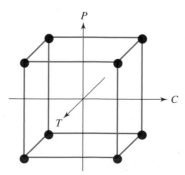

crucial information on the effect of the factors, in this case C, T, and P, on the response. Enlightened technical managers recognize the need to balance the loss of usable product against the long-term benefit from the information gained by the experiment. Engineers often use designed experiments in process development, improvement, and optimization, particularly during the startup of new processes.

Some engineers, particularly older ones, do not use statistically based, planned experiments. These engineers complain that designed experiments require too many runs, cost too much, and take too long. In some cases, they complain that the designed experiment takes away their "creativity". Instead, they prefer to "tweak" the process until they achieve acceptable results.

These engineers fail to realize that every time they tweak the process, they are conducting an experimental run. As a consequence, they often conduct far more experimental runs than a well planned experiment would require, always in the spirit that the "right" set of conditions will be the next they try. Even worse, a well planned experiment reveals significant insight as to the basic nature of the process. Ultimately, tweaking is inefficient, undisciplined, and uninformative. A well planned experiment simply provides structure to the experimental process. A designed experiment never interferes with true creativity. True creativity lies in picking the appropriate factors and correct levels. A well planned experimental strategy simply makes the truly creative engineer more effective.

1.6 Sampling

Observational studies require samples in order to learn about the underlying engineering process. The better these samples represent the true behavior of the process, the more we can learn. Engineers and statisticians have developed many good sampling strategies that guarantee representative samples and minimize any systematic biases. Three commonly recommended sampling schemes are:

1. Simple random sampling
2. Stratified random sampling
3. Systematic random sampling

In most cases, engineers sample from an ongoing process (a dynamic situation) where the process continues to generate items during the sampling. In this situation, we sample to determine the current state of the process. In other cases, engineers sample from fixed "lots" (a static situation) where all of the items have been generated prior to the actual sampling. In this situation, we sample mostly to characterize the lot.

Simple Random Samples

Definition 1.2	**Simple Random Sample (Finite Populations)**
	Consider a sample of n observations taken from a finite population (a much larger group of observations that are of interest). We call this sample a *simple random sample* if every possible sample of n observations taken from this population has the same chance of being selected.

Strictly speaking, this definition applies only to samples taken from finite populations. In Chapter 3, we shall give a more general definition of a random sample.

Example 1.11	**Monitoring the Porosity of Nickel Battery Plates**
	Nickel–hydrogen (Ni-H) batteries use a nickel plate as the anode. The electrode deposition (ED) process determines the specific electrical properties of these plates. In the ED process, sets of 40 plates are placed in a special bath and subjected to an electrical load for a specified period of time. During each shift, production supervision picks a random sample of five plates from a set of 40 for a destructive stress test. In this case, we treat the set of 40 plates as the population of interest. The manufacturer numbers the exact position of each plate within the ED bath. To ensure a random sample, the operator uses a computer based protocol to generate five random integers from the interval 1 to 40. The plates at these five specific locations then form the random sample used for the stress test.

Stratified Random Samples

In many engineering situations, we are interested in populations that are composed of several distinct, nonoverlapping subpopulations. For example, management needs to know the overall nonconformance rate (i.e., the overall proportion of items that fail to meet specifications) for an injection molding process. The materials manager purchases large batches of polypropylene from three different suppliers; all go into inventory. Production randomly selects material from the inventory. Consequently, some of the time production uses supplier A's material, some of the time B's, and some of the time C's. In this case, the overall population, which is all the production from this injection molding process, consists of three distinct, nonoverlapping subpopulations, the production from each source of polypropylene. Figure 1.6 illustrates this situation. Since each supplier's material runs slightly differently on the equipment, we should expect the nonconformance rates to differ by each source. Stratified sampling allows us to take this fact into account as we collect data to estimate the overall nonconformance rate.

VOICE OF EXPERIENCE

When planning studies, it is important to recognize the potential sources of variation and to account for them.

Figure 1.6 | Polypropylene Sources for an Injection Molding Process

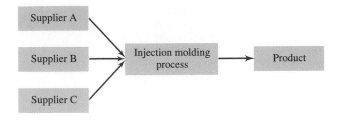

Definition 1.3 | **Stratified Random Sample**

Suppose that a population can be divided into m different, nonoverlapping subpopulations. A **stratified random sample** is one where we take a simple random sample within each subpopulation.

Let n_i be the size of the simple random sample taken from the ith subpopulation. If we know the relative sizes of the different subpopulations, then we often choose the n_i's proportional to the individual subpopulation sizes.

Example 1.12 | **Rate of Nonconformances for an Automobile Assembly Line**

A major manufacturer of automobiles operates an assembly line two shifts a day. The first shift accounts for approximately two-thirds of the overall production. Management routinely monitors the overall average number of nonconformances per automobile produced on this line by closely inspecting nine automobiles a day and counting the number of imperfections. The manufacturer gives each car produced a specific identification number and keeps a precise count of the number of cars made during each shift. As a result, the inspector can use a random-number generator at the end of the shift to obtain a simple random sample. To obtain a stratified random sample, the inspector randomly selects six cars from the first shift and three cars from the second shift each day for the detailed inspection. Figure 1.7 illustrates this sampling scheme.

Systematic Random Samples

Often engineers find it difficult to take true random samples; however, many engineering settings lend themselves very nicely to sampling every mth item. This is especially true in high-speed part-manufacturing operations, that now have equipment to take such samples automatically. In these situations, systematic random sampling works quite well.

Figure 1.7	Stratified Random Sample from a Two-Shift Automobile Assembly Process

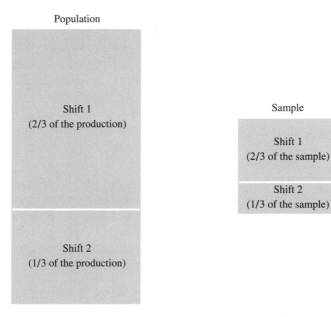

Definition 1.4	**Systematic Random Sample**

Suppose we wish to take as our sample every *m*th item. A **systematic random sample** is one that starts this process with an item randomly selected from the first *m*.

Example 1.13	**Sampling from a Sintering Process**

A manufacturer of nickel–hydrogen batteries uses a sintering process to produce its nickel anodes. This process, which uses a high-temperature furnace, controls the porosity of the plates. Production knows the precise sintering order of each plate. Typically, this company sinters 300 plates per shift. Production monitors the porosity of the plates by taking a systematic sample of five plates per day. The operator uses a calculator to obtain a random number between 1 and 60, inclusive. She selects this plate as the first item in the sample and then picks every 60th plate thereafter to complete the sample.

Nonadvisable Sampling Techniques

Two other common sampling strategies are

1. Quota sampling
2. Convenience sampling

In quota sampling, the sampler collects data until a specific quota is filled. For example, in the automobile assembly line example, the inspector might use some ad hoc method to select the six automobiles from the first shift by his lunch time. In so doing, he fills his quota of six automobiles. Of course, he also ignores the quality of the automobiles produced after lunch! Consequently, his sample at best represents the quality of the automobiles produced before lunch, which is not the same thing as the quality of the automobiles produced on the first shift.

In convenience sampling, the sampler uses an ad hoc technique "convenient" to him or her. For example, in the sintering process example, the operator might always select the first plate through the furnace and the others just before her breaks through the rest of the day. Although these times may be convenient for her, they also may correspond to some periodic effect in the sintering process. A periodic effect would bias the resulting sample, and we would not get an accurate representation of this process.

In general, we always prefer to use some type of random sampling scheme to obtain data. Such an approach ensures a representative sample and minimizes potential biases. In addition, formal statistical analysis depends on the assumption that the data come from a random sample.

1.7 Basic Principles of Experimental Design

Experimental and Observational Units

Two primary concepts underlie experimental design:

1. The experimental unit with its associated error
2. The observational unit with its associated error

Too often, engineers confuse these two concepts, which can seriously impair the resulting analysis!

Definition 1.5 | **Experimental Unit**

The *experimental unit* is the smallest unit to which we apply a treatment combination.

The *experimental error* is a measure of the variability among the experimental units used in the experiment.

Definition 1.6	**Observational Unit**

The *observational unit* is the unit upon which we make the measurement.

The *observational error* is a measure of the variability among the observational units used in the experiment. The observational error forms a part of the experimental error.

Example 1.14	**Strength of Sewer Pipe**

A major manufacturer of sewer pipe experienced a problem in the pipe's breaking strength. Since this pipe is made of a ceramic material, the firing process controls the final strength. The furnace used is as long as a football field and is divided up into several firing zones. Production can control the speed at which the pipe travels through the furnace. The firing regime is the specific combination of firing temperatures and speed. Depending on the size of the pipe, this firing process can require up to four weeks!

The engineers assigned as technical support to this process decided to compare three different firing regimes. Since most engineers consider the firing process to be "black box," they plan to treat the firing regime as a single factor with three levels. In this particular case, a treatment combination is a specific firing regime.

R&D has a test furnace available for this experiment that can fire three sample pipes at a time. Figure 1.8 considers two firing regimes in the test furnace. Since the firing regime controls the conditions within the furnace, each of the pipes within the furnace receives the same treatment. Within a specific set of three pipes, we cannot apply firing regime 2 to one of the pipes and firing regime 1 to the other two. Instead, when we apply firing regime 1 to the furnace, we are applying firing regime 1 to the entire set of three pipes in the furnace. Thus, the experimental unit is the set of three pipes in the test furnace. Since R&D tests each pipe, the observational units are the individual pipes.

The experimental error in this example is the variability that results from trying to reproduce the firing regimes from run to run. Although these firing regimes are well defined, we cannot exactly reproduce the firing conditions each time we run the experiment. The experimental error quantifies this variation. The observational

Figure 1.8	Experimental Setup for the Sewer Pipe Experiment

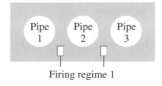
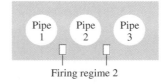

error is the variability among the three pipes within each furnace run. Although the furnace is designed to provide as uniform a heat distribution as possible throughout the furnace, we still expect to see some differences. Pipe 1 sees a slightly different heat profile than pipe 2, and pipe 2 sees a slightly different profile than pipe 3. The observational error quantifies this variation.

Basic Principles of Experimental Design

The three basic principles of experimental design are

1. Replication
2. Randomization
3. Local control of error

Replication means that we apply at least one of the treatment combinations to more than one experimental unit. Replication allows us to estimate the experimental error and to perform formal statistical analysis. Replicating observational units, which is often called repeat runs, minimizes the impact of measurement errors, but it does not provide an estimate of the experimental error. *Randomization* means that we perform the specific experimental runs in a random order, which minimizes the impact of any systematic bias over the course of the entire experiment. Formal statistical analysis presupposes randomization. *Local control of error* seeks to reduce the random error among the experimental units. The basic idea is to control anything other than the factors that might affect the response. With proper local control of error, we make the experiment more sensitive for detecting the differences due to the factors by eliminating extraneous sources of variability. We often use pairing or *blocking* to reduce the impact of these extraneous sources of variability. Engineers use blocking when they suspect significant differences among the experimental units available for the study. We can minimize the impact of this variability by dividing the experimental units into groups, called blocks, which are reasonably alike (homogeneous). We then apply each treatment combination to an experimental unit within the block. The next examples illustrate these principles.

> **VOICE OF EXPERIENCE**
>
> *The estimate of true experimental error requires replicates, not repeats! Experimental error deals with the duplication of results, not the reproducibility.*

Example 1.15 | **Strength of Sewer Pipe—Revisited**

After some preliminary discussion, the engineers suggested that we run 12 sets of 3 pipes each through R&D's test furnace. Because we were testing three different firing regimes, we could run each regime four times and thus *replicate* the experimental units. To minimize the impact of any systematic bias, we *randomly* allocated the specific batches to the three firing regimes. We actually put 12 pieces of paper in a hat: four with the number 1, four with the number 2, and four with

Figure 1.9	The Sewer Pipe Experimental Design in Random Order

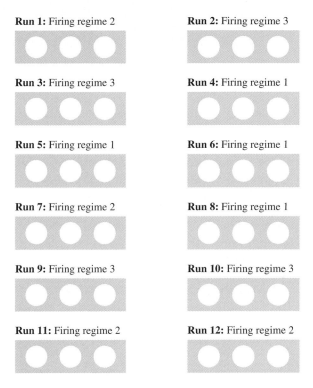

Run 1: Firing regime 2

Run 2: Firing regime 3

Run 3: Firing regime 3

Run 4: Firing regime 1

Run 5: Firing regime 1

Run 6: Firing regime 1

Run 7: Firing regime 2

Run 8: Firing regime 1

Run 9: Firing regime 3

Run 10: Firing regime 3

Run 11: Firing regime 2

Run 12: Firing regime 2

the number 3. We shook the hat and then drew the pieces out. Each time we drew out a piece of paper, we recorded the number on it. In this way, we generated a random order for the firing regimes. As a final measure, the engineers noted that sufficient materials were available so that we could make all 36 (12 × 3) batches of pipe from one large production unit. In this manner, we reduced extraneous sources of variability that may have obscured the effects of the firing regime (*local control of error*). Figure 1.9 illustrates the design in the randomized run order.

Example 1.16	**Testing Octane Blends**

Snee (1981) conducted an experiment to determine whether two methods for measuring the octane ratings of gasoline blends produced different results. This experiment thus had one factor, method, with two levels. The smallest unit to which we can apply the treatment is the particular sample of gasoline for which we obtain an octane rating. Each blend could produce several samples for testing.

Figure 1.10 | The Experimental Design for the Gasoline Blends

As a result, Snee could use two homogeneous experimental units from each blend.

Petroleum engineers used these test methods on gasoline blends that had a wide range of target octane ratings. Snee was interested in detecting the differences in the test methods over this particular range. As a result, this study required many gasoline blends with widely differing target octane ratings. Samples taken from different blends had significantly different octane ratings that were not due to only the method of measurement. These blend-to-blend differences represented a significant source of extraneous variability in the comparison of the two methods.

Snee had available 32 different blends over an appropriate spectrum of target octane ratings. To remove the blend-to-blend variability, Snee took two samples from each blend. He used one method on the first sample and the second method on the other sample. He thus used 32 blocks, each with two experimental units. By allocating one experimental unit to each method in each block, he made each method have 32 experimental units, which provided *replication*. He *randomized* the experiment in two ways. First, the blends were tested in a random order, and then the two samples from each blend were randomly allocated to the two test methods. Figure 1.10 illustrates this design. In this experiment, Snee used the blends as blocks, which provided *local control of error*. By pairing the data on the specific gasoline blends, he removed the effect of the target octane rating from the study.

Completely Randomized and Randomized Complete Block Designs

We mentioned that randomization is one of the three basic principles of experimental design. Two common randomization strategies are

1. completely randomized designs
2. randomized complete block designs

The completely randomized design randomly allocates all of the treatment combinations to the experimental units. This approach ensures that each experimental unit has the same chance of receiving any specific treatment combination. Example 1.15 illustrates this approach. The completely randomized design is a generalization of the two-independent-samples *t*-test, which we discuss in Section 4.5. The randomized complete block design randomly allocates the treatment combinations to the experimental units *within the blocks*. This approach requires that every block must consist of sufficient experimental material so that every treatment combination can occur within each block. Example 1.16 illustrates this approach. The randomized complete block design is a generalization of the paired *t*-test, which we discuss in Section 4.6.

This book usually assumes that the engineer uses a completely randomized design, which is the most common way engineers run experiments. We do not have time in this text to fully discuss randomized complete block designs. The interested student may read about this strategy, particularly within an RSM framework, in Montgomery (2004) and in Myers and Montgomery (2002).

1.8   Examples of Engineering Experiments

Engineers run planned experiments for many reasons, including these:

- To *screen* the factors to determine which truly influence the response
- To *predict* the behavior of the response over a specified range of the factors
- To *optimize* the response—that is, to find the specific settings of the factors that produce the best value for the response
- To make products and processes *robust* to known sources of variability

The underlying purpose of the experiment can greatly influence the specific experimental strategy.

Experiments for Screening Factors

Screening experiments consider a moderate to large number of factors, all usually at two levels. These experiments attempt to separate those factors that primarily influence the behavior of the response from those that have little or no influence. Engineers typically use screening experiments during the first phases of process improvement. These experiments use as few runs as possible in order to save resources for later experimentation. In general for screening experiments, the range between the levels should be large.

Example 1.17 | **Growing Silicon Layers on Integrated Circuits**

Shoemaker, Tsui, and Wu (1991) study the process that grows silicon wafers for integrated circuits. These wafers consist of a series of silicon layers. These layers need to be as uniform as possible because later processing steps form electrical devices within these layers. The smallest unit to which the engineers can apply the processing steps is a batch of wafers. The individual wafers are the observational units.

In their initial experiment, the engineers seek to determine which of six possible factors truly influence the uniformity of the silicon layer. They intend to conduct follow-up experimentation using the significant factors found in this initial experiment. To minimize the size of the design, they use only two levels for each factor. The factors and their levels are listed here:

Factor	Low	High
Deposition temperature	1210° F	1220° F
Deposition time	Low	High
Argon flow rate	55%	59%
HCl etch temperature	1180	1215
HCl flow rate	10%	14%
Nozzle position	2	4

An appropriate screening design is the eight-run fractional factorial in Table 1.1. The engineers randomize the actual order of these runs before they perform the experiment. With each setting of the process conditions, they allow the process to reach equilibrium, produce a "batch" of wafers, and measure the uniformity of the silicon layers. In Chapter 7, we shall perform the appropriate analysis to identify the factors that truly influence the response. Follow-up experimentation can provide a basis for optimizing the uniformity of the silicon layer.

Experiments for Prediction

In some cases, engineers believe that a specific model should explain the behavior of the data. They plan the experiment to estimate the proposed model as efficiently as possible with the resources available. Engineers use the data from these experiments to determine the adequacy of the proposed model. If they

Table 1.1	The Experimental Design Settings for the Silicon Wafer Experiment

Run	Deposition Temperature	Deposition Time	Argon Flow Rate	HCl Etch Temperature	HCl Flow Rate	Nozzle Position
1	1210	Low	55%	1180	14%	4
2	1220	Low	55%	1215	10%	4
3	1210	High	55%	1215	10%	2
4	1220	High	55%	1180	14%	2
5	1210	Low	59%	1215	14%	2
6	1220	Low	59%	1180	10%	2
7	1210	High	59%	1180	10%	4
8	1220	High	59%	1215	14%	4

are satisfied with the model, they then can use it to predict the behavior of the response over the ranges of the factors studied in the experiment.

Example 1.18 | **Springs with Cracks**

Box and Bisgaard (1987) discuss a manufacturing operation for carbon-steel springs that has a severe problem with cracks. Basic metallurgy suggests that the cracking depends on these factors:

- The temperature of the steel before quenching
- The amount of carbon in the formulation
- The temperature of the quenching oil

In this case, the engineers apply each treatment combination to an entire production lot of springs. The engineers use the percent of springs that do not exhibit cracking as their response. They seek to determine the basic relationships among the three factors and the cracking. In the process, they hope to find conditions that will reduce or even eliminate the problem.

They decide to pursue a two-level experiment involving these three factors. The following actual levels are used:

Factor	Low	High
Steel temperature	1450° F	1600° F
Carbon	0.50%	0.70%
Oil temperature	70° F	120° F

An appropriate design for this situation uses every combination of the two levels for each of the three factors and is called a 2^3 *factorial* design. The 2 indicates the number of levels used for each factor, and the 3 exponent indicates the total number of factors. The actual design is shown in Table 1.2.

The engineers run the actual design in random order to minimize the impact of any potential biases. With each setting of the process conditions, they allow the process to reach equilibrium, produce a lot of springs, and determine the percent that exhibit cracking. In Chapter 7, we will lay the foundation for performing the appropriate analysis to produce a model that allows the engineers to predict the amount of cracking within a production lot given a specific combination of the processing conditions.

Experiments for Optimization

Experiments for optimizing a process or system tend to be rather large and should be pursued only after previous experimentation has identified the important factors and an appropriate region in the factors for exploration. In Chapter 8, we shall outline a sequential philosophy of experimentation that more fully addresses these issues.

Table 1.2	The Experimental Design Settings for the Springs-with-Cracks Experiment

Run	Steel Temperature	Carbon	Oil Temperature
1	1450° F	0.50%	70° F
2	1600° F	0.50%	70° F
3	1450° F	0.70%	70° F
4	1600° F	0.70%	70° F
5	1450° F	0.50%	120° F
6	1600° F	0.50%	120° F
7	1450° F	0.70%	120° F
8	1600° F	0.70%	120° F

Example 1.19	**Optimizing a Process to Etch Silicon Gates**

Preuninger and colleagues (1993) used an experiment to optimize the uniformity of line size of an etching process for polycrystalline silicon gates used in the manufacture of integrated circuits. Initially, they used a screening experiment to determine which of the following factors influence the uniformity of the line size:

- RF power
- Pressure
- Percent hydrogen bromide (HBr)
- Temperature
- Magnetic field

The screening experiment indicates that only RF power, pressure, and percent HBr are important. The engineers then run a five-level experiment, called a *central composite design*, in these three factors to determine the optimum operating conditions. Table 1.3 gives the design. The engineers perform the actual experiment in a randomized order. Each run consists of a single wafer, which is the experimental unit. An inspector measures the line size at five different locations on each wafer. The last four runs of the experiment provide replication. The engineers use the same etch chamber for each experimental run to provide local control of error. In Chapter 8, we present the appropriate analysis for finding the optimum combinations for the factors of a designed experiment.

Experiments for Robust Design

Engineers are beginning to realize that to optimize a process does not always mean to maximize or to minimize. Consider the manufacture of a ballpoint pen. The fit between the cap and the barrel is a primary quality characteristic. The caps

Table 1.3	The Uniformity Experiment

Run	RF Power	Pressure	Percent HBr
1	180 W	30 mT	20
2	180 W	30 mT	10
3	180 W	20 mT	20
4	180 W	20 mT	10
5	120 W	30 mT	20
6	120 W	30 mT	10
7	120 W	20 mT	20
8	120 W	20 mT	10
9	150 W	25 mT	23
10	150 W	25 mT	7
11	150 W	33 mT	15
12	150 W	17 mT	15
13	200 W	25 mT	15
14	100 W	25 mT	15
15	150 W	25 mT	15
16	150 W	25 mT	15
17	150 W	25 mT	15
18	150 W	25 mT	15

and the barrels are made by separate injection molding processes. We seek to find the conditions for the injection molding process that will produce the diameters as close to a stated nominal as possible. We call a process that achieves a target condition for a characteristic of interest with minimal variability *robust* because it can achieve the target over a wide range of operating conditions.

Example 1.20	**Cake Mix Experiment**

Box and Jones (1992) discuss an experiment involving a cake mix. The manufacturer has direct control over the amounts of these ingredients:

- Flour (x_1)
- Shortening (x_2)
- Sugar (x_3)

Table 1.4	The Cake Mix Experiment

x_1	x_2	x_3	z_1	z_2
−1	−1	−1	−1	−1
−1	−1	−1	1	1
1	−1	−1	1	−1
1	−1	−1	−1	1
−1	1	−1	1	−1
−1	1	−1	−1	1
1	1	−1	−1	−1
1	1	−1	1	1
−1	−1	1	1	−1
−1	−1	1	−1	1
1	−1	1	−1	−1
1	−1	1	1	1
−1	1	1	−1	−1
−1	1	1	1	1
1	1	1	1	−1
1	1	1	−1	1

The quality of the cake also depends on these factors:

- Oven temperature (z_1)
- Cooking time (z_2)

Both of these are controlled by the consumer. The manufacturer seeks the proper combination of flour, shortening, and sugar that will produce a good-tasting cake over a wide range of oven temperatures and cooking times. For proprietary reasons, Box and Jones report the levels for each factor as only ±1. Table 1.4 summarizes the design. The manufacturer ran the actual design in random order.

1.9 Purpose of Engineering Statistics

The proper application of statistics should guide the engineer to the proper use and analysis of data. This chapter has outlined these topics:

- The engineering method
- Statistical models

- The importance of data collection
- The basic issues and concepts underlying data collection
- Some basic sampling techniques
- Some basic experimental design strategies

Well-thought-out data collection provides a good start toward solving engineering problems; however, it is only a start. We must develop tools that elicit as much information as we can from our data.

As this book unfolds, we first look at appropriate graphical displays of data. Often, these displays provide more than enough information to solve the problem at hand. In many cases, however, we need more formal methodologies. As the problems become more complex, we require more sophisticated techniques. Ultimately, we see that how we plan to perform our analysis can significantly influence how we collect our data. We actually complete the circle in Chapters 7 and 8 when we return to the data-collection issue. Then we are armed with an arsenal of data analytic tools to aid our process.

1.10 Case Study | *Manufacture of Writing Instruments*

This text concludes each chapter by examining the proper application of statistics to actual engineering problems associated with the manufacture of writing instruments. These case studies reflect the work experience of one of the authors at the Faber-Castell Corporation. The manufacture of writing instruments uses a wide array of engineering disciplines, including industrial, chemical, ceramic, mechanical, and electrical engineering. The major purposes of this case study are to illustrate these ideas:

- The proper application of statistics in its broadest sense, which includes the engineering method and statistical thinking
- Real engineering problems within their specific context
- The broad array of real engineering problems associated with manufacturing, even with such common products as pencils, pens, and markers
- The challenges of modern engineering and proper engineering data analysis

The unifying theme for each example in this case study is the need for sound data analysis to solve real engineering problems. The examples come from the core plant, where the pencil lead is made; injection molding, where the pen and marker casings are made; the automated pencil line, where the final pencil is assembled; quality control; and the laboratories. In the process, you, the reader, will get a deeper appreciation for a wide array of engineering problems and the role of data analysis in their solution.

⮞ 1.11 Ideas for Projects

1. As a class, perform the "catapult" experiment. Many instructors use a rubber band catapult to generate experimental data. The ones we use allow you to change the tension on the rubber band in several ways. Perform the experiment to generate a prediction for how far the catapult throws the ball. Use the prediction equation to try to hit a target a known distance away.

2. Perform the "paper helicopter" experiment. Many instructors make paper helicopters and drop them in class. We prefer to drop them from the fourth floor of a building on campus. The goal is to design a paper helicopter that flies as long as possible.

⮞ References

1. Box, G. E. P., and Bisgaard, S. (1987). The scientific context of quality improvement. *Quality Progress, 22*(6), 54–61.

2. Box, G. E. P., and Jones, S. (1992). Designing products that are robust to the environment. *Total Quality Management, 3,* 265–282.

3. Montgomery, D. C. (2004). *Design and Analysis of Experiments.* 6th edition. New York: John Wiley.

4. Myers, R. H., and Montgomery, D. C. (2002). *Response Surface Methodology: Process and Product Optimization Using Designed Experiments.* 2nd edition. New York: John Wiley.

5. Preuninger, F., Blasko, J., Meester, S., and Kook, T. (1993). Use of experimental design to optimize a process for etching polycrystalline silicon gates. *Statistics in the Semiconductor Industry.* Austin, TX: SEMATECH. Technology Transfer No. 92051125A-GEN, Vol. II, pp. 4-129–4-150.

6. Shoemaker, A. C., Tsui, K., and Wu, C. F. J. (1991). Economical experimentation methods for robust design. *Technometrics, 33,* 415–427.

7. Snee, R. D. (1981). Developing blending models for gasoline and other mixtures. *Technometrics, 23,* 119–130.

⮞ Exercises

1.1 A study was carried out on the coating thickness of a panel produced by a paint operation. Viscosity is thought to impact the coating thickness. For simplicity, only two levels of viscosity (low and high) are used. The experimenters randomly selected 12 panels from each level of viscosity. The goal of the study was to determine if higher viscosity leads to thicker coatings.

a. What is the response of interest?

b. List the factors of interest and whether they are categorical or continuous.

c. List the levels for the factors.

d. List out all possible treatment combinations.

e. What is the experimental unit? What is the observational unit?

f. How would the experiment be randomized?

1.2 A company compares two temperature instruments daily: one coupled to a process computer and the other used for visual control. Day-to-day variation is expected. On each of seven days, the two instruments are used to measure the temperature. Ideally, these two instruments should agree.

a. What is the response of interest?

b. List the factors of interest and whether they are categorical or continuous.

c. List the levels for the factors.

d. List out all possible treatment combinations.

e. What is the experimental unit? What is the observational unit?

f. How would the experiment be randomized?

1.3 Nickel-hydrogen batteries use nickel plates as the anode. After the plates are sintered or fired in a high-temperature furnace, they are grouped into lots of 40 plates each and then placed into an "electrode deposition" (ED) bath where they are placed under an electrical load. This bath controls the electrical properties of the cell. An important characteristic of the nickel plates batteries is "stress growth." As the battery cell undergoes its charge–discharge cycle, the plates actually begin to expand due to the stress. One of the engineers believes that the more porous the plate, the greater the stress growth. He wants to conduct a test to confirm this belief. The engineer knows that the specific conditions of the ED bath have a major impact on stress growth. Since no two ED bath runs are identical, the engineer expects a lot of variability in the stress growth purely from the ED baths. To minimize the impact of the ED baths, he has set up each ED lot so that 20 plates have "low" porosity and 20 plates have "high" porosity. After the ED run (15 are used), the engineer randomly selects five low-porosity plates to make a test battery cell and five high-porosity plates to form a second test cell. The data are the average percent increases in the plates' thicknesses after 200 charge–discharge cycles.

a. What is the response of interest?

b. List the factors of interest and whether they are categorical or continuous.

c. List the levels for the factors.

d. List out all possible treatment combinations.

e. What is the experimental unit? What is the observational unit?

f. How would the experiment be randomized?

1.4 A company that makes calculator cases is interested in the breaking strength of the plastic. Small plastic pellets are pressed into a test specimen and then measured for breaking strength. There are two suppliers of the plastic pellets. Interest lies in determining if the mean breaking strength differs for the two suppliers. Fifteen batches of pellets are selected at random from each supplier.

a. What is the response of interest?

b. List the factors of interest and whether they are categorical or continuous.

c. List the levels for the factors.
d. List out all possible treatment combinations.
e. What is the experimental unit? What is the observational unit?
f. How would the experiment be randomized?

1.5 Octane ratings for 32 gasoline blends are determined by two methods: motor and research. An important question is whether one method tends to produce higher ratings than the other. The sample consists of 32 gasoline blends covering a wide range of target octane ratings. Each blend is divided into two samples so that each blend can be tested by both methods.

a. What is the response of interest?
b. List the factors of interest and whether they are categorical or continuous.
c. List the levels for the factors.
d. List out all possible treatment combinations.
e. What is the experimental unit? What is the observational unit?
f. How would the experiment be randomized?

1.6 An independent consumer group tested radial tires from two major brands to determine whether there were any differences in the expected tread life. Twelve cars are assigned Brand 1 and twelve are assigned Brand 2. The data are collected by running the tires until the tread wears and recording the number of miles driven.

a. What is the response of interest?
b. List the factors of interest and whether they are categorical or continuous.
c. List the levels for the factors.
d. List out all possible treatment combinations.
e. What is the experimental unit? What is the observational unit?
f. How would the experiment be randomized?

1.7 Two brands of ultrasonic humidifiers are compared with respect to the rate at which they output moisture. The company chose eight humidifiers from brand A and eight from brand B. They measured the maximum outputs (in fluid ounces) per hour in a chamber controlled at a temperature of 70° F and a relative humidity of 30%.

a. What is the response of interest?
b. List the factors of interest and whether they are categorical or continuous.
c. List the levels for the factors.
d. List out all possible treatment combinations.
e. What is the experimental unit? What is the observational unit?
f. How would the experiment be randomized?

1.8 A study was conducted on the running times for 20 fuses. The goal was to verify that the two operators are giving consistent measurements. Since the fuses can vary, the experiment is carried out by having both operators measure each fuse.

a. What is the response of interest?
b. List the factors of interest and whether they are categorical or continuous.

   c. List the levels for the factors.

   d. List out all possible treatment combinations.

   e. What is the experimental unit? What is the observational unit?

   f. How would the experiment be randomized?

1.9 A manufacturer requires a certain type of zirconium sponge. They are in the process of awarding a new contract for the production of these sponges. One requirement is that the level of chlorine be held as low as possible. Five vendors (A, B, C, D, E) have applied for the contract. Each vendor makes four sponges using their own compositions. These are sent to a central testing laboratory where the amount of chlorine is measured.

   a. What is the response of interest?

   b. List the factors of interest and whether they are categorical or continuous.

   c. List the levels for the factors.

   d. List out all possible treatment combinations.

   e. What is the experimental unit? What is the observational unit?

   f. How would the experiment be randomized?

1.10 Fuel components in nuclear cores consist of rectangular fuel wafers made from natural uranium dioxide (UO_2). A study of factors involved in the compaction and sintering processes has been conducted to determine the levels required to achieve a high density (enhances fission gas retention). The study considered four factors: compaction pressure (40 tons/in^2 and 50 tons/in^2), compaction rate (20 wafers/min and 30 wafers/min), sintering temperature (1500° C and 1700° C), and sintering time (6 hrs. and 8 hrs.). Three batches of uranium dioxide were available, and variation was expected between batches. From each batch 100 wafers of each treatment combination were produced and the percent theoretical density was determined.

   a. What is the response of interest?

   b. List the factors of interest and whether they are categorical or continuous.

   c. List the levels for the factors.

   d. List out all possible treatment combinations.

   e. What is the experimental unit? What is the observational unit?

   f. How would the experiment be randomized?

1.11 A manufacturer of plastic containers for the beverage industry is interested in the pressure required to separate the cap from the bottle. The containers involve an injection molding process. The engineers have identified three important factors of the process: injection speed (40 and 75), mold temperature (25° C and 45° C), and cooling time (10 sec. and 25 sec.). The experiment involved using an 8-cavity injection mold.

   a. What is the response of interest?

   b. List the factors of interest and whether they are categorical or continuous.

   c. List the levels for the factors.

   d. List out all possible treatment combinations.

  e. Four different shots of the mold are conducted, and suppose that the experiment involves randomly assigning a treatment to a cavity.
   (1) What is the experimental unit? What is the observational unit?
   (2) How would the experiment be randomized?
  f. Suppose now that sixteen different shots of the mold are conducted, and suppose that the experiment involves randomly assigning a treatment to all cavities in the mold.
   (1) What is the experimental unit? What is the observational unit?
   (2) How would the experiment be randomized?

1.12 An experimenter is planning to study the control of citrus blight. It is believed that a vector of this disease is the leafhopper. To control this pest, three new chemical sprays (A, B, C) are being considered. Each chemical spray will be applied in two amounts (Low and High). A small grove containing five rows of six trees each is available for this study. Since this grove was once used as a demonstration project, each row of trees is on a different rootstock. There is a strong possibility of large differences between rootstocks. The experiment is carried out by selecting a row (rootstock) and then randomly assigning the treatments to the six trees. After a period of time, ten leaves are taken from each tree to measure citrus blight.

  a. What is the response of interest?
  b. List the factors of interest and whether they are categorical or continuous.
  c. List the levels for the factors.
  d. List out all possible treatment combinations.
  e. What is the experimental unit? What is the observational unit?
  f. How would the experiment be randomized?

1.13 An experiment has been performed to determine the power requirements for cutting metal with ceramic tools. The response of interest is the vertical component of a dynamometric reading. Two factors were examined: angle of edge level ($15°$ and $30°$) and type of cut (continuous and interrupted). Nine pieces of metal were available for the experiment. Each piece of metal is cut with all treatments. The pieces of metal are expected to vary.

  a. What is the response of interest?
  b. List the factors of interest and whether they are categorical or continuous.
  c. List the levels for the factors.
  d. List out all possible treatment combinations.
  e. What is the experimental unit? What is the observational unit?
  f. How would the experiment be randomized?

1.14 An automotive engineer wishes to improve the burst pressure of radiators for the next generation of vehicle. The engineer focuses on the brazing process used to join the aluminum radiator components together. The engineer runs an experiment in which radiators are fabricated using 3 different types of filler (1,2,3) material and 2 different brands of flux (A,B). The flux prepares the surfaces to be joined and protects them from oxidation while the filler material joins the components.

The radiators are brazed in an oven at the same temperature setting. The oven holds twelve radiators so two of each treatment combination are used.

a. What is the response of interest?
b. List the factors of interest and whether they are categorical or continuous.
c. List the levels for the factors.
d. List out all possible treatment combinations.
e. What is the experimental unit? What is the observational unit?
f. How would the experiment be randomized?

1.15 A food company is considering a new recipe for their cake mix. They have identified three factors that affect texture of the cake: amount of flour (low and high), amount of egg powder (low and high), and amount of oil added when mixing (low and high). They are also interested in varying the temperature of the oven when baking the cake (375° and 400°). Changing the temperature for each individual cake would be a very time-consuming process. The experiment is conducted by choosing a temperature, then mixing and baking all eight cakes together. Then the temperature is changed and again all eight cakes are baked together.

a. What is the response of interest?
b. List the factors of interest and whether they are categorical or continuous.
c. List the levels for the factors.
d. List out all possible treatment combinations.
e. Do all factors have the same experimental unit?
f. Identify the experimental unit and observational unit for the temperature factor.
g. Identify the experimental unit and observational unit for the other factors.

2

Data Displays

⫶ 2.1 Importance of Data Displays

Communicating with Data

The success of the engineering method hinges on the ability to convert data into information, usually in the form of a probabilistic model. This model must be able to explain the data observed and must account for the variation encountered. The key to developing good probabilistic models is "listening" to what the data have to say.

For example, an aircraft engineer may be very concerned about the wall thicknesses of the coolant jackets on the cylinder heads for an aluminum engine designed for high-altitude applications. This thickness must be at least 0.190 in. Many questions, which require data to answer, immediately come to mind. How often is this thickness less than 0.190 in.? What target value should reasonably guarantee that all the thicknesses will be at least 0.190 in.? Obviously, the larger this target value is, the more aluminum is required to make the engine, which in turn makes the engine heavier (a major drawback) and more expensive (possibly a major drawback). Can the target thickness be reduced and still reasonably guarantee that all the thicknesses will be at least 0.190 in.? Only by listening to properly collected and analyzed data can we hope to answer these questions.

Consider another example. A safety engineer must study the time between accidents at a major chemical facility over a ten-year period. Over this time, the facility has instituted several major safety programs. Again, several reasonable questions leap to mind. What is the typical time between accidents? How often do accidents happen very close together? Are there times when this facility has very few accidents? Have the safety initiatives been effective? We must analyze data to answer these questions, and the key to good data analysis is listening to what the data have to say.

Listening requires us to enter into a conversation with data. Most of us find our first encounters with data to be very one-sided, just like dealing with a very shy person. Unless properly prodded, data sets can be very unresponsive. As we become more skilled in starting a dialogue, we find that data sets can gush with information.

The Explanatory Power of Simple Graphs and Displays

Clever plots of data provide a powerful basis for starting a good conversation with data. For novice data analysts, data displays are ideal "ice breakers" with data. Engineers, by their background and training, are very comfortable with graphs and plots. Many engineers find interpreting plots of data more intuitive and more informative than doing formal statistical analyses. In many cases, clever plots of the data provide more than enough information to answer the questions at hand. In such situations, performing formal statistical analyses actually would be a waste of time and effort.

VOICE OF EXPERIENCE

Whenever possible, graph your data.

Data displays can provide powerful insights about the nature of the data and about the appropriate probabilistic model. For example, many measurements tend to follow bell-shaped patterns, which we should expect for the aircraft engine wall thicknesses. The times between events often follow a more J-shaped pattern, which we should expect for the times between accidents. These insights into the nature of the data are crucial to the success of formal analyses. Formal statistical analyses depend on the specific probabilistic model chosen. We express our expectations about the behavior of the data through specific *assumptions*. The resulting analyses are valid only to the extent that the model's underlying assumptions are met. Data displays provide a quick, easy, and effective method for checking the validity of these assumptions.

This chapter develops in detail:

- Stem-and-leaf displays
- Boxplots

Both are very easy to construct by hand for small data sets. We also introduce:

- Histograms
- Time plots

They are powerful tools for analyzing moderate to large data sets and are best left to appropriate software to generate. In later chapters, we shall introduce:

- Q–Q plots, especially the normal probability plot (Chapter 3)
- Scatter plots (Chapter 6)
- Contour plots (Chapter 8)
- Surface plots (Chapters 7 and 8)

Engineers commonly use all of these plots to analyze data. The plots not presented in this chapter all require some statistical foundations best left to later chapters.

⸫ **2.2** Stem-and-Leaf Displays

The Basic Stem-and-Leaf Display

The stem-and-leaf display provides a basis for evaluating the "shape" of a data set with minimal loss of the original information. The basic idea is to let the data themselves suggest natural groupings, which then may be exploited to produce a plot that displays the data's shape. This process loses only the time order of the data. It is best illustrated through an example.

Example 2.1 | **Wall Thicknesses of Aircraft Parts**

Eck Industries, Inc. (see Mee 1990) manufactures cast aluminum cylinder heads that are used for liquid-cooled aircraft engines. The wall thicknesses of the coolant jackets are critical, particularly in high-altitude applications. Engineering specifications require that this thickness must be at least 0.190 in. The thicknesses (in inches) for 18 cylinder heads as measured by ultrasound, which is a nondestructive technique, are listed here:

0.223	0.193	0.218	0.201	0.231	0.204
0.228	0.223	0.215	0.223	0.237	0.226
0.214	0.213	0.233	0.224	0.217	0.210

A natural basis for grouping the data is the first two digits of each measurement. For example, we shall group together in some fashion all the measurements that are between 0.220 and 0.229 in., inclusive. We shall call the first two digits of each measurement the *stem*. The last digit is the *leaf*. We notice that the smallest measurement is 0.193 in., and the largest is 0.237 in. Thus, the smallest stem is 0.19, and the largest is 0.23. We begin the stem-and-leaf display by plotting the stems as illustrated in Figure 2.1. Note that we must list all possible stems between 0.19 and 0.23 even if some of these stems do not contain any observations.

Figure 2.1 | Starting the Stem-and-Leaf Display

Stem	Leaves
0.19:	
0.20:	
0.21:	
0.22:	
0.23:	

Figure 2.2 | Adding the First Leaf to the Stem-and-Leaf Display

Stem	Leaves
0.19:	
0.20:	
0.21:	
0.22:	3
0.23:	

Blank stems provide powerful information about potential unusual observations. Next, we add the leaves to the plot. In Figure 2.2, we add the leaf associated with the first measurement, 0.223. Figure 2.3 gives the stem-and-leaf display for the first six measurements, and Figure 2.4 gives the "raw" stem-and-leaf display.

Depth   Many textbooks stop at this point because this raw display clearly shows the basic shape of the data set. Often, however, we use the stem-and-leaf as a tool for generating other data displays. Certain "refinements" are thus helpful. In Figure 2.5, we rearrange the leaves in ascending order, add a count of the number of leaves in each stem, and, finally, add the depth information. In the process of rearranging the leaves in ascending order, we actually rearrange the entire data set in ascending order, which will be important for constructing boxplots later. The depth information provides a quick and easy basis for locating critical values within the ordered data set. We shall see the true value of the depth when we construct boxplots in the next section. The *depth* represents how far the underlined observation is from the appropriate end of the ordered data set. For example, the value 0.204, which is the last value on the 0.20 stem, is the third observation from the beginning of the ordered data set. The value 0.223, which is the first value on the 0.22 stem, is the ninth value from the end of the ordered data set. The counts per stem help us to determine the depth. The depth from the beginning of the ordered data set is always the depth from the previous stem plus the count on the current stem. Similarly, the depth from the end of the ordered data set is always the depth from the stem immediately below plus the count on the current stem.

Figure 2.3 | The Stem-and-Leaf Display for the First Line of the Data

Stem	Leaves
0.19:	3
0.20:	14
0.21:	8
0.22:	3
0.23:	1

| Figure 2.4 | The Raw Stem-and-Leaf Display for the Aircraft Thicknesses |

Stem	Leaves
0.19:	3
0.20:	14
0.21:	854370
0.22:	383364
0.23:	173

When the middle value of the ordered data set falls on a stem, we do not give the depth information for that stem because it would be ambiguous whether the depth is from the beginning or the end of the ordered data set.

Reading a Stem-and-Leaf Display

As we said earlier, the purpose of data displays is to engage in a conversation with the data. Like any good conversation, the key to getting good information is to ask good questions. Some questions are obvious, but many others are more subtle. Statistical analysis distills a data set into its essential elements. Often, we can well describe the data by just two measures: the "center" and the "spread." Thus, two obvious questions are

- What is a typical value (the "center") for the data set?
- What is the variability ("spread") of the data?

More subtle questions are:

- Do the data follow some pattern?
- If the data follow a pattern, is it symmetric (see Figure 2.6), skewed toward large values (sometimes called *skewed right* or *right-tailed*, see Figure 2.7), or skewed toward small values (sometimes called *skewed left* or *left-tailed*, see Figure 2.8)?

| Figure 2.5 | The Stem-and-Leaf Display for the Aircraft Thicknesses |

Stem	Leaves	Number	Depth
0.19:	3	1	1
0.20:	14	2	3
0.21:	034578	6	
0.22:	333468	6	9
0.23:	137	3	3

Figure 2.6 | A Symmetric Data Pattern

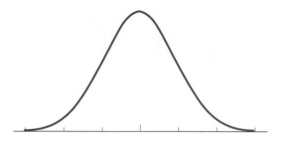

Figure 2.7 | A Right-Tailed or Right-Skewed Data Pattern

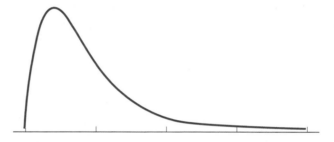

- Are there multiple peaks?
- Are there outliers?
- Are there any other interesting features?

Detecting Patterns   Virtually all data sets follow some pattern. Failure to discern a pattern usually indicates that the analyst has chosen a poor stem structure for the display. If the analyst uses stems that have too many significant digits, then the resulting display has too many stems to permit any reasonable pattern. Rounding the data corrects this problem. We shall see other ways to correct poor stem structures in the following pages. Most engineering data follow roughly symmetric patterns, but not all, as we shall see in other examples.

When the data follow a skewed pattern, the engineer needs to ask why and to seek out possible explanations. *Multiple peaks* in the data are almost always dead giveaways of contamination from another population. *Outliers* are data values that clearly are out of sync with the other values in the data set. It is crucial that the engineer determine the exact cause of these values. In some cases, the outlier is the result of a transcription error and is easy to correct. In other cases, something went wrong with the engineering process at the time the value was collected. Clearly, such a data value should not be used in subsequent analysis. Sometimes, however, nothing appears to be wrong. Then the analyst must presume that the data value is valid and that the underlying engineering process produces some odd values at least intermittently.

| Figure 2.8 | A Left-Tailed or Left-Skewed Data Pattern |

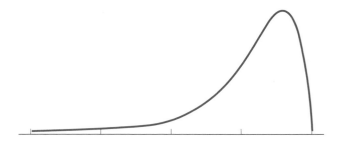

The best way to illustrate a conversation with a data set is through an example. Consider again the wall thickness data for aircraft engines.

| Example 2.2 | **Wall Thicknesses of Aircraft Parts—Revisited** |

Looking at the stem-and-leaf display in Figure 2.5, we see that the typical wall thickness is somewhere around 0.21–0.22 in., which is quite a bit larger than the minimum specification of 0.190 in. The data range from 0.193 to 0.237 in., with most somewhere between 0.210 and 0.230 in. The data appear to follow a fairly symmetric pattern. The pattern would be perfectly symmetric if the value 0.193 were a little larger. The value 0.193 looks out of line as a result, but the stem-and-leaf display does not give us much evidence to suggest that it is an outlier. The data suggest that the manufacturer has no real problem meeting the given specification. On the whole, this data set appears fairly typical with no extraordinary features.

Extensions to the Basic Stem-and-Leaf Display

The number of stems used in a stem-and-leaf display plays an important role in our ability to see interesting features in the data. If too few stems are used, then all of the leaves appear on just two or three stems, and the analyst really cannot discern much usable information. On the other hand, if too many stems are used, the data appear to have no pattern whatsoever. Extensions to the basic stem-and-leaf display help to avoid these problems.

| Example 2.3 | **Ambient Levels of Peroxyacyl Nitrates** |

Williams, Grosjean, and Grosjean (1993) studied the ambient levels of peroxyacyl nitrates in Atlanta, Georgia, during the period July 22 through August 26, 1992. Peroxyacyl nitrates are eye irritants and possible skin cancer agents. Since they have no known direct source, they are excellent indicators of photochemical air pollution. As a result, these compounds are of direct interest to environmental engineers. During severe smog episodes, the ambient level of PAN, which is the

---

Figure 2.9 | The Stem-and-Leaf Display for the PAN Data

Stem	Leaves	Number	Depth
0.:	34477789	8	8
1.:	0011111123333556899	19	
2.:	12444789	8	8

---

most common form of peroxyacyl nitrate, can reach 30–50 parts per billion (ppb). A previous study of the ambient levels of PAN in Atlanta during a comparable period in 1981 found the typical maximum daily level to be approximately 4 ppb.

The daily maximum ambient levels of PAN (in ppb) during the period studied in 1992 are listed here. One day's value was dropped because there were no data after 9:02 A.M.

0.3	1.0	0.8	1.1	1.3	1.1	2.4
2.9	1.6	1.3	0.4	0.7	1.5	1.3
2.1	1.9	1.0	1.8	1.1	2.2	1.9
0.7	0.4	1.5	0.7	2.4	2.4	2.8
2.7	1.1	1.2	1.1	1.1	0.9	1.3

Figure 2.9 gives the stem-and-leaf display using the natural stems from the data. Since there are only three stems, we really cannot see the patterns in the data very well. A logical extension is to split the stems into two. For example, the stem 0. can be split into the two stems 0∗. and 0•., where the stem 0∗. represents the values 0.0–0.4 and the stem 0•. represents the values 0.5–0.9. Notice that there are five possible values for each stem. Thus, if the stem 0•. in our stem-and-leaf display contains more of the data than the stem 0∗., it is not an artifact of how we choose to split our stems. The plot that results from splitting the original stems into two is called a *stretched* or *double* stem-and-leaf display. Figure 2.10 shows the stretched stem-and-leaf for the PAN data.

Figure 2.10 | The Stretched Stem-and-Leaf Display for the PAN Data

Stem	Leaves	Number	Depth
0∗.:	344	3	3
0•.:	77789	5	8
1∗.:	0011111123333	13	
1•.:	556899	6	14
2∗.:	12444	5	8
2•.:	789	3	3

The stretched display now allows us to talk with the data set. Typical values are 1.0–1.4. The data range from 0.3 to 2.9. The pattern looks almost symmetric. If the data pattern is skewed, it is toward the larger values; thus, it may be a little right tailed. There appear to be no outliers. Recall that the study from 1981 had found the typical maximum PAN value to be approximately 4 ppb. If anything, it appears that the maximum PAN value has dropped from 1981 to 1992, which would indicate that smog has decreased in the Atlanta area over that period of time.

> **VOICE OF EXPERIENCE**
>
> *Choosing the proper stem structure requires careful thought.*

In some cases, splitting the stems into two parts is not sufficient to reveal the general shape of the data. The critical idea that underlies splitting stems is that each resulting stem must have the same number of possible values. Thus, the next level of splitting is five. The resulting plot is called a *squeezed* stem-and-leaf display. Again, the best way to illustrate this technique is through an example, in this case one to which all students should be able to relate.

Example 2.4 | **Scores on a Homework Assignment**

These homework scores on the first assignment were given to an honors statistics class that consisted of 11 freshmen and 9 upperclassmen:

19	16	23	22	24
25	15	19	23	23
23	17	25	23	20
18	24	17	18	18

This course was given in the fall semester, so the freshmen were just making the transition from high school to college. The entire assignment was worth 25 points. We define our stems in this way:

- **1**∗ represents the values 10 and 11.
- **1t** represents the values 12 and 13 (the t stands for two and three).
- **1f** represents the values 14 and 15 (the f stands for four and five).
- **1s** represents the values 16 and 17 (the s stands for six and seven).
- **1•** represents the values 18 and 19.

Figure 2.11 gives the resulting squeezed stem-and-leaf display. This plot shows two clear peaks: one at 1• and the other at 2t. Is there a reasonable explanation for these two peaks? Of course there is!

Table 2.1 gives the scores broken down by freshmen and upperclassmen. How should we plot a stem-and-leaf display that will allow us to compare and contrast two or more data sets? Figure 2.12 gives a *side-by-side* stem-and-leaf display for these data sets that uses the same stem structure for both plots. This plot helps us to understand why two peaks appeared in the original stem-and-leaf display. A typical score for the freshmen was just under 20, whereas the upperclassmen

Figure 2.11 | The Stem-and-Leaf Display for the Homework Scores

Stem	Leaves	Number	Depth
1f:	5	1	1
1s:	677	3	4
1•:	88899	5	9
2*:	0	1	10
2t:	233333	6	10
2f:	4455	4	4

Table 2.1 | Homework Scores by Classification

Freshmen				Upperclassmen		
19	16	22	24	23	25	15
19	23	17	20	23	23	25
18	18	18		23	24	17

Figure 2.12 | The Side-by-Side Stem-and-Leaf Display for the Homework Scores

Stem	Freshmen			Upperclassmen		
	Leaves	Number	Depth	Leaves	Number	Depth
1f:				5	1	1
1s:	67	2	2	7	1	2
1•:	88899	5				
2*:	0	1	4			
2t:	23	2	3	3333	4	
2f:	4	1	1	455	3	3

typically scored just under 25. The freshmen scores are skewed to the right or to the larger values. It is interesting that two of the upperclassmen performed poorly on this assignment and appear as possible outliers.

The instructor conducted a brief investigation to determine the source of these differences. He discovered that most of the freshmen put forth the same level of effort as they did in high school. Unfortunately, this particular instructor expected more. The upperclassmen, familiar with college-level expectations, typically put in more effort as reflected by the grades. The instructor used this stem-and-leaf display to highlight the differences between the freshmen and the upperclassmen. The problem was corrected.

Stratification   In Example 2.4 we broke the original data set into two parts based on the class of the students. We call breaking apart a data set into subparts based on attendant information *stratification*. Often, multiple peaks in a stem-and-leaf display indicate a need for stratification.

Properly done, stratification can lead to profound insights into the nature of the data and is a key element to statistical thinking. Stratification often provides a basis for isolating why strange features appear in a data display. It also allows us to identify extraneous sources of variability. Removing these sources leads to better engineering processes and to more powerful analyses. Whenever possible, engineers should consider stratifying their data.

Stratification and side-by-side stem-and-leaf displays provide a powerful basis for analyzing data, as we shall see this chapter's case study. We can extend stratification and side-by-side stem-and-leaf displays naturally to three or more groups.

⁖ *Exercises*

2.1   These are the outside diameters (in inches) of the barrels of a popular felt-tip marker:

0.379	0.376	0.379	0.379	0.378
0.378	0.377	0.378	0.377	0.379
0.378	0.377	0.377	0.379	0.378
0.377	0.377	0.378	0.379	0.378
0.379	0.380	0.379	0.378	0.379
0.380	0.378	0.379	0.379	0.379
0.379	0.380	0.380	0.381	0.379

Plot an appropriate stem-and-leaf display and comment on your results. Discuss any interesting features. If possible, propose reasons for these features.

2.2   Yashchin (1992) studies the thicknesses of metal wires produced in a chip-manufacturing process. Ideally, these wires should have a target thickness of 8 microns. These are the thicknesses (in microns) of a sample of wires:

8.4	8.0	7.8	8.0	7.9	7.7	8.0	7.9	8.2	7.9
7.9	8.2	7.9	7.8	7.9	7.9	8.0	8.0	7.6	8.2
8.1	8.1	8.0	8.0	8.3	7.8	8.2	8.3	8.0	8.0
7.8	7.9	8.4	7.7	8.0	7.9	8.0	7.7	7.7	7.8
7.8	8.2	7.7	8.3	7.8	8.3	7.8	8.0	8.2	7.8

Plot an appropriate stem-and-leaf display and comment on the results. How well do these wires meet the specified target thickness? Discuss any interesting features. If possible, propose reasons for these features.

2.3 Cryer and Ryan (1990) discuss the following chemical process data, where the measurement variable is a color property:

0.67	0.63	0.76	0.66	0.69	0.71	0.72
0.71	0.72	0.72	0.83	0.87	0.76	0.79
0.74	0.81	0.76	0.77	0.68	0.68	0.74
0.68	0.69	0.75	0.80	0.81	0.86	0.86
0.79	0.78	0.77	0.77	0.80	0.76	0.67

Plot an appropriate stem-and-leaf display and comment on the results. Discuss any interesting features. If possible, propose reasons for these features.

2.4 Padgett and Spurrier (1990) analyze the breaking strengths of carbon fibers used in fibrous composite materials. These fibers measure 50 mm in length and 7–8 microns in diameter. Periodically, the manufacturer selects random samples of five fibers and tests their breaking stresses. Specifications require that 99% of the fibers must have a breaking stress of at least 1.2 GPa (gigapascals). These are the breaking stresses (in GPa) from 20 such samples:

3.7	2.7	2.7	2.5	3.6	3.1	3.3	2.9	1.5	3.1
4.4	2.4	3.2	3.2	1.7	3.3	3.1	1.8	3.2	4.9
3.8	2.4	3.0	3.0	3.4	3.0	2.5	2.7	2.9	3.2
3.4	2.8	4.2	3.3	2.6	3.3	3.3	2.9	2.6	3.6
3.2	2.4	2.6	2.6	2.4	2.8	2.8	2.2	2.8	1.9
1.4	3.7	3.0	1.4	1.0	2.8	4.9	3.7	1.8	1.6
3.2	1.6	0.8	5.6	1.7	1.6	2.0	1.2	1.1	1.7
2.2	1.2	5.1	2.5	1.2	3.5	2.2	1.7	1.3	4.4
1.8	0.4	3.7	2.5	0.9	1.6	2.8	4.7	2.0	1.8
1.6	1.1	2.0	1.6	2.1	1.9	2.9	2.8	2.1	3.7

Plot an appropriate stem-and-leaf display and comment on the results. How well do these breaking stresses compare to the minimum acceptable stress? Discuss any interesting features. If possible, propose reasons for these features.

2.5 The National Bureau of Standards (see Mulrow et al., 1988) tests 14 samples of biphenyl measured on a differential calorimeter calibrated with two standards in order to establish this substance's melting point in °C. Here are the data:

343.0	342.4	343.4	343.1	343.3	343.7	343.5
343.1	343.3	343.4	343.8	343.3	343.3	343.3

Plot an appropriate stem-and-leaf display and comment on the results. Discuss any interesting features. If possible, propose reasons for these features.

2.6 Montgomery, Peck, and Vining (2001) look at the time required to deliver cases of a popular soft drink to vending machines. The following data are the times (in minutes) required by the driver to stock a machine:

16.7	11.5	12.0	14.9	13.8
18.1	8.0	17.8	79.2	21.5
40.3	21.0	13.5	19.8	24.0
29.0	15.4	19.0	9.5	35.1
17.9	52.3	18.8	19.8	10.8

Plot an appropriate stem-and-leaf display and comment on the results. Discuss any interesting features. If possible, propose reasons for these features.

2.7 A two-component distillation column produced these yields for 25 successive batches:

0.99	0.93	0.95	0.99	0.89
0.96	0.94	0.96	0.99	0.99
0.98	0.81	0.97	0.92	0.99
0.97	0.99	0.96	0.94	0.99
0.99	0.99	0.99	0.80	0.95

Plot an appropriate stem-and-leaf display and comment on the results. Discuss any interesting features. If possible, propose reasons for these features.

2.8 Albin (1990) studied aluminum contamination in recycled PET plastic from a pilot plant operation at Rutgers University. She collected 26 samples and measured, in parts per million (ppm), the amount of aluminum contamination. The maximum acceptable level of aluminum contamination, on the average, is 220 ppm. These are the data:

291	222	125	79	145	119	244	118	182
63	30	140	101	102	87	183	60	191
119	511	120	172	70	30	90	115	

Plot an appropriate stem-and-leaf display and comment on the results. How well does this pilot plant perform relative to the maximum acceptable level of aluminum contamination? Discuss any interesting features. If possible, propose reasons for these features.

2.9 Good and Gaskins (1980) studied the silica in meteors. The calculated percentage of silica in each of 22 chondrites meteors are listed here:

20.77	22.56	22.71	22.99	26.39	27.08	27.32	27.33	27.57	27.81	28.69
29.36	30.25	31.89	32.88	33.23	33.28	33.40	33.52	33.83	33.95	34.82

Plot an appropriate stem-and-leaf display and comment on the results. Discuss any interesting features. If possible, propose any reasons for these features.

2.10 A chemist at the National Institute of Standards & Technology (NIST) conducted a study to determine the certified transmittance value that may be attached to the particular filter under study. The sample consisted of the following 50 transmittance measurements (at a sampling rate of 10 observations per second) from a filter with a nominal value of 2:

2.0018	2.0017	2.0017	2.0021	2.0013	2.0014	2.0015	2.0020	2.0024	2.0026
2.0017	2.0015	2.0018	2.0020	2.0015	2.0013	2.0016	2.0020	2.0025	2.0027
2.0018	2.0014	2.0018	2.0016	2.0015	2.0014	2.0015	2.0021	2.0027	2.0026
2.0019	2.0015	2.0019	2.0014	2.0016	2.0015	2.0016	2.0022	2.0026	2.0025
2.0018	2.0015	2.0019	2.0013	2.0015	2.0014	2.0019	2.0023	2.0026	2.0024

Plot an appropriate stem-and-leaf display and comment on the results. Discuss any interesting features. If possible, propose any reasons for these features.

2.11 Juran and Gryna (1980) provide an example involving an electronic component. These are the failure times in hours for 84 of the electronic components:

1.0	6.4	19.2	54.2	88.4	114.8
1.2	6.8	28.1	55.6	89.9	115.1
1.3	6.9	28.2	56.4	90.8	117.4
2.0	7.2	29.0	58.3	91.1	118.3
2.9	8.3	30.6	63.7	92.1	120.6
3.0	8.7	32.4	64.6	97.9	121.0
3.1	9.2	33.0	65.3	100.8	122.9
3.5	10.2	36.1	70.1	103.2	124.5
3.8	10.4	40.1	71.0	104.0	125.8
4.3	11.9	42.8	75.1	104.3	126.6
4.7	14.4	44.5	78.4	105.8	128.4
4.8	15.6	50.4	79.2	106.5	129.2
5.2	16.2	51.2	84.1	110.7	129.5
5.4	17.0	52.0	86.0	112.6	129.9

Plot an appropriate stem-and-leaf display and comment on the results. Discuss any interesting features. If possible, propose any reasons for these features.

2.12 Daily rainfall (in millimeters) was recorded over a 47-year period in Turramurra, Sydney, Australia (see Rayner and Best (1989)). For each year the day with the greatest rainfall was identified.

1468	909	841	475	846	452
3830	1397	556	978	1715	747
909	2002	1331	1227	2543	2649
1781	1717	2718	584	1859	1138
2675	1872	1359	1544	1372	1334
955	1849	719	1737	1389	681
1565	701	994	1188	962	1564
1800	580	1106	880	850	

Plot an appropriate stem-and-leaf display and comment on the results. Discuss any interesting features. If possible, propose any reasons for these features.

2.13 Smith and Naylor (1987) studied the strengths of 1.5-cm glass fibers measured at the National Physical Laboratory, England. The data are given below.

0.55	0.74	0.77	0.81	0.84
0.93	1.04	1.11	1.13	1.24
1.25	1.27	1.28	1.29	1.30
1.36	1.39	1.42	1.48	1.48
1.49	1.49	1.50	1.50	1.51
1.52	1.53	1.54	1.55	1.55
1.58	1.59	1.60	1.61	1.61
1.61	1.61	1.62	1.62	1.63
1.64	1.66	1.66	1.66	1.67
1.68	1.68	1.69	1.70	1.70
1.73	1.76	1.76	1.77	1.78
1.81	1.82	1.84	1.84	1.89
2.00	2.01	2.24		

Plot an appropriate stem-and-leaf display and comment on the results. Discuss any interesting features. If possible, propose any reasons for these features.

2.14 A pharmaceutical plant believes the variability of a synthetic process of a drug substance (measured in assay percentage) is related to quality differences in the starting material from their suppliers. The assay percentages are:

Supplier 1			Supplier 2		
85.9	85.0	82.6	83.5	83.2	83.1
84.5	85.3	87.2	86.6	85.7	88.1
86.2	84.3	84.7	83.9	84.8	87.9
86.3	88.4	85.4	83.8	87.5	87.7

Plot an appropriate stem-and-leaf display and comment on the results. Explicitly compare the two groups. Discuss any interesting features. If possible, propose any reasons for these features.

2.15 Pignatiello and Ramberg (1985) studied the impact of several factors involving the heat treatment of leaf springs. In this process, a conveyor system transports leaf spring assemblies through a high-temperature furnace. After the spring leaves a high-pressure press, an oil quench cools it to near ambient temperature. An important quality characteristic of this process is the resulting free height of the spring, which has a target value of 8 inches. An engineer assigned to this process strongly believes that the heating time affects the free height. She has chosen two times: 23 sec. and 25 sec. Here are the data:

23 seconds						25 seconds					
7.5	7.6	7.5	7.5	7.6	7.5	7.8	7.8	7.8	7.5	7.3	7.1
7.6	7.6	7.8	7.6	7.8	7.6	8.2	8.2	7.9	7.9	7.9	7.9
7.6	7.6	7.4	7.2	7.2	7.3	7.9	8.0	7.9	7.3	7.4	7.4
7.6	7.8	7.7	7.8	7.5	7.6	7.7	8.1	8.1	7.6	7.7	7.6

Plot an appropriate stem-and-leaf display and comment on the results. Explicitly compare the two groups. Discuss any interesting features. If possible, propose any reasons for these features.

2.16 Lucas (1985) studied the rate of accidents at a DuPont facility over a ten-year period. The following data are the number of industrial accidents per calendar quarter for the first five-year period (period I) and for the second (period II):

Period I				Period II			
5	5	10	8	3	4	2	0
4	5	7	3	1	3	2	2
2	8	6	9	7	7	1	4
5	6	5	10	1	2	2	1
6	3	3	10	4	4	4	4

Plot an appropriate stem-and-leaf display and comment on the results. Explicitly compare these two periods. Discuss any interesting features. If possible, propose reasons for these features.

2.17   Eibl, Kess, and Pukelsheim (1992) studied the impact of viscosity on the observed coating thicknesses produced by a paint operation. Ideally, this process should produce a coating thickness of 0.8 mm. For simplicity, they chose to study only two viscosities: "low" and "high." The coating thicknesses (in mm) are listed here:

"Low" Viscosity							
1.09	1.12	0.83	0.88	1.62	1.49	1.48	1.59
0.88	1.29	1.04	1.31	1.83	1.65	1.71	1.76

"High" Viscosity							
1.46	1.51	1.59	1.40	0.74	0.98	0.79	0.83
2.05	2.17	2.36	2.12	1.51	1.46	1.42	1.40

Plot an appropriate stem-and-leaf display and comment on the results. Explicitly compare the two groups. Discuss any interesting features. If possible, propose reasons for these features.

2.18   A manufacturer of aircraft monitors viscosity of primer paint (see Montgomery 2004, pp. 232–236). These are the viscosities for two different time periods:

Time Period 1						Time Period 2				
33.8	33.1	34.0	33.8	33.5		33.5	33.3	33.4	33.3	34.7
34.0	33.7	33.3	33.5	33.2		34.8	34.6	35.0	34.8	34.5
33.6	33.0	33.5	33.1	33.8		34.7	34.3	34.6	34.5	35.0

Plot an appropriate stem-and-leaf display and comment on the results. Explicitly compare the two periods. Discuss any interesting features. If possible, propose reasons for these features.

2.19   A chemical engineer studied the effects of two different reflux rates on the yields from a distillation column. Here are the yields:

Reflux Rate of 70						Reflux Rate of 80				
0.99	0.92	0.94	0.92	0.87		0.81	0.93	0.95	0.97	0.89
0.96	0.92	0.95	0.99	0.95		0.97	0.93	0.95	0.76	0.99
0.99	0.79	0.97	0.91	0.99		0.90	0.80	0.97	0.92	0.81
0.97	0.99	0.96	0.93	0.95		0.97	0.99	0.96	0.94	0.94
0.99	0.95	0.96	0.78	0.97		0.81	0.94	0.90	0.79	0.95

Plot an appropriate stem-and-leaf display and comment on the results. Explicitly compare the two groups. Discuss any interesting features. If possible, propose reasons for these features.

⊳ 2.3 Boxplots

The boxplot provides a quick display of some important features of the data. Unlike the stem-and-leaf display, which retains virtually all of the information in the data set, the boxplot "distills" the data set down to its most important features. As a result, the boxplot does lose some information contained in the data.

The boxplot gives the analyst a formal tool for discriminating outliers during preliminary data analysis. Since we expect to encounter outliers, we construct the boxplot by using measures of the center and the spread based on the *median* and the *quartiles,* which are *resistant* or insensitive to the presence of outliers. Both the median and the quartiles require the data set to be arranged in ascending order.

The Median

Definition 2.1 | **Median**

The median, $\tilde{y}$, is the middle value of the data set once it has been arranged in ascending order.

To find the median, we first must order the data set. Suppose we have n observations in our data set. Let $y_1, y_2, \ldots, y_n$ denote this data set. We rearrange the data set in ascending order. Let

$$y_{(1)} \le y_{(2)} \le \cdots \le y_{(n)}$$

be the rearranged data set. Sometimes this rearranged data set is called the set of *order statistics,* which are extremely important in the field of nonparametric statistics. The refined stem-and-leaf display directly yields this rearranged data set and serves as an excellent precursor for developing the boxplot.

The median gives a measure of the "center" of the data. It literally splits the data set into two equal parts. Let ℓ_m denote the "location" of the median within the rearranged data set. Thus,

$$\ell_m = \frac{n+1}{2}. \tag{2.1}$$

If n is odd, then ℓ_m is an integer and

$$\tilde{y} = y_{(\ell_m)}. \tag{2.2}$$

If n is even, then ℓ_m contains the fraction $\frac{1}{2}$, which presents a problem. We want the "middle" value, but there is no unique one. By convention, we take as the median the average of the two values closest to the middle. For those who like formulas, this is

$$\tilde{y} = \frac{y_{(\ell_m-1/2)} + y_{(\ell_m+1/2)}}{2}. \qquad (2.3)$$

The next two examples illustrate how easy it is to find the median. They also illustrate how we can use the depth information from the refined stem-and-leaf display to our advantage.

Example 2.5 | **Median of the Ambient Levels of PAN**

Recall Example 2.3, which discussed the ambient levels of PAN in Atlanta, Georgia, during the summer of 1992. We now wish to find the median value. We first must rearrange the data in ascending order. Figure 2.10 gives the appropriate stem-and-leaf display. Our method for constructing this display ensures that the values are in ascending order. We next need to find the location of the median. From equation (2.1), we have

$$\ell_m = \frac{n+1}{2} = \frac{35+1}{2} = 18.$$

Since n in this case is odd, the resulting location of the median is an integer, which makes life a little easier for us. From equation (2.2), we have

$$\tilde{y} = y_{(\ell_m)} = y_{(18)}.$$

We thus must locate the 18th value in the rearranged data set. Looking down the depth column of Figure 2.10, we see that the last value on the stem 0•. is the 8th value and that the last value on the stem 1∗. is the 21st value. Since the median is the 18th value, it must be the 10th value on the 1∗. stem. Thus,

$$\tilde{y} = 1.3.$$

Example 2.6 | **Median of the Wall Thicknesses of Aircraft Parts**

Recall Example 2.1, in which we examined the wall thicknesses of an aircraft engine part. We now wish to find the median wall thickness. Once again, we first must rearrange the data in ascending order. Figure 2.5 gives the refined stem-and-leaf display with the observations in ascending order. We next must find the location of the median. From equation (2.1), we note

$$\ell_m = \frac{n+1}{2} = \frac{18+1}{2} = 9\frac{1}{2}.$$

In this case, since n is even, the location of the median is not an integer. We therefore must locate the two values that "surround" this location and take their average. In this case, we must locate the 9th ($\ell_m - \frac{1}{2}$) and the 10th ($\ell_m + \frac{1}{2}$) values. Looking down the depth column of Figure 2.5, we see that last value on the 0.20 stem is the 3rd value. We next note that there are six values on the 0.21 stem. Thus, the last value on the 0.21 stem is the 9th value. The first value on the next stem (the 0.22 stem) then must be the 10th value. The median is the average of these two values. Using equation (2.3), we have

$$\tilde{y} = \frac{y_{(\ell_m - 1/2)} + y_{(\ell_m + 1/2)}}{2} = \frac{y_{(9)} + y_{(10)}}{2} = \frac{0.218 + 0.223}{2} = 0.2205.$$

The Quartiles

Definition 2.2 | **Quartiles**

Consider a data set rearranged in ascending order. The quartiles are those values that divide the data set into four equal parts.

Let Q_1 be the *first* or *lower* quartile, and let Q_3 be the *third* or *upper* quartile. The median turns out to be the *second* or *middle* quartile. Q_1 can be thought of as the "median" of the data below the actual median of the data. Likewise, Q_3 can be thought of as the "median" of the data above the actual median of the data.

Let ℓ_q be the location for the quartiles. Various authors define this location differently, but most use

$$\ell_q = \begin{cases} \dfrac{n+3}{4} & \text{if } n \text{ is odd} \\[2ex] \dfrac{n+2}{4} & \text{if } n \text{ is even.} \end{cases} \tag{2.4}$$

If ℓ_q is an integer, then we find the first quartile by counting ℓ_q values from the beginning of the ordered data set. To find the third quartile, we count ℓ_q values from the end of the ordered data set. In terms of formal formulas,

$$Q_1 = y_{(l_q)}$$
$$Q_3 = y_{(n+1-\ell_q)}. \tag{2.5}$$

If ℓ_q is not an integer, then the appropriate quartile is the average of the two values that surround the appropriate location. In formulas, we have

$$Q_1 = \frac{y_{(\ell_q - 1/2)} + y_{(\ell_q + 1/2)}}{2}$$

$$Q_3 = \frac{y_{(n+1-\ell_q - 1/2)} + y_{(n+1-\ell_q + 1/2)}}{2}. \tag{2.6}$$

We illustrate how to find the quartiles by continuing our previous two examples.

Example 2.7 | **Quartiles for the Wall Thicknesses of Aircraft Parts**

To find the quartiles, we need to determine the appropriate location. Since n is even, equation (2.4) gives us

$$\ell_q = \frac{n+2}{4} = \frac{18+2}{4} = 5,$$

which is an integer and makes life a little easier for us. The first quartile is the fifth value from the beginning of the ordered data set, and the third quartile is the fifth value from the end. We can find both of these quantities easily from the depth information on the stem-and-leaf display in Figure 2.5. According to the depth column, the last value on the 0.20 stem is the third value from the beginning of the ordered data set. Thus, the first quartile is the second value on the next or 0.21 stem, or 0.213. Again, according to the depth column, the first value on the 0.23 stem is the third value from the end of the ordered data set. Thus, the third quartile is the second value from the end of the previous or 0.22 stem, or 0.226. In terms of equation (2.5), we have

$$Q_1 = y_{(\ell_q)} = y_{(5)} = 0.213$$
$$Q_3 = y_{(n+1-\ell_q)} = y_{(14)} = 0.226.$$

Example 2.8 | **Quartiles for the Ambient Levels of PAN**

To find the quartiles, we again need to determine the appropriate location. Since n is odd, equation (2.4) gives

$$\ell_q = \frac{n+3}{4} = \frac{35+3}{4} = 9\frac{1}{2},$$

which clearly is not an integer and presents some complications. The first quartile is the average of the ninth and the tenth values from the beginning of the ordered data set. According to the depth information from the stem-and-leaf display in

Figure 2.10, the last value on the 0●. stem is the eighth from the beginning. As a result, the first two values on the next (1∗.) stem are the ninth and tenth. Using equation (2.6), we obtain the first quartile by

$$Q_1 = \frac{y_{(\ell_q - 1/2)} + y_{(\ell_q + 1/2)}}{2}$$

$$= \frac{y_{(9)} + y_{(10)}}{2}$$

$$= \frac{1.0 + 1.0}{2}$$

$$= 1.0.$$

(2.7)

Similarly, the depth information tells us that the first value on the 2∗. stem is the eighth from the end of the ordered data set. As a result, the last two values on the previous (1●.) stem are the ninth and tenth. Thus, the third quartile is the average of the ninth and tenth values from the end of the ordered data set. Again, using equation (2.6), we obtain the third quartile by

$$Q_3 = \frac{y_{(n+1-\ell_q-1/2)} + y_{(n+1-\ell_q+1/2)}}{2}$$

$$= \frac{y_{(26)} + y_{(27)}}{2}$$

$$= \frac{1.9 + 1.9}{2}$$

$$= 1.9.$$

Resistance

Statisticians often call the median and the quartiles *resistant* statistics because they are relatively insensitive to the presence of outliers. Consider the wall thickness example. We obtain the same values for the median and the quartiles whether the smallest value is the original 0.193 or 0.100. Making the smallest value even smaller does not affect the location of the median or the quartiles. Similarly, making the largest value even larger does not affect the median or the quartiles.

We typically use boxplots to identify possible outliers. We compromise our ability to identify outliers if extreme values distort our definition of a possible outlier. We thus need to use resistant measures of the center and the spread when we construct boxplots, which explains why we use the median and the quartiles.

The Basic Boxplot

The boxplot provides the analyst at a glance:

- The center of the data set
- Where most of the data fall
- The spread of the unquestionably "good" data
- Possible outliers

The next example illustrates a seven-step procedure for constructing boxplots.

Example 2.9 | **Boxplot for the Wall Thicknesses of Aircraft Parts**

In Example 2.2, we noted that the wall thickness 0.193 in. looks a little different from the other data values. We now construct the boxplot, which provides a more formal basis for determining whether this point is in fact "different."

1. Construct a horizontal scale, marked conveniently, that covers at least the range of the data. For this example, we use a scale from 0.190 to 0.240 in. (see Figure 2.13).
2. Find the median and the quartiles. For this example, we have already found these values:

$$\tilde{y} = 0.2205$$
$$Q_1 = 0.213$$
$$Q_3 = 0.226.$$

Use Q_1 and Q_3 to make a rectangular box above the horizontal scale. Draw a vertical line through the box at the median (see Figure 2.14).

Figure 2.13 | Horizontal Scale for the Wall Thickness Boxplot

Figure 2.14 | The "Box" for the Wall Thickness Boxplot

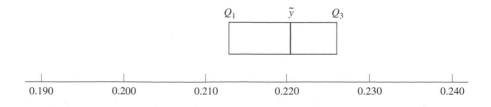

3. Find the step size to identify any outliers. Let *step* be this step size. For theoretical reasons, most authors define the step size by

$$step = 1.5 \cdot (Q_3 - Q_1). \tag{2.8}$$

Some authors call the quantity $Q_3 - Q_1$ the *interquartile range*. The interquartile range is sometimes used as a measure of the variability in the data. In our case,

$$step = 1.5 \cdot (0.226 - 0.213) = 0.0195.$$

4. Find the *inner fences*, which define the bounds for unquestionably good data. By "unquestionably good," we mean that we cannot classify the data values that fall within the inner fences as outliers. We calculate the upper inner fence (*UIF*) and the lower inner fence (*LIF*) by

$$\begin{aligned} UIF &= Q_3 + step \\ LIF &= Q_1 - step. \end{aligned} \tag{2.9}$$

In our example,

$$UIF = 0.226 + 0.0195 = 0.2455$$
$$LIF = 0.213 - 0.0195 = 0.1935.$$

5. Locate the most extreme data values on or within the inner fences. In our specific case, the *LIF* is 0.1935. The smallest value on or within the *LIF* is 0.201, so this value is the lower adjacent. The *UIF* is 0.2455. The largest value on or within the *UIF* is 0.237, so it is the upper adjacent. Draw vertical lines at these points, which are sometimes called the *adjacents*. The adjacents represent the most extreme observed data values considered unquestionably good. Connect these points to the box with a horizontal line, which is often called the *whisker* (see Figure 2.15). At this point, we have displayed a measure of the typical value (the median), where the bulk of the data fall (the box defined by the quartiles), and the observed bounds for the unquestionably good data (the whiskers).

6. Find the *outer fences*, which provide a basis for discriminating between mild and extreme outliers. *Extreme outliers* are data values that have virtually no

Figure 2.15 | The "Box and Whiskers" Portion of the Wall Thickness Boxplot

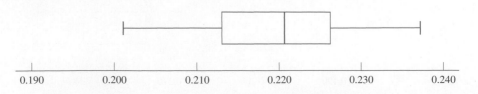

Figure 2.16 | The Complete Boxplot for the Wall Thickness Data

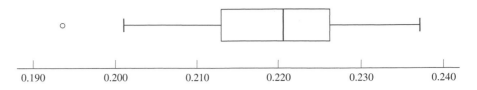

chance of coming from the same population as the bulk of the data. We define the upper outer fence (*UOF*) and the lower outer fence (*LOF*) by

$$UOF = Q_3 + 2 \cdot step$$
$$LOF = Q_1 - 2 \cdot step.$$

(2.10)

In our example,

$$UOF = 0.226 + 2 \cdot (0.0195) = 0.265$$
$$LOF = 0.213 - 2 \cdot (0.0195) = 0.174.$$

7. Mark possible outliers. We use a ○ to denote *mild* outliers, which are those data points between the inner and the outer fences. We use a ● to denote *extreme* outliers, which are those points on or beyond the outer fences. In this example, the thickness 0.193 in. is a mild outlier (see Figure 2.16).

In summary, the boxplot tells us this information at a glance:

- A typical value is 0.2205 in.
- Much of the data fall between 0.213 and 0.226 in.
- The unquestionably good data—that is, the data we do not consider even potentially outliers—fall between 0.201 and 0.237 in.
- The value 0.193 in. appears to be a possible outlier and deserves further investigation.

Parallel Boxplots

Parallel or side-by-side boxplots are a powerful tool for comparing two or more data sets simultaneously. We actually have seen more than 30 data sets easily analyzed by such a plot! The key to this plot is using the same scale for all of the boxplots. The next example illustrates this technique.

Example 2.10 | **Sensory Modalities**

Galinsky and colleagues (1993) studied the impact of sensory modalities (either aural or visual) on people's ability to monitor a specific display for critical events to which they must respond. Such tasks are critical components of jobs like air traffic control, industrial quality control, robotic manufacturing operations, and nuclear power plant monitoring. One aspect of the study focused on the difference in response to aural and visual stimuli. In particular, the researchers monitored the motor activity of the subject's dominant wrist as a measure of "restlessness" or "fidgeting"; the greater the activity, the more restless the subject. Galinsky and her colleagues recorded the number of wrist movements over 10-minute periods of time. The data are listed in Table 2.2. Table 2.3 summarizes the medians and the quartiles for these two data sets. Figure 2.17 shows the parallel boxplots. We clearly see no outliers in either data set. Visual stimuli appear to lead to more wrist movement, thus indicating more restlessness, than aural stimuli. For example, 100% of the auditory measurements were below 700 movements while only 50% of the visual were below 700 movements. Both data sets reveal similar amounts of variability.

Table 2.2 | The Sensory Modality Data

	Auditory					Visual			
418	236	281	416	578	386	517	617	870	892
329	197	397	677	698	416	574	782	838	885

Table 2.3 | The Medians and Quartiles for the Sensory Modality Data

	Auditory	Visual
$\tilde{y}$	406.5	699.6
Q_1	281	517
Q_3	578	870

Figure 2.17 | Parallel Boxplots Comparing Visual and Aural Stimuli for Restlessness

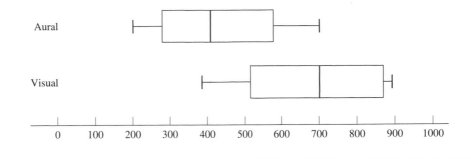

⫶ *Exercises*

2.20 Construct a boxplot for the felt-tip marker data given in Exercise **2.1** and comment on your results. Discuss any interesting features. What insights does the boxplot offer above the stem-and-leaf display?

2.21 Construct a boxplot for the thicknesses of metal wires given in Exercise **2.2** and comment on your results. Discuss any interesting features. What insights does the boxplot offer above the stem-and-leaf display?

2.22 Construct a boxplot of the color property data given in Exercise **2.3** and comment on your results. Discuss any interesting features. What insights does the boxplot offer above the stem-and-leaf display?

2.23 Construct a boxplot for the breaking strengths of carbon fibers given in Exercise **2.4** and comment on your results. Discuss any interesting features. What insights does the boxplot offer above the stem-and-leaf display?

2.24 Construct a boxplot for the melting points of biphenyl given in Exercise **2.5** and comment on your results. Discuss any interesting features. What insights does the boxplot offer above the stem-and-leaf display?

2.25 Construct a boxplot for the delivery time data given in Exercise **2.6** and comment on your results. Discuss any interesting features. What insights does the boxplot offer above the stem-and-leaf display?

2.26 Construct a boxplot for the yields given in Exercise **2.7** and comment on your results. Discuss any interesting features. What insights does the boxplot offer above the stem-and-leaf display?

2.27 Construct a boxplot for the amounts of aluminum contamination given in Exercise **2.8** and comment on your results. Discuss any interesting features. What insights does the boxplot offer above the stem-and-leaf display?

2.28 Construct a boxplot for the percentage of silica in chondrites meteors given in Exercise **2.9** and comment on your results. Discuss any interesting features. What insights does the boxplot offer above the stem-and-lead display?

2.29 Construct a boxplot for the transmittance measurements given in Exercise **2.10** and comment on your results. Discuss any interesting features. What insights does the boxplot offer above the stem-and-leaf display?

2.30 Construct a boxplot for the failure times given in Exercise **2.11** and comment on your results. Discuss any interesting features. What insights does the boxplot offer above the stem-and-leaf display?

2.31 Construct a boxplot for the annual maximum rainfall given in Exercise **2.12** and comment on your results. Discuss any interesting features. What insights does the boxplot offer above the stem-and-leaf display?

2.32 Construct a boxplot for the strength of glass fibers given in Exercise **2.13** and comment on your results. Discuss any interesting features. What insights does the boxplot offer above the stem-and-leaf display?

2.33 Construct parallel boxplots for the assay percentages given in Exercise **2.14** and comment on your results. Discuss any interesting features. What insights do the boxplots offer above the stem-and-leaf display?

2.34 Construct parallel boxplots for the free heights given in Exercise **2.15** and comment on your results. Discuss any interesting features. What insights do the boxplots offer above the stem-and-leaf display?

2.35 Construct parallel boxplots for the accident data given in Exercise **2.16** and comment on your results. Discuss any interesting features. What insights do the boxplots offer above the stem-and-leaf display?

2.36 Construct parallel boxplots for the coating thicknesses given in Exercise **2.17** and comment on your results. Discuss any interesting features. What insights do the boxplots offer above the stem-and-leaf display?

2.37 Construct parallel boxplots for the paint viscosities given in Exercise **2.18** and comment on your results. Discuss any interesting features. What insights do the boxplots offer above the stem-and-leaf display?

2.38 Construct parallel boxplots for the yields given in Exercise **2.19** and comment on your results. Discuss any interesting features. What insights do the boxplots offer above the stem-and-leaf display?

2.39 Construct parallel boxplots for the homework scores given in Example 2.4 and comment on your results. Discuss any interesting features. What insights do the boxplots offer above the stem-and-leaf display?

2.40 George, Sullivan, and Park (1994) studied three different thermoplastic starches: waxy maize, native corn, and high amylose corn. They injection molded 2-mm-thick test specimens at a melt temperature of $350\,°F$, an injection speed of 3 in./sec, a hold pressure of 5000 psi, a hold time of 5 sec, and a cool time of 10 sec. For each sample, they recorded the minimum injection pressure, which measures the processability of the thermoplastic starch. A lower minimum injection pressure indicates easier processability. The pressures (in thousands of psi) for each thermoplastic starch are listed here:

Waxy Maize			
13.0	9.0	10.0	10.0
10.0	6.0	7.0	7.0

Native Corn			
22.5	18.0	9.0	9.0
9.0	6.0	10.0	10.0

High Amylose Corn			
15.0	13.0	18.0	14.5
12.0	11.0	8.9	8.0

Construct parallel boxplots for these data and comment on your results.

2.4 Using Computer Software

Many standard statistical software packages produce stem-and-leaf displays and boxplots. In addition, these packages can produce other graphical displays such as *histograms* and *time plots*. The best way to introduce these packages is with specific examples.

Example 2.11	**Times Between Industrial Accidents**

Lucas (1985) studied the times between accidents for a ten-year period at a DuPont facility. DuPont historically has strongly emphasized the importance of safety in its operations and has always striven to reduce the number of accidents over time. During this period, 178 accidents occurred. Table 2.4 gives the data, which are the numbers of days since the previous accident.

Basic Stem-and-Leaf Displays and Boxplots

The standard statistical software packages all produce similar-looking plots. Figure 2.18 gives the stem-and-leaf display from Splus for the data in Example 2.11. Such data as times between accidents, equipment lifetimes, times to repair, and incomes often display shapes such as the one in Figure 2.18, which is extremely skewed to the right. Why? The time between accidents cannot be less

Table 2.4	Times Between Accidents at a Major Chemical Facility								
15	16	33	4	51	3	15	19	14	7
13	2	4	0	7	20	0	18	1	30
0	12	15	9	0	2	6	56	26	20
13	55	10	1	6	31	18	22	7	9
36	0	4	55	7	11	23	53	53	0
14	22	12	3	20	0	6	24	4	14
16	9	34	23	3	18	3	13	8	2
3	38	1	9	10	28	36	11	32	3
5	7	39	27	18	3	7	27	2	14
30	10	1	1	8	2	3	15	5	6
24	13	18	48	12	27	46	7	7	22
2	13	14	19	3	0	6	14	8	34
21	19	36	14	8	1	98	20	173	49
15	40	60	35	34	66	44	3	7	39
0	1	11	29	11	3	22	7	0	14
67	58	4	28	22	72	53	43	86	26
72	43	35	36	2	2	68	9	7	23
	9	20	14	60	21	11	25		

than zero, so we encounter a natural lower bound for these times. On the other hand, it is entirely possible to go significant periods of time without an accident, so extreme values tend to be large—hence, the skew in our data. Splus has identified three points as potential outliers: the values 86, 98, and 173. Figure 2.19 shows the boxplot, which confirms two of these three points as being extreme. In addition, it identifies other times in the 70–80 day range as mild outliers.

Histograms

Most statistical software packages generate *histograms*, which provide another way to see the shape of the data. The software divides the range of the data into a suitably chosen number of intervals, all with the same length. The software then plots either the count for the number of observations that fall within each interval (for small data sets) or the relative proportion of observations that fall within each interval (for large data sets). The intervals chosen by the software package may or may not correspond with the stems it chooses for the corresponding stem-and-leaf display. In addition, since the histogram plots either the count or

Figure 2.18 | Stem-and-Leaf Display of the Times Between Accidents

```
N = 177   Median = 14
Quartiles = 6, 28
Decimal point is 1 place to the right of the colon
    41   41    0 : 00000000001111111222222223333333333344444
    69   28    0 : 5566666777777777778888999999
         25    1 : 000111112223333344444444444
    83   15    1 : 555556688888999
    68   17    2 : 0000011222223334444
    51    9    2 : 566777889
    42    8    3 : 00123444
    34    9    3 : 556666899
    25    4    4 : 0334
    21    3    4 : 689
    18    4    5 : 1333
    14    4    5 : 5568
    10    2    6 : 00
     8    3    6 : 678
     5    2    7 : 22
High:   86   98    173
```

the relative proportion of data values that fall in the interval, we actually lose the individual data values. In general, we should prefer stem-and-leaf displays for small data sets because then we do not lose the individual data values. For larger data sets, we usually prefer histograms because we can scale them to fit on our page or screen without losing any further information.

Figure 2.19 | Boxplot of the Times Between Accidents

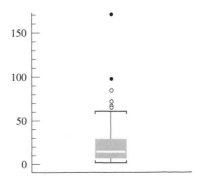

Figure 2.20 | Histogram of the Times Between Accidents

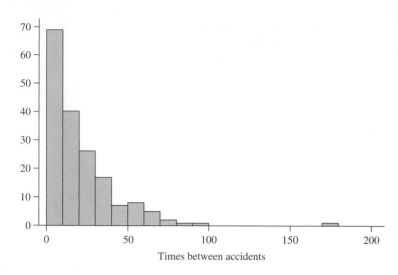

Times between accidents

Statistical software packages use different rules for determining the number of intervals for histograms as well as the number of stems for stem-and-leaf displays. Some texts suggest that the maximum number of stems we should use is $10 \log_{10}(n)$, where n is the number of data values. Whenever a statistical package gives a display that appears to use too many intervals or too many stems, we should check to see whether the package observed this maximum number. If it did not, usually the package will allow us to override the default number of intervals or stems.

Figure 2.20 shows the histogram for the times between accidents. We again see a definite right-tailed pattern. Also, note that because we plot the counts per interval, we can scale the plot in any convenient manner, unlike the stem-and-leaf display, which must record each data value.

> **VOICE OF EXPERIENCE**
>
> *Be very careful with stem-and-leaf displays and histograms. It is very easy to drastically change the appearance of the shape of the distribution by how the data are split or how the bin size is chosen.*

Time Plots

In many cases, we observe data over time, like the times between accidents. In such cases, we should plot the data over time and look for trends. Such plots always put the characteristic of interest on the y-axis. Sometimes, we plot the actual time as the x-axis. In other cases, the x-axis simply represents the sequence in which we observed the data. In this latter case, if $x = 3$, then the value plotted on the y-axis is the third value observed in the sequence. In either case, we are plotting the data in time order, hence the name *time plot*. Some quality engineers call this plot a *run chart*.

> **VOICE OF EXPERIENCE**
>
> *Boxplots and stem-and-leaf displays do not account for trends over time.*

Figure 2.21 | Time Plot of the Times Between Accidents

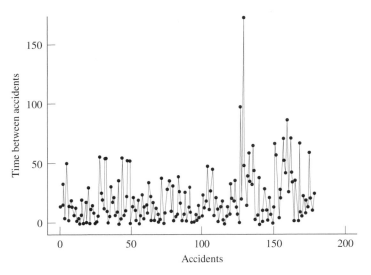

Engineers use run charts as a simple tool for monitoring processes. Typically, a process remains stable until it is acted upon by some force that results in some change in the characteristic of interest. For example, consider an ethanol–water distillation column. The yields generally tend to range within a narrow interval around 94% (ethanol and water form an azeotrope that prevents the yield from exceeding 95%). However, if a problem in the column occurs, these yields will begin to drop, which the time plot or run chart will reflect.

Figure 2.21 shows the time plot for the times between accidents. For each accident, we plot the time elapsed since the last accident. Thus, the y-axis is the time between accidents and the x-axis is the accident number. On careful inspection, we see that the times seem to be longer during the last several years of the study. This increase in the times suggests that accidents were beginning to occur less frequently, a very good trend.

Parallel Boxplots

Most statistical software packages also generate parallel or side-by-side boxplots. To create these plots, we need to identify the group to which each data value belongs. In this particular example, Lucas points out that the accident rate at this facility seems to have dropped over the last five years as compared with the first five. As a result, the times between accidents seem to be longer in the last five years, which the time plot of the data appears to confirm. We thus should consider stratifying our data based on this extra information. The first 120 times come from the first five years, and the last 57 times come from the last five years

Figure 2.22 | Parallel Boxplots Comparing the Times Between Accidents for the First Five Years Versus the Second Five Years

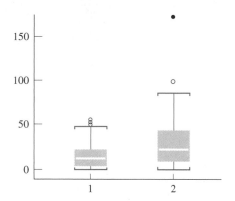

of the study. To generate the parallel boxplots, each software package requires an extra column of data identifying which five-year period the time comes from. The value 1 represents the first five years; the value 2 represents the second five years.

The standard software packages produce similar parallel boxplots. Figure 2.22 gives the Splus output, which clearly shows that the times between accidents during the second five years do tend to be longer than those in the first five years. Obviously, the plant's safety efforts have borne fruit.

Example 2.12 | **Comparing Two Temperature Instruments—Paired Data**

Van Nuland (1992) uses two different temperature instruments for controlling a chemical process: one coupled to a process computer and the other used for visual control. Ideally, at any given point in time, these two instruments should agree. Each day, he collects the two temperature measurements, both taken at the same point in time. The temperatures for 30 successive days are listed in Table 2.5. A naive analysis treats the two sets of temperatures separately. Figure 2.23 gives the parallel boxplots for the two temperature instruments. Instrument 2 appears to register slightly hotter temperatures than instrument 1; however, the plots do not indicate a pronounced difference. We see a definite overlap of the two boxes. Many of the temperatures registered by the first instrument fall well within the typical temperatures registered by the second instrument. The parallel boxplots do not give conclusive evidence that the two instruments disagree.

The true process temperature at the time of the two measurements clouds our ability to see a possible difference. The true process temperature fluctuates over some narrow band during normal operation. Each day, when the operator takes the two measurements, we expect the true process temperature to be different. This day-to-day variability in the true process temperature obscures our ability to see the true differences between the two temperature instruments.

Table 2.5	The Temperature Instrument Data									
Day	1	2	3	4	5	6	7	8	9	10
Instrument 1	84.3	84.3	84.5	84.4	84.3	84.1	84.7	84.5	84.2	84.7
Instrument 2	84.6	84.3	84.6	84.7	84.6	84.6	84.9	84.6	84.5	84.7
Day	11	12	13	14	15	16	17	18	19	20
Instrument 1	84.5	84.2	84.3	84.3	84.4	84.8	84.0	84.4	84.3	84.4
Instrument 2	84.4	84.7	84.7	84.4	84.5	84.7	84.4	84.5	84.5	84.4
Day	21	22	23	24	25	26	27	28	29	30
Instrument 1	84.4	84.2	84.4	84.6	84.1	84.3	84.4	84.6	84.4	84.5
Instrument 2	84.5	84.5	84.6	84.7	84.5	84.5	84.8	84.6	84.6	84.9

A better analysis eliminates the day-to-day variability by looking at the observed *differences*. Let y_{i1} be the temperature registered by instrument 1 on day i, and let y_{i2} be the temperature registered by instrument 2 on day i. We can define the difference, d_i, for the ith day by $d_i = y_{i1} - y_{i2}$. If $d_i > 0$, then instrument 1 registered a hotter temperature than instrument 2 on the ith day. By focusing on the daily differences, we remove the effect of the true process temperature on our analysis. If the process was hotter than normal on day i, then both instruments should register hotter than normal conditions. On the other hand, if one instrument registers a hotter than normal temperature but the other does not, we know that the two instruments do not agree, which indicates a problem. This pairing of the data is the simplest form of an important statistical principle called blocking.

Figure 2.23	Side-by-Side Boxplots for the Two Temperature Instruments

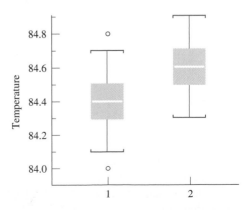

Figure 2.24 | The Stem-and-Leaf Display of the Daily Temperature Differences

```
N = 30    Median = -0.2
Quartiles = -0.3, -0.1
Decimal point is 1 place to the left of the colon
    2     2   -5 : 00
    7     5   -4 : 00000
   12     5   -3 : 00000
          5   -2 : 00000
   13     7   -1 : 0000000
    6     4    0 : 0000
    2     2    1 : 00
```

Figure 2.25 | The Boxplot of the Daily Temperature Differences

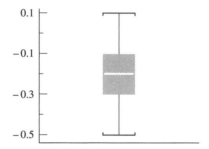

Figure 2.24 gives the stem-and-leaf display for the daily differences. We see a somewhat skewed pattern toward the negative differences. Most of the differences are less than zero. Figure 2.25 shows the boxplot for the daily differences. We see that the typical difference is −0.2 degree, which indicates that instrument 2 typically registers a slightly hotter temperature than instrument 1. In addition, almost all of the differences are less than zero, strongly suggesting that instrument 2 consistently registers a hotter temperature. The two instruments do not seem to agree.

⊁ Exercises

2.41 Use appropriate software to construct a stem-and-leaf display, a boxplot, and a histogram for the felt-tip marker data given in Exercise **2.1**. Give a thorough discussion of your results, including the relative advantages and disadvantages of each display.

2.42 Use appropriate software to construct a stem-and-leaf display, a boxplot, a histogram, and a time plot (the data are in order) for the thicknesses of metal wires given in Exercise **2.2**. How well do these data meet the target of 8.0 microns? Give a thorough discussion of your results, including the relative advantages and disadvantages of each display.

2.43 Use appropriate software to construct a stem-and-leaf display, a boxplot, a histogram, and a time plot (the data are in order) for the color property data given in Exercise **2.3**. Give a thorough discussion of your results, including the relative advantages and disadvantages of each display.

2.44 Use appropriate software to construct a stem-and-leaf display, a boxplot, a histogram, and a time plot for the breaking strengths of carbon fibers given in Exercise **2.4**. How well do these data compare to the minimum acceptable stress? Give a thorough discussion of your results, including the relative advantages and disadvantages of each display.

2.45 Use appropriate software to construct a stem-and-leaf display, a boxplot, and a histogram for the melting points of biphenyl given in Exercise **2.5**. Give a thorough discussion of your results, including the relative advantages and disadvantages of each display.

2.46 Use appropriate software to construct a stem-and-leaf display, a boxplot, and a histogram for the delivery time data given in Exercise **2.6**. Give a thorough discussion of your results, including the relative advantages and disadvantages of each display.

2.47 Use appropriate software to construct a stem-and-leaf display, a boxplot, a histogram, and a time plot for the yields given in Exercise **2.7**. Give a thorough discussion of your results, including the relative advantages and disadvantages of each display.

2.48 Use appropriate software to construct a stem-and-leaf display, a boxplot, and a histogram for the amounts of aluminum contamination given in Exercise **2.8**. How do these data compare to the maximum allowable concentration? Give a thorough discussion of your results, including the relative advantages and disadvantages of each display.

2.49 Use appropriate software to construct a stem-and-leaf display, a boxplot, and a histogram for the percentage of silica in chondrites meteors given in Exercise **2.9**. Give a thorough discussion of your results, including the relative advantages and disadvantages of each display.

2.50 Use appropriate software to construct a stem-and-leaf display, a boxplot, and a histogram for the transmittance measurements given in Exercise **2.10**. Give a thorough discussion of your results, including the relative advantages and disadvantages of each display.

2.51 Use appropriate software to construct a stem-and-leaf display, a boxplot, and a histogram for the failure times given in Exercise **2.11**. Give a thorough discussion of your results, including the relative advantages and disadvantages of each display.

2.52 Use appropriate software to construct a stem-and-leaf display, a boxplot, and a histogram for the annual maximum rainfall given in Exercise **2.12**. Give a thorough discussion of your results, including the relative advantages and disadvantages of each display.

2.53 Use appropriate software to construct a stem-and-leaf display, a boxplot, and a histogram for the strength of glass fibers given in Exercise **2.13**. Give a thorough discussion of your results, including the relative advantages and disadvantages of each display.

2.54 McGrath et al. (1995) describe materials prepared by selective functionalization of olefin-containing polymers to produce polybutadiene polyols. The effect on the ultimate strength of adding 1,4-butanediol to the formulation was studied. The data are:

1,4-Butanediol		
1:1	**1:2**	**1:3**
513	1278	2332
1415	2528	2688
619	758	1238
1699	2332	2477

Use appropriate software to plot appropriate parallel boxplots and thoroughly discuss your results. Give a possible reason for the large variation in the strength at each 1-4 butanediol ratio.

2.55 Use appropriate software to construct parallel boxplots for the assay percentages given in Exercise **2.14**. Give a thorough discussion of your results.

2.56 Use appropriate software to construct parallel boxplots for the free heights given in Exercise **2.15**. Give a thorough discussion of your results.

2.57 Use appropriate software to construct parallel boxplots for the accident data given in Exercise **2.16**. Give a thorough discussion of your results.

2.58 Use appropriate software to construct parallel boxplots for the coating thicknesses given in Exercise **2.17**. Give a thorough discussion of your results.

2.59 Use appropriate software to construct parallel boxplots for the paint viscosities given in Exercise **2.18**. Give a thorough discussion of your results.

2.60 Use appropriate software to construct parallel boxplots for the homework scores given in Example 2.4. Give a thorough discussion of your results.

2.61 Use appropriate software to construct parallel boxplots for the minimum injection pressures given in Exercise **2.40**. Give a thorough discussion of your results.

2.62 Snee (1983) examined the thicknesses of paint can ears. Periodically, the manufacturer took random samples of five cans each and measured the thicknesses of the ears. The data (in units of 0.001 in.) are listed here:

29	36	39	34	34	29	29	28	32	31
34	34	39	38	37	35	37	33	38	41
30	29	31	38	29	34	31	37	39	36
30	35	33	40	36	28	28	31	34	30
32	36	38	38	35	35	30	37	35	31
35	30	35	38	35	38	34	35	35	31
34	35	33	30	34	40	35	34	33	35
34	35	38	35	30	35	30	35	29	37
40	31	38	35	31	35	36	30	33	32
35	34	35	30	36	35	35	31	38	36
32	36	36	32	36	36	37	32	34	34
29	34	33	37	35	36	36	35	37	37
36	30	35	33	31	35	30	29	38	35
35	36	30	34	36	35	30	36	29	35
38	36	35	31	31	30	34	40	28	30

Use a computer software package to construct an appropriate stem-and-leaf display, a histogram, and a boxplot. Give a thorough discussion of your results, including the relative advantages and disadvantages of each display.

2.63 Snee (1981) conducted an experiment to determine whether two methods for measuring the octane rating of gasoline blends produced different results. Each blend could produce two samples for testing. Snee had available 32 different blends over an appropriate spectrum of target octane ratings. The following ratings summarize the results:

Method 1	105.0	81.4	91.4	84.0	88.1	91.4	98.0	90.2
Method 2	106.6	83.3	99.4	94.7	99.7	94.1	101.9	98.6
Method 1	94.7	105.5	86.5	83.1	86.2	87.7	84.7	83.8
Method 2	103.1	106.2	92.3	89.2	93.6	97.4	88.8	85.9
Method 1	86.8	90.2	92.4	85.9	84.8	89.3	91.7	87.7
Method 2	96.5	99.5	99.8	97.0	95.3	100.2	96.3	93.9
Method 1	91.3	90.7	93.7	90.0	85.0	87.9	85.2	87.4
Method 2	97.4	98.4	101.3	99.1	92.8	95.7	93.5	97.5

Use a computer software package to construct an appropriate stem-and-leaf display, a histogram, and a boxplot. Give a thorough discussion of the results, including the relative advantages and disadvantages of each display.

2.64 Grubbs (1983) obtained data on the running times of 20 fuses. Two operators, acting independently, measured these times for each fuse:

Operator 1	4.85	4.93	4.75	4.77	4.67	4.87	4.67	4.94	4.85	4.75
Operator 2	5.09	5.04	4.95	5.02	4.90	5.05	4.90	5.15	5.08	4.98
Operator 1	4.83	4.92	4.74	4.99	4.88	4.95	4.95	4.93	4.92	4.89
Operator 2	5.04	5.12	4.95	5.23	5.07	5.23	5.16	5.11	5.11	5.08

Use a computer software package to construct an appropriate stem-and-leaf display, a histogram, and a boxplot. Give a thorough discussion of your results, including the relative advantages and disadvantages of each display.

2.65 Measuring the actual dimensions of a manufactured part is a classical problem that faces many different disciplines, especially mechanical and industrial engineering. A mechanical engineer must grapple with the thicknesses of nickel plates for a nickel–hydrogen battery. By the way the plate is made, he can consistently identify specific locations on each plate. Thus, location A on the first plate measured is the same as location A on the second plate. He believes that one specific location, A, of each plate is consistently thicker than another specific location, B. The actual thicknesses (in mm) of ten plates are listed here:

Location A	31.10	31.10	30.90	30.80	32.20	30.40	29.65	29.85	29.85	30.65
Location B	29.75	29.75	30.15	30.80	30.20	30.40	30.35	29.75	29.15	30.50

Use a computer software package to construct an appropriate stem-and-leaf display, a histogram, and a boxplot. Give a thorough discussion of your results, including the relative advantages and disadvantages of each display.

⋗ 2.5 Using Boxplots to Analyze Designed Experiments

Engineers often use parallel boxplots to analyze results from a planned experiment, especially if the factors are categorical. The general idea is to determine whether or not there are differences. For example, Nelson (1989) examined the cold cranking power of car batteries. This study used five different models and

Table 2.6	The Cranking Power Data

Model 1	Model 2	Model 3	Model 4	Model 5
41	42	27	48	28
43	43	26	45	32
42	46	28	51	37
46	38	27	46	25

measured how many seconds an individual battery provided its rated amperage without falling below 7.2 V at 0° F. The data are given in Table 2.6.

Figure 2.26 gives the parallel boxplots for the five models. Models 1, 2, and 4 appear better than models 3 and 5 with model 4 looking the best. A more formal procedure for comparing several groups is called the analysis of variance (ANOVA). In general, ANOVA compares the variation *between* groups to the variation *within* groups. Although ANOVA will not be discussed in this book, there is a direct relationship between ANOVA and the regression methods presented in Chapters 6, 7, and 8.

We also can do this for factorial experiments. The groups become the different treatment combinations. Doing the analysis on the treatment combinations is called unfolding the factorial structure.

Figure 2.26	The Boxplot of the Cranking Power Times

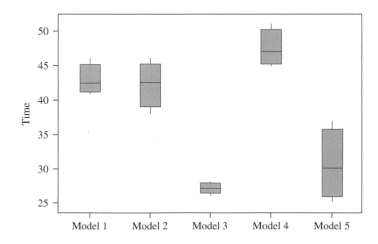

Table 2.7 | The Machine-Cutting Tool Lifetime Data

Feed	Depth	Life
0.010	0.05	23.7
0.010	0.05	20.8
0.010	0.05	21.7
0.010	0.05	21.9
0.010	0.05	22.6
0.010	0.20	15.1
0.010	0.20	15.4
0.010	0.20	16.3
0.010	0.20	13.8
0.010	0.20	13.5
0.026	0.05	6.6
0.026	0.05	7.4
0.026	0.05	8.7
0.026	0.05	7.0
0.026	0.05	7.0
0.026	0.20	1.2
0.026	0.20	0.8
0.026	0.20	0.8
0.026	0.20	0.8
0.026	0.20	1.1

Example 2.13 | **Lifetime of a Machine Cutting Tool**

An experiment is conducted to study the effect of the feed rate and the depth of the cut on the lifetime of a machine-cutting tool. The four combinations of these two factors were replicated 5 times. The data are shown in Table 2.7. The boxplots for the unfolded data (the four distinct combinations of feed rate and depth) are shown in Figure 2.27. There is a negative effect for both feed rate and depth (the lifetime decreases as you move from the low level of the factor to the high level of the factor). The best setting of feed rate at 0.01 and depth at 0.05 produces an average lifetime around 22 months.

The rest of the book will be spent on more formally dealing with this kind of analysis, culminating in Chapter 8.

| Figure 2.27 | The Boxplot of the Machine-Cutting Tool Lifetimes |

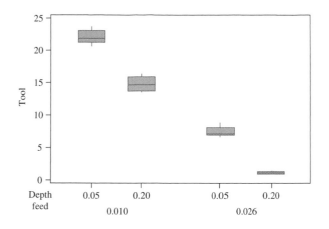

Exercises

2.66 An engineering statistics class ran a catapult experiment to develop a prediction equation for how far a catapult can throw a plastic ball. The class decided to manipulate two factors each at two levels: how far back the operator draws the arm (angle), measured in degrees, and the height of the pin that supports the rubber band, measured in equally spaced locations. The class used four replicates of the four combinations. The data are:

Angle	Height	Distance (inches)			
140	2	27	27	32	26
180	2	81	67	75	79
140	4	67	82	74	81
180	4	137	158	140	154

Use appropriate software to construct parallel boxplots for the unfolded data. Give a thorough discussion of your results.

2.67 Liu, Kan and Chen (1993) conducted an experiment to determine the effect of gas velocity and fluid viscosity on a dimensional factor, K, related to the pressure drop across a screen plate used in bubble columns. Four replicates of the four combinations were used. The data are:

Velocity	Viscosity	K			
2.14	2.63	24.2	17.6	14.0	33.8
8.15	2.63	20.9	15.8	18.3	28.1
2.14	10	28.9	27.2	19.7	29.2
8.15	10	26.4	23.2	22.8	23.6

Use appropriate software to construct parallel boxplots for the unfolded data. Give a thorough discussion of your results.

2.68   Ng and Aspinwall (2002) studied the effect of workpiece hardness and cutting speed on the resultant force of hotwork die steel. Four replicates of the six combinations based on the article are:

Workpiece Hardness	Cutting Speed	Resultant Force			
28	75	1810	1820	1790	1795
28	150	1622	1608	1595	1597
28	200	1451	1448	1463	1447
42	75	1603	1598	1587	1612
42	150	1550	1542	1546	1553
42	200	1500	1492	1496	1511

Use appropriate software to construct parallel boxplots for the unfolded data. Give a thorough discussion of your results.

2.6 Case Study

Pencil lead is a ceramic material consisting of clay and graphite. The clay provides the ceramic matrix that supports the graphite. The ratio of clay to graphite and the "firing" process primarily determine the ceramic properties. A gas-fired kiln converts the clay into the ceramic matrix via a fairly complex chemical reaction. Essentially, the firing process drives out the clay's water of hydration, which then hardens the clay's structure. This loss of the water of hydration causes the resulting ceramic body, in this case the pencil lead, to be porous. After all the water of hydration leaves the body, the firing process actually begins to close the pores. Thus, the porosity of the pencil lead provides a reasonable measure of the quality of the firing process. In general, for a given clay–graphite formulation, the more porous the pencil lead, the softer it writes. Similarly, for a given firing regimen, the more graphite in the formulation, the more porous the lead, and the softer it writes.

An inspector takes a sample of pencil lead from each batch or "lot" of pencil lead produced, determines the average porosity, and records the results in a log to provide process documentation. These logs prove especially important when problems in writing quality occur. Table 2.8 gives the porosities for the basic #2 pencil lead from one such problem period.

The stem-and-leaf display in Figure 2.28 indicates that most of the data lie between 11.5 and 14.5. However, this plot also reveals a possible second peak centered between 15.0 and 15.5. The pattern for the bulk of the data appears to be fairly symmetric but not particularly bell-shaped, which causes some concern.

Table 2.8	The Pencil Lead Porosity Data

12.1	13.5	11.7	12.5	12.5	12.7
13.3	13.8	12.3	12.7	12.5	12.8
12.3	13.5	14.3	13.5	13.2	14.2
13.7	13.9	13.6	13.6	13.3	13.3
13.0	12.6	13.2	14.2	14.1	13.7
12.0	14.2	11.7	12.3	14.5	11.8
12.1	12.7	12.6	13.2	12.3	12.8
13.1	13.4	12.7	13.2	11.8	11.9
14.8	12.4	13.9	15.2	15.8	15.8
15.3	15.4	15.4	15.0	14.3	15.0

In too many situations, such a broad pattern indicates that we have missed an important source of variability. The boxplot in Figure 2.29 cannot pick up the second peak. It does confirm that typical porosities are between 12.5 and 14.0, however. The boxplot does not indicate any outliers.

The presence of the second peak clearly indicates a need for further analysis. A second peak almost always indicates contamination from some source that we need to identify and remove, if possible. In this particular case, quality control has records on which lots were accepted and which were rejected. We thus can stratify our data set. Figure 2.30 gives the side-by-side stem-and-leaf displays comparing

Figure 2.28	The Original Stem-and-Leaf Display for the Pencil Lead Porosities

Stem	Leaves	Number	Depth
11.•:	77889	5	5
12.*:	01133334	8	13
12.•:	55566777788	11	24
13.*:	0122223334	10	34
13.•:	5556677899	10	
14.*:	122233	6	16
14.•:	58	2	10
15.*:	002344	6	8
15.•:	88	2	2

Figure 2.29 | The Original Boxplot for the Pencil Lead Porosities

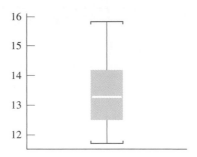

the lots determined acceptable by quality control with the lots rejected. The good lots appear to be slightly skewed to the lower porosities. Typical porosities fall between 13.0 and 14.0. The rejected lots clearly show two peaks: one centered between 12.0 and 13.0 and the other between 15.0 and 16.0. Figure 2.31 gives the corresponding parallel boxplots. The porosities of the good lots form group 1; the porosities of the rejected lots form group 2. Again, the boxplot cannot pick up the two peaks in the rejected data. It does indicate that the rejected lots have much more variability in their porosities even though the "center" seems fairly close to the typical porosity of the good lots.

The two peaks in the porosities for the rejected lots demand further attention because there should be a rational explanation for their existence. Quality control

---

Figure 2.30 | Side-by-Side Stem-and-Leaf Displays Comparing the Porosities of Good and Rejected Lots of Pencil Lead

	Good Lots			Rejected Lots		
Stem	Leaves	Number	Depth	Leaves	Number	Depth
11.•:	7	1	1	7889	4	4
12.*:	133	3	4	01334	5	9
12.•:	5556778	7	11	6778	4	13
13.*:	022333	6		1224	4	17
13.•:	555667789	9	13	9		
14.*:	1223	4	4	23	2	12
14.•:				58	2	10
15.*:				002344	6	8
15.•:				88	2	2

Figure 2.31 | Parallel Boxplots Comparing the Porosities of Good and Rejected Lots of Pencil Lead

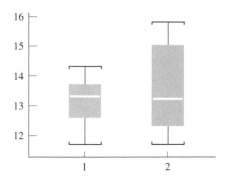

records the reason for rejection. These lots were rejected because they were either too firm or too weak. We can use this information to stratify the rejected lots into these two groups. Figure 2.32 gives the side-by-side stem-and-leaf displays comparing the firm lots to the weak lots. Clearly, the firm lots tend to have lower porosities, which are consistent with slightly too much clay in the formulation, slightly "overfired," or some combination of both. The weak lots tend to have much higher porosities, which are consistent with slightly too much graphite in the formulation, slightly "underfired," or some combination. Figure 2.33 shows the parallel boxplots comparing the good lots (group 1), firm lots (group 2), and weak lots (group 3). Clearly, the firm lots tend to have slightly lower porosities

Figure 2.32 | Side-by-Side Stem-and-Leaf Displays Comparing the Porosities of Firm and Weak Lots of Pencil Lead

	Firm Lots			Weak Lots		
Stem	Leaves	Number	Depth	Leaves	Number	Depth
11.●:	889	3	3	7	1	1
12.*:	01334	5	8			
12.●:	678	3		7	1	2
13.*:	1224	4	8			
13.●:	9	1	4			
14.*:	2	1	3			
14.●:				58	2	4
15.*:				002344	6	
15.●:				88	2	2

Figure 2.33 | Parallel Boxplots Comparing the Porosities of Good (1), Firm (2), and Weak (3) Lots of Pencil Lead

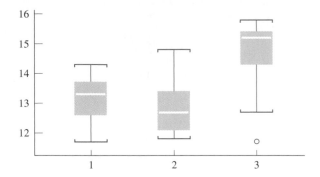

than the good lots. The weak lots tend to have significantly higher porosities. Both phenomena are perfectly consistent with what the clay chemistry predicts. Variations in the moisture content of the clay and variations in the carbon content of the graphite caused the problems with the writing quality of the lead.

2.7 Need for Probability and Distributions

Graphical displays provide a powerful and intuitive basis for studying data and serve as a springboard for further data analysis. All good statistical analyses begin with some graphical examination of the data. In some cases, this graphical examination completely addresses the questions of interest. More typically, the analyst requires more sophisticated techniques.

We have seen how the boxplot uses the median and the quartiles to distill the data down to more fundamental quantities. More sophisticated statistical analyses require us to distill the data down to their most essential elements. Classical statistics distills the data into two quantities: the mean and the variance. *Under certain assumptions*, these two statistics completely describe the behavior of the data and thus provide a basis for more sophisticated analyses. These analyses are useful only to the extent that the assumptions about the behavior of these two statistics hold. Data analysts thus must appreciate these assumptions, why they are important, and how to check their validity.

We can only begin to appreciate the behavior of data and statistics calculated from data through a basic understanding of *probability, random variables,* and the *distributions* of these random variables. These three concepts lie at the heart of statistical modeling and analysis and provide the basis for understanding the behavior of real data. Our next chapter on modeling random behavior provides the foundation for classical statistical analysis.

❯ 2.8   Ideas for Projects

All good projects illustrating the utility of data displays involve collecting real data, as we discussed in Chapter 1.

1. Get data from a laboratory class in which a large number of measurements are taken. For example, in organic chemistry laboratory classes, many students perform the same organic reaction experiment and record their yields. Ideally, the basic chemical reaction produces a specific yield. Use data displays to analyze the actual yields relative to the theoretical value. Discuss differences and any interesting features. Other good sources are surveying classes and unit operations laboratories.

2. Use data displays to analyze the scores from either a homework assignment or an examination. Without getting personal, discuss the results, particularly any outliers.

3. Many instructors use a catapult to teach basic statistical concepts. If you have access to one, fix the throwing conditions and launch the ball eight to ten times. Record the distances and use data displays to analyze the results. Change the throwing conditions, and repeat the exercise. Compare and contrast the two sets of throwing conditions using data displays.

4. Weigh yourself daily for at least a week and record the results. Use data displays to analyze these data. Consider alternative plots that may help to analyze these data.

5. Collect data that will address some personal question of interest. Use data displays to analyze these data. For example, record your daily expenditures for food or entertainment for two weeks. Analyze these data to determine your typical spending patterns. Even better, have one or two friends also record their expenditures. Compare and contrast the results.

❯ *References*

1. Albin, S. L. (1990). The lognormal distribution for modeling quality data when the mean is near zero. *Journal of Quality Technology, 22,* 105–110.
2. Cryer, J. D., and Ryan, T. P. (1990). The estimation of sigma for an X chart: MR/d2 or S/c4. *Journal of Quality Technology, 22,* 187–192.
3. Eibl, S., Kess, U., and Pukelsheim, F. (1992). Achieving a target value for a manufacturing process: A case study. *Journal of Quality Technology, 24,* 22–26.
4. Galinsky, T. L., Rosa, R. R., Warm, J. S., and Dember, W. N. (1993). Psychophysical determinants of stress in sustained attention. *Human Factors, 35,* 603–614.
5. George, E. R., Sullivan, T. M., and Park, E. H. (1994). Thermoplastic starch blends with a poly (ethylene-co-vinyl alcohol): Processability and physical properties. *Polymer Engineering and Science, 34,* 17–23.
6. Good, I. J., and Gaskins, R. A. (1980). Density estimation and bumphunting by the penalized likelihood method explemplified by scattering and meteorite data. *Journal of the American Statistical Association, 75,* 42–56.

7. Grubbs, F. E. (1983). Grubbs' estimators (precision and accuracy of measurement). In S. Kotz & N. L. Johnson (Eds.), *Encyclopedia of Statistical Sciences,* Vol. 3. New York: Wiley.

8. Juran, J., and Gryna, F. (1980). *Quality Planning and Analysis.* New York: McGraw-Hill.

9. Liu, C. H., Kan, M., and Chen, B. H. (1993). A correlation of two-phase pressure drops in screen-plate bubble column. *Canadian Journal of Chemical Engineering, 71,* 460–463.

10. Lucas, J. M. (1985). Counted data CUSUM's. *Technometrics, 27,* 129–144.

11. McGrath, M. P. et al. (1995). Novel polymeric alcohols by controlled catalytic polymer functionalization. *Journal of Applied Polymer Science, 56,* 533–543.

12. Mee, R. W. (1990). An improved procedure for screening based on a correlated, normally distributed variable. *Technometrics, 32,* 331–337.

13. Montgomery, D. C. (2004). *Introduction to statistical quality control, 5th ed.* New York: John Wiley.

14. Montgomery, D. C., Peck, E. A., and Vining, G. G. (2001). *Introduction to linear regression analysis, 3rd ed.* New York: John Wiley.

15. Mulrow, J. M., Vecchia, D. F., Buonaccorsi, J. P., and Iyer, H. K. (1988). Problems with interval estimation when data are adjusted via calibration. *Journal of Quality Technology, 20,* 233–247.

16. Nelson, P. R. (1989). Multiple comparisons of means using simultaneous confidence intervals. *Journal of Quality Technology, 21,* 283–288.

17. Ng, E., and Aspinwall, D. K. (2002). The effect of workpiece hardness and cutting speed on the machinability of AISI H13 hotwork ore steel when using PCBN toolry. *Journal of Manufacturing Science and Engineering, 124,* 588–594.

18. Padgett, W. J., and Spurrier, J. D. (1990). Shewart-type charts for percentiles of strength distributions. *Journal of Quality Technology, 22,* 283–288.

19. Pignatiello, J. J., Jr., and Ramberg, J. S. (1985). Discussion of "Off-line quality control, parameter design, and the Taguchi method" by R. N. Kackar. *Journal of Quality Technology, 17,* 198–206.

20. Rayner, J., and Best, D. J. (1989). *Smooth Tests of Goodness of Fit.* Oxford: Oxford University Press.

21. Smith, R. L., and Naylor, J. C. (1987). A comparison of maximum likelihood and Bayesian estimators for the three-parameter Werbull distribution. *Applied Statistics, 36,* 358–369.

22. Snee, R. D. (1981). Developing blending models for gasoline and other mixtures. *Technometrics, 23,* 119–130.

23. Snee, R. D. (1983). Graphical analysis of process variation studies. *Journal of Quality Technology, 15,* 76–88.

24. Van Nuland, Y. (1992). Maintaining calibration control with a control chart. *Quality Progress, 25*(3), 152.

25. Willams, E. L. II, Grosjean, E., and Grosjean, D. (1993). Ambient levels of the peroxyacyl nitrates PAN, PPN, and MPAN in Atlanta, Georgia. *Air & Waste, 43,* 873–879.

26. Yashchin, E. (1992). Analysis of CUSUM and other Markov-type control schemes by using empirical distributions. *Technometrics, 34,* 54–63.

3

Modeling Random Behavior

▷ ## 3.1 Probability

The Language of Statistics

By now, we should appreciate that real data exhibit variability and that variability entails uncertainty. Statistics uses probability to model this variability and to quantify the resulting uncertainty. Statisticians use probability in much the same way as chemical engineers use chemistry. Chemistry provides a context and a language to the chemical engineer. In a similar way, probability gives us a language for describing uncertainty. The specific probabilities that describe our uncertainty require certain assumptions. How well we model real engineering variability depends on how well these specific assumptions are met.

Three probability statements are

1. The probability that a lot of pencil lead is unacceptable is 0.05.
2. There is a 30% chance that our engineering design firm will get the Nissan contract.
3. There is a 50–50 chance that we will move our office to Nashville.

In each case, we may view the "probability" as the "size" of the chances that something will occur from a set of possibilities of interest relative to the "size" of the chances something will occur from the set of all possible outcomes. Ultimately, *probability represents a standardized measure of chance.*

Defining Probability   An *event* is a set of possible outcomes of interest to us. Let S represent the *sample space,* which is the set of all possible outcomes for a situation of interest. If A is an event, then the *probability* of A, denoted by $P(A)$, is

$$P(A) = \frac{\text{size of the event } A}{\text{size of the sample space } S}.$$

In this case, *size* refers to an appropriate measure of chance. Essentially, $P(A)$ represents the size of the event A relative to the size of the sample space.

Too often, statisticians immediately jump to the conclusion that the most appropriate way to measure size is by counting. In some cases, this approach works; in many, it does not. Engineers often use different approaches to measure size. To quantify the size of a particular problem, they may count how many times it occurs. To quantify the size of a roll of aluminum, they may report its length. To quantify the size of a given day's production, engineers may report a count, a weight, or even a volume. The ways we determine these sizes range from extremely objective to fairly subjective. For example, the engineer may accurately measure a size, roughly measure it, or even give a best guess. The same is true when we measure the chances that something will occur. Note that even the most objective methods for measuring the chances associated with engineering events require certain assumptions. As a result, these probabilities are truly objective only to the extent that the underlying assumptions hold.

Some statisticians view probabilities as *the measures* of chance or likelihood. For these people, the size of the chance that an event will occur is its probability, and thus the definition is circular. They prefer to view probability as a measure of chance that conforms to certain rules. Such a view is legitimate and has merit. On the other hand, we can best see the probability of certain engineering events by determining, by whatever appropriate method, the sizes of the chances for the set of interest and for the set of all possibilities. At least conceptually, we can use this approach for any engineering problem. We have found pedagogical advantages to the relative size approach, which is why we pursue it in this text.

From this definition, we see that

$$0 \le P(A) \le 1.$$

Texts on mathematical statistics use certain axioms plus some basic definitions to derive all the basic theorems of probability.

| Example 3.1 | **Maintenance of Spinning Machines** |

A major manufacturer of textile fibers has several spinning "plants" at a single location. The central maintenance shop provides support for all major repairs and overhauls. Plant 2A with 60 spinning machines and plant 3 with 18 spinning machines produce the same product, tire cord. Since both plants run similar processes, it is reasonable to assume that the chances are the same that any given spinning machine requires attention by the central maintenance shop.

Consider the probability that the next spinning machine requiring attention is in plant 3. We now have this information:

- The set of possible outcomes of interest in this case consists of the spinning machines in plant 3.
- The set of all possible outcomes consists of all the spinning machines in the two plants.
- The "size" of the set of interest is the number of spinning machines in plant 3, which is 18.

- The "size" of the set of all possible outcomes is the total number of spinning machines, which is 18 + 60 or 78.

In this case, our measure of size requires that each spinning machine be equally likely to require attention. If the spinning machines in each plant are of the same make and of similar age, then this assumption is probably reasonable. On the other hand, if the machines in plant 3 are older and historically display more problems, then we may need to determine the measure of size by some other method.

Under the assumption that each spinning machine is equally likely to require repair, the probability, P, that the next machine to require attention is in plant 3 is given by

$$P = \frac{\text{size of the set of interest}}{\text{size of the set of all possible outcomes}}$$

$$= \frac{18}{78} = \frac{3}{13}.$$

This probability provides a basis for modeling the random behavior of which plant's machines require attention. Until the next request comes into the central maintenance shop, we have no idea whether it will come from plant 2A or plant 3. However, if we truly can assume that each spinning machine is equally likely to require repair, then we have solid reason to believe that 3/13 of the time, the next request will come from plant 3. If we see a proportion of requests from plant 3 significantly greater than 3/13, we should begin to question whether this plant's spinning machines behave the same as those in plant 2A.

Using Probability to Make Inferences   We can use probability to make *inferences* about a population of interest. Consider the spinning machine example again. Suppose that the next 50 requests to the central maintenance shop all come from plant 3. Would you believe that the spinning machines in plant 3 require the same attention as the spinning machines in plant 2A? Of course not! Does this conclusion mean that it is impossible for the next 50 requests to come from plant 3 when the spinning machines are truly equally likely to need the same attention? Not at all. Rather, we are making our decision based on the fact that so many requests should be an extremely *rare event*.

Examine this same problem in a slightly different context. Consider each request sequentially:

1. First request comes from plant 3.
2. Second request comes from plant 3.
3. And so on.

At what point do we begin to suspect that the machines in plant 3 need more attention than the machines in plant 2A? To answer this question, we must look at the probabilities associated with these events. We tend to suspect that the machines in plant 3 require more attention once the probability of seeing Y requests in a row is sufficiently small under the assumption that they require the same attention. We shall use a similar approach when we make decisions about populations.

Conditional Probability

Often, two events are related. If we know the relationship between the two events, then we can model the behavior of one event in terms of the other. *Conditional probability* quantifies the chances of one event occurring given that the other occurs. It thus provides a basis for explaining the relationship between the two events. For example, a major manufacturing facility for titanium dioxide uses two production lines. We might believe that the operation of the two lines is related. Thus, knowing that one line is running may give us information about whether the other is running. We then can predict whether one line is operating based on whether the other is running.

The overlap or the intersection of the two events defines the nature of their relationship. If A and B are events, then the *intersection* of A and B, denoted by $A \cap B$, is the set of outcomes that are both in A and in B. Thus, the event $A \cap B$ is the event that both A and B occur.

A useful tool for visualizing the relationships among sets is the *Venn diagram*. The universal set, S, is represented by a rectangle. Particular sets within S are denoted by geometric figures. Figure 3.1 gives the Venn diagram for the intersection of two events.

In some cases, knowing that one event has occurred gives us information about the chances that the other event will occur. For example, suppose we select two silicon wafers at random from a set of ten, of which four happen to be defective. By selecting these two wafers at random, we assume that each wafer is equally likely to be selected. Under this assumption, the appropriate way to measure size is by counting. Let A be the event that the first wafer selected is defective, and let B be the event that the second wafer selected is defective. The probability that A occurs is

$$P(A) = \frac{\text{size of the event } A}{\text{size of the sample space } S}$$

$$= \frac{4}{10} = .4.$$

Figure 3.1 | The Venn Diagram Illustrating the Intersection of Two Events

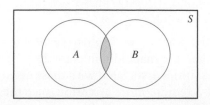

Once we know that A has occurred, there are only nine wafers left in the sample space, of which only three are defective. Thus, the probability that B occurs given that we know that A has occurred is

$$\frac{3}{9} = \frac{1}{3}.$$

We call the probability that the event B will occur given that the event A has occurred the *conditional probability of B given A*. We denote such a probability by $P(B \mid A)$. Figure 3.1 helps to illustrate this concept. Once we know that line B is operating (that event B has occurred), then the set of all possible outcomes becomes B. Furthermore, the set of interest now may be thought of as only that part of A that resides in B. Thus, once we know that the event B has occurred, the set of interest is $A \cap B$.

This insight provides a basis for establishing "mathematically" the definition for conditional probability:

$$P(A \mid B) = \frac{\text{size of the set of interest } (A \cap B)}{\text{size of the set of all possible outcomes } (B)}$$

$$= \frac{\text{size of the set of interest } (A \cap B)}{\text{size of the set of all possible outcomes } (B)} \cdot 1$$

$$= \frac{\text{size of the set of interest } (A \cap B)}{\text{size of the set of all possible outcomes } (B)} \cdot \frac{\frac{1}{\text{size of } S}}{\frac{1}{\text{size of } S}}$$

$$= \frac{\frac{\text{size of the set of interest } (A \cap B)}{\text{size of } S}}{\frac{\text{size of the set of all possible outcomes } (B)}{\text{size of } S}}$$

$$= \frac{P(A \cap B)}{P(B)}.$$

Definition 3.1 | **Conditional Probability**

Let A and B be events in S. The **conditional probability** of B given that A has occurred is

$$P(B \mid A) = \frac{P(A \cap B)}{P(A)}$$

if $P(A) > 0$. Similarly, the conditional probability of A given that B has occurred is

$$P(A \mid B) = \frac{P(A \cap B)}{P(B)}$$

if $P(B) > 0$.

An example helps to illustrate this concept.

Example 3.2 | **Titanium Dioxide Production Lines**

A major manufacturer of titanium dioxide, which is the white pigment in paint, reacts titanium ore with chlorine to form titanium tetrachloride. It then reacts the titanium tetrachloride with oxygen to form the final product. This particular facility uses two separate production lines (A and B) of approximately the same size. The process, which requires highly corrosive materials and high temperatures, is a maintenance manager's nightmare. Historically, the probability that line A is operating is 0.85, the probability that line B is operating is 0.85, and the probability that they are both operating is 0.71. Let A be the event line A is operating, and let B be the event line B is operating. Thus,

$$P(A) = 0.85$$
$$P(B) = 0.85$$
$$P(A \cap B) = 0.71.$$

The conditional probability that line A is operating given that line B is operating is

$$P(A \mid B) = \frac{P(A \cap B)}{P(B)}$$
$$= \frac{0.71}{0.85} = 0.835.$$

Thus, if we know that line B is operating, then the chances that line A is also operating are approximately 83.5%.

Conditional probability is extremely useful in cases of unequal sample sizes. Consider a case of 100 parts made by two operators. Suppose that operator 1 made 5 defects and operator 2 made 10 defects, which appears to imply operator 2 is much worse (assuming they each made 50). However, suppose that operator 1 made only 20 of the parts. Then operator 1's production contained $5/20 = 25\%$ defective parts. However, operator 2's production contained $10/80 = 12.5\%$ defective parts. As a result, operator 1 clearly has a higher defective rate than operator 2.

Independence

Conditional probability allows us to see the nature of the relationship between two events. In many engineering settings, two or more events have no real relationship. Thus, information about the event B provides no new information about A. When this occurs, we say that the two events are statistically *independent*. Independent events are important for several reasons:

- Many engineering events are either independent or close enough that the assumption of independence is a very reasonable first approximation.
- Independence provides a powerful basis for modeling the joint behavior of engineering events.
- The formal concept of a *random sample* assumes that the individual observations are independent of one another.

Definition 3.2 | **Independence**

Let A and B be events in S. A and B are said to be **independent** if

$$P(A \mid B) = P(A).$$

Similarly, if A and B are independent, then

$$P(B \mid A) = P(B).$$

If two events are not independent, then they are said to be *dependent*. The basic idea underlying independence is that information about the event A provides no new information about B. Thus, the probability that A will occur given that B has occurred is still $P(A)$.

Example 3.3 | **Titanium Dioxide Production Lines—Revisited**

Consider our last example involving the two titanium dioxide production lines. To determine whether these two events are independent, we need to compare either $P(A \mid B)$ to $P(A)$ or $P(B \mid A)$ to $P(B)$. In the last example, we found

$$P(A) = 0.85 \quad \text{and} \quad P(A \mid B) = 0.835.$$

Thus, these two events are not independent. If we know that one of the lines is operating, then we have some information about the chances that the other line is operating. In this particular case, both lines undergo regular preventive maintenance where the line is taken out of operation for inspection and repairs. Production management clearly wants to have at least part of the plant running as much of the time as possible. As a result, production management will not shut down an operating line for preventive maintenance when the other line is already down for unscheduled maintenance. Only when the unscheduled repairs are made and that line is operating again will production allow the other line to go down for its scheduled maintenance. Consequently, the two events are truly dependent.

Basic Rules of Probability

The following rules of probability often prove useful to engineers. Occasionally in this text, we make use of these rules to establish some important results.

1. $0 \leq P(A) \leq 1$.
2. If $\emptyset$ is the empty set, then $P(\emptyset) = 0$.
3. **The probability of complements:** If A is an event in some sample space S, then the *complement* of the set A relative to S is the set of outcomes in S that are not in A. The complement of A is denoted by $\overline{A}$. It is straightforward to establish that

    $$P(\overline{A}) = 1 - P(A).$$

4. **The additive law of probability:** This result deals with the union of events. If A and B are events in S, then the *union* of A and B, denoted by $A \cup B$, is the set of outcomes either in A or in B. In this context, we use the inclusive sense of "or." Thus, $A \cup B$ is the event that either A, B, or both occur. The additive law of probability is

    $$P(A \cup B) = P(A) + P(B) - P(A \cap B).$$

    If A and B are *mutually exclusive*, which means that A and B cannot both happen, then $A \cap B = \emptyset$ and $P(A \cap B) = 0$. In this case;

    $$P(A \cup B) = P(A) + P(B).$$

5. **The multiplicative law of probability:** This result follows directly from the concept of conditional probability. If A and B are events in S, with $P(A) > 0$ and $P(B) > 0$, then

    $$P(B \mid A) = \frac{P(A \cap B)}{P(A)}.$$

    Thus,

    $$P(A \cap B) = P(A) \cdot P(B \mid A).$$

    Similarly,

    $$P(A \cap B) = P(B) \cdot P(A \mid B).$$

    If A and B are independent, then

    $$P(A \mid B) = P(A) \quad \text{and} \quad P(B \mid A) = P(B).$$

Thus, if A and B are independent, then

$$P(A \cap B) = P(A) \cdot P(B).$$

This property is a very powerful result, making independence quite important for finding the probabilities associated with the intersections of events.

6. **Simplest form of the law of total probability:** Let A and B be events in S. We may partition B into two parts: the part that overlaps A, $A \cap B$, and the part that overlaps $\overline{A}$, $\overline{A} \cap B$. Thus,

$$P(B) = P(A \cap B) + P(\overline{A} \cap B)$$
$$= P(A) \cdot P(B \mid A) + P(\overline{A}) \cdot P(B \mid \overline{A}).$$

7. **The simplest form of Bayes rule:** Let A and B be events in S. Suppose we are given $P(A)$, $P(B \mid A)$, and $P(B \mid \overline{A})$. We can find $P(A \mid B)$ by *Bayes rule*, which in this case is

$$P(A \mid B) = \frac{P(A \cap B)}{P(B)}$$
$$= \frac{P(A) \cdot P(B \mid A)}{P(A) \cdot P(B \mid A) + P(\overline{A}) \cdot P(B \mid \overline{A})}.$$

Example 3.4	**Engineering Design Project Bids**

Many engineering design and construction firms fluctuate between feast (having too many contracts at one time) and famine (not enough contracts). Of course, these firms must make their bids within a highly competitive environment. At any given time, a typical project manager may have several bids under consideration in the hopes that one or two "hit." Clearly, the project manager must consider the likelihood of each bid being accepted.

Recently, a project manager for an engineering firm submitted bids to provide multidisciplinary engineering design on four projects. The firm formally evaluates the chances that each bid will be accepted. The following table summarizes the firm's beliefs by project.

Company	Probability
A	0.3
B	0.8
C	0.1
D	0.5

In this situation, we can reasonably assume that the individual bids are truly *independent*. Unless the parties involved discuss the bids among themselves, which really is unlikely, then knowing that company A has accepted the bid gives us no real information about whether company B will accept.

Project managers worry most about no bids being accepted. If at least one is accepted, they still have a job the next week! The situation "at least one of the bids is accepted" corresponds to the event

$$A \cup B \cup C \cup D.$$

Finding the union of four or more events (often, three or more events) can be a major challenge. Under the assumption of independence, however, finding the probability of intersections is easy.

Our approach begins by observing that the complement of either A or B or C or D accepting their bids is

$$\overline{(A \cup B \cup C \cup D)}$$

and is equivalent to none of them accepting their bids. Since company A not accepting its bid is the complement of A, none of the parties accepting their bids is

$$\overline{A} \cap \overline{B} \cap \overline{C} \cap \overline{D}.$$

Thus,

$$\overline{(A \cup B \cup C \cup D)} = \overline{A} \cap \overline{B} \cap \overline{C} \cap \overline{D},$$

which is an application of *DeMorgan's law* for sets. Putting all of this together, we have

$$P(A \cup B \cup C \cup D) = 1 - P[\overline{(A \cup B \cup C \cup D)}]$$

$$= 1 - P(\overline{A} \cap \overline{B} \cap \overline{C} \cap \overline{D}).$$

By independence, the probability of the intersection is the product of the individual probabilities. In this case,

$$P(A \cup B \cup C \cup D) = 1 - P(\overline{A}) \cdot P(\overline{B}) \cdot P(\overline{C}) \cdot P(\overline{D})$$

$$= 1 - (0.7)(0.2)(0.9)(0.5)$$

$$= 1 - 0.063$$

$$= 0.937.$$

As a result, the project manager should feel quite comfortable that his firm will have work in the near future; there is a 93.7% chance.

Of course, the firm would have great trouble completing all four projects at the same time. It thus must consider the chances that all four bids are accepted, which is $P(A \cap B \cap C \cap D)$. Again, if we can assume independence, then

$$P(A \cap B \cap C \cap D) = P(A) \cdot P(B) \cdot P(C) \cdot P(D)$$

$$= (0.3)(0.8)(0.1)(0.5)$$

$$= 0.012.$$

The firm should not worry too much about having to fulfill all four contracts and thus being completely overwhelmed.

Finally, consider the probability that only company A accepts the bid, which is $P(A \cap \overline{B} \cap \overline{C} \cap \overline{D})$ and is given by

$$P(A \cap \overline{B} \cap \overline{C} \cap \overline{D}) = P(A) \cdot P(\overline{B}) \cdot P(\overline{C}) \cdot P(\overline{D})$$
$$= (0.3)(0.2)(0.9)(0.5)$$
$$= 0.027.$$

⋙ **3.2** Random Variables and Distributions

Random variables and their distributions provide a basis for modeling and describing the behavior of important characteristics of interest. We can best illustrate the concept of a random variable by an example.

Example 3.5	**Titanium Dioxide Production Lines—Continued**

Production management really needs to know *how many* lines are running at any given time rather than which lines are running. Let Y be the number of lines running at any given time. Since there are two lines, the possible values for Y are 0, 1, or 2. Each of these values for Y corresponds to some *event* involving the two production lines:

- $Y = 0$ corresponds to the event that both lines are not running.
- $Y = 1$ corresponds to the event that either one of the lines is running, but not both.
- $Y = 2$ corresponds to the event that both lines are running.

Since each possible value for Y corresponds to an event, we call Y a *random variable*.

Definition 3.3	**Random Variable**

We call Y a **random variable** if Y is a function that assigns a real numbered value to every possible event in a sample space of interest.

VOICE OF EXPERIENCE

Random variables provide a way to model the behavior of real data.

Statisticians generally use uppercase Roman letters to denote random variables and lowercase Roman letters to denote specific values for these variables. Statistical tradition occasionally supports the use of lowercase Roman letters for both to avoid unnecessary complication and confusion.

In Example 3.5, Y is the random variable representing the number of lines operating at any given time. The value y represents a specific value for Y and must be either 0, 1, or 2.

Distributions

Distributions describe the random behavior of random variables and provide a basis for developing probabilistic models for important characteristics of interest. Since every possible set of values for a random variable Y corresponds to some event, it has a probability associated with it. A random variable's distribution details in a meaningful way the probabilities associated with these sets of values. In so doing, we can model the behavior of the random variable.

We can model the behavior of every random variable by its *cumulative distribution function*.

Definition 3.4 | **Cumulative Distribution Function**

If Y is a random variable, then the *cumulative distribution function* (cdf), denoted by $F(y)$, is given by

$$F(y) = P(Y \leq y)$$

for all real numbers y.

All random variables must have a cdf. For some random variables, we can describe their random behavior only by their cdf's.

Example 3.6 | **Titanium Dioxide Production Lines—Continued**

From historical data, we know the probabilities associated with the number of lines operating at any given time, Y. We thus can summarize Y's distribution with Table 3.1 We can generate the values for the column $P(Y = y_i)$ from the information given in Example 3.3 and the basic rules of probability.

We shall discuss two types of random variables: discrete and continuous.

Table 3.1 | The Distribution Function for the Titanium Dioxide Production Lines

y_i	$P(Y = y_i)$	$P(Y \leq y_i)$
0	.01	.01
1	.28	.29
2	.71	1.00

⊳ 3.3 Discrete Random Variables

Definition 3.5	**Discrete Random Variable**
	A **discrete random variable** is one that can assume at most a countable number of values.

By *countable,* we mean that the possible values, at least theoretically, are able to be counted. Three examples of discrete random variables are the number of defective silicon chips in a lot, the number of bids on contracts accepted, and the number of defects in a new car.

Every discrete random variable has a *probability function, $p(y)$,* defined by

$$p(y) = P(Y = y).$$

Often, we use the probability function to outline the distribution of a discrete random variable. Two properties of $p(y)$ are

1. $0 \le p(y) \le 1$, for every possible value of y.
2. $\sum_y p(y) = 1$.

Property 1 follows from the fact that $p(y)$ is a probability. Property 2 follows from the fact that $P(S) = 1$; that is, something from the sample space must occur.

The next example illustrates how we can model the behavior of a discrete random variable by the probability function, $p(y)$, and the cumulative distribution function, $F(y)$.

Example 3.7	**Engineering Design Project Bids—Revisited**
	In Example 3.4, we discussed a real engineering design firm that submitted bids to provide multidisciplinary engineering design on four projects. Let Y be the random variable representing the number of bids accepted. By following the techniques illustrated in that example, we can generate this random variable's probability function and cdf based on the firm's beliefs, which Table 3.2 summarizes.

It is interesting that three of the firm's bids were actually accepted. The distribution provides us with a basis for modeling the number of bids accepted. In so doing, we can evaluate the chances of being at least this successful. Based on the firm's beliefs, the chances of having three or more bids accepted is

$$P(Y \ge 3) = 1 - P(\overline{Y \ge 3})$$
$$= 1 - P(Y < 3)$$
$$= 1 - P(Y \le 2)$$

because one cannot accept a fraction of a bid. Thus,

$$P(Y \ge 3) = 1 - F(2) = 1 - 0.837 = 0.163.$$

| Table 3.2 | The Distribution Function for the Engineering Design Project Bids |

y	$p(y)$	$F(y)$
0	0.063	0.063
1	0.349	0.412
2	0.425	0.837
3	0.151	0.988
4	0.012	1.000

As a result, *if the firm's beliefs really were true,* then the chances of three or more bids being accepted were about 16%. In this particular case, the firm had been seriously concerned about not having enough work, which probably caused them to be rather pessimistic about their chances for these bids being accepted.

Expected Values for Discrete Distributions

Random variables and their distributions provide a basis for modeling and for describing populations. The preceding example used a discrete distribution as an appropriate *probabilistic model*. This model provides a basis for identifying such important characteristics as the expected number of bids accepted and the possible variability in the number of bids accepted. A reasonable definition of the expected number of bids accepted is the "long-run average" number of bids accepted if we could repeat this specific bidding process an infinite number of times.

In general, if Y is a random variable, then its long-run average is called the *expected value*, denoted by $E(Y)$.

| Definition 3.6 | **The Population Mean** |

We define the *population mean*, μ, by

$$\mu = E(Y).$$

Obviously, we cannot physically repeat this specific bidding process an infinite number of times. However, for a discrete random variable, we can use its probability function, $p(y)$, to describe what would happen if we could physically repeat the bidding process. Under the assumption that $p(y)$ adequately describes the long-run behavior of a random variable, Y, we can obtain the expected value by

$$\mu = E(Y) = \sum_y y \cdot p(y),$$

which is nothing more than a weighted average of the possible values.

Population Variance and Standard Deviation   Variability implies that the data *deviate* from one another. The classical definitions of variability all look at how the data deviate from a typical value. For example, let y_i be a specific data value from some distribution with mean μ. We can define its *deviation* from the population mean by $y_i - \mu$. We can easily show that

$$E(y_i - \mu) = 0$$

because the negative deviations cancel the positive ones. The classical measure of the population variability avoids this problem by squaring the deviation.

Definition 3.7 | **Population Variance**

We define the *population variance*, σ^2, by

$$\sigma^2 = \text{var}(Y) = E[(Y - \mu)^2].$$

We may view the population variance as the long-run average squared deviation from the population mean. The computational formula for the population variance is

$$\sigma^2 = E(y^2) - \mu^2.$$

For a discrete random variable,

$$E(y^2) = \sum_y y^2 \cdot p(y).$$

Definition 3.8 | **Population Standard Deviation**

We define the *population standard deviation, σ,* by

$$\sigma = \sqrt{\sigma^2}.$$

Why do we have two measures of variability? Consider the engineering design bids example. The "units" associated with the population variance, σ^2, are bids^2! The units for σ, however, are bids. Thus, the scale for σ is the same as the scale of the data.

Some textbooks motivate the value of the population standard deviation by the so-called *empirical rule*, which states that virtually all of the data for a particular distribution should fall within the interval $\mu \pm 3\sigma$. In general, we take this recommendation with a grain of salt. Very skewed or heavy-tailed distributions do have valid data values outside this interval. However, the rule points out that we can begin to describe the behavior of many distributions with just two measures: the population mean and the population standard deviation. Many engineers commonly evaluate their data using this notion of the mean plus or minus three standard deviations because even for most skewed distributions, the probability of seeing any single data value outside of this range is remote.

| Example 3.8 | **Engineering Design Project Bids—Continued** |

We have discussed an engineering design firm that submitted bids to provide multidisciplinary engineering design on four projects. Management would like to know the expected number of bids that will be accepted and some measure of the variability around this expected value. Let Y be the random variable representing the number of bids accepted. Table 3.3 lists the information relevant for finding the appropriate expected values based on this firm's beliefs.

First, consider the population mean, which is given by

$$\mu = E(Y) = \sum_y y \cdot p(y) = 1.7.$$

The engineering firm should expect 1.7 of the bids to be accepted. To find the variance, we first must find $E(y^2)$, which is given by

$$E(Y^2) = \sum_y y^2 \cdot p(y) = 3.6.$$

The variance is then

$$\sigma^2 = E(y^2) - \mu^2$$
$$= 3.6 - (1.7)^2$$
$$= 3.6 - 2.89$$
$$= 0.71.$$

| Table 3.3 | Summary of Information for Finding the Population Mean and Variance for the Engineering Design Project Bid Example |

y	$p(y)$	$y \cdot p(y)$	y^2	$y^2 \cdot p(y)$
0	0.063	0.0	0	0.000
1	0.349	0.349	1	0.349
2	0.425	0.850	4	1.700
3	0.151	0.453	9	1.359
4	0.012	0.048	16	0.192
	1.000	1.700		3.600

The population standard deviation is

$$\sigma = \sqrt{\sigma^2} = \sqrt{0.71} = 0.843.$$

By the empirical rule, the actual number of bids accepted should fall within the interval

$$\mu \pm 3\sigma = 1.7 \pm 3(0.843)$$
$$= (-0.829, 4.229).$$

Since the entire range of possible bids accepted is (0, 4), we see that the empirical rule actually does contain all of the possible values.

The Binomial Distribution

We often must model the random behavior of data that we can classify as either a *success* or a *failure*. For example, we may need to model one of these variables:

- The number of battery plates in a lot that fail to meet specifications, where each plate either meets the specification or does not.
- The number of people in a sample who prefer our new automobile model, where each person either prefers our new model or does not.

In these situations, we typically use the *binomial distribution* as our probability model. Since we encounter these dichotomous situations so often, the binomial is the single most important discrete distribution. We can best illustrate this distribution through an example.

Example 3.9 | **A Single Operator Making Battery Plates**

The manufacturer of nickel battery plates has imposed a tight initial weight specification that is difficult to meet. Consider the next three attempts made by an operator who has a 40% chance of achieving the specification on any given attempt. These three attempts take approximately 10 minutes of production time. Let S represent a successful attempt, let F represent a failed attempt, and let Y represent the number of successful attempts she makes.

Consider the probability that exactly two out of these three attempts are successful, or $P(Y = 2)$. The possible ways she can get exactly two successful attempts are

$$(SSF) \quad (SFS) \quad (FSS).$$

Since these events are *mutually exclusive* (no two of these events can happen at the same time), the probability of exactly two successful attempts is

$$P(Y = 2) = P(SSF) + P(SFS) + P(FSS).$$

In this situation, we can reasonably assume that each attempt is *independent* of the others (knowing that she succeeded on the first attempt gives us no useful information about whether she will be successful on the next attempt). Let p be the probability that she succeeds in meeting the weight specification on any given attempt. Thus, $p = .4$. Let q be the probability that she fails. In general, $q = 1 - p$. In this case, $q = .6$. Since each attempt is independent of the others, we have

$$P(SSF) = P(S) \cdot P(S) \cdot P(F) = p \cdot p \cdot q = p^2 \cdot q = 0.096$$
$$P(SFS) = P(S) \cdot P(F) \cdot P(S) = p \cdot q \cdot p = p^2 \cdot q = 0.096$$
$$P(FSS) = P(F) \cdot P(S) \cdot P(S) = q \cdot p \cdot p = p^2 \cdot q = 0.096.$$

As a result,

$$\begin{aligned}
P(Y = 2) &= P(SSF) + P(SFS) + P(FSS) \\
&= p^2 \cdot q + p^2 \cdot q + p^2 \cdot q \\
&= 3 \cdot p^2 \cdot q \\
&= (\text{number of ways to get two successes}) \cdot p^2 \cdot q \\
&= 3 \cdot (0.096) = 0.288.
\end{aligned}$$

In general, if she makes n total attempts, the probability that she succeeds exactly y times is

$$P(Y = y) = (\text{number of ways to get } y \text{ successes out of } n) \cdot p^y \cdot q^{n-y},$$

which is a form for the probability function of a binomial random variable.

We commonly use the binomial coefficient $\binom{n}{y}$, read "n choose y," to denote the number of ways to get y successes from n total attempts. We define $\binom{n}{y}$ by

$$\binom{n}{y} = \frac{n!}{y!(n-y)!}$$

where $n! = n(n-1)(n-2)\cdots(2)(1)$ and by definition

$$0! = 1.$$

We now can write the probability of obtaining exactly y successes out of n total attempts as

$$P(Y = y) = (\text{number of ways to get } y \text{ successes out of } n) \cdot p^y \cdot q^{n-y}$$

$$= \binom{n}{y} p^y \cdot q^{n-y}$$

$$= \frac{n!}{y!(n-y)!} p^y \cdot q^{n-y}.$$

The General Form of the Binomial Distribution   Consider an experiment that meets these five conditions:

1. The experiment consists of n identical attempts or "trials."
2. Each trial results in one of two outcomes: one outcome that is considered a "success" and the other considered a "failure."
3. The probability of a success on a single trial is p and remains the same from trial to trial. As a result, the probability of a failure is $q = 1 - p$.
4. The trials are independent.
5. We are interested in Y, which represents the total number of successes among the n trials.

These conditions describe a binomial experiment. The variable, Y, is called a binomial random variable and is said to follow a *binomial distribution*. The probability function for a binomial random variable is

$$p(y) = \begin{cases} \binom{n}{y} p^y \cdot q^{n-y} = \dfrac{n!}{y!(n-y)!} p^y \cdot q^{n-y} & \text{for } y = 0, 1, 2, \ldots, n \\ 0 & \text{otherwise.} \end{cases}$$

If Y follows a binomial distribution, then we can show that

$$\mu = E(Y) = np$$

$$\sigma^2 = npq = np(1 - p)$$

$$\sigma = \sqrt{np\,q}.$$

Note that while the contract bids example has two possible outcomes it does not follow a binomial distribution because the probability of getting the project is *not* constant across the four contract opportunities.

Example 3.10 | **Nonconforming Brick**

Marcucci (1985) reports on a brick manufacturing process that classifies the product into one of these categories:

- Suitable for all purposes (standard)
- Structurally sound but not suitable for all uses (chipped face)
- Unacceptable for use (cull)

The latter two categories may be viewed as not meeting the standard, or as non-conforming. Historically, the process has produced 5% nonconforming bricks. Under normal circumstances, this facility makes 25 bricks per hour.

First, consider the probability that this facility produces exactly two nonconforming bricks in the next hour. Let Y be the random variable associated with the number of nonconforming bricks produced. In this case, $n = 25$ and we seek $P(Y = 2)$, which is

$$P(Y = 2) = \binom{n}{y} p^y \cdot q^{n-y}$$

$$= \frac{25!}{2! \cdot 23!} (0.05)^2 \cdot (0.95)^{23}$$

$$= 300 \cdot (0.0007684)$$

$$= 0.23.$$

As a result, we should not be surprised to see two nonconforming bricks in the next hour's production; there is a 23% chance.

Management is really more concerned about the probability that at least one brick in the next hour's production fails to conform to the standards. We thus seek $P(Y \geq 1)$. The easiest way to approach this problem notes that

$$P(Y \geq 1) = 1 - P(\overline{Y \geq 1})$$

$$= 1 - P(Y < 1)$$

$$= 1 - P(Y = 0)$$

because the only possible number of nonconforming bricks less than 1 is 0. As a result,

$$P(Y \geq 1) = 1 - \binom{25}{0} \cdot p^0 \cdot q^{25-0}$$

$$= 1 - \frac{25!}{0!25!} \cdot (0.95)^{25}$$

$$= 1 - (0.95)^{25} = 0.723.$$

Consequently, management should expect at least one nonconforming brick an hour more than 70% of the time!

Finally, consider the typical number, μ, of nonconforming bricks per hour and the variation (σ^2 and σ) in the number of nonconforming bricks per hour:

$$\mu = np = 25(.05) = 1.25$$

$$\sigma^2 = npq = 25(0.05)(0.95) = 1.1875$$

$$\sigma = \sqrt{npq} = \sqrt{1.1875} = 1.09.$$

Engineers use the *hypergeometric distribution* to model the number of successes out of *n* trials when sampling without replacement from a finite population of N objects that contains exactly *r* successes. As N gets large, we often can well approximate the hypergeometric distribution by the binomial. Historically, engineers have used the hypergeometric distribution to develop acceptance sampling plans for small lots where we cannot model the behavior by the binomial distribution.

⤷ Exercises

3.1 Consider a process for making nickel battery plates that has an operator who successfully meets the weight specification only 20% of the time. Let Y be the number of times she successfully meets the specification on her next three attempts. If we assume that each attempt is independent of the other attempts, then the following table summarizes the probability function describing Y:

y_i	0	1	2	3
$p(y_i)$	0.512	0.384	0.096	0.008

   a. Find the probability that she is successful fewer than two times.
   b. Find the probability that she is successful more than one time.
   c. Find the expected number of times she is successful.
   d. Find the variance and the standard deviation for the number of times she is successful.

3.2 A pencil company has four extruders for making pencil lead. The maintenance manager has determined from historical data that the following table describes the distribution of the number of extruders down (out of operation) on any given day:

y	0	1	2	3	4
$p(y)$	0.5	0.3	0.1	0.05	0.05

   a. Find the probability that three or more extruders are down.
   b. Find the probability that fewer than one extruder is down.
   c. Find the expected number of extruders down.
   d. Find the population variance and the population standard deviation for the number of extruders down.

3.3 An injection molding process for making detergent bottles uses four different machines. The table gives the distribution for the number of machines operating at any given time:

y_i	0	1	2	3	4
$p(y_i)$	0.005	0.010	0.035	0.050	0.900

a. Find the probability that two or fewer machines are running.
b. Find the probability that at least one machine is running.
c. Find the expected number of machines running.
d. Find the variance and the standard deviation for the number of machines running.

3.4 A sales engineer for a manufacturer of high-speed grinding equipment has just returned from visiting five possible clients. She believes that the following table describes the distribution for the number of sales she will make:

y_i	0	1	2	3	4	5
$p(y_i)$	0.05	0.30	0.30	0.20	0.10	0.05

a. Find the probability that she makes more than three sales.
b. Find the probability that she makes two or more sales.
c. Find the expected number of sales.
d. Find the variance and the standard deviation for the number of sales.

3.5 A manufacturer of silicon wafers has encountered an operating problem wherein too many of the chips made are unacceptable. An inspector selects a sample of five wafers from each shift. The following table, based on the recent performance of this process, describes the distribution of the number of unacceptable wafers in the sample:

y_i	0	1	2	3	4	5
$p(y_i)$	0.3106	0.4313	0.2098	0.0442	0.0040	0.0001

a. Find the probability that exactly two wafers in the sample are unacceptable.
b. Find the probability that at least one wafer is unacceptable.
c. Find the mean, the variance, and the standard deviation for the number of unacceptable wafers in the sample.

3.6 A metallurgical engineer believes that too many steel specimens produced by an older process will fail to meet a new, more stringent strength standard imposed by a customer. An inspector selects four specimens for destructive testing. The table, based on the engineer's best guess, describes the distribution of the number of specimens in the sample that fail to meet the specification:

y_i	0	1	2	3	4
$p(y_i)$	0.1022	0.3633	0.3814	0.1387	0.0144

a. Find the probability that at least three specimens fail to meet the new specification.
b. Find the probability that fewer than two fail to meet the new specification.
c. Find the mean, the variance, and the standard deviation for the number of specimens that fail to meet the new specification.

3.7 A metal cutting plant is concerned with accidents. The table gives the distribution for the number of accidents per month:

y_i	0	1	2	3	4	5	6	7	8
$p(y_i)$	0.716	0.180	0.060	0.020	0.010	0.010	0.002	0.000	0.002

a. Find the probability that one or fewer accidents will occur.
b. Find the probability that more than one but fewer than 5 accidents will occur.
c. Find the expected number of accidents.
d. Find the variance and standard deviation for the number of accidents.

3.8 A software company has experienced an increase in programming errors. Four pages are randomly selected. The following table, based on prior process knowledge, describes the distribution of programming errors in the sample:

y_i	0	1	2	3	4
$p(y_i)$	0.52	0.24	0.12	0.09	0.03

a. Find the probability that no programming errors are found in the sample.
b. Find the probability that more than two programming errors are found in the sample.
c. Find the mean, variance, and standard deviation for the programming errors in the sample.

3.9 One quality characteristic for cabinet manufacturers is easy sliding drawers. A drawer is considered easy sliding if it does not get stuck when opened. Historically, only 2% of the drawers get stuck. Ten drawers are randomly selected from a new lot. Each drawer is tested and all 10 must be easy sliding before the new lot is shipped.
a. Find the probability that the lot will be shipped.
b. Find the probability that at least one drawer will get stuck.

    c. Find the expected number of stuck drawers, the variance, and the standard deviation.

3.10 Airplanes approaching the runway for landing are required to stay within the localizer (a certain distance left and right of the runway). When an airplane deviates from the localizer, it is sometimes referred to as an exceedence. Consider one airline at a small airport with six daily arrivals and an exceedence rate of 7%.

    a. Find the probability that on a particular day exactly one plane has an exceedence.

    b. Find the probability that on a particular day no planes experience an exceedence.

    c. Find the expected number of planes that have an exceedence, the variance, and the standard deviation.

3.11 A manufacturer of water filters for refrigerators monitors the process for defective filters (the filter leaks). Historically, this process averages 5% defective filters. Suppose 5 filters are selected for testing.

    a. Find the probability that all five filters are defective.

    b. Find the probability that at least 1 but no more than 3 filters are defective.

    c. Find the expected number of defective filters, the variance, and the standard deviation.

3.12 A manufacturer of nickel–hydrogen batteries discovered a problem with "blisters" on its nickel plates. These blisters cause the resulting battery cell to short out prematurely. During a specific production period, 8.5% of the plates exhibited blisters within 50 test cycles. During this period, the company made a series of test cells, each using ten plates.

    a. Find the probability that none of the ten plates blisters.

    b. Find the probability that exactly one blisters.

    c. Find the expected number of plates that blister, the variance, and the standard deviation.

3.13 Atwood (1986) studied the failure of pumps used in standby safety systems for commercial nuclear power plants. The number of pumps in a safety system ranged from two to eight, depending on the specific nature of the system. He found that the probability that a randomly selected pump failed to run after starting was 0.16.

    a. Typically, low-pressure coolant injection systems use four pumps. Consider a periodic inspection of this system that tests each pump.

       (1) Find the probability that all four fail a periodic inspection.

       (2) Find the probability that at least one fails.

       (3) Find the expected number of pumps that fail, the variance, and the standard deviation.

    b. Suppose a highly critical system uses eight pumps. Consider a periodic inspection of this system that tests each pump.

       (1) Find the probability that none of the pumps fails.

(2) Find the probability that exactly two pumps fail.

(3) Find the expected number of pumps that fail, the variance, and the standard deviation.

3.14 Metal casting processes are notoriously slow and expensive. A common problem facing many older casting processes is "flashing." In casting, liquid metal is shot into a mold and rapidly cooled. A flash commonly forms on the piece at the spot in the mold where the metal flows. Sanding corrects minor problems with flashing, but severe problems may require scrapping the part. A particular casting operation historically has scrapped 10% of its parts due to severe problems with flashing. Consider the next ten parts cast by this process, which represents about one hour of production.

a. Find the probability that this company must scrap exactly two of these parts.

b. Find the probability that it must scrap at least one part.

c. Find the expected number of parts scrapped, the variance, and the standard deviation.

3.15 An automobile manufacturer gives a 5-year/60,000-mile warranty on its drive train. Historically, 7% of this manufacturer's automobiles have required service under this warranty. Consider a random sample of 15 cars.

a. Find the probability that exactly one car requires service under the warranty.

b. Find the probability that more than one car requires service under the warranty.

c. Find the expected number of cars that require service, the variance, and the standard deviation.

3.16 Historically, 10% of the homes in Florida have radon levels higher than recommended by the Environmental Protection Agency. Radon is a weakly radioactive gas known to contribute to health problems. A city in north central Florida has hired an environmental consulting firm to check a random sample of 20 homes. Assume that this city is typical for the state of Florida.

a. Find the probability that exactly three homes have radon levels that exceed EPA recommendations.

b. Find the probability that one or fewer homes has radon levels that exceed EPA recommendations.

c. Find the expected number of homes with excessive radon levels, the variance, and the standard deviation.

3.17 A chemical engineer monitors a dyeing process for polyester yarn used in clothing by comparing a sample of the yarn against a standard color chart. He accepts or rejects the entire dyeing batch based on the sample's results. Historically, this process averages 5% rejected batches. Each shift, the process dyes eight batches. Assume that the next shift forms a random sample.

a. Find the probability that at least seven batches are accepted.

b. Find the probability that at least one batch is rejected.

c. Find the expected number of batches accepted, the variance, and the standard deviation.

    d. Find the expected number of batches rejected, the variance, and the standard deviation.

    e. Comment on your results from parts **c** and **d**.

3.18 A civil engineering professor assigns a bridge building project using Popsicle sticks each semester. To get a passing grade, the structure must support at least 20 lb. Historically, 10% of the student bridges fail to support 20 lb. Assume that the current class of 15 students forms a random sample.

    a. Find the probability that everyone passes the project.

    b. Find the probability that at least one person fails the project.

    c. Find the expected number of students who pass, the variance, and the standard deviation.

    d. Find the expected number of students who fail, the variance, and the standard deviation.

    e. Comment on your results from parts **c** and **d**.

3.19 An automobile manufacturer has just increased the strength standard for an aluminum alloy commonly used in car bodies. Recent history suggests that one supplier fails to meet this new specification 20% of the time. Assume that the next 15 batches of this alloy are a random sample.

    a. Find the probability that exactly three shipments fail to meet the new specifications.

    b. Find the probability that all of the shipments meet the new specifications.

    c. Find the expected number of shipments that fail to meet the new specifications, the variance, and the standard deviation.

    d. Find the expected number of shipments that do meet the new specifications, the variance, and the standard deviation.

    e. Comment on your results from parts **c** and **d**.

3.20 An industrial engineering class has studied the number of customers who drop off prescriptions at the outpatient pharmacy of a major teaching hospital. From their study, this pharmacy averages 20 customers an hour. Consider a randomly selected hour of operation.

    a. Find the probability that exactly 10 customers drop off prescriptions.

    b. Find the probability that exactly 20 customers drop off prescriptions.

    c. Find the expected number of customers, the variance, and the standard deviation.

3.21 If we reduce the data on the times between industrial accidents to the number of accidents each month, then we can well model the data by a Poisson distribution with an accident rate of 1.5 per month.

    a. Find the probability that no accidents occur in a given month.

    b. Find the probability that at least one accident occurs in a given month.

    c. Historically, DuPont management has placed the highest priority on safety and typically will reassign any plant manager whose facility has an excessive number of accidents. Suppose that management begins to consider reassignment when a facility has five accidents in a month. Find the probability that this facility has exactly five accidents in a given month.

d. Find the mean number of accidents per month, the variance, and the standard deviation.

3.22 Nelson (1987) discusses a process that historically has averaged 2.6 flaws per 1000-m length of wire.

a. Find the probability that a 1000-m length of wire has one or fewer flaws.

b. Find the probability that a 1000-m length of wire has more than two flaws.

c. Find the mean number of flaws per 1000-m length of wire, the variance, and the standard deviation.

d. Consider a 500-m length of wire.
    (1) Find the probability that it has no flaws.
    (2) Find the expected number of flaws, the variance, and the standard deviation.

3.23 Kalbfleisch, Lawless, and Robinson (1991) modeled the number of warranty claims within one year of purchase for a particular system on a single car model with a Poisson distribution and a rate of 0.75 claim per vehicle. Consider a randomly selected automobile.

a. Find the probability that this automobile has no claims within one year.

b. Find the probability that this automobile has exactly three claims within one year.

c. Find the expected number of claims, the variance, and the standard deviation.

3.24 The manufacture of silicon wafers used in integrated circuits requires the removal of contaminating particles of a certain size. Yashchin (1995) studied a rinsing process for these wafers. This process rinses batches of 20 wafers with deionized water. The process then dries these wafers by spinning off the water droplets. Prior to loading the wafers in the rinser/dryer, production personnel count the number of contaminating particles. This count provides feedback on the cleanliness of the manufacturing environment. The counts are well modeled by a Poisson distribution with a rate of six particles per wafer. Consider a randomly selected wafer.

a. Find the probability that this wafer has at least one particle.

b. Find the probability that this wafer has exactly six particles.

c. Find the expected number of particles, the variance, and the standard deviation.

3.25 A major automobile manufacturing company is known to average five defects per car. Consider the next car made.

a. Find the probability that it has exactly seven defects.

b. Find the population mean, the population variance, and the population standard deviation for the number of defects.

3.26 An industrial engineer who is studying the clerical staffing of a physicians' group practice must decide whether it requires a full-time phone operator. From historical data, she has determined that the group receives phone calls at the rate of 15 per hour.

a. Find the probability that the group receives exactly 15 calls in a given hour.
b. Find the population mean, the population variance, and the population standard deviation for the number of calls.

3.27   A highway engineer who is studying the number of accidents at a busy intersection has determined that accidents occur at the rate of 2.5 per month.

a. Find the probability that none occurs in a given month.
b. Find the probability that more than one occur in a given month.
c. Find the population mean, the population variance, and the population standard deviation for the number of accidents in a given month.

3.28   A server at a large university is known to average 2.5 hours of downtime per week. Consider one week.

a. Find the probability that the server is down exactly four hours.
b. Find the population mean, the population variance, and the population standard deviation for the number of hours the server is down.

3.29   A printing company is considering a new machine. They would like the probability of seeing no defects per 100 pages to exceed 50%. Historically, the current machine produces 1 defect per 100 pages of text. The new machine is touted as producing 0.6 defect per 100 pages of text.

a. Find the probability that a run of 100 pages of text from the current machine has no defects.
b. Find the probability that a run of 100 pages of text from the new machine has no defects.
c. Comment on your answers from parts **a** and **b**. Should they purchase the new machine?
d. Find the population mean, the population variance, and the population standard deviation for the number of defects per 100 pages of text for the current machine.

3.30   On average, a particular type of wood has 1.2 knots per 10 cubic feet of wood.

a. Find the probability that a 10-cubic-foot block of wood has fewer than two knots.
b. Find the probability that a 10-cubic-foot block of wood has exactly four knots.
c. Find the population mean, the population variance, and the population standard deviation for the number of knots per 10-cubic-foot block of wood.

3.31   Suppose that 30% of the applicants for a certain engineering job possess advanced statistical training. Applicants are randomly selected and are interviewed sequentially.

a. Find the probability that the first applicant with advanced statistical training is found on the fourth interview.

b. Find the probability that the first applicant with advanced statistical training is found on the first interview.

c. Find the expected number of applicants interviewed until the first one with advanced statistical training is found, the variance, and the standard deviation.

d. Find the probability that the seventh applicant interviewed is the second person with advanced statistical training.

3.32 Consider a process for making nickel battery plates that has an operator who successfully meets the weight specification 40% of the time. Let Y be the number of plates she makes until she is successful.

a. Find the probability that she requires four attempts to make her first successful plate.

b. Find the probability that she requires more than one attempt to make her first successful plate.

c. Find the expected number of attempts until her first success, the variance, and the standard deviation.

d. Find the probability that she requires ten attempts to make her third successful plate.

e. Find the expected number of attempts until her third success, the variance, and the standard deviation.

3.33 Metal casting processes are notoriously slow and expensive. A common problem facing many older casting processes is "flashing." In casting, liquid metal is shot into a mold and rapidly cooled. A flash commonly forms on the piece at the spot in the mold where the metal flows. Sanding corrects minor problems with flashing, but severe problems may require scrapping the part. A particular casting operation historically has scrapped 10% of its parts due to severe problems with flashing.

a. Find the probability that the first part scrapped is the tenth one made.

b. Find the probability that at least two parts are made before the first one is scrapped.

c. Find the expected number of parts until the first one is scrapped, the variance, and the standard deviation.

d. Find the probability that the fourth part scrapped is the 20th made.

e. Find the expected number of parts until the fourth part is scrapped, the variance, and the standard deviation.

3.34 A clothing store chain has hired an industrial engineer to study the buying habits of its customers. After a period of study, she has concluded that 30% of the customers who enter the store actually buy something.

a. What is the probability that the fifth customer who enters the store is the first to buy something?

b. What is the probability that the first customer who enters the store buys something?

    c. Find the expected number of customers until someone buys something, the variance, and the standard deviation.

    d. Find the probability that the fifth customer who enters is the second to buy something.

    e. Find the expected number of customers until the second to buy something, the variance, and the standard deviation.

3.35 An automobile manufacturer gives a 5-year/60,000-mile warranty on its drive train. Historically, 7% of this manufacturer's automobiles have required service under this warranty. Consider a new dealer.

    a. Find the probability that the first claim under this warranty is the tenth car sold.

    b. Find the expected number of cars sold until the first claim, the variance, and the standard deviation.

    c. Find the probability that the 20th car sold is the third to require service under the warranty.

    d. Find the expected number of cars sold until the third claim, the variance, and the standard deviation.

3.36 A manufacturer of silicon wafers has encountered an operating problem where 20% of the chips made are unacceptable. Consider a lot of 50 wafers, of which ten (exactly 20%) are unacceptable. An inspector selects a sample of five wafers from the lot of 50.

    a. Find the probability that exactly two wafers in the sample are unacceptable.

    b. Find the probability that at least one wafer is unacceptable.

3.37 At a particular golf resort, if a person hits a hole-in-one, the player gets to select 4 slips of paper from a bin of 100. Of the 100 slips of paper, 2 are good for a brand new driver valued at $300. Consider a person who hit one hole-in-one.

    a. Find the probability that the person will win one driver.

    b. Find the probability that the person will not win a driver.

    c. Find the mean, the variance, and the standard deviation for the number of winning slips of paper in the sample.

3.4 Continuous Random Variables

In contrast to discrete random variables, continuous random variables can assume an uncountable number of values. Consider the length of a pen barrel. At first glance, it appears to be about 6 in. long. If we measure it with a ruler, we may discover that it is about 5.75 in. long. If we send it down to an appropriate lab, we may discover that it is approximately 5.7443286 in. long. Ultimately, however, we shall never exactly know its true *specific* length.

Next, consider someone's height. He claims that he is 6 ft, 1 in. tall. Does that mean that he is exactly 6 ft, 1 in. tall? Of course not. He is simply claiming that his height is somewhere between 6 ft, $\frac{1}{2}$ in. and 6 ft, $1\frac{1}{2}$ in.

Let Y be a random variable that is measured over a continuum, and let y be a specific value. If Y is truly continuous, then we can show that $P(Y = y) = 0$. As a result, it really does not make sense to talk about $P(Y = y)$ for continuous random variables. Instead, we talk about the probabilities that the random variable falls within some interval. For example, $P(Y \leq y) = F(y)$ or $P(y_1 \leq Y \leq y_2)$.

Definition 3.9 | **Continuous Random Variable**

Let Y be a random variable. Y is said to be a **continuous random variable** if, at least theoretically, Y can assume any possible real value over some interval.

Some examples of continuous random variables are the outside diameter of a pen barrel, the times between accidents, the net weight of milk in an 8-oz container, and a reaction temperature over time.

Every continuous random variable we shall use has a probability density function (pdf), denoted by $f(y)$. The pdf is a mathematical expression that defines the shape of the distribution. We often use $f(y)$ to describe the distribution and to model the behavior of the random variable. Five properties of $f(y)$ are

1. $f(y) \geq 0$ for all possible values of y, because probabilities must be nonnegative.
2. $\int_{-\infty}^{\infty} f(y) \, dy = 1.0$ because the range $(-\infty, \infty)$ covers the entire sample space, and the probability that something from the sample space occurs is 1.
3. If y_0 is a specific value of interest, then we find the cdf by

$$F(y_0) = P(Y \leq y_0) = \int_{-\infty}^{y_0} f(y) dy.$$

4. If y_1 and y_2 are specific values of interest, then

$$P(y_1 \leq Y \leq y_2) = \int_{y_1}^{y_2} f(y) dy = F(y_2) - F(y_1).$$

Thus, the probabilities for specific intervals may be viewed as areas under the curve defined by $f(y)$. For example, suppose that Y is a continuous random variable with the pdf shown in Figure 3.2. Then $P(1.50 \leq Y \leq 2.25)$ is the shaded area under the curve.

5. If y_0 is a specific value, then $P(Y = y_0) = 0$. Since $P(Y = y) = 0$, we have

$$P(y_1 \leq Y \leq y_2) = P(y_1 < Y \leq y_2) = P(y_1 \leq Y < y_2)$$
$$= P(y_1 < Y < y_2).$$

We may view $f(y)$ as a "weight" or "mass" function that describes the weight we can ascribe to particular intervals for Y. In this particular case, we scale the weight function such that the total weight over all the possible values for Y is 1.

Figure 3.2 | The pdf for an Arbitrary Random Variable

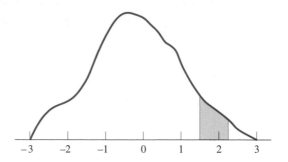

Example 3.12 | Times Between Industrial Accidents

Lucas (1985) studied the times between accidents for a ten-year period at a DuPont facility (see Example 2.12). We can well model these times by an *exponential distribution* that has the following pdf:

$$f(y) = \begin{cases} \lambda \exp(-\lambda y) & y > 0 \text{ and } \lambda > 0 \\ 0 & \text{otherwise} \end{cases}$$

where λ is the accident *rate* (in this case, the expected number of accidents per day).

The exponential distribution is an important example of a continuous distribution that engineers often use to model lifetimes and times between incidents. It thus has wide applications in reliability studies. There are many other continuous distributions. For example, the normal distribution is used for the majority of standard statistical analyses, and the Weibull distribution is used extensively in reliability analysis. The cdf for the exponential distribution is

$$F(y_0) = P(Y \le y_0)$$
$$= \int_{-\infty}^{y_0} f(y)\, dy$$
$$= \int_{0}^{y_0} \lambda \exp(-\lambda y)\, dy$$
$$= 1.0 - \exp(-\lambda y_0).$$

For the accident data, $\lambda = 0.05$ accident per day; thus, the cdf is

$$F(y) = 1.0 - \exp(-0.05 y).$$

The chances that this facility goes fewer than ten days between accidents is given by

$$P(Y < 10) = \int_0^{10} 0.05 \exp(-0.05y)\, dy$$
$$= 1.0 - \exp[-0.05(10)]$$
$$= 0.393.$$

As a result, this facility has a reasonably large chance, approximately 39.3%, of going fewer than ten days between accidents.

In general, the probability that an exponential random variable is greater than some specific value y_0 is given by

$$P(Y > y_0) = \int_{y_0}^{\infty} \lambda \exp(-\lambda y)\, dy$$
$$= \exp(-\lambda y_0).$$

Thus, the chances that this facility goes longer than 80 days without an accident is given by

$$P(Y > 80) = \exp[-0.05(80)]$$
$$= 0.018.$$

We should not expect this facility to go longer than 80 days between accidents very often!

Finally, for an exponential random variable, the probability of falling within the interval (y_1, y_2) is given by

$$P(Y_1 < Y < Y_2) = \int_{y_1}^{y_2} \lambda \exp(-\lambda y)\, dy$$
$$= \exp(-\lambda y_1) - \exp(-\lambda y_2).$$

Thus, the likelihood that this facility goes between 10 and 80 days between accidents is given by

$$P(10 < Y < 80) = \exp[-0.05(10)] - \exp[-0.05(80)]$$
$$= 0.607 - 0.018$$
$$= 0.589.$$

As a result, the chances are almost 59% that the facility will go between 10 and 80 days between accidents.

The Relationship of p(y), f(y), and a Stem-and-Leaf Display

Distributions can provide a powerful basis for modeling the random behavior of important characteristics of interest as long as certain assumptions hold. Later, we shall see that formal statistical analyses require certain assumptions about the underlying distribution from which the data come. Typically, these assumptions focus on the "shape" of the data. Appropriate data displays provide a quick and easy way to check these assumptions, especially the stem-and-leaf display. For a discrete random variable, the probability function, $p(y)$, and for a continuous random variable, the pdf, $f(y)$, define the theoretical shape for the stem-and-leaf display if the data actually come from the assumed distribution. The next example illustrates this point.

Example 3.13 | **Times Between Industrial Accidents—Revisited**

We claimed earlier that we could well model the times between industrial accidents by an exponential distribution with $\lambda = 0.05$. Figure 3.3 is a graph of the pdf for this specific distribution. To see whether this distribution provides a reasonable model for the data, consider Figure 3.4, which overlays an appropriately scaled plot of the pdf on the stem-and-leaf display (which we have converted to a histogram). This plot indicates that the exponential distribution does provide a reasonable basis for modeling these times. As a result, we should be fairly comfortable about the probabilities we calculated earlier based on this assumption.

Expected Values for Continuous Random Variables

For a continuous random variable, we can use integration over the possible values of the random variable to find the expected value. Once again, the expected value is a weighted average. For a continuous variable, the pdf, $f(y)$, defines the

---

Figure 3.3 | The pdf for the Theoretical Distribution for the Times Between Industrial Accidents

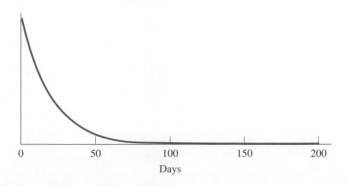

Figure 3.4 | The Overlay of the pdf and the Actual Data for the Times Between Industrial Accidents

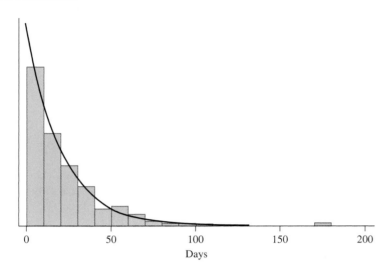

weights. We thus find the expected value by

$$\mu = E(Y) = \int_{-\infty}^{\infty} yf(y)dy.$$

Like discrete random variables, the population variance for a continuous random variable is

$$\sigma^2 = \text{var}(Y) = E(y^2) - \mu^2.$$

In this case,

$$E(Y^2) = \int_{-\infty}^{\infty} y^2 f(y)dy.$$

Once again, the population standard deviation is

$$\sigma = \sqrt{\sigma^2}.$$

Example 3.14 | **Times Between Industrial Accidents—Continued**

For an exponential distribution, the population mean is given by

$$\mu = \int_{-\infty}^{\infty} y\, f(y)\, dy$$

$$= \int_{0}^{\infty} y\lambda \exp(-\lambda y)\, dy.$$

We must use integration by parts to obtain

$$\mu = \left[-y\exp(-\lambda y) - \frac{1}{\lambda}\exp(-\lambda y) \right]_0^\infty$$

$$= \frac{1}{\lambda}.$$

We thus see that the expected time between accidents is the inverse of the rate, λ. In our particular case, $\lambda = 0.05$; thus,

$$\mu = \frac{1}{0.05} = 20.$$

As a result, this facility, on the average, goes 20 days between accidents, which makes sense if the expected number of accidents is 0.05 per day.

To find the variance, we first need to find $E(y^2)$. Again, we require integration by parts, which gives

$$E(y^2) = \int_{-\infty}^{\infty} y^2 f(y)\, dy$$

$$= \int_0^{\infty} y^2 \lambda \exp(-\lambda y)\, dy$$

$$= \left[-y^2 \exp(-\lambda y) - \frac{2y}{\lambda}\exp(-\lambda y) - \frac{2}{\lambda^2}\exp(-\lambda y) \right]_0^\infty$$

$$= \frac{2}{\lambda^2}.$$

The variance is

$$\sigma^2 = E(y^2) - \mu^2$$

$$= \frac{2}{\lambda^2} - \left(\frac{1}{\lambda} \right)^2$$

$$= \frac{2}{\lambda^2} - \frac{1}{\lambda^2}$$

$$= \frac{1}{\lambda^2}.$$

The population standard deviation is

$$\sigma = \sqrt{\sigma^2} = \sqrt{\frac{1}{\lambda^2}} = \frac{1}{\lambda}.$$

It is interesting that for the exponential distribution, the population standard deviation equals the population mean. Such a result implies that data following this distribution are highly variable. For this particular case,

$$\sigma^2 = \frac{1}{(0.05)^2} = 400$$

$$\sigma = \frac{1}{0.05} = 20.$$

By the empirical rule, we should expect the actual number of accidents to fall somewhere within the interval

$$\mu \pm 3\sigma = 20 \pm 3(20)$$
$$= (-40, 80).$$

In this case, the lower bound makes no sense because of the large skew in this distribution. On the other hand, the upper bound suggests that we should not expect to go longer than 80 days between accidents. Any time that we do is evidence that we have improved the safety of this facility.

In Example 3.12, we found that $P(Y > 80) = 0.018$, which is not essentially zero as the empirical rule implies. Why do we see the discrepancy? The empirical rule does not work particularly well for very skewed distributions such as the exponential. Nonetheless, even for the exponential distribution, the probability of being more than three standard deviations from the population mean is quite small even if it is not essentially zero.

Some Important Continuous Distributions

Table 3.5 lists the probability density function, the expected value, and the population variance for three important continuous distributions:

1. Uniform
2. Weibull
3. Gamma

Engineers often use these distributions to model times between events (such as failures) and interarrival times. Thus, these distributions play an important role in reliability and queuing studies. In Exercise 3.45, we show that the cdf for the Weibull distribution is given by

$$F(y) = 1 - \exp[-(\lambda y)^{\beta}].$$

The exponential distribution is a special case of both the gamma ($\alpha = 1$) and the Weibull distributions ($\beta = 1$). Both distributions make use of the gamma function defined by

$$\Gamma(\alpha) = \int_0^{\infty} y^{\alpha-1} \exp(-y).$$

Table 3.5 | Summary of Some Important Continuous Distributions

Distribution	Density Function		Expected Value	Population Variance
Uniform	$\dfrac{1}{b-a}$ 0	for $a \leq y \leq b$ otherwise	$\dfrac{a+b}{2}$	$\dfrac{(b-a)^2}{12}$
Weibull	$\lambda\beta(\lambda y)^{\beta-1}\exp[-(\lambda y)^\beta]$ 0	$y > 0, \lambda > 0, \beta > 0$ otherwise	$\dfrac{\Gamma[(\beta+1)/\beta]}{\lambda}$	$\dfrac{\Gamma[(\beta+2)/\beta] - [\Gamma[(\beta+1)/\beta]]^2}{\lambda^2}$
Gamma	$\dfrac{\lambda^\alpha y^{\alpha-1}}{\Gamma(\alpha)}\exp(-\lambda y)$ 0	$y > 0, \lambda > 0, \alpha > 0$ otherwise	$\dfrac{\alpha}{\lambda}$	$\dfrac{\alpha}{\lambda^2}$

Tables exist that summarize values for this integral for various α's. In addition, many software packages compute this integral directly.

⇨ Exercises

3.38 Davis and Lawrance (1989) present time-to-failure data for 171 automobile tires. Baltazar-Aban and Pena (1995) modeled these data with an exponential distribution with a failure rate of 0.004 tire per hour.

a. Find the probability that a randomly selected tire fails during the first 100 hours of testing.

b. Find the probability that a randomly selected tire fails between 50 and 150 hours of testing.

c. Find the probability that a randomly selected tire survives more than 200 hours.

d. Find the expected time to failure for a randomly selected tire.

e. Find the variance and the standard deviation for the time to failure for a randomly selected tire.

3.39 Miyamura (1982) modeled the lifetimes of the electromagnetic valve used for starting the idle-up actuator of an air conditioner by an exponential distribution with a rate of 0.05 failure per million revolutions. Consider a randomly selected valve.

a. Find the probability that this valve fails within the first half million revolutions.

b. Find the probability that this valve lasts longer than 3 million revolutions.

c. Find the expected time to failure for this valve.

d. Find the variance and the standard deviation for the time to failure for this valve.

3.40 The maintenance manager at a chemical facility knows that the times between repairs, Y, for a specific chemical reactor are well modeled by this distribution:

$$f(y) = 0.01e^{-0.01y}$$

for $y > 0$ and 0 otherwise.

a. Find the probability that Y is less than 30.
b. Find the probability that Y is greater than 15.
c. Find the probability that Y is exactly equal to 100.
d. Find the probability that Y is between 50 and 150.

3.41 Lewis (1986) studied the times (in seconds) between successive vehicles travelling northwards along the M1 motorway in England as they passed a fixed point near Bedfordshire. Suppose these times can be well modeled by an exponential distribution with a rate of 0.03. Consider a randomly selected car.

a. Find the probability that the time until the next car is less than 20 seconds.
b. Find the probability that the time until the next car is more than 30 seconds.
c. Find the expected time until the next car.
d. Find the variance and standard deviation for the time between cars.

3.42 According to Zimmels (1983), the sizes of particles used in sedimentation experiments often have a uniform distribution. Suppose the particles have diameters that are uniformly distributed between 0.01 and 0.05 cm.

a. Derive the mean for this distribution.
b. Derive the variance and standard deviation for this distribution.
c. Find the probability that a particle has a diameter bigger than 0.042 cm.

3.43 Rather than collecting the actual times between accidents, Lucas could have reduced the data to the number of accidents per month. Suppose that we know only that one accident occurred during a given month and that we can well model the actual times between accidents by an exponential distribution. In this case, given that we know an accident occurred at some time during the month, the time within the month it happened follows a *uniform distribution*. In general, the pdf for a uniform distribution is

$$f(y) = \begin{cases} \frac{1}{b-a} & a \leq y \leq b \\ 0 & \text{otherwise.} \end{cases}$$

This pdf essentially says that all the values within this interval are equally likely to occur.

a. Derive the cdf for this distribution.
b. Derive the mean for this distribution.
c. Derive the variance and the standard deviation for this distribution.

  d. For the time between accidents data, assume that a month has 30 days and that an accident occurs during that month.
    (1) Find the probability that the accident occurred on or before the 15th.
    (2) Find the probability that the accident occurred after the 25th.
    (3) Find the expected time it occurred.
    (4) Find the variance and the standard deviation for the time it occurred.

3.44 Engineers often use the uniform distribution to model the arrival time of some event given that the event did occur within some interval. For example, production knows that a particular pump failed at some time between 1:00 and 3:00 P.M. Given that we know it failed at some time during this period, the pdf for the specific time within the period is

$$f(y) = \begin{cases} \frac{1}{b-a} & a \leq y \leq b \\ 0 & \text{otherwise} \end{cases}$$

where $a = 1$ and $b = 3$. This pdf essentially says that all the times within this interval are equally likely to occur.

  a. Derive the mean for this distribution.
  b. Derive the variance and the standard deviation for this distribution.
  c. Find the probability that the pump failed after 1:30 P.M.

3.45 The Weibull is probably the most widely used distribution in reliability studies. The pdf for the Weibull is given by

$$f(y) = \begin{cases} \lambda\beta(\lambda y)^{\beta-1} \exp[-(\lambda y)^\beta] & y > 0, \lambda > 0, \beta > 0 \\ 0 & \text{otherwise.} \end{cases}$$

For $\beta = 1.0$, the Weibull simplifies to the exponential distribution.

  Padgett and Spurrier (1990) used a Weibull with $\lambda = 0.4$ and $\beta = 2$ to model the breaking strengths of carbon fibers used in fibrous composite materials. These fibers measure 50 mm in length and 7–8 microns in diameter. Periodically, the manufacturer selects random samples of five fibers and tests their breaking stresses.

  a. Derive the cdf for a Weibull distribution.
  b. Specifications suggest that 99% of the fibers must have a breaking strength of at least 1.2 GPa (gigapascals). Find the probability that the breaking strength is less than 1.2 GPa.
  c. Find the expected strength.
  d. Find the variance and the standard deviation of the breaking strengths.
  e. The breaking stresses (in GPa) for the samples are listed here. Compare the plot of the pdf with a stem-and-leaf display of the actual strengths and comment on your results.

1.4	3.7	3.0	1.4	1.0	2.8	4.9	3.7	1.8	1.6
3.2	1.6	0.8	5.6	1.7	1.6	2.0	1.2	1.1	1.7
2.2	1.2	5.1	2.5	1.2	3.5	2.2	1.7	1.3	4.4
1.8	0.4	3.7	2.5	0.9	1.6	2.8	4.7	2.0	1.8
1.6	1.1	2.0	1.6	2.1	1.9	2.9	2.8	2.1	3.7

3.46 All batteries have definite "shelf" lives; that is, batteries begin to deteriorate in storage. Morris (1987) used a Weibull distribution (see Exercise 3.45) with $\beta = 2$ and $\lambda = 0.1$ to model the storage time required for long-life Li/SO$_4$ batteries to become unacceptable for use.

a. Show that the cdf of a Weibull distribution is

$$F(y) = 1 - \exp[-(\lambda y)^{\beta}].$$

b. Find the probability that a randomly selected battery becomes unacceptable after between 12 and 18 months.

c. Find the expected number of months until a randomly selected battery becomes unacceptable.

d. Find the variance and the standard deviation for the time until a randomly selected battery becomes unacceptable.

⁖ 3.5 The Normal Distribution

Among all of the distributions used in classical statistics, the single most important is the *normal distribution,* which is often described as the "bell-shaped" distribution or curve. Two major reasons for the importance of the normal distribution are

1. We can well model the behavior of many phenomena by this distribution.
2. Under certain conditions, we can well model the behavior of averages by this distribution.

The normal distribution is actually a family of distributions with each member characterized by the population mean, μ, and the population variance, σ^2.

The pdf for a normal random variable is given by

$$f(y) = \frac{1}{\sqrt{2 \cdot \pi \cdot \sigma^2}} \exp\left[-\frac{1}{2}\left(\frac{y - \mu}{\sigma}\right)^2\right] \qquad -\infty \leq y \leq \infty.$$

For a normal distribution, $\mu \pm \sigma$ represent the points of inflection for the pdf. Figure 3.5, which gives a plot of the normal pdf, illustrates why we often call it a bell-shaped curve. For a normal distribution,

---

Figure 3.5 | The Normal Probability Density Function

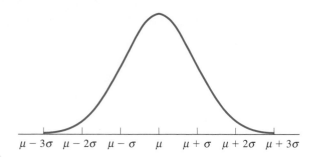

$$\mu - 3\sigma \quad \mu - 2\sigma \quad \mu - \sigma \quad \mu \quad \mu + \sigma \quad \mu + 2\sigma \quad \mu + 3\sigma$$

- Approximately 68% of the observations fall within the interval $\mu \pm \sigma$.
- Approximately 95% fall within the interval $\mu \pm 2\sigma$.
- Virtually all (approximately 99.7%) fall within the interval $\mu \pm 3\sigma$.

It is important to note that the normal distribution is the fundamental basis for the empirical rule.

The Standard Normal Distribution

Of all the normal distributions, the single most important is the *standard normal distribution* because we can convert any normal random variable into a standard normal. Why is this fact important? We cannot integrate analytically the pdf for the normal distribution, so we must use numerical integration techniques to find the probabilities associated with normal random variables. Since we can convert any normal random variable to a standard normal, we can summarize the cdf for any normal random variable by the cdf for the standard normal random variable.

The standard normal distribution has $\mu = 0$ and $\sigma^2 = 1$. We denote a standard normal random variable by Z. The probabilities associated with this random variable are given in Table 1 of the appendix. For a specific value z_0, the table gives $P(Z \leq z_0)$, which is the shaded area under the standard normal curve (see Figure 3.6). For example, the table gives $P(Z \leq 1.96) = 0.9750$.

Figure 3.6 |

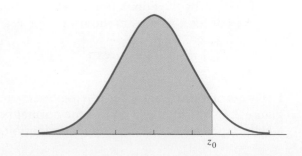

z_0

Figure 3.7

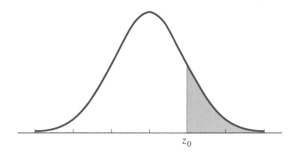

Figure 3.7 considers $P(Z > z_0)$. Thus, we can find $P(Z > z_0)$ by

$$P(Z > z_0) = 1 - P(\overline{Z > z_0}) = 1 - P(Z \leq z_0).$$

For example, we may find $P(Z > 2.33)$ by

$$P(Z > 2.33) = 1 - P(Z \leq 2.33) = 1 - 0.9901 = 0.0099.$$

Finally, consider the interval defined by $z_1 < z_2$. Figure 3.8 illustrates $P(z_1 < Z \leq z_2)$. We find this probability by

$$P(z_1 < Z \leq z_2) = P(Z \leq z_2) - P(Z \leq z_1).$$

For example, $P(1.00 < Z \leq 1.96)$ is given by

$$P(1.00 < Z \leq 1.96) = P(Z \leq 1.96) - P(Z \leq 1.00)$$
$$= 0.9750 - 0.8413$$
$$= 0.1337.$$

We often need to use the Z-value associated with specific "tail" areas of the standard normal distribution. Let z_α represent the Z-value associated with a right-hand "tail area" of α (see Figure 3.9). z_α is that value for Z such that

$$P(Z > z_\alpha) = \alpha.$$

Figure 3.8

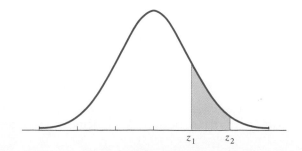

Figure 3.9

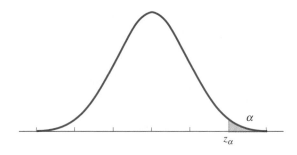

As a result, z_α is that value from the table that satisfies

$$1.0 - P(Z \le z_\alpha) = \alpha$$

or

$$P(Z \le z_\alpha) = 1.0 - \alpha$$

$$\text{value from table} = 1.0 - \alpha.$$

For example, $z_{.025}$ is that Z such that

$$P(Z \le z_{.025}) = 1.0 - 0.025 = 0.975.$$

Looking into the body of the table, we get

$$z_{.025} = 1.96.$$

Transforming a General Normal Random Variable to a Standard Normal

If Y is a normal random variable with mean μ and variance σ^2, we can transform it into a standard normal by the equation

$$Z = \frac{Y - \mu}{\sigma}.$$

By subtracting μ, we recenter the random variable around 0. By dividing by σ, we rescale the random variable so that the variance is 1. In the process, we convert the original normal random variable into a standard normal.

Figure 3.10 illustrates what this transformation actually does. By subtracting μ, which is the expected value of Y, the expected value of Z is 0. Dividing by σ rescales the random variable so that the resulting Z-value represents how many standard deviations a value of a random variable lies from its mean.

Example 3.15 | **Production at a Titanium Dioxide Facility**

A major titanium dioxide (white pigment) facility uses two production lines, each with a nominal designed capacity of 300 tons per day. Typically, engineers

Figure 3.10 | Transforming a General Normal Random Variable into a Standard Normal Random Variable

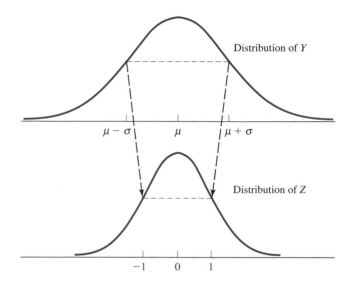

"overdesign" chemical facilities, which means that both of these lines can actually produce somewhat more than 300 tons on given days. Historically, the total daily production of this facility approximately follows a normal distribution with a mean of 500 tons and a standard deviation of 50 tons.

The corporate office can sell everything this facility can make. As a result, corporate management really would like to know the probability that this facility will exceed its nominal total capacity of 600 tons on any given day. Let Y be the total production on a given day. We thus seek $P(Y > 600)$:

$$P(Y > 600) = P\left(\frac{Y - \mu}{\sigma} > \frac{600 - \mu}{\sigma}\right)$$
$$= P\left(Z > \frac{600 - 500}{50}\right)$$
$$= P(Z > 2.0)$$
$$= 1 - P(Z \leq 2.0)$$
$$= 1 - 0.9772$$
$$= 0.0228.$$

This facility should exceed the nominal capacity only about 2% of the time.

The production superintendent believes that the "typical" performance is between 480 and 520 tons. The probability that production on any given day falls within that interval is given by

$$P(480 \leq Y \leq 520) = P\left(\frac{480 - \mu}{\sigma} \leq \frac{Y - \mu}{\sigma} \leq \frac{520 - \mu}{\sigma}\right)$$

$$= P\left(\frac{480 - 500}{50} \leq Z \leq \frac{520 - 500}{50}\right)$$

$$= P(-0.40 \leq Z \leq 0.40)$$

$$= P(Z \leq 0.40) - P(Z \leq -0.40)$$

$$= 0.6554 - 0.3446$$

$$= 0.3108.$$

The facility meets this definition of "typical" about 31% of the time.

Example 3.16 | **Setting the Appropriate Mean for a Dairy Packaging Line**

Maxcy and Lowry (1984) report on a packaging process for 8-oz milk cartons. Since the specific gravity of milk is approximately 1.033, one can measure the volume by measuring the weight. In this case, 8 oz of milk should weigh 245 g. Historically, these weights follow a normal distribution with a standard deviation of 1.65 g. Federal inspectors require that the mean amount of milk packaged must be significantly greater than 245 g in order to minimize any underages. Clearly, plant management would like the mean amount to be no more than necessary to meet the inspectors' standards. They argue that a reasonable mean weight should produce less than 1% of the cartons underweight. Let Y be the weight of a specific carton of milk. Plant management seeks to find μ such that $P(Y < 245) < 0.01$. To find this true mean, we first observe

$$P(Y < 245) = P\left(\frac{Y - \mu}{\sigma} < \frac{245 - \mu}{\sigma}\right)$$

$$= P(Z < z_0)$$

where $z_0 = (245 - \mu)/\sigma$. If $P(Y < 245) < 0.01$, then $P(Z < z_0) < 0.01$, and from the table, $z_0 < -2.33$. As a result, we need to choose μ such that

$$\frac{245 - \mu}{\sigma} < -2.33$$

$$\frac{245 - \mu}{1.65} < -2.33$$

$$245 - \mu < -2.33(1.65)$$

$$\mu > 245 + 2.33(1.65)$$

$$\mu > 248.83.$$

So plant management would like to deliver, on the average, 248.83 g per carton. In so doing, less than 1% of the cartons would contain less than 8 oz.

⸭ *Exercises*

3.47 Wasserman and Wadsworth (1989) discuss a process for manufacturing steel bolts that continuously feed an assembly line downstream. Historically, the thicknesses of these bolts follow a normal distribution with a mean of 10.0 mm and a standard deviation of 1.6 mm. Process supervision becomes concerned about the process if the thicknesses begin to get larger than 10.8 mm or smaller than 9.2 mm. Assume that the current process mean is 10.0 mm, and consider a randomly selected bolt.

    a. Find the probability that the thickness of this bolt is between 9.2 and 10.8 mm.

    b. Find the probability that the thickness of this bolt is smaller than 9.2 mm.

3.48 Canning (1993) studied the performance of a new photolithography process. This process deposits layers of material on silicon wafers. The thicknesses of one of the layers deposited followed a normal distribution with a standard deviation of 0.057. Since all thicknesses were measured relative to the nominal, the mean was 0. The lower and upper specifications were −0.15 and 0.15, respectively.

    a. Find the probability that any given layer deposited by the new process falls within the specifications.

    b. Find the value for the mean thickness required to make the probability of exceeding the upper specification limit be less than 1%.

3.49 Runger and Pignatiello (1991) consider a plastic injection molding process for a part that has a critical width dimension that follows a normal distribution with a historic mean of 100 and historic standard deviation of 8. They monitor this process with a control chart (which we shall discuss in detail in Chapter 5) with limits of 90 and 110. Assume that the mean width is 100, and consider a randomly selected part.

    a. Find the probability that the width of this part is greater than 110.

    b. Find the probability that the width is smaller than 90.

    c. Runger and Pignatiello also use "warning" limits of 99 and 101 to help monitor this process. Find the probability that the width is between 99 and 101.

    d. Find the value for the mean thickness required to make the probability of exceeding 101 be less than 10%.

3.50 The strength of a chemical paste product delivered in casks is known to follow a normal distribution with a mean of 59 and a standard deviation of 3.4.

a. Find the probability that a cask will have a strength between 55 and 65.
b. Find the probability that a cask will have a strength greater than 53.
c. Find the mean strength needed for the casks such that the probability that a cask strength is below 55 is less than 3%.

3.51 Sokal and Rohlf (1981) studied purebred Canadian dairy cattle. The butterfat percentages from 2-year-old cows follow a normal distribution with a mean of 4.51 and a standard deviation of 0.348.

a. Find the probability that a 2-year-old cow will have a butterfat percentage between 4.25 and 4.60.
b. Find the probability that a 2-year-old cow will have a butterfat percentage less than 4.75.

3.52 Buxton (1991) investigated the welding properties of a high-density polyethylene used in an injection mold and then cut into two pieces. The quality of the weld was then measured by the ratio of the yield stress of the welded bar to the mean yield stress of unwelded bars. Assume the weld ratios follow a normal distribution with a mean of 0.83 and a standard deviation of 0.039.

a. Find the probability that a welded bar will have a weld ratio greater than 0.9.
b. Find the probability that a welded bar will have a weld ratio between 0.8 and 0.9.
c. Find the mean weld ratio needed for the welded bars such that the probability that a welded bar's ratio is above 0.80 is greater than 95%.

3.53 An ethanol–water distillation column historically produces yields that are well modeled by a normal distribution with a mean of 70 volume percent and a standard deviation of 2.

a. Find the probability that a yield exceeds 75%.
b. Find the probability that a yield is between 67% and 73%.
c. Find the mean yield for this process such that the probability that a yield is below 70% is less than 0.02.

3.54 The volumes delivered by a nominal 20-oz soft drink bottling process follow a normal distribution with a mean of 20.2 oz and a standard deviation of 0.07.

a. Find the probability that this process underfills a bottle.
b. The bottle will overflow if the volume delivered exceeds 20.35 oz. Find the probability of an overflow.

3.55 The daily production of a sulfuric acid process is known to follow a normal distribution with a mean of 400 tons per day and a variance of 225 tons per day.

a. Find the probability that today's production will be between 375 and 425 tons.
b. Find the probability that today's production will be less than 360 tons.

⋗ 3.6   Random Behavior of Means

Sampling Distributions

We have already seen that the engineering method requires the collection of data. Often, we summarize the data with quantities calculated from the observed data. For example, we calculated medians and quartiles in Chapter 2. We

VOICE OF EXPERIENCE

All statistics follow a distribution.

often call such quantities calculated from data *statistics*. By the appropriate use of probabilistic models for modeling the data, we can also model the random behavior of these statistics.

Definition 3.10 | **Sampling Distribution**

The probability distribution of a statistic is its *sampling distribution*.

Under certain assumptions, statistics follow well-known distributions. The sampling distribution of a statistic allows us to model in a probabilistic way the statistic's behavior. With this information, we can begin to make inferences back to populations!

The Sample Mean

We often use the arithmetic average or the *sample mean* to describe the data's typical value.

Definition 3.11 | **Sample Mean**

Let y_1, y_2, y_3, ..., y_n denote a sample of interest. The sample mean, denoted by $\overline{y}$, is given by

$$\overline{y} = \frac{\sum_{i=1}^{n} y_i}{n} = \frac{1}{n} \sum_{i=1}^{n} y_i,$$

where n represents the total number of observations in our sample.

Example 3.17 | **Wall Thicknesses of Aircraft Parts**

Eck Industries, Inc. (see Mee 1990), manufactures cast aluminum cylinder heads that are used for liquid-cooled aircraft engines. The wall thicknesses of the coolant

| Figure 3.11 | The Histogram of the Thickness Data |

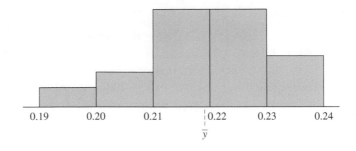

jackets are critical, particularly in high-altitude applications. Engineering specifications require that this thickness must be at least 0.190 in. The values are the thicknesses (in inches) of 18 cylinder heads as measured by ultrasound, which is a nondestructive technique:

0.223	0.193	0.218	0.201	0.231	0.204
0.228	0.223	0.215	0.223	0.237	0.226
0.214	0.213	0.233	0.224	0.217	0.210

For these data, $\sum_{i=1}^{n} y_i = 3.933$. Thus,

$$\overline{y} = \frac{1}{n} \sum_{i=1}^{n} y_i = \frac{1}{18}(3.933) = 0.2185.$$

The sample mean, $\overline{y}$, represents the "center of gravity" for the data set. Figure 3.11 is the histogram with the sample mean located for the data. $\overline{y}$ represents the point that would "balance," in a loose sense, the histogram. If we think of the data placed on a seesaw or a lever, $\overline{y}$ serves as the fulcrum. Just like a small weight placed on the far end of a lever can balance much heavier weights closer to the center but on the other side, a single observation quite distant from the center of the data can balance out the influence of a large number of points near but on the other side of the sample mean. We say that a point has *leverage* if it is distant from the center. Such a point tends to "draw" the sample mean to itself. As a result, a single outlier can often distort the sample mean. In Example 2.9, we constructed the boxplot for these data and saw that the observation 0.193 was a mild outlier. As a result, this observation is somewhat distant from the center of the data and has leverage. If we drop this value and recalculate the sample mean, we get $\overline{y} = 0.220$. This single observation moves the sample mean 0.0015 unit toward itself, which, within the framework of these data, is quite a bit. In Chapter 2, we mentioned that the sample median is a *resistant* measure of the typical value. The sample mean, on the other hand, is not because of the leverage imparted by outliers.

The Central Limit Theorem

Classical statistics almost always assumes that the data come from a random sample.

Definition 3.12 | **Random Sample (Formal Definition)**

A *random sample* is one in which all the observations are statistically independent and follow exactly the same distribution.

Statisticians commonly say a random sample is one where the observations are *independent and identically distributed*, abbreviated as iid.

Consider taking a series of random samples, all of size n, from some population and calculating $\overline{y}$ for each one. Since the y's are random variables, the individual $\overline{y}$'s are also random variables and follow a distribution. Let σ denote the population standard deviation for the original population (for the y's), and let $\sigma_{\overline{y}}$ be the standard deviation for the sampling distribution of the $\overline{y}$'s. We often call $\sigma_{\overline{y}}$ the *standard error of the mean*. We can show that

$$\sigma_{\overline{y}} = \frac{\sigma}{\sqrt{n}}.$$

We can also show that

$$E[\overline{Y}] = \mu.$$

Thus, if the original population (*underlying* or *parent* population) follows a normal distribution, then $\overline{y}$ also follows a normal distribution with mean μ and standard error $\sigma_{\overline{y}} = \sigma/\sqrt{n}$.

Unfortunately, we rarely sample from truly normal distributions. However, we often can appeal to the Central Limit Theorem.

Definition 3.13 | **Central Limit Theorem**

Consider a series of random samples all containing n observations, each from an underlying population with mean μ and variance σ^2. If n is sufficiently large, then the sampling distribution of $\overline{y}$ is approximately normal with mean μ and standard error $\sigma/\sqrt{n}$.

Thus, if n is sufficiently large, then

$$Z = \frac{\overline{Y} - \mu}{\sigma_{\overline{y}}} = \frac{\overline{Y} - \mu}{\sigma/\sqrt{n}}$$

follows a standard normal distribution.

Figure 3.12 | A "Well-Behaved" Distribution

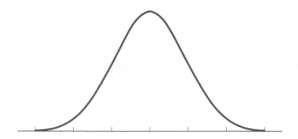

Of course, we need to consider how large is sufficiently large. If the underlying population actually follows a normal distribution, then $n = 1$ is sufficiently large! If the underlying population is symmetric and single-peaked and the tails die out rapidly, as in Figure 3.12, then many statisticians suggest that n's of 3 to 5 are sufficiently large. The literature suggests that n's of 6 to 12 are sufficiently large for the *uniform distribution,* which is illustrated in Figure 3.13. Clearly, the uniform distribution does not have a specific peak; however, it is symmetric and its tails in some sense do die rapidly. As a result, we can apply the Central Limit Theorem with relatively small sample sizes in this case.

> **VOICE OF EXPERIENCE**
>
> *The applicability of the Central Limit Theorem depends on the sample size used for the specific distribution.*

Many texts suggest that n is sufficiently large to assume the Central Limit Theorem whenever $n \geq 30$. Although this holds true for a number of parent distributions, we can name many counterexamples. In general, the more the shape of the parent distribution deviates from the bell-shaped curve, the larger the sample size, n, is required to assume the Central Limit Theorem. Whenever possible, we should use graphical displays, such as the stem-and-leaf, to check whether the sample size is sufficiently large to assume the Central Limit Theorem.

Example 3.18 | **Thicknesses of Silicon Wafers**

Hurwitz and Spagon (1993) analyzed the performance of a planarization device that polishes silicon wafers to a high degree of smoothness. Historically, the thicknesses at the dead center of the wafer have a mean of 3200 angstroms with a standard

Figure 3.13 | The Uniform Distribution

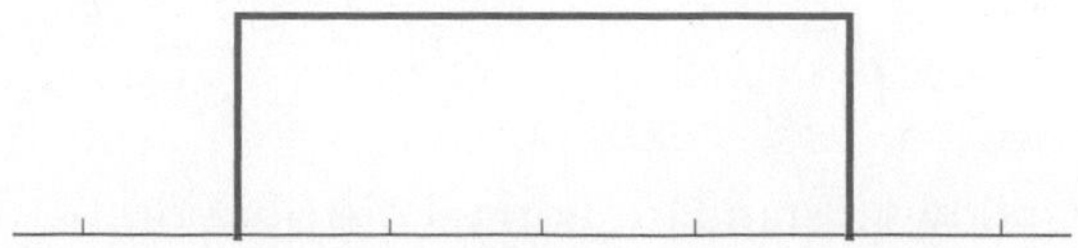

deviation of 80 angstroms. The following thicknesses are for 23 wafers from a single lot, all measured at the center of the wafer:

3240	3200	3220	3210	3250	3220
3190	3190	3150	3160	3270	3180
3200	3270	3180	3300	3250	3330
3300	3280	3270	3270	3200	

Let y_i be the thickness of the ith wafer. For these data, $\sum_{i=1}^{n} y_i = 74{,}330$. Thus,

$$\bar{y} = \frac{1}{n} \sum_{i=1}^{n} y_i = \frac{1}{23}(74{,}330) = 3232.$$

Production management would like to know whether there is evidence to suggest that this lot is thicker than normal. Assume that these data form a random sample. We can address this question by looking at the probability of obtaining a sample mean of 3232 or greater when the true process mean is 3200 with a standard deviation of 80. We therefore seek $P(\bar{Y} > 3232)$. If we assume the Central Limit Theorem, then

$$P(\bar{Y} > 3232) = P\left(\frac{\bar{Y} - \mu}{\sigma/\sqrt{n}} > \frac{3232 - \mu}{\sigma/\sqrt{n}}\right)$$

$$= P\left(Z > \frac{3232 - 3200}{80/\sqrt{23}}\right)$$

$$= P(Z > 1.92)$$

$$= 0.0274.$$

A sample mean of 3232 is 1.92 standard errors from the true process mean. As a result, the chances of observing a sample mean of 3232 or greater if the true process mean is 3200 are less than 3%, which is fairly remote. We may conclude that there is some evidence to suggest that this particular lot is thicker than normal.

The validity of this analysis depends on whether we really can assume the Central Limit Theorem. In other words, for the distribution of these thicknesses, can we well model the sample mean of 23 thicknesses by a normal distribution? Consider the stem-and-leaf display of the sample data in Figure 3.14, which gives us a reasonable "picture" of the distribution for these thicknesses. Although the data do not follow a clear bell-shaped pattern, they do seem to come from a distribution with a single peak and with rapidly dying tails. As a result, we should feel comfortable that with a sample size of 23, we can well model the sample mean by a normal distribution.

Figure 3.14	The Stem-and-Leaf Display for the Thickness of Silicon Wafers

Stem	Leaves	Number	Depth
31•.:	568899	6	6
32*.:	0001224	7	
32•.:	5577778	7	10
33*.:	003	3	3

Normal Probability and Q–Q Plots

Many software packages generate *normal probability plots* to help us determine whether we can reasonably model our data by a normal distribution. To construct a normal probability plot, we first must rearrange the data in ascending order. Following our notation in Chapter 2, let $y_{(1)} \leq y_{(2)} \leq \cdots \leq y_{(n)}$ represent this reordered data set. The subscript (i) represents the *rank* of the particular data value. For each value in the reordered data set, we compute the estimated cumulative probability point, $P_{(i)}$, defined by

$$P_{(i)} = \frac{i - 0.5}{n}.$$

Some software packages plot each pair of $y_{(i)}$ and $P_{(i)}$ on normal probability paper. Usually, these packages plot the $y_{(i)}$'s on the x-axis and the $P_{(i)}$'s on the y-axis. Other software packages avoid the need to use normal probability paper and plot the normal quantile associated with $P_{(i)}$ on the y-axis. What do we mean by this normal quantile? For a given $P_{(i)}$, the corresponding normal quantile is that value from the body of the standard normal table that corresponds to a left-hand tail area of $P_{(i)}$. In other words, the quantile is that value, z_0, for the standard normal random variable such that the cdf $[P(Z \leq z_0)]$ equals $P_{(i)}$. Consider the thicknesses of silicon wafers in Example 3.18. The data set consists of 23 observations; hence, $n = 23$. The smallest observed data value is 3150; thus, $y_{(1)} = 3150$. The corresponding $P_{(i)}$ is

$$P_{(i)} = \frac{i - 0.5}{n} = \frac{1 - 0.5}{23} = 0.0217.$$

The normal quantile associated with this data value is -2.02 because $P(Z \leq -2.02) = 0.0217$.

For either plot, if the data roughly follow a normal distribution, then we should see roughly a straight line on our plot: The more the plot deviates from a straight line, the less likely the data follow a normal distribution. We use the data values closest to the median value to define the appropriate straight line.

The value on the data axis that corresponds to the normal quantile of zero is actually an estimate of the true mean for the data set. The slope of the line is a function of the standard deviation. If we plot the data on the x-axis and the normal quantiles on the y-axis, then the slope of the line is inversely proportional to the standard deviation. As a result, a steep slope for the line, when plotted this way, indicates a small standard deviation.

A natural question considers how straight is straight. The answer depends on the size of the data set and how closely the true distribution follows a normal distribution. Some of our colleagues suggest the "fat pencil" rule. If we can cover the plotted values with a "fat" pencil, then we feel fairly comfortable that the data roughly follow a normal distribution. Although this approach sounds rather ad hoc, it does seem to work well in practice, particularly for moderate size samples ($n \geq 15$) and for the size plots generated by most software packages. For smaller data sets, we need to make some allowances. Some analysts do not rely on normal probability plots for data sets smaller than 15 observations. Other analysts use them cautiously. Typical violations of the fat pencil rule include data sets that have distinct outliers—that is, individual data values that fall off the straight line—and data sets whose true distribution is fairly skewed, in which case, the pattern follows more of an S-shape.

The next two examples illustrate the use of this plot.

Example 3.19	**Thicknesses of Silicon Wafers—Continued**

Figure 3.15 shows the normal probability plot for the thicknesses of silicon wafers. The data appear roughly to follow a straight line. We thus feel reasonably

Figure 3.15	The Normal Probability Plot of the Thicknesses

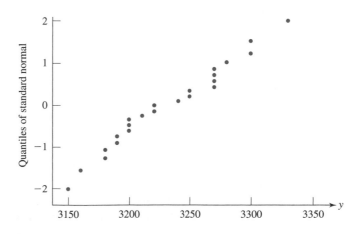

Figure 3.16 | The Normal Probability Plot of the Times Between Accidents

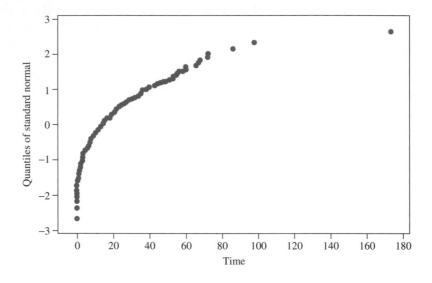

comfortable that the data come from a well-behaved distribution. Especially with a sample size of 23, we feel comfortable that the sample mean follows a normal distribution by the Central Limit Theorem.

Example 3.20 | **Times Between Industrial Accidents—Continued**

Figure 3.16 gives the normal probability plot for the times between industrial accidents. Earlier, we noted that the exponential distribution provides a reasonable model for these data. The exponential distribution looks nothing like a normal distribution. As a result, we expect this plot to look quite different from a straight line. In this case, the data appear to follow a curved shape, which indicates that we cannot model these data by a normal distribution.

VOICE OF EXPERIENCE

Whenever possible, use graphical techniques to check the reasonableness of the Central Limit Theorem.

The normal probability plot is an example of a quantile–quantile or *Q–Q plot*. Many standard statistical software packages generate Q–Q plots as a way to help us to determine whether the data roughly follow a specified distribution. The basic idea of these plots is to graph the data against the quantiles, assuming that these data follow the specified distribution. The only difference between these other Q–Q plots and the normal probability plot is the cdf used to determine the actual quantiles. Some other common distributions for which we often can generate Q–Q plots are the uniform, gamma, and Weibull.

∷ Computer Exercises

3.1 Use a statistical software package to illustrate the Central Limit Theorem when sampling from a normal distribution.

    a. Generate 1000 random samples, each of size 4, from a normal distribution with a mean of 10 and a standard deviation of 10. Have the software calculate the sample mean for each sample. Plot a histogram of the sample means, and comment on the plot.

    b. Generate another 1000 random samples, each of size 10, from the same normal distribution. Have the software calculate the sample mean for each sample and plot a histogram of the results. Comment on your plot.

    c. Generate another 1000 random samples, each of size 30, from the same normal distribution. Have the software calculate the sample mean for each sample and plot a histogram of the results. Comment on the plot.

3.2 Use a statistical software package to illustrate the Central Limit Theorem when sampling from a uniform distribution.

    a. Generate 1000 random samples, each of size 4, from a uniform distribution over the interval -7.5 to 27.5 that has a mean of 10 and a standard deviation of 10.1. Have the software calculate the sample mean for each sample and plot a histogram of the results. Comment on the plot.

    b. Generate another 1000 random samples, each of size 10, from the same uniform distribution. Have the software calculate the sample mean for each sample and plot a histogram of the results. Comment on the plot.

    c. Generate another 1000 random samples, each of size 30, from the same uniform distribution. Have the software calculate the sample mean for each sample and plot a histogram of the results. Comment on the plot.

3.3 Use a statistical software package to illustrate the Central Limit Theorem when sampling from an exponential distribution.

3.4 Use a statistical software package to illustrate the Central Limit Theorem when sampling from a Weibull distribution.

    a. Generate 1000 random samples, each of size 4, from a Weibull distribution with $\lambda = 3$ and $\beta = 2$. Have the software calculate the sample mean for each sample and plot a histogram of the results. Comment on the plot.

    b. Generate another 1000 random samples, each of size 10, from the same Weibull distribution. Have the software calculate the sample mean for each sample and plot a histogram of the results. Comment on the plot.

c. Generate another 1000 random samples, each of size 30, from the same Weibull distribution. Have the software calculate the sample mean for each sample and plot a histogram of the results. Comment on the plot.

⁖ *Exercises*

3.56 Kane (1986) discusses the concentricity of an engine oil seal groove. Concentricity measures the cross-sectional coaxial relationship of two cylindrical features. In this case, he studies the concentricity of an oil seal groove and a base cylinder in the interior of the groove. He measures the concentricity as a positive deviation using a dial indicator gauge. Historically, this process has produced an average concentricity of 5.6 with a standard deviation of 0.7. To monitor this process, he periodically takes a random sample of three measurements. If the average is greater than 6.8 or less than 4.3, he concludes that the process mean has shifted. Assume that the process mean is 5.6.

  a. Find the probability that on the next sample he concludes the process mean has shifted purely due to random chance.

  b. What did you assume in order to find this probability?

3.57 Runger and Pignatiello (1991) consider a plastic injection molding process for a part with a critical width dimension that has a historic mean of 100 and a historic standard deviation of 8. Periodically, clogs form in one of the feeder lines, causing the mean width to change. As a result, the operator periodically takes random samples of size four. If the sample mean width of these four parts is either larger than 101.0 or smaller than 99.0, then he must immediately take another sample. Consider the next sample taken. Assume that the actual mean width is 100.

  a. Find the probability that the operator must take another sample immediately.

  b. What did you assume in order to find this probability?

3.58 Yashchin (1995) discusses a process for the chemical etching of silicon wafers used in integrated circuits. This process etches the layer of silicon dioxide until the layer of metal beneath is reached. This company monitors the thickness of the silicon oxide layer because thicker layers require longer etching times. Historically, the layer has a true mean thickness of 1 micron and a standard deviation of 0.06 micron.

  a. A recent random sample of four wafers yielded a sample mean of 1.134. Find the probability of observing such a mean or something smaller assuming the historic mean and standard deviation.

  b. What did you assume in order to find this probability?

3.59 A supplier of plastic pellets claims that when the plastic is pressed into wafers of uniform thickness, they will have a mean breaking strength of 165 newtons and a standard deviation of 5 newtons. An engineer at the company randomly selects wafers from 12 batches of pellets.

a. Find the probability that the sample mean is less than 162 newtons.
b. Find the probability that the sample mean is between 164 and 166 newtons, inclusive.
c. What did you assume in order to do this analysis? How should you check this assumption?

3.60 The average modulus of rupture (MOR) for a particular grade of pencil lead is known to be 6500 psi with a standard deviation of 250 psi.

a. Find the probability that a random sample of 16 pencil leads will have an average MOR between 6400 and 6550 psi.
b. What did you assume in order to find this probability?

3.61 Historically, the chemical reactor for a polyester polymer has averaged 65% with a standard deviation of 1.5%. Consider a random sample of ten batches.

a. Find the probability that the sample mean is greater than 67%.
b. Find the probability that the sample mean is less than 64%.
c. Find the probability that the sample mean is between 62% and 66%, inclusive.
d. What did you assume in order to do this analysis? How should you check these assumptions?

3.62 An ethanol–water distillation column has an average yield of 93% with a standard deviation of 1%. A random sample of eight recent batches produced these yields:

0.90	0.93	0.95	0.86
0.90	0.87	0.93	0.92

a. Find the sample mean for these data.
b. Find the probability that we observe this sample mean or something smaller given the assumed information about this column.
c. If we could remove any single observation from this data set, which one has the most influence on the sample mean? Which one has the least influence? Justify your answer.
d. What did you assume in order to do this analysis? How comfortable are you with these assumptions?

3.63 Snee (1983) examined the thicknesses (in 0.001 in.) of paint can ears, which have a historic mean of 34.2 and a standard deviation of 2.8. The manufacturer took random samples of five cans each and measured the thicknesses of the ears in order to monitor the process. A recent sample yielded these thicknesses: 34, 31, 37, 39, and 36.

Table 3.6

29	36	39	34	34	29	29	28	32	31
34	34	39	38	37	35	37	33	38	41
30	29	31	38	29	34	31	37	39	36
30	35	33	40	36	28	28	31	34	30
32	36	38	38	35	35	30	37	35	31
35	30	35	38	35	38	34	35	35	31
34	35	33	30	34	40	35	34	33	35
34	35	38	35	30	35	30	35	29	37
40	31	38	35	31	35	36	30	33	32
35	34	35	30	36	35	35	31	38	36
32	36	36	32	36	36	37	32	34	34
29	34	33	37	35	36	36	35	37	37
36	30	35	33	31	35	30	29	38	35
35	36	30	34	36	35	30	36	29	35
38	36	35	31	31	30	34	40	28	30

a. Find the sample mean for these data.
b. Assume that the true population mean is 34.2. Find the probability that we observe the sample mean we calculated in part **a** or something larger.
c. If you could remove any single observation from this data set, which one has the most influence on the sample mean? Which one has the least influence? Justify your answer.
d. What assumptions are necessary to perform this analysis?
e. The data in Table 3.6 are the thicknesses from 30 samples. In light of these data, how comfortable are you with the assumptions required to do the above analysis? Justify your answer.

3.7 Random Behavior of Means When the Variance Is Unknown

The Sample Variance

The Central Limit Theorem requires that we know the variance, σ^2, which we rarely do. In such situations, we require an estimate of σ^2. The classical estimator of σ^2 is s^2.

Definition 3.14 | **Sample Variance**

We define the sample variance, s^2, by

$$s^2 = \frac{1}{n-1} \sum_{i=1}^{n} (y_i - \overline{y})^2.$$

This statistic looks like an "average" of the squared deviations from the sample mean. In this case, we use $n - 1$ for the denominator instead of n. We can show that by defining s^2 this way, $E(s^2) = \sigma^2$, which is a desirable property we shall discuss in Chapter 4. Clearly from the definitional formula,

$$s^2 \geq 0.$$

Although the definitional formula is very important for conceptual purposes, it should not be used for actually computing s^2. Frankly, most spreadsheet and engineering calculators find s^2 directly. In the few cases where we need to calculate s^2 by hand, we should use the *computational formula*:

$$s^2 = \frac{n \sum_{i=1}^{n} y_i^2 - \left(\sum_{i=1}^{n} y_i \right)^2}{n \cdot (n-1)}.$$

The classical estimator of σ is the sample standard deviation, s, given by

$$s = \sqrt{s^2}.$$

The t-Distribution

Since we calculate s^2 from the observed data, it is a random variable. In Section 4.7, we shall discuss its sampling distribution. The Central Limit Theorem tells us that if the sample size is sufficiently large, then

$$Z = \frac{\overline{Y} - \mu}{\sigma / \sqrt{n}}$$

follows a standard normal distribution. This Z-statistic, however, involves only a single random variable, $\overline{Y}$. When we substitute s for σ, we introduce a new source of variability. As a result, we have no basis for modeling the resulting statistic by a standard normal distribution.

If the y's follow a normal distribution, we can show that

$$t = \frac{\overline{Y} - \mu}{s / \sqrt{n}}$$

VOICE OF EXPERIENCE

We rarely know s^2; hence, the t-distribution is extremely important.

follows a t-distribution with $n - 1$ *degrees of freedom*. Essentially, the t-statistic represents how many estimated standard deviations a given value of a random variable lies relative to its mean. The degrees of freedom represent the amount of information present in our estimate of the standard error. In general,

$$\text{degrees of freedom} = \text{number of observations}$$
$$- \text{number of parameters estimated.}$$

It is well known that this result is quite robust to the normality assumption. By that, we mean that if the y's follow a well-behaved distribution (single peaked and relatively symmetric, and the tails die rapidly) and if n is "moderately" large, then we can well model

$$t = \frac{\overline{Y} - \mu}{s/\sqrt{n}}$$

by a t-distribution with $n - 1$ degrees of freedom. The larger the degrees of freedom associated with our t-statistic, the more the parent population can depart from a normal distribution. We typically use a stem-and-leaf display (if we have more than roughly 15 observations), a normal probability plot, or both to check the "well-behaved" assumption.

The t-distribution is actually a family of distributions indexed by the degrees of freedom. Essentially, the t-distribution is a "squatter" version of the Z, as illustrated in Figure 3.17. As the degrees of freedom increase, the t-distribution approaches the standard normal. In other words, as the amount of information associated with our estimate of the standard error gets larger, the closer this estimate comes to the true value. As a result, the t-distribution comes closer and closer to the Z-distribution.

Table 2 of the appendix gives values of $P(t \leq t_\alpha)$ for various important choices for α and the degrees of freedom. We shall use this table extensively in Chapter 4, so we save a detailed discussion of its use until then. Nonetheless, we can begin to interpret an observed value of the t-statistic as the number of estimated standard errors $\overline{y}$ lies relative to its presumed true mean. As a result,

Figure 3.17 | Comparison of the Z and t-Distributions

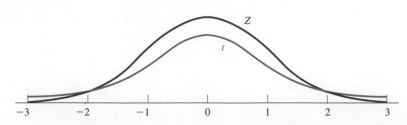

we have these two rough rules of thumb:

1. It is rare to observe $|t| > 3$ due to random chance.
2. It is unlikely, though possible, to observe $|t| > 2$ due to random chance.

The exact probabilities of observing such results depend on the number of degrees of freedom, which explains the formal need for Table 2 of the Appendix.

Example 3.21 | **Packaged Weights**

King (1992) discusses the net weights of a nominally 16-oz packaged product. An inspector collected a sample of 20 packages and accurately measured their net contents. Let y_i denote the net contents of the ith package. Table 3.7 lists the necessary information to calculate the sample mean and the sample variance. For these data, $\sum_{i=1}^{n} y_i = 330.4$ and $\sum_{i=1}^{n} y_i^2 = 5458.68$. The sample mean, the sample variance, and the sample standard deviation for these data are

$$\bar{y} = 16.52$$

$$s^2 = \frac{n \sum_{i=1}^{n} y_i^2 - \left(\sum_{i=1}^{n} y_i\right)^2}{n \cdot (n-1)}$$

$$= \frac{20(5458.68) - (330.4)^2}{20 \cdot (20-1)}$$

$$= 0.0248$$

$$s = \sqrt{s^2} = \sqrt{0.0248} = 0.16.$$

Table 3.7 | The Packaged Weights Data

i	y_i	y_i^2	i	y_i	y_i^2
1	16.4	268.96	11	16.6	275.56
2	16.4	268.96	12	16.6	275.56
3	16.5	272.25	13	16.8	282.24
4	16.5	272.25	14	16.3	265.69
5	16.6	275.56	15	16.4	268.96
6	16.7	278.89	16	16.5	272.25
7	16.2	262.44	17	16.5	272.25
8	16.4	268.96	18	16.6	275.56
9	16.4	268.96	19	16.7	278.89
10	16.5	272.25	20	16.8	282.24

Figure 3.18 | The Stem-and-Leaf Display for the Packaged Weights Data

Stem	Leaves	Number	Depth
16t:	23	2	2
16f:	4444455555	10	
16s:	666677	6	8
16●.:	88	2	2

Assume that these data form a random sample. Management must ensure that the true amount packaged is something greater than 16 oz. We shall discuss formal approaches to this kind of problem in Chapter 4. However, if $\mu = 16.0$, then the value of the observed t-statistic is

$$t = \frac{\overline{y} - \mu}{s/\sqrt{n}}$$

$$= \frac{16.52 - 16.0}{0.16/\sqrt{20}}$$

$$= 14.53.$$

Thus, a sample mean of 16.52 is more than 14.5 estimated standard errors from 16.0. The chances of seeing such an extreme value due to random chance are extremely remote and, in fact, are approximately zero. We are quite confident that the true process mean is something greater than 16.0 oz.

Of course, this informal analysis depends on how well we can model

$$\frac{\overline{Y} - \mu}{s/\sqrt{n}}$$

by a t-distribution. Consider the stem-and-leaf display of the data in Figure 3.18. It indicates that the data appear to come from a single-peaked, roughly symmetric distribution with tails that die fairly rapidly. Figure 3.19 gives the normal probability plot for these data. This is a typical plot for a moderate number of observations that assume a relatively few number of distinct values. Although the data do not form a perfectly straight line, we are satisfied that the data come from a well-behaved distribution, especially because we have a sample size of 20. As a result, we should feel reasonably comfortable with our analysis of these data.

Figure 3.19	The Normal Probability Plot for the Packaged Weights Data

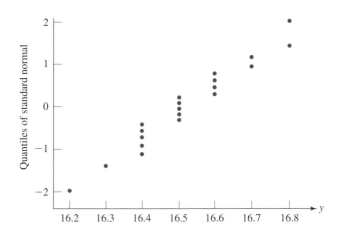

∷ *Computer Exercises*

3.5   Use a statistical software package to illustrate the distribution of the *t*-statistic when sampling is from a normal distribution.

a.   Generate 1000 random samples, each of size 4, from a normal distribution with a mean of 10 and a standard deviation of 10. Have the software calculate the sample mean, the sample standard deviation, and the *t*-statistic for each sample. Plot a histogram of the *t*-statistics and comment on the plot.

b.   Generate another 1000 random samples, each of size 10, from the same normal distribution. Have the software calculate the sample mean, the sample standard deviation, and the *t*-statistic for each sample. Plot a histogram of the results and comment on your plot.

c.   Generate another 1000 random samples, each of size 30, from the same normal distribution. Have the software calculate the sample mean, the sample standard deviation, and the *t*-statistic for each sample. Plot a histogram of the results and comment on the plot.

3.6   Use a statistical software package to illustrate the distribution of the *t*-statistic when sampling is from a uniform distribution.

a.   Generate 1000 random samples, each of size 4, from a uniform distribution over the interval −7.5 to 27.5 that has a population mean of 10 and a population standard deviation of 10.1. Have the software calculate the sample mean, the sample standard deviation, and the *t*-statistic for each sample. Plot a histogram of the results and comment on the plot.

b.   Generate another 1000 random samples, each of size 10, from the same uniform distribution. Have the software calculate the sample mean, the sample standard deviation, and the t-statistic for each sample. Plot a histogram of the results and comment on the plot.

c. Generate another 1000 random samples, each of size 30, from the same uniform distribution. Have the software calculate the sample mean, the sample standard deviation, and the t-statistic for each sample. Plot a histogram of the results and comment on the plot.

3.7   Use a statistical software package to illustrate the distribution of the t-statistic when sampling is from an exponential distribution.

a. Generate 1000 random samples, each of size 4, from an exponential distribution with $\lambda = 0.1$ that has a population mean and a population standard deviation of 10. Have the software calculate the sample mean, the sample standard deviation, and the t-statistic for each sample. Plot a histogram of the results and comment on the plot.

b. Generate another 1000 random samples, each of size 10, from the same exponential distribution. Have the software calculate the sample mean, the sample standard deviation, and the t-statistic for each sample. Plot a histogram of the results and comment on the plot.

c. Generate another 1000 random samples, each of size 30, from the same exponential distribution. Have the software calculate the sample mean, the sample standard deviation, and the t-statistic for each sample. Plot a histogram of the results and comment on the plot.

3.8   Use a statistical software package to illustrate the distribution of the t-statistic when sampling from a Weibull distribution.

a. Generate 1000 random samples, each of size 4, from a Weibull distribution with $\lambda = 3$ and $\beta = 2$. Have the software calculate the sample mean, the sample standard deviation, and the t-statistic for each sample. Plot a histogram of the results and comment on the plot.

b. Generate another 1000 random samples, each of size 10, from the same Weibull distribution. Have the software calculate the sample mean, the sample standard deviation, and the t-statistic for each sample. Plot a histogram of the results and comment on the plot.

c. Generate another 1000 random samples, each of size 30, from the same Weibull distribution. Have the software calculate the sample mean, the sample standard deviation, and the t-statistic for each sample. Plot a histogram of the results and comment on the plot.

⋗ Exercises

3.64   Pignatiello and Ramberg (1985) studied the heat treatment of leaf springs. In this process, a conveyor system transports leaf spring assemblies through a high-temperature furnace. After this heat treatment, a high-pressure press induces the curvature. After the spring leaves the press, an oil quench cools it to near ambient temperature. An important quality characteristic of this process is the resulting free height of the spring, which has a target value of exactly 8 in. The following table lists the resulting leaf spring heights (in inches) for a heating time of 23 sec. Assume these data form a random sample.

7.5	7.6	7.5	7.5	7.6	7.5
7.6	7.6	7.8	7.6	7.8	7.6
7.6	7.6	7.4	7.2	7.2	7.3
7.6	7.8	7.7	7.8	7.5	7.6

a. Find the sample mean, the sample variance, and the sample standard deviation for these data.

b. Assume that the true population mean is 8 in. Find the observed value of the t-statistic. Use the rough rules of thumb to interpret your result.

c. What did you assume to do the informal analysis? Can you evaluate how well these data meet the assumptions? If yes, determine how comfortable you are with them. If not, explain why.

3.65 An ethanol–water distillation column has an average yield of 93%. A random sample of eight recent batches produced these yields:

0.90	0.93	0.95	0.86
0.90	0.87	0.93	0.92

a. Find the sample mean for these data.

b. Assume that the historical average is true. Calculate the observed value of the t-statistic. Use the rough rules of thumb to interpret your result.

c. What did you assume to do the informal analysis? Can you evaluate how well these data meet the assumptions? If yes, determine how comfortable you are with them. If not, explain why.

3.66 Consider the filling operation for 20-oz bottles of a popular soft drink. Historically, this operation averages 20.2 oz. A recent random sample of 12 bottles yielded these volumes:

20.0	20.1	20.0	19.9	20.5	20.9
20.1	20.4	20.2	19.1	20.1	20.0

a. Find the sample mean for these data.

b. Assume that the historical average is true. Calculate the observed value of the t-statistic. Use the rough rules of thumb to interpret your result.

c.   What did you assume to do the informal analysis? Can you evaluate how well these data meet the assumptions? If yes, determine how comfortable you are with them. If not, explain why.

3.67   The nominal power produced by a student-designed internal combustion engine is 100 hp. The student team that designed this engine conducted ten tests to determine the actual power. The data follow:

98	101	102	97	101
98	100	92	98	100

a.   Find the sample mean, the sample variance, and the sample standard deviation for these data.

b.   Assume that the nominal average is true. Calculate the observed value for the t-statistic. Use the rough rules of thumb to interpret your result.

c.   What did you assume to do this analysis? How comfortable are you with these assumptions?

3.68   A cheese manufacturer is concerned that a supplier is adding water to their milk to increase profits. Adding water to milk raises its freezing temperature, which is $-0.545°$ C. A random sample of 10 batches of the supplier's milk yielded the following freezing temperatures:

−0.54140	−0.53873	−0.52649	−0.53203	−0.53340
−0.54351	−0.53729	−0.52877	−0.53872	−0.54827

a.   Find the sample mean for these data.

b.   Assume the supplier is not adding water to the milk. Calculate the observed value for the t-statistic. Use the rough rules of thumb to interpret your result.

c.   What did you assume to do the informal analysis? Can you evaluate how well these data meet the assumption? If yes, determine how comfortable you are with them. If not, explain why.

3.69   A maintenance manager believes that the true mean time between repairs for a major piece of packaging equipment is 20 days. Assume that the last 15 repairs form a random sample. The times between repairs (in days) are listed here:

10	23	12	18	20
17	11	17	23	54
12	14	18	22	11

a.   Find the sample mean for these data.

b.   Assume that the maintenance manager's belief is true. Calculate the observed value of the t-statistic. Use the rough rules of thumb to interpret your result.

c. What did you assume to do the informal analysis? Can you evaluate how well these data meet the assumptions? If yes, determine how comfortable you are with them. If not, explain why.

3.70 Farnum (1994, p. 195) discusses a chrome plating process. Small electric currents are run through a chemical plate that contains nickel, resulting in a thin plating of the metal on the part. Since the bath loses nickel as the plating proceeds, the operators periodically add more nickel to the bath. The process runs three shifts per day. The following are the bath concentrations at the beginning of the day for five days: 4.8, 4.5, 4.4, 4.2, and 4.4. Assume that these concentrations form a random sample.

a. Find the sample mean, the sample variance, and the sample standard deviation for these data.

b. Assume that the true mean nickel concentration is 4.5 oz/gal, which is the operating standard. Calculate the observed value for the t-statistic. Use the rough rules of thumb to interpret your result.

c. What did you assume to do the informal analysis? Can you evaluate how well these data meet the assumptions? If yes, determine how comfortable you are with them. If not, explain why.

d. If you add any single value from the range 4.0–5.0, which one produces the largest sample variance? Which one produces the smallest? Justify your answer statistically. (*Hint:* What is the definitional formula for the sample variance?)

3.71 Albin (1990) studied aluminum contamination in recycled PET plastic from a pilot plant operation at Rutgers University. She collected 26 samples and measured, in parts per million (ppm), the amount of aluminum contamination. The maximum acceptable level of aluminum contamination, on the average, is 220 ppm. The data follow:

291	222	125	79	145	119	244	118	182
63	30	140	101	102	87	183	60	191
119	511	120	172	70	30	90	115	

a. Find the sample mean, the sample variance, and the sample standard deviation for these data.

b. Assume that the true mean amount of aluminum contamination is 220 ppm, which is the maximum acceptable level. Calculate the observed value for the t-statistic. Use the rough rules of thumb to interpret your result.

c. What did you assume to do the informal analysis? Can you evaluate how well these data meet the assumptions? If yes, determine how comfortable you are with them. If not, explain why.

d. If you could add any single observation from the range 0–500, which one produces the largest sample variance? Which one produces the smallest? Justify your answer statistically. (*Hint:* What is the definitional formula for the sample variance?)

⋗ **3.8** Normal Approximation to the Binomial

Engineers often require the binomial distribution to model data when the sample size, n, is too large to make direct calculation of the probabilities convenient. We can use the Central Limit Theorem to show that if the sample size, n, is sufficiently large, then we can well model the distribution of a binomial random variable, Y, by a normally distributed random variable, Y^*, that has the same mean and variance as Y. Thus,

$$E(Y^*) = np$$

$$\mathrm{var}[Y^*] = npq = np \cdot (1 - p).$$

Typically, we consider n to be sufficiently large if both

$$np \geq 5 \quad \text{and} \quad nq \geq 5.$$

The approximation works better if both np and nq are greater than or equal to 10.

Example 3.22	**Packaging 50-lb Graphite Bags**

People have found many uses for graphite, from brake linings to pencil lead. A major graphite company commonly distributes graphite in bags that have a nominal net weight of 50 lb. Most users require that the true mean net weight really be 50 lb. This company defines a correctly filled bag to contain between 48 and 52 lb. Historically, 1% of the bags filled by this company's packaging process fall outside this range and thus do not conform to the specifications.

Each week, this company carefully weighs the net contents of 1000 randomly selected bags to determine the proportion that fails to conform to the specifications. Let Y be the number of nonconforming bags. We can well model Y by a binomial distribution with mean and variance

$$\mu = E(Y) = np = 1000(0.01) = 10$$

$$\sigma^2 = npq = 1000(0.01)(0.99) = 9.9,$$

respectively. We could use the probability function for a binomial random variable to compute specific probabilities; however, with such a small p and such a large n, we probably should seek an easier solution. Since both np and nq are greater than 5, we can approximate Y by a normal random variable, which we will call Y^*, that has the same mean and variance as Y.

Consider now the probability that we observe exactly ten nonconforming bags. Although it makes perfect sense to talk about having exactly ten nonconforming bags, for a continuous random variable, $P(Y^* = 10) = 0$. Since Y^* has meaning only over some interval of interest, we must define intervals for Y^* that correspond

Figure 3.20

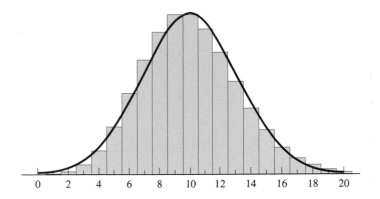

to the discrete values of Y. Consider the plot in Figure 3.20. A reasonable interval corresponding to $Y = 10$ is $9.5 \leq Y^* \leq 10.5$ (see Figure 3.21). Thus, we can approximate $P(Y = 10)$ by $P(9.5 \leq Y^* \leq 10.5)$. We can approximate $P(Y \leq 10)$ by $P(Y^* \leq 10.5)$, and we can approximate $P(Y < 10)$ by $P(Y^* \leq 9.5)$. Using the constant 0.5 in this manner is called the *correction for continuity.*

In general, if Y really follows a binomial distribution with parameters n and p and we wish to approximate its distribution by Y^*, which is a normal random variable with mean np and variance npq, then (see Figure 3.22)

$$P(Y = a) \approx P(a - 0.5 \leq Y^* \leq a + 0.5)$$

$$P(Y \leq a) \approx P(Y^* \leq a + 0.5)$$

$$P(Y < a) \approx P(Y^* \leq a - 0.5)$$

$$P(Y > a) \approx P(Y^* \geq a + 0.5)$$

$$P(Y \geq a) \approx P(Y^* \geq a - 0.5).$$

Figure 3.21

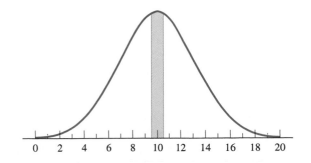

Figure 3.22

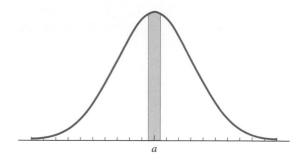

a

With this information, the probability that exactly ten bags fail to meet the specifications is given by

$$P(Y = 10) \approx P(9.5 \leq Y^* \leq 10.5)$$

$$= P\left(\frac{9.5 - np}{\sqrt{npq}} \leq \frac{Y^* - np}{\sqrt{npq}} \leq \frac{10.5 - np}{\sqrt{npq}}\right)$$

$$= P\left(\frac{9.5 - 10}{\sqrt{9.9}} \leq Z \leq \frac{10.5 - 10}{\sqrt{9.9}}\right)$$

$$= P(-0.16 \leq Z \leq 0.16)$$

$$= P(Z \leq 0.16) - P(Z \leq -0.16)$$

$$= 0.5636 - 0.4364$$

$$= 0.1272.$$

Thus, we should expect approximately 13% of the time to have ten bags that fail to meet the specifications.

A major customer of this company, an automobile manufacturer, rejects shipments of graphite when more than 1.5% of the bags fail to meet the specifications. Thus, consider the probability that more than 1.5% of the random sample of 1000 bags fail to meet the specifications, which means that they find more than 15 nonconforming bags. We thus seek $P(Y > 15)$, which is given by

$$P(Y > 15) \approx P(Y^* \geq 15.5)$$

$$= P\left(\frac{Y^* - np}{\sqrt{npq}} \geq \frac{15.5 - np}{\sqrt{npq}}\right)$$

$$= P\left(Z \geq \frac{15.5 - 10}{\sqrt{9.9}}\right)$$

$$= P(Z \geq 1.75)$$

$$= 1.0 - P(Z \leq 1.75)$$
$$= 1.0 - 0.9599 = 0.0401.$$

Approximately 4% of the time we should expect to see more than 15 nonconforming bags out of 1000 inspected.

:> Exercises

3.72 A manufacturer of nickel–hydrogen batteries discovered a problem with "blisters" on its nickel plates. These blisters cause the resulting battery cell to short out prematurely. During a specific production period, 8.5% of the plates exhibited blisters within 50 test cycles. A normal production cell uses 60 nickel plates. The customer strongly believes that the cell will short out prematurely if two or more plates blister within 50 cycles.

   a. Find the probability that two or more plates in this cell blister.
   b. Find the probability that one or fewer fails.

3.73 Atwood (1986) studied the failure of pumps used in standby safety systems for commercial nuclear power plants and found that the probability that a randomly selected pump failed to run after starting was 0.16. Consider a nuclear power plant with 200 standby safety pumps.

   a. Find the probability that more than 10% of the pumps fail to run after starting.
   b. Find the probability that 32 or fewer fail to run after starting.

3.74 Felt-tip markers have shelf lives of approximately two years. Ideally, at least 95% of the felt-tip markers should still write after sitting on a shelf for two years. Most manufacturers use accelerated life testing whereby markers are placed in an oven at elevated temperatures for a given period of time, usually about six weeks. The proportion that survive the elevated temperatures provides a good estimate of the proportion that should survive two years on a shelf. Suppose that historically 95% of the markers tested survive the accelerated life test. Consider a random sample of 300 markers.

   a. Find the probability that more than 15 markers fail to survive the test.
   b. Find the probability that fewer than ten fail to survive the test.
   c. Suppose a random sample of 300 markers yielded 30 that failed to survive. What would you begin to conclude about the true proportion that should survive two years on the shelf? Justify your answer statistically.

3.75 Marketing studies for many consumer goods, including automobiles, often ask 600 people whether they prefer one product or brand over another. For example, an automobile manufacturer wants to market its latest luxury model to those who currently drive the leading selling model. Consider a random sample of people who all drive the leading selling luxury automobile. Each person in this group drives the manufacturer's latest luxury automobile and then gives his or her preference.

a. Suppose that 50% of *all* people who drive the leading selling luxury automobile would actually prefer the manufacturer's latest model. Find the probability that more than 550 people in the random sample of 600 prefer the manufacturer's new model.

b. Suppose that 25% of *all* people who drive the leading selling luxury automobile would actually prefer the manufacturer's latest model. Find the probability that fewer than 135 people in the random sample of 600 prefer the manufacturer's new model.

3.76 A major bottler of soft drinks historically averages 0.5% underfills. Consider a random sample of 2000 bottles with carefully measured volumes.

a. Find the probability that ten or fewer bottles are underfilled.
b. Find the probability that 15 or more bottles are underfilled.
c. Find the probability that 20 or more bottles are underfilled.
d. Suppose that a consumer group carefully measures the volumes of 2000 bottles and finds 20 or more underfilled. What should they conclude? Justify your answer statistically.

3.77 As explained earlier, airplanes approaching the runway for landing are required to stay within the localizer (a certain distance left and right of the runway). When an airplane deviates from the localizer, it is sometimes referred to as an exceedence. Consider one airline at a medium sized airport with 30 daily arrivals and an exceedence rate of 12%.

a. Find the probability that on a particular day more than 10 planes have an exceedence.
b. Find the probability that on a particular day 5 planes or fewer experience an exceedence.
c. Find the probability that on a particular day more than 2 but fewer than 5 experience an exceedence.
d. Find the exact probability that on a particular day more than 2 but fewer than 5 experience an exceedence.
e. Do your answers to parts **c** and **d** differ? Why or why not?

3.78 A manufacturer of CDs claims that 99.2% of its CDs are defect-free. A software company that buys these CDs wants to verify the manufacturer's claim. They randomly select 1500 CDs.

a. Find the probability that 25 or fewer CDs are defective.
b. Find the probability that 4 or fewer CDs are defective.
c. Suppose a random sample of 1500 CDs yielded 4 defective. What would you begin to conclude about the manufacturer's claim? Justify your answer statistically.

3.9 Case Study

Generally, Faber-Castell sells its products either in a polypropylene film package or in a blister pack. Both packaging processes are high volume, so the company seeks to minimize downtime due to repair. One way to achieve this goal is to develop appropriate probability models for the time between repairs for the key pieces of equipment. Maintenance then can allocate its mechanics in an optimal way to meet the anticipated repairs.

Faber uses an older piece of equipment for a significant proportion of the polypropylene film packaging. From historical data, Faber can well model the time between repairs for this piece of equipment by an exponential distribution with $\lambda = 0.15$. As a result, on the average, this piece of equipment goes $1/\lambda = 6.7$ days between repairs with a standard deviation of 6.7 days. The probability that it goes longer than y_0 days between repairs is $\exp(-0.15 \cdot y_0)$. The following values are the days between repairs for this piece of equipment during a four-month period in 1995:

4	6	1	9	6
0	1	4	2	29
20	1	4	7	1

Figure 3.23 gives the stem-and-leaf display, which shows a pattern quite similar to the times between industrial accidents data, also modeled by an exponential distribution. As an initial impression, we may think that the times 20 and 29 are outliers. However, according to our exponential model, the probability of this equipment going 20 or more days between repairs is

$$\exp(-0.15 \cdot 20) = 0.050$$

Figure 3.23 | The Stem-and-Leaf Display for the Film Packaging Equipment Data

Stem	Leaves	Number	Depth
0*:	011112444	9	
0•:	6679	4	6
1*:			
1•:			
2*:	0	1	2
2•:	9	1	1

or one in 20. As a result, these values (20 and 29) are perfectly consistent with an exponential distribution that has a very heavy right tail. We would be making a major mistake to conclude that these times are unusual. Periodically, we expect this piece of equipment to go a significant time between repairs.

We also must consider the mean time between repairs (6.7 days) in light of the shape of the actual times. We usually think that roughly half of the data values should fall below and roughly half above the mean. This expectation is quite reasonable when the shape of the distribution is roughly symmetric. However, the exponential distribution is highly skewed to the right. As a result, most of the times actually fall well below the mean. The occasional extreme value distorts the average to make it larger.

Faber uses a blister package to sell individual items such as erasers. A critical piece of equipment for this process forms the plastic bubble or blister. From historical data, Faber can well model the times between repairs for this piece of equipment by a Weibull distribution with $\lambda = 0.075$ and $\beta = 0.75$. As a result, on the average, this piece of equipment goes 15.9 days between repairs with a standard deviation of 21.5 days. The probability that it goes longer than y_0 days between repairs is $\exp(-0.075 \cdot y_0)^{0.75}$. The following values are the days between repairs for this piece of equipment during a six-month period in 1995:

1	25	50	1	7
12	2	8	1	1
0	2	22	68	29

The stem-and-leaf display in Figure 3.24 shows a highly skewed pattern. As an initial impression, we may think that the times 50 and 68 are outliers. However,

---

Figure 3.24   The Stem-and-Leaf Display for the Blister Package Equipment Data

Stem	Leaves	Number	Depth
0:	011112278	9	
1:	22	2	6
2:	5	1	4
3:			
4:			
5:	0	1	2
6:	8	1	1

according to the Weibull model, the probability of this equipment going 50 or more days between repairs is

$$\exp - [(0.075 \cdot 50)^{0.75}] = 0.068$$

or more than one in 20. As a result, these values (50 and 68) are perfectly consistent with a Weibull distribution that has a very heavy right tail. We would be making a major mistake to conclude that these times are unusual. Periodically, we expect this piece of equipment to go a significant time between repairs.

Again, we must consider the mean time between repairs (15.9 days) in light of the shape of the actual times. We usually think that roughly half of the data values should fall below and roughly half above the mean. Such an expectation is quite reasonable when the shape of the distribution is roughly symmetric. However, this Weibull distribution is highly skewed to the right. As a result, most of the times actually fall well below the mean. The occasional extreme value distorts the average to make it larger.

The packaging department has many pieces of equipment requiring similar frequency of repair. Since these pieces of equipment require frequent attention, maintenance keeps one full-time mechanic in this area during the first shift to handle the minor repairs.

⁘ References

1. Albin, S. L. (1990). The lognormal distribution for modeling quality data when the mean is near zero. *Journal of Quality Technology, 22,* 105–110.
2. Atwood, C. L. (1986). The binomial failure rate common cause model. *Technometrics, 28,* 139–148.
3. Baltazar-Aban, I., and Pena, E. A. (1995). Properties of hazard-based residuals and implications in model diagnostics. *Journal of the American Statistical Association, 90,* 185–197.
4. Buxton, J. R. (1991). Some comments on the use of response variable transformations in empirical modelling. *Applied Statistics, 40,* 391–400.
5. Canning, J. T. (1993). Marathon report for a photolithography exposure tool. *Statistics in the Semiconductor Industry.* Austin, TX: SEMATECH. Technology Transfer No. 92051125A-GEN, Vol. III, pp. 6-10–6-35.
6. Davis, T. P., and Lawrance, A. J. (1989). The likelihood for competing risk survival analysis. *Scandinavian Journal of Statistics, 16,* 23–28.
7. Farnum, N. R. (1994). *Modern statistical quality control and improvement.* Pacific Grove, CA: Duxbury Press.
8. Hurwitz, A. M., and Spagon, P. D. (1993). Identifying sources of variation in a wafer planarization process. *Statistics in the Semiconductor Industry.* Austin, TX: SEMATECH. Technology Transfer No. 92051125A-GEN, Vol. I, pp. 3-45–3-71.
9. Kalbfleisch, J. D., Lawless, J. F., and Robinson, J. A. (1991). Methods for the analysis and prediction of warranty claims. *Technometrics, 33,* 273–285.

10. Kane, V. E. (1986). Process capability indices. *Journal of Quality Technology, 18,* 41–52.

11. King, J. R. (1992). Tutorial comments on Lehrman manual [QE 4(1)]. *Quality Engineering, 5,* 107–122.

12. Lewis, T., and the M345 Course Team (1986). *M345 statistical methods, unit 2: basic methods: testing and estimation.* Milton Keynes: The Open University.

13. Lucas, J. M. (1985). Counted data CUSUM's. *Technometrics, 27,* 129–144.

14. Marcucci, M. (1985). Monitoring multinomial processes. *Journal of Quality Technology, 17,* 86–91.

15. Maxcy, R. B., and Lowry, S. R. (1984). Evaluating variability of filling operations. *Food Technology, 38*(12), 51–55.

16. Mee, R. W. (1990). An improved procedure for screening based on a correlated, normally distributed variable. *Technometrics, 32,* 331–337.

17. Miyamura, T. (1982). Estimating component failure rates for combined component and systems data: Exponentially distributed component lifetimes. *Technometrics, 24,* 313–318.

18. Moore, L. M., and Beckman, R. J. (1988). Approximate one-sided tolerance bounds on the number of failures using Poisson regression. *Technometrics, 30,* 283–290.

19. Morris, M. D. (1987). A sequential experimental design for estimating a scale parameter from quantal life testing data. *Technometrics, 29,* 173–181.

20. Nelson, L. S. (1987). Comparison of Poisson means: The general case. *Journal of Quality Technology, 19,* 173–179.

21. Padgett, W. J., and Spurrier, J. D. (1990). Shewhart-type charts for percentiles of strength distributions. *Journal of Quality Technology, 22,* 283–288.

22. Pignatiello, J. J., Jr., and Ramberg, J. S. (1985). Discussion of "Off-line quality control, parameter design, and the Taguchi method" by R. N. Kackar. *Journal of Quality Technology, 17,* 198–206.

23. Runger, G. C., and Pignatiello, J. J., Jr. (1991). Adaptive sampling for process control. *Journal of Quality Technology, 23,* 135–155.

24. Snee, R. D. (1983). Graphical analysis of process variation studies. *Journal of Quality Technology, 15,* 76–88.

25. Sokal, R. R., and Rohlf, F. J. (1981). *Biometry,* 2nd ed. San Francisco: W. H. Freeman.

26. Wasserman, G. S., and Wadsworth, H. M. (1989). A modified Beattie procedure for process modeling. *Technometrics, 31,* 415–421.

27. Yashchin, E. (1995). Estimating the current mean of a process subject to abrupt changes. *Technometrics, 37,* 311–323.

28. Zimmels, Y. (1983). Theory of kindred sedimentation of polydisperse mixtures. *American Institute of Chemical Engineers Journal, 29,* 669–676.

Estimation and Testing

⊳ 4.1 Estimation

Populations and Samples

In Chapter 3, we developed the foundations for modeling the random behavior of engineering data. These models provide the basis for predicting physical phenomena. From a statistician's perspective, these models actually describe *populations*.

Most engineering data come from processes that change over time. Within this context, the population corresponds to the state of the process at a specific interval of time. Three examples of populations are

- The true mean outside diameter of the pen barrels produced by an injection molding process
- The differences in the elastic strengths of polymer yarn produced by two different machines
- The coefficients that relate the effect of catalyst, temperature, and pressure to a polymer filament's strength

Certain features of populations are of real interest to engineers. The outside diameter of the pen barrel must be as close as possible to a specific nominal value to guarantee the proper fit between the barrel and the pen cap. Thus, the true mean diameter and some appropriate measure of the diameters' variability are of real interest. Often we use the standard deviation as our measure of variability. We should be vitally concerned about the difference among the typical strengths of the yarns spun on the two different machines. Clearly, there must be an engineering reason to explain any difference that exists. The coefficients in our model for the filament strength determine the influence that the three factors (catalyst, temperature, and pressure) exert on the filament strength. If a given coefficient is actually zero, then the particular factor associated with that coefficient has no influence on the filament's strength. Important terms in our models lead to the concept of parameters.

Definition 4.1	**Parameter**

A **parameter** is a numeric quantity that describes an important feature of a population.

Parameters always depend, either implicitly or explicitly, on the model selected to describe the population, and they summarize the relevant information about the population. Statistical thinking recognizes that we often can "distill" all the important information about a population of interest into just a few quantities. Examples of parameters are

- The true average or *population mean*, μ, outside diameter of a pen barrel
- The *population standard deviation*, σ, which is an important measure of variability, for the strengths of the yarns spun on a specific machine
- The true value of the *coefficient*, β, that relates the effect of changes in the polymerization temperature and a polymer filament's strength

In general, we shall use lowercase Greek letters to denote parameters. Occasionally, we shall use θ to denote some arbitrary parameter.

The Basic Estimation Problem   The engineering method requires us to estimate the parameters in our model and to address "interesting questions" about these parameters. In real life, we never see populations, at least not in their entirety. As a result, the parameters that describe the important features of these populations are never truly known. We thus have a problem. Consider the outside diameters of the pen barrels. Of real concern to an engineer is the true mean outside diameter (μ) and the population standard deviation (σ), but we are unable to ever know their true values. Do you begin to see the dilemma? The commonsense approach to this quandary is to *sample* from the population and then to use this sample to *estimate* the appropriate parameters.

Definition 4.2	**Statistic**

A *statistic* is a quantity calculated from a sample that describes an important feature of that sample.

In general, we shall use either a lowercase Roman letter or a lowercase Greek letter with the symbol $\wedge$, called a "hat," over it to denote a statistic. Three examples of statistics are

- The sample average or *sample mean*, $\bar{y}$, of the outside diameters of ten pen barrels produced by an injection molding process

- The *sample standard deviation, s*, for the strengths of five sample yarns spun on a specific machine
- The *estimated coefficient, $\hat{\beta}$*, that relates the effect of changes in the polymerization temperature and a polymer filament's strength

---

Definition 4.3 | **Estimator**

An **estimator** is a statistic used to estimate an unknown parameter of a population.

VOICE OF EXPERIENCE

Both statistics and estimators are random variables.

We shall let $\hat{\theta}$ represent an estimator of the arbitrary parameter θ of some population of interest. Examples of estimators are $\bar{y}$, s, and $\hat{\beta}$.

Example 4.1 | **Filling Milk Cartons**

A common industrial engineering problem deals with packaging equipment. Consider the process that fills 8-oz milk cartons. Clearly, if we put too much in the carton, we waste milk through spillage. A more subtle issue, though, is giving away "free product." Government regulations require that, on the average, the distributor must put something more than 8 oz in the carton to prevent underages. From an economic standpoint, we would like to make sure that the true mean amount is no more than absolutely necessary to meet these regulations. Two questions are very important for the operation of this process:

- What is the true *current* mean amount of milk delivered to these cartons?
- Is the current mean too large or too small?

Given the natural variability in the amounts delivered to each individual carton, these questions are nontrivial and demand sound statistical analysis.

Figure 4.1 illustrates the relationship between a population and a sample. Our real interest is in the population—in this case, in the amounts of milk delivered by the process. Often, we may assume that the population produces observations that either follow some model or are well approximated by some model. For example, we may model the amounts delivered by a normal distribution. Thus, there are two important parameters of characteristics of this population:

- μ, the population mean amount of milk delivered
- σ, the population standard deviation for these amounts

Under the assumption that the amounts follow a normal distribution, knowledge of μ and σ provides complete information about the probabilistic behavior of the data. Unfortunately, we would need to observe the amount delivered to every carton over the life of this process to actually know both μ and σ, which is clearly not realistic. Thus, μ and σ are truly unknowable, and this presents a problem.

Figure 4.1 | The Relationships Among Populations, Parameters, Samples, and Statistics

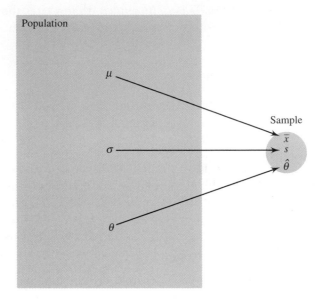

Our solution is to take a *sample* from our population and use the sample data to estimate μ and σ.

We have already encountered two potential estimators of μ:

- The sample mean
- The sample median

The two possible estimators of σ are

- s, the *sample standard deviation*
- Some constant multiple of the *interquartile range*

Now, which estimators should we use to estimate μ and σ? In this context, how should we estimate the true current mean amount of milk delivered and the true standard deviation for these amounts?

Choosing Estimators

To lay a foundation for addressing how to select appropriate estimators, we need to pursue a slight digression. Consider the purchase of a new personal computer. What factors should we consider: price, microprocessor, speed, RAM, storage on the hard drive, peripherals, or others? Clearly, we should make our decision based on criteria that weigh the relative values of these factors.

Consider a second digression. Let θ be the true time at this very instant. Do we really know θ? Of course not. However, our watch or a clock provides an estimate of θ. How do we define a "good" estimate of θ? Most people tend to define *good* in terms of these criteria:

- *accuracy,* where our watch is set to the correct time
- *precision,* where we prefer a watch with a second hand to one with only a minute hand and we prefer one that neither gains nor loses time

These same two issues arise in statistical estimation. Classical statistics tends to measure accuracy through the concept of unbiasedness.

Definition 4.4 | **Unbiased Estimator**

An *unbiased estimator* of an unknown parameter is one whose expected value is equal to the parameter of interest.

We call $\hat{\theta}$ an unbiased estimator of θ if

$$E[\hat{\theta}] = \theta.$$

Thus, the estimator yields, on the average, an estimate close to the true value. A biased estimator is one where $E[\hat{\theta}] \neq \theta$. Figure 4.2 shows the sampling distributions for two possible estimators, $\hat{\theta}_1$ and $\hat{\theta}_2$, of some parameter θ. The mean value of $\hat{\theta}_1$ is θ, whereas the mean value of $\hat{\theta}_2$ is not. In terms of accuracy, we prefer $\hat{\theta}_1$ because its long-run average is θ—that is, because it is an unbiased estimator of θ. The concept of precision looks at the variances of the estimators.

Definition 4.5 | **Precision**

An estimator is more *precise* if its sampling distribution has a smaller standard error.

Figure 4.2 | Biased and Unbiased Estimators

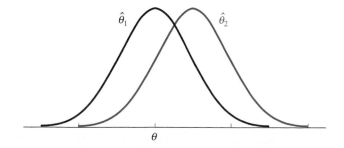

Figure 4.3 | Illustration of Precision

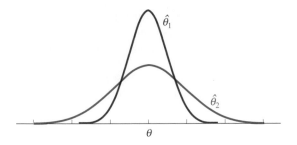

Figure 4.3 shows the sampling distributions for two unbiased estimators, $\hat{\theta}_1$ and $\hat{\theta}_2$, of some parameter θ. Both estimators appear to be unbiased; however, $\hat{\theta}_1$ has a much smaller variance than $\hat{\theta}_2$. Consequently, we expect $\hat{\theta}_1$ to give us more estimates closer to the true value of θ than $\hat{\theta}_2$. Hence, we prefer to use $\hat{\theta}_1$.

Back to the question at hand: Which estimators should we use to estimate the true current mean amount of milk delivered by this process, μ, and the variance of the amounts delivered, σ^2? Under the assumption of a random sample from a normal distribution, it can be shown that the sample mean, $\overline{y}$, and the sample variance, s^2, are the most precise unbiased estimators of μ and σ^2, respectively.

Point and Interval Estimates

We call $\overline{y}$ and s^2 *point estimators* because they estimate the specific values for μ and σ^2, respectively. Since they estimate only the specific "point," they possess grave limitations. For example, consider $\overline{y}$. Because $\overline{y}$ is a continuous random variable, we have

$$P(\overline{y} = \mu) = 0.$$

Thus, we know that this point estimate has no chance of being correct! As a result, statisticians prefer to talk about *interval estimates* for parameters. Ideally, the width of the interval should reflect two factors:

1. Our confidence in the interval
2. The variability of the estimator

How can we generate such intervals? From the Central Limit Theorem, if the sample size, n, is sufficiently large, then $\overline{y}$ follows a normal distribution with mean μ and standard error $\sigma/\sqrt{n}$. Suppose we randomly select two $\overline{y}$'s from this distribution and construct intervals of the form

$$\overline{y} \pm z_{\alpha/2} \frac{\sigma}{\sqrt{n}}.$$

Figure 4.4 shows two possibilities. The lower interval does not contain μ, which sometimes happens, but we expect it to happen only $\alpha \cdot 100\%$ of the time.

Figure 4.4 | Illustration of Confidence Intervals

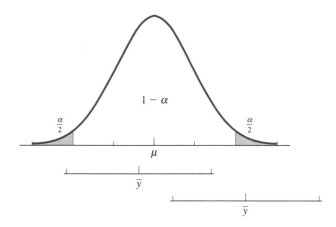

In Chapter 3, we observed that we rarely know the population variance; however,

$$\frac{\overline{y} - \mu}{s/\sqrt{n}}$$

follows a t-distribution under a wide range of conditions. We can use this insight to construct the basic $(1 - \alpha) \cdot 100\%$ confidence interval for μ by the following argument:

$$
\begin{aligned}
P(-t_{df,\alpha/2} \leq t \leq t_{df,\alpha/2}) &= P\left(-t_{df,\alpha/2} \leq \frac{\overline{y} - \mu}{s/\sqrt{n}} \leq t_{df,\alpha/2}\right) \\
&= P\left(-t_{df,\alpha/2}\frac{s}{\sqrt{n}} \leq \overline{y} - \mu \leq t_{df,\alpha/2}\frac{s}{\sqrt{n}}\right) \\
&= P\left(\overline{y} - t_{df,\alpha/2}\frac{s}{\sqrt{n}} \leq \mu \leq \overline{y} + t_{df,\alpha/2}\frac{s}{\sqrt{n}}\right).
\end{aligned}
$$

VOICE OF EXPERIENCE

"Confidence" refers to the reliability of the procedure, not the specific interval.

The factor $t_{df,\alpha/2}$ is the value from the t-table (Table 2 in the appendix) that corresponds to a right-hand tail area of $\alpha/2$ for a t-distribution with df degrees of freedom (see Figure 4.5).

If we could repeat this process an infinite number of times, the interval

$$\overline{y} \pm t_{df,\alpha/2}\frac{s}{\sqrt{n}}$$

Figure 4.5

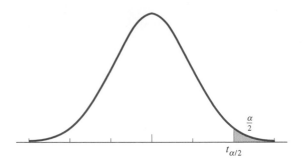

would contain the true value for $(1-\alpha)\cdot 100\%$ of the time. Thus, a $(1-\alpha)\cdot 100\%$ confidence interval for μ is

$$\overline{y} \pm t_{df,\alpha/2}\,\frac{s}{\sqrt{n}}. \tag{4.1}$$

The width of the interval reflects these two factors:

1. Our confidence in the interval through $t_{df,\alpha/2}$
2. The variability of the estimator through $s/\sqrt{n}$

In those rare cases where we essentially know σ^2 (we have an independent estimate with a very large number of degrees of freedom), $t_{df,\alpha/2}$ is essentially $z_{\alpha/2}$ and an appropriate interval is

$$\overline{y} \pm z_{\alpha/2}\,\frac{\sigma}{\sqrt{n}}.$$

It is important to note that the true value, μ, is either in this interval or not. It is from the way we construct our confidence interval that we are highly confident our procedure has produced an interval that actually contains the true value of the parameter. A confidence interval thus represents the range of values for the parameter of interest that we feel is reasonable or plausible. Finally, this confidence interval is for the population mean, not individual values.

Example 4.2 | **Filling Milk Cartons—Revisited**

Maxcy and Lowry (1984) report on a filling process for milk cartons that nominally contain 8 fluid ounces (fl oz). Since the specific gravity of milk is approximately 1.033, one can measure the volume by measuring the weight. In this case, 8 fl oz should weigh 245 g. Historically, the standard deviation for these weights has been 1.65 g. For most engineering processes, the variance is much more stable over time than the mean, so we may treat this standard deviation as known.

The operators monitor the process by accurately weighing five cartons of milk each day. Historically, these weights tend to follow a well-behaved distribution, so a sample size of five is large enough to assume the Central Limit Theorem. The weights for day 15 were 263.9, 266.2, 266.3, 266.8, and 265.0 g. The sample mean, $\bar{y}$, is given by

$$\bar{y} = \frac{1}{n} \sum_{i=1}^{n} y_i = \frac{1328.2}{5} = 265.64.$$

The 95% confidence interval is then

$$\bar{y} \pm z_{\alpha/2} \frac{\sigma}{\sqrt{n}} = 265.64 \pm 1.96 \frac{1.65}{\sqrt{5}}$$
$$= 265.64 \pm 1.45$$
$$= (264.19, 267.09).$$

Thus, we feel reasonably comfortable that the true current mean amount of milk delivered to each carton is somewhere between 264.19 and 267.09 g. More technically, this interval was generated by a procedure that actually contains the true current mean 95% of the time. We are thus 95% confident that the interval 264.19 to 267.09 g really does contain the true current mean amount, and we believe that the true current mean amount is somewhere in this interval.

Why should we focus so much attention on the mean when the real question of interest is whether the individual cartons meet the minimum specification of 245 g? If we can reasonably model the amounts delivered by a normal distribution, then we completely describe the amounts' probabilistic behavior by μ and σ. In this case, σ is known and constant, so we need to consider only the population mean. The lower bound for our confidence interval, 264.19, is more than 11.6 standard deviations from the specification limit! At least for day 15, we have no reason to worry about this process's performance relative to the specification.

Calculating Sample Sizes

Engineers often need to collect data to estimate a parameter of interest within a given precision. For example, we may need to estimate the mean force required to break a polymer filament to within ±100 lb. A reasonable question is: How many polymer filaments do we require in order to obtain the given precision?

Suppose we wish to estimate a population mean when we have at least a reasonable idea of the population standard deviation, σ. In this case, we may treat σ as known; thus, the form of the appropriate confidence interval is

$$\bar{y} \pm z_{\alpha/2} \frac{\sigma}{\sqrt{n}}.$$

We observe that $\sigma/\sqrt{n}$ controls the width of this interval. Since we essentially know σ, we can make the width of the interval as small as we wish by an appropriate choice of n. Suppose we wish to estimate this mean to within $\pm B$ units. We thus need to choose n such that

$$z_{\alpha/2}\,\frac{\sigma}{\sqrt{n}} \leq B.$$

Performing the algebra, we see that

$$B \cdot \sqrt{n} \geq z_{\alpha/2} \cdot \sigma$$
$$\sqrt{n} \geq \frac{z_{\alpha/2} \cdot \sigma}{B} \tag{4.2}$$
$$n \geq \left(\frac{z_{\alpha/2} \cdot \sigma}{B}\right)^2.$$

Since n must be an integer, we should always round up.

    This entire procedure assumes that n is sufficiently large for the Central Limit Theorem to apply. As a result, we need to have some basic idea about the shape of the underlying distribution before we can determine the appropriate sample size. Often, we base our knowledge of σ on some baseline data. These baseline data can provide a basis for generating either a stem-and-leaf display or a normal probability plot. The plots then provide us with some insight as to an appropriate minimum sample size in order to assume the Central Limit Theorem.

| Example 4.3 | **Filling Milk Cartons—Continued** |

Suppose a government inspector requires this company to generate a 95% confidence interval with a width of ± 0.5 g. We already know that the population standard deviation is 1.65 g and that the data follow a well-behaved distribution such that sample sizes as small as five are appropriate for assuming the Central Limit Theorem. By applying equation (4.2), we obtain

$$n \geq \left(\frac{z_{\alpha/2} \cdot \sigma}{B}\right)^2$$
$$\geq \left(\frac{1.96 \cdot (1.65)}{0.5}\right)^2$$
$$\geq 41.83$$
$$\geq 42$$

because the sample size must be an integer.

⸭ *Computer Exercises*

4.1   Use a statistical software package to illustrate confidence intervals when sampling is from a normal distribution.

a. Generate 1000 random samples, each of size 4, from a normal distribution with a mean of 6 and a variance of 12. Have the software calculate the sample mean and a 95% confidence interval for each sample. Count how many samples fail to contain the true mean value. Comment on your results.

b. Generate another 1000 random samples, each of size 10, from the same normal distribution. Have the software calculate the sample mean and a 95% confidence interval for each sample. Count how many samples fail to contain the true mean value. Comment on your results.

c. Generate another 1000 random samples, each of size 30, from the same normal distribution. Have the software calculate the sample mean and a 95% confidence interval for each sample. Count how many samples fail to contain the true mean value. Comment on your results.

4.2   Use a statistical software package to illustrate confidence intervals when sampling is from a uniform distribution.

a. Generate 1000 random samples, each of size 4, from a uniform distribution over the interval 0 to 12 that has a mean of 6 and a variance of 12. Have the software calculate the sample mean and a 95% confidence interval for each sample. Count how many samples fail to contain the true mean value. Comment on your results.

b. Generate another 1000 random samples, each of size 10, from the same uniform distribution. Have the software calculate the sample mean and a 95% confidence interval for each sample. Count how many samples fail to contain the true mean value. Comment on your results.

c. Generate another 1000 random samples, each of size 30, from the same uniform distribution. Have the software calculate the sample mean and a 95% confidence interval for each sample. Count how many samples fail to contain the true mean value. Comment on your results.

4.3   Use a statistical software package to illustrate confidence intervals when sampling is from a χ^2 distribution.

a. Generate 1000 random samples, each of size 4, from a χ^2 distribution with 6 degrees of freedom that has a mean of 6 and a variance of 12. Have the software calculate the sample mean and a 95% confidence interval for each sample. Count how many samples fail to contain the true mean value. Comment on your results.

b. Generate another 1000 random samples, each of size 10, from the same χ^2 distribution. Have the software calculate the sample mean and a 95% confidence interval for each sample. Count how many samples fail to contain the true mean value. Comment on your results.

c. Generate another 1000 random samples, each of size 30, from the same χ^2 distribution. Have the software calculate the sample mean and a 95% confidence interval for each sample. Count how many samples fail to contain the true mean value. Comment on your results.

4.4 Use a statistical software package to illustrate confidence intervals when sampling is from a Weibull distribution.

a. Generate 1000 random samples, each of size 4, from a Weibull distribution with $\lambda = 3$ and $\beta = 2$. Have the software calculate the sample mean and a 95% confidence interval for each sample. Count how many samples fail to contain the true mean value. Comment on your results.

b. Generate another 1000 random samples, each of size 10, from the same Weibull distribution. Have the software calculate the sample mean and a 95% confidence interval for each sample. Count how many samples fail to contain the true mean value. Comment on your results.

c. Generate another 1000 random samples, each of size 30, from the same Weibull distribution. Have the software calculate the sample mean and a 95% confidence interval for each sample. Count how many samples fail to contain the true mean value. Comment on your results.

⫸ Exercises

4.1 Kane (1986) discusses the concentricity of an engine oil seal groove. Concentricity measures the cross-sectional coaxial relationship of two cylindrical features. In this case, he studied the concentricity of an oil seal groove and a base cylinder in the interior of the groove. He measures the concentricity as a positive deviation using a dial indicator gauge. Historically, the standard deviation for the concentricity is 0.7. To monitor this process, he periodically takes a random sample of three measurements.

a. A recent sample yielded a sample mean of 5.8. Construct a 95% confidence interval for the true mean concentricity.

b. Find the sample size required to estimate the true mean concentricity to within ±0.2 using a 95% confidence interval.

c. What did you assume to do these analyses? How can you check these assumptions?

4.2 Runger and Pignatiello (1991) consider a plastic injection molding process for a part that has a critical width dimension with a historic standard deviation of 8. Periodically, clogs form in one of the feeder lines, causing the mean width to change. As a result, the operator periodically takes random samples of size four.

a. A recent sample yielded a sample mean of 101.4. Construct a 99% confidence interval for the true mean width.

b. Find the sample size required to estimate the true mean width to within ± 2 units using a 99% confidence interval.

c. What did you assume to do these analyses? How can you check these assumptions?

4.3 Yashchin (1995) discusses a process for the chemical etching of silicon wafers used in integrated circuits. This process etches the layer of silicon dioxide until the layer of metal beneath is reached. This company monitors the thickness of the silicon oxide layer because thicker layers require longer etching times. Historically, the layer thicknesses have a standard deviation of 0.06 micron.

a. A recent random sample of four wafers yielded a sample mean of 1.134 microns. Construct a 95% confidence interval for the true mean thickness.

b. Find the sample size required to estimate the true mean thickness to within ± 0.01 micron using a 95% confidence interval.

c. What did you assume to do these analyses? How can you check these assumptions?

4.4 The modulus of rupture (MOR) for a particular grade of pencil lead is known to have a standard deviation of 250 psi.

a. A random sample of 16 pencil leads yielded a sample mean of 6490. Construct a 90% confidence interval for the true mean MOR.

b. Find the sample size required to estimate the true mean MOR to within ± 100 using a 90% confidence interval.

c. What did you assume to do these analyses? How can you check these assumptions?

4.5 The yields from an ethanol–water distillation column have a standard deviation of 1%. A random sample of eight recent batches produced these yields:

0.90	0.93	0.95	0.86
0.90	0.87	0.93	0.92

a. Construct a 99% confidence interval for the true mean yield.

b. Find the sample size required to estimate the true mean yield to within $\pm 0.5\%$ using a 99% confidence interval.

c. What did you assume to do these analyses? How can you check these assumptions?

4.6 Wasserman and Wadsworth (1989) discuss a process for the manufacture of steel bolts that continuously feed an assembly line downstream. Historically, the thicknesses of these bolts follow a normal distribution with a standard deviation of 1.6 mm. A recent random sample yielded these thicknesses:

9.7	9.9	10.3	10.1	10.5
9.4	9.9	10.1	9.7	10.3

a. Construct a 95% confidence interval for the true mean thickness.
b. Find the sample size required to estimate the true mean thickness to within 0.2 mm using a 95% confidence interval.
c. What did you assume to do these analyses? How can you check these assumptions?

4.7 The ocean swell produces spectacular eruptions of water through a hole in the cliff at Kiama, Australia, known as the Blowhole. Historically, the time between eruptions has a standard deviation of 10 seconds. The times at which 10 successive eruptions occurred was measured in seconds:

60	55	77	56	68
89	73	61	69	61

a. Construct a 99% confidence interval for the true mean time between eruptions.
b. Find the sample size required to estimate the true mean time between eruptions to within ±5 seconds using a 99% confidence interval.
c. What did you assume to do these analyses? How can you check these assumptions?

4.8 Data on the concentration of polychlorinated biphenyl (PCB) residues in a series of lake trout from Cayuga Lake, NY, can be found in Bates and Watts (1988). Each whole fish was mechanically chopped, ground and thoroughly mixed, and 5-gram samples taken. The samples were treated and PCB residues in parts per million (ppm) were estimated using column chromatography. Historically, the PCB residues have a standard deviation of 0.7 ppm. A subset of the data is shown in the table.

0.6	1.6	0.5	1.2	2.0
1.3	2.5	2.2	2.4	1.2

a. Construct a 90% confidence interval for the true mean PCB residues.
b. Find the sample size required to estimate the true mean PCB residues to within ±0.075 using a 90% confidence interval.
c. What did you assume to do these analyses? How can you check these assumptions?

4.9 A study was done on the propagation of an ultrasonic stress wave through a substance. In particular, attenuation (the decrease in amplitude of the stress wave measured in neper/cm) in fiber-glass-reinforced polyester composites was measured. Attenuation values are known to have a standard deviation of 0.25.

2.6	2.5	2.2	2.0	2.1	2.3	2.6

a. Construct a 95% confidence interval for the true mean attenuation values.
b. Find the sample size required to estimate the true mean attenuation values to within ±0.20 using a 95% confidence interval.
c. What did you assume to do these analyses? How can you check these assumptions?

⟫ 4.2 Hypothesis Testing

Overview

Consider the milk carton example in more detail. Since the specific gravity of milk is approximately 1.033, 8 fl oz of milk weigh 245 g. The milk filling process has a target mean amount of 260 g in order to guarantee no underages. This target is more than *nine standard deviations* from the minimum requirement! The population mean can actually drop three standard deviations (almost 5 g) and the probability that the process underfills a given carton would still be virtually zero. As a result, the industrial engineer assigned to this process is primarily concerned about overfilling the cartons. In Example 4.2, the random sample of five cartons on day 15 had a sample mean weight of 265.64 g. Is this sufficient evidence to conclude that the process is beginning to overfill the cartons?

- If it is, then the engineer is responsible for taking appropriate action.
- If it is not, then the best thing to do is to leave the process alone. Taking unnecessary action is a waste of effort and can add needless variability to the process.

Just like this engineer, all of us constantly must make decisions under risk, whether we are in the business world or in a laboratory class. How should we make these decisions? To address this question, we first require a digression.

Consider a criminal court case—a capital murder case where the defendant faces a death sentence if found guilty. Standard American judicial procedure presumes that the defendant is innocent until proven guilty, and the jurors must be convinced "beyond a reasonable doubt" that the defendant is guilty. The criminal court situation is illustrated by Table 4.1.

Table 4.1

		Jury's Decision	
		Convict	**Acquit**
Defendant's State	Innocent	Type I error	OK
	Guilty	OK	Type II error

Convicting an innocent person is called a *Type I error.* Acquitting a guilty person is called a *Type II error.* We can actually describe this whole process as testing two hypotheses. The nominal claim, which we shall call the *null hypothesis,* denoted by H_0, is

$$H_0: \quad \text{The defendant is innocent.}$$

The alternative claim, which we shall call the *alternative hypothesis,* denoted by H_a, is

$$H_a: \quad \text{The defendant is guilty.}$$

Note that the alternative hypothesis is what the prosecutor wishes to establish.

The Error Probabilities

Let α be the largest probability of rejecting H_0 when the null hypothesis is true. Thus, α represents the maximum probability of making a Type I error. In the

VOICE OF EXPERIENCE

Rejecting the nominal claim is a strong statement.

court example, α is the probability that we convict an innocent person. The actual size of α used by an individual juror defines what he or she considers to be "beyond a reasonable doubt," and, ideally, should be rather small particularly in a capital murder case, where the defendant faces the prospect of a death

sentence. As the result of using a small α, rejecting the null hypothesis, H_0, is a strong claim. Not rejecting the null hypothesis is a weak claim. Why? Does failing to convict the defendant imply that the defendant is really innocent? Of course not! It simply means that there was insufficient evidence to convict. The jury may honestly believe that the defendant is probably guilty and still acquit him or her. They were simply not convinced beyond a reasonable doubt.

A Type II error is failing to reject the null hypothesis when the alternative is true. In the court case, failing to convict a guilty defendant is a Type II error. If α is small—that is, we require a large amount of evidence to convince us beyond a reasonable doubt—then the chances that we will make a Type II error tend to be large.

Typically, for a fixed situation, decreasing α will cause the probability of making a Type II error to increase. α is often called the *significance level* for our test. The significance level for a hypothesis test, α, is exactly the same α we used in confidence intervals. We often call the probability that we reject the null hypothesis, when a specific alternative hypothesis is true, the *power* of our test. Ideally, we would like the power to be as large as possible. For a fixed sample size, once α is fixed, then so is the power and vice versa. The tradeoffs between α and the power are extremely important in determining sample size, which we shall discuss later in this section.

Table 4.2

		Engineer's Decision	
		Correct process	Don't correct
Actual Weight	= 260 g	Type I error	OK
	> 260 g	OK	Type II error

One-Sided Alternatives

What does all of this have to do with our industrial engineer? We may consider filling the milk cartons as a test of two hypotheses:

$$H_0: \quad \mu = 260 \text{ g}$$
$$H_a: \quad \mu > 260 \text{ g}.$$

Why do we make $\mu > 260$ g the alternative hypothesis? We need substantial evidence before we should conclude that the process needs correction. We must be certain "beyond a reasonable doubt." Note that $H_a: \mu > 260$ g is an example of a "one-sided" alternative because the engineer is really interested only when the process puts in more than 260 g of milk.

We summarize this test in Table 4.2. A Type I error causes the engineer to look for a change in the process when none truly is present. Thus, he or she will waste time and effort needlessly. On the other hand, a Type II error means that the engineer should have searched for the cause of a process change but did not. Consequently, the process has a greater risk of overfilling the cartons.

In this particular situation, the engineer believes that a reasonable α or significance level is .05, which represents a good balance between the chances of making a Type I and a Type II error when we sample five cartons per day. Statisticians often use this level for conducting tests because it represents a reasonable compromise between Type I and Type II errors for many practical sample sizes. The most commonly used values for α are 0.10, 0.05, and 0.01.

The Two-Sided Alternative

Consider the filling process from a general management perspective. Since the population variance is known and remains constant, we need to focus only on the population mean. If $\mu < 260$ g, then the government regulators may intervene. On the other hand, if $\mu > 260$ g, then we are giving away more product than we should and we risk spilling milk as the result of the overfilling. From this perspective, the engineer should seek to keep the filling process set to precisely 260 g.

Consider the following monitoring procedure that determines when the filling process needs to be adjusted. Each hour, five cartons are weighed and a test of the following hypotheses is performed:

$$H_0: \quad \mu = 260 \text{ g}$$

$$H_a: \quad \mu \neq 260 \text{ g}.$$

In this case, we have a "two-sided" alternative because we are interested in means both larger and smaller than the nominal value. If we obtain sufficient evidence to reject the nominal claim, then we are justified in adjusting the process. We thus are able to bring the filling process back into line before a major quality problem occurs. This procedure is actually very common in manufacturing! Since we sample the process frequently, we are less concerned about a Type II error—not seeing a process shift when one is present. If we do not see a shift on this sample, we should see it in a subsequent sample within a few hours. On the other hand, a Type I error requires the engineer to search for a cause of a possible change when none is present, which from a management perspective should be minimized. Actual industrial practice uses an α of 0.0027. In Chapter 5, we shall discuss why.

Rejecting the null hypothesis is easier in a one-sided hypothesis test than in a two-sided hypothesis test. Mathematically, this is true because putting all of α on one side makes the critical value smaller. Conceptually, this is true because by focusing on only one side of the distribution, the shift does not need to be as extreme since it can happen in only one direction.

The Five-Step Hypothesis Testing Procedure

We now are in a position to outline the general five-step hypothesis testing procedure:

1. **State the appropriate hypotheses:**

$$H_0: \quad \text{the nominal claim}$$

$$H_a: \quad \text{the alternative claim}$$

VOICE OF EXPERIENCE

All of our test statistics may be viewed as signal-to-noise ratios.

Usually the null hypothesis is that some parameter, θ, is at some nominal value, θ_0. The alternative claim is what we wish to establish and should be chosen before collecting data.

2. **State the appropriate test statistic:** If the variance is unknown, which is typical, then the test statistic has the form

$$t = \frac{\hat{\theta} - \theta_0}{\hat{\sigma}_{\hat{\theta}}},$$

where θ_0 is the nominal value for the parameter of interest and $\hat{\sigma}_{\hat{\theta}}$ is an appropriate estimate of the standard error of $\hat{\theta}$. The primary exception to this rule is when we test variances. If the variance is known, then the test statistic has the form

$$Z = \frac{\hat{\theta} - \theta_0}{\sigma_{\hat{\theta}}},$$

where $\sigma_{\hat{\theta}}$ is the standard error of $\hat{\theta}$. Again, the primary exception to this rule is when we test variances.

3. **State the critical region for the test statistic:** Determine the values for the test statistic that constitute sufficient evidence to reject the nominal claim. Essentially, we define those values for the test statistic that are "beyond a reasonable doubt."

4. **Conduct the experiment and find the specific value for the test statistic:** In proper statistical analysis, the first three steps should be done *before any data are collected*. This point is often lost in textbooks because the exercises give the data results in the basic problem statements. However, a well-planned experiment should have clear criteria for making decisions before the data collection in order to ensure objectivity.

5. **Reach appropriate conclusions and state them in English:** A hypothesis test has two possible conclusions:

   - Fail to reject the null hypothesis
   - Reject the null hypothesis

   We never "accept the null hypothesis"; rather, we just did not have sufficient evidence to establish the alternative. In many cases, quite a bit of evidence supports the alternative hypothesis but not enough to convince us "beyond a reasonable doubt." The conclusions "fail to reject H_0" and "reject H_0" are statistical jargon and really should not be used in reporting the final conclusions. Instead, we always interpret these results in light of the engineering context of the problem. We should always give a well-phrased answer in English. If we reject the null hypothesis, we should give an appropriate confidence interval to quantify the range of values for the parameter that we consider plausible in light of our data.

Finally, whenever possible, we should check to make sure that the assumptions we make are reasonable. Most of our tests and confidence intervals are based on the *t*-distribution. Strictly speaking, the *t*-distribution assumes that the data themselves follow a normal distribution; however, it is well known that the *t*-distribution is very robust to this requirement. For very small sample sizes, we shall require that the data follow a distribution quite close to a normal. As the sample size increases, we may relax that assumption. In general, we are looking for the data to follow a distribution that is single peaked, roughly symmetric, and with tails that die rapidly.

| Example 4.4 | **Filling Milk Cartons—Continued** |

As stated in example 4.2, Maxcy and Lowry (1984) report on a filling process for milk cartons that nominally contain 8 fluid ounces (fl oz, 245 g). To ensure that the process does not underfill the cartons, management wants to maintain a true mean amount of 260 g. Historically, the standard deviation for these weights has been 1.65 g. For most engineering processes, the variance is much more stable over time than the mean, so we may treat this standard deviation as known. The operators monitor this process by accurately weighing five cartons of milk each day. Historically, these weights tend to follow a well-behaved distribution, so a sample size of five is large enough to assume the Central Limit Theorem.

1. **State hypotheses:** Let μ_0 be the nominal value for μ for our test. In general, these are the possible sets of hypotheses:

$$H_0: \ \mu = \mu_0 \qquad H_0: \ \mu = \mu_0 \qquad H_0: \ \mu = \mu_0$$
$$H_a: \ \mu < \mu_0 \qquad H_a: \ \mu > \mu_0 \qquad H_a: \ \mu \neq \mu_0$$

In this case, $\mu_0 = 260$. Suppose that management is primarily concerned with overfilling the cartons. Then the appropriate hypotheses are

$$H_0: \ \mu = 260$$
$$H_a: \ \mu > 260.$$

2. **State the test statistic:** Since the amounts delivered come from a well-behaved distribution and since we know the population variance, we may use the Central Limit Theorem. Our test statistic is

$$Z = \frac{\overline{y} - \mu_0}{\sigma/\sqrt{n}}.$$

3. **State the critical region:** The critical regions depend on the alternative hypotheses. Consider $H_a: \mu < \mu_0$. Figure 4.6 illustrates that we reject H_0 if $Z < -z_\alpha$. Now consider $H_a: \mu > \mu_0$. Figure 4.7 illustrates that we reject H_0 if

| Figure 4.6 | Critical Region for $H_a: \mu < \mu_0$ |

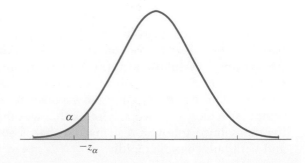

Figure 4.7 | Critical Region for H_a: $\mu > \mu_0$

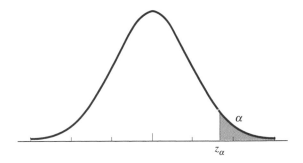

$Z > z_\alpha$. Finally, consider H_a: $\mu \neq \mu_0$. Figure 4.8 illustrates that we reject H_0 if $|Z| > z_{\alpha/2}$. We need to determine an appropriate significance level. Common choices for α are 0.10, 0.05, and 0.01. Since we conduct this test often, we really are more concerned about a Type I error, rejecting the null hypothesis when it is true, than a Type II error, failing to detect a change in the true amount of milk delivered. Why? If we make a Type I error, we must search for the cause of a change when none is present. Since we are conducting this test so frequently, we run the risk of constantly searching for problems that do not exist. On the other hand, since we are conducting this test so frequently, if we do not detect a change in the true mean on any given sample, we should pick it up on a subsequent sample. Thus, we shall use $\alpha = 0.01$. Since we have H_a: $\mu > \mu_0$, we reject the null hypothesis if

$$Z > z_\alpha$$
$$> z_{.01}$$
$$> 2.326.$$

4. **Conduct the experiment and find Z:** The weights for day 15 were 263.9, 266.2, 266.3, 266.8, and 265.0 g. For these data, the sample mean

Figure 4.8 | Critical Region for H_a: $\mu \neq \mu_0$

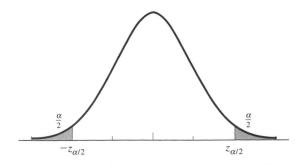

$\overline{y} = 265.64$, and we have

$$Z = \frac{\overline{y} - \mu_0}{\sigma / \sqrt{n}}$$

$$= \frac{265.64 - 260}{1.65 / \sqrt{5}}$$

$$= 7.64.$$

5. **Reach conclusions and state them in English:** Since $7.64 > 2.326$, we have sufficient evidence to reject the null hypothesis. As a result, we have sufficient evidence to conclude that the packaging process is overfilling the cartons. Thus, we should search out the presence of any changes to our process. Rejecting the nominal claim simply means that the data do not support the claim that the true mean amount delivered is 260 g. Once we reject this nominal claim, we should give a range of plausible values for the true mean. A 99% confidence interval for the true mean is

$$\overline{y} \pm z_{\alpha/2} \frac{\sigma}{\sqrt{n}} = 265.64 \pm 2.576 \frac{1.65}{\sqrt{5}}$$

$$= 265.64 \pm 1.90$$

$$= (263.74, 267.54).$$

Thus, the range of plausible values for the true mean amount delivered is 263.74 to 267.54 g. This entire range lies more than two standard deviations above the target value for the mean! At least on day 15, this process appears to be overfilling the cartons by a substantial amount.

One may ask why we constructed a two-sided interval when we conducted a one-sided test. In this example, once we reject the null hypothesis, we no longer believe the true mean amount delivered is 260 g. We then face the question, What is the true mean value? Within a hypothesis testing framework, we use the confidence interval to quantify the extent of the difference in order to determine whether the difference is of real practical concern. We use the confidence interval as a powerful diagnostic tool. We could construct a one-sided interval, in which case the upper bound for the true mean is ∞. For diagnostic purposes, many engineers do not find such a bound meaningful. The two-sided interval often gives a better sense of the plausible values for the true mean once we have rejected the null hypothesis.

Practical Versus Statistical Significance

Too many people, not just engineers, confuse *statistical significance* and *practical significance*. When a statistical test rejects the nominal claim that $\theta = \theta_0$, the data suggest that the true value for θ really is something other than θ_0: nothing more, nothing less. Reconsider the milk carton example where we rejected the claim that the true mean amount is 260 g. In this particular example and at that particular point, we can conclude only that the true mean amount is something greater than 260 g. At that particular point, for all we know, the true mean amount may actually be 260.001 g, which is greater than 260 but only trivially so. It is for this very reason that we should always give a confidence interval when we reject the nominal claim. The confidence interval gives us a range of plausible values so we can determine whether the difference between the true value of the parameter and the nominal value is of real, practical significance. In the milk carton example, the confidence interval seems to indicate that the difference between the true mean amount and the nominal claim is truly of practical concern.

The issue of statistical versus practical significance is not purely academic. With a large enough sample size, we can show with statistical significance any minute difference from the nominal value for the parameter of interest. Ideally, we plan our studies so that our statistical procedure can detect differences of practical importance with reasonably high probability but do not detect marginal differences very often.

Relationship Between Confidence Intervals and Tests

A *two-sided* hypothesis test with a significance level of α is equivalent to constructing a $(1 - \alpha) \cdot 100\%$ confidence interval and checking to see whether the interval contains the nominal value of the parameter. The α we use for the hypothesis test is exactly the same α we use for the confidence interval. Consider the possibilities:

- If the interval does contain this value, then we fail to reject H_0.
- If the interval does not contain this value, then we reject H_0.

We constructed our interval in such a manner that we are highly confident that the true value lies somewhere within it. In some sense, each value in the interval is a plausible candidate for the true value. Thus, if the nominal value of the parameter of interest falls within the confidence interval, then we have no evidence to conclude that it is not a plausible value for the parameter. Hence, we cannot reject the null hypothesis. On the other hand, if our interval does not contain the nominal value, then the nominal value is not plausible, and we do have sufficient evidence to reject the nominal claim.

Many engineers and statisticians prefer to concentrate solely on confidence intervals because they both clearly estimate the parameter of interest and address the interesting questions for which hypothesis tests are designed. These people point out that by concentrating purely on the confidence interval, we can completely skip the hypothesis test. *Confidence intervals provide a simple, powerful, and direct basis for addressing both practical and statistical significance.* In many of the examples given in this chapter, we shall illustrate these basic points. Do not be misled by the brevity of these examples. *Actually, their brevity should point out their very advantage!*

Personally, we are sympathetic to this use of confidence intervals, although we do believe that hypothesis tests play an important role in statistical analysis, particularly for control charts (see Chapter 5), regression analysis (see Chapter 6), and the formal analysis of designed experiments (see Chapters 7 and 8), especially because most statistical software emphasizes hypothesis testing over confidence intervals. Once we discuss *p*-values, we shall begin to see why the software tends to prefer hypothesis tests. Ultimately, the instructor and, later, the data analyst are free to emphasize either confidence intervals or hypothesis tests, depending on philosophical slant.

p-Values

Recall that α is often called the *significance level* for a hypothesis test. In some sense, α represents our standard of evidence. Once we decide on α, we determine the appropriate critical region for our test. Any value of the test statistic that is more extreme than the "critical value" is considered sufficient evidence to reject the null hypothesis or nominal claim. For a fixed sample size, the smaller our α, the more evidence is required to reject the null hypothesis.

An alternative method looks at the *observed significance level*, sometimes called the *attained significance level*, which is the smallest Type I error rate that would allow us to reject the null hypothesis. The observed significance level is the probability of seeing the particular value of our test statistic, or something more extreme, if H_0 is true. This probability is usually called a *p-value*. Most statistical software packages report *p*-values because they do not know what the researcher wishes to use for α. One rejects H_0 whenever the *p*-value is less than α. In Chapters 6, 7, and 8 we shall make extensive use of *p*-values when performing regression analysis with statistical software.

We have pointed out that statistical software packages prefer hypothesis testing over confidence intervals. We have seen that confidence intervals provide a simple and powerful way to conduct our analyses. They depend, however, on using a specified, predetermined α. Historically, since the software developers recognize that different analysts need to use different α's, they have relied on *p*-values and hypothesis testing. The software developers prefer the greater flexibility *p*-values provide.

p-Values for Tests Based on the Standard Normal Random Variable

The *p*-value depends on the specific alternative used for the test. Let z_0 be the observed value for the test statistic.

- For $H_a: \mu < \mu_0$, the *p*-value is $P(Z < z_0)$.
- For $H_a: \mu > \mu_0$, the *p*-value is $P(Z > z_0)$.
- For $H_a: \mu \neq \mu_0$, we must consider both tails of the standard normal distribution and the *p*-value is $2 \cdot P(Z > |z_0|)$.

| Example 4.5 | **Breaking Strengths of Carbon Fibers—Revisited** |

In Example 4.4, we tested these hypotheses:

$$H_0: \quad \mu = 260$$
$$H_a: \quad \mu > 260.$$

The data produced a test statistic value of $z_0 = 7.64$. Thus, for this test,

$$p\text{-value} = P(Z > z_0)$$
$$= P(Z > 7.64)$$
$$= 0.0000$$

In this example, $\alpha = 0.01$. Since our p-value is less than 0.01, we would reject the null hypothesis. Once again, we have sufficient evidence to conclude that the packaging process is overfilling the cartons.

p-Values for t- and χ^2 Tests

Statistical software packages give exact p-values for this situation. Unfortunately, when we use the t- or the χ^2 tables, we must give *intervals* within which the p-value falls. The p-value depends on the specific alternative used for our test. For the one-sided alternatives, we find the two values of α, α_1 and α_2, that have critical values that "straddle" the observed t-statistic. We then say that the p-value is between α_1 and α_2. For the two-sided alternative, we again find the two values of α, α_1 and α_2, that have critical values that "straddle" the observed t-statistic. We then say that the p-value is between $2 \cdot \alpha_1$ and $2 \cdot \alpha_2$. If the observed test statistic is more extreme than the t-value associated with the smallest α, then our p-value is simply less than this smallest α.

Power and Sample Size (Optional)

The Formal Concept of Power   The power of a hypothesis test is the probability that we reject the null hypothesis when the alternative hypothesis is true. We borrow the notation of conditional probability to express this concept mathematically:

$$P(\text{reject } H_0 \mid H_a \text{ is true}).$$

Determining the power of a hypothesis test for a parameter θ always requires that we know these values:

- The significance level, α
- The specific alternative hypothesis, H_a

- A specific alternative value, θ_a, that we wish to detect with high probability if the alternative hypothesis is true

For example, consider the following one-sided hypotheses for the population mean, μ, when we know the population standard deviation, σ:

$$H_0: \quad \mu = \mu_0$$

$$H_a: \quad \mu > \mu_0.$$

Let α be the stated significance level for this test, and let μ_1 be the specific alternative value for μ that we wish to detect with a stated power. For this particular situation, we reject H_0 if the value of the test statistic exceeds the critical value, z_α, which in this case means

$$\frac{\overline{y} - \mu_0}{\sigma/\sqrt{n}} > z_\alpha.$$

As a result, the power of our test is

$$\text{power} = P\left(\frac{\overline{y} - \mu_0}{\sigma/\sqrt{n}} > z_\alpha \mid \mu = \mu_1\right).$$

Under the alternative hypothesis, the test statistic, $(\overline{y} - \mu_0)/(\sigma/\sqrt{n})$, does not follow a standard normal distribution because $\overline{y}$ is centered around μ_1 rather than around μ_0. As a result, to find the power, we must restandardize $\overline{y}$. Performing the algebra, we get

$$\text{power} = P\left(\frac{\overline{y} - \mu_0}{\sigma/\sqrt{n}} > z_\alpha \mid \mu = \mu_1\right)$$

$$= P\left(\overline{y} - \mu_0 > z_\alpha \cdot \frac{\sigma}{\sqrt{n}} \mid \mu = \mu_1\right)$$

$$= P\left(\overline{y} > \mu_0 + z_\alpha \cdot \frac{\sigma}{\sqrt{n}} \mid \mu = \mu_1\right).$$

Under the specific alternative hypothesis, $\overline{y}$ has a mean of μ_1 and a standard error of $\sigma/\sqrt{n}$. We then find the power by

$$\text{power} = P\left(\overline{y} - \mu_1 > \mu_0 - \mu_1 + z_\alpha \cdot \frac{\sigma}{\sqrt{n}}\right)$$

$$= P\left(\frac{\overline{y} - \mu_1}{\sigma/\sqrt{n}} > \frac{\mu_0 - \mu_1 + z_\alpha \cdot \sigma/\sqrt{n}}{\sigma/\sqrt{n}}\right)$$

$$= P\left(Z > z_\alpha - \frac{\mu_1 - \mu_0}{\sigma/\sqrt{n}}\right). \tag{4.3}$$

Figure 4.9 | Distribution of the Test Statistic Under the Null Hypothesis

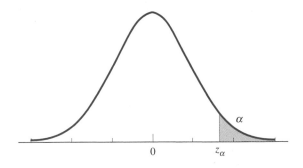

We often find it convenient to use γ, which is defined by

$$\gamma = \frac{\mu_1 - \mu_0}{\sigma/\sqrt{n}}.$$

Under H_a, $(\bar{y} - \mu_0)/(\sigma/\sqrt{n})$ follows a normal distribution with a mean γ and a standard error of $\sigma/\sqrt{n}$. Since the test statistic is now centered at γ rather than zero, we call γ the *noncentrality parameter*. We can rewrite equation (4.3) as

$$P(Z > z_\alpha - \gamma).$$

Figures 4.9 and 4.10 illustrate the concept of power. Figure 4.9 shows the distribution of the test statistic, $(\bar{y} - \mu_0)/(\sigma/\sqrt{n})$, under the null hypothesis, H_0. We reject H_0 if the test statistic exceeds z_α. If the null hypothesis is true, we reject H_0 $\alpha \cdot 100\%$ of the time, which is our Type I error rate. Figure 4.10 shows the situation when the alternative hypothesis, H_a, is true. Under H_a, the test statistic

Figure 4.10 | Distribution of the Test Statistic Under the Alternative Hypothesis

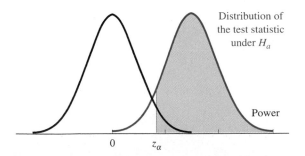

Distribution of the test statistic under H_a

Power

is actually centered at γ, but we still reject H_0 if the test statistic exceeds z_α. As a result, the probability of rejecting H_0, which is the shaded area under the curve, is much greater when H_a is true.

In a similar manner, we can show that the power for the hypotheses

$$H_0: \quad \mu = \mu_0$$
$$H_a: \quad \mu < \mu_0$$

is

$$P\left(Z < -z_\alpha - \frac{\mu_1 - \mu_0}{\sigma/\sqrt{n}}\right) = P(Z < -z_\alpha - \gamma) \qquad (4.4)$$

where $\gamma < 0$. For the two-sided hypotheses

$$H_0: \quad \mu = \mu_0$$
$$H_a: \quad \mu \neq \mu_0$$

the power is

$$P\left(Z < -z_{\alpha/2} - \frac{\mu_1 - \mu_0}{\sigma/\sqrt{n}}\right) + P\left(Z > z_{\alpha/2} - \frac{\mu_1 - \mu_0}{\sigma/\sqrt{n}}\right)$$
$$= P(Z < -z_{\alpha/2} - \gamma) + P(Z > z_{\alpha/2} - \gamma). \qquad (4.5)$$

For most practical situations either $P(Z < -z_{\alpha/2} - \gamma)$ or $P(Z > z_{\alpha/2} - \gamma)$ is zero, which simplifies matters.

Finding Sample Sizes   We can use equations (4.3), (4.4), and (4.5) to find sample sizes for a specific testing situation. The procedure requires a specific alternative value of interest. Consider the hypotheses

$$H_0: \quad \mu = \mu_0$$
$$H_a: \quad \mu > \mu_0.$$

Let p_1 be the desired power when $\mu = \mu_1$, where $\mu_1 > \mu_0$. μ_1 is the specific alternative value of interest. In this case, it represents the smallest value for μ that we wish to detect with probability p_1. Applying equation (4.3) to this situation, we obtain

$$P\left(Z > z_\alpha - \frac{\mu_1 - \mu_0}{\sigma/\sqrt{n}}\right) \geq p_1.$$

We note that we can achieve an exact power of p_1 if

$$z_\alpha - \frac{\mu_1 - \mu_0}{\sigma/\sqrt{n}} = z_{p_1}$$

where z_{p_1} is the Z-value associated with a right-hand tail area of p_1. In general, $p_1 > 0.5$, which in this case implies that $z_{p_1} < 0$. We thus need to find the smallest n that satisfies

$$z_\alpha - \frac{\mu_1 - \mu_0}{\sigma/\sqrt{n}} \leq z_{p_1}.$$

Performing the algebra, we have

$$z_\alpha - \frac{\mu_1 - \mu_0}{\sigma/\sqrt{n}} \leq z_{p_1}$$

$$\frac{\mu_1 - \mu_0}{\sigma/\sqrt{n}} \geq z_\alpha - z_{p_1}$$

$$\frac{\sqrt{n}\,(\mu_1 - \mu_0)}{\sigma} \geq z_\alpha - z_{p_1} \qquad (4.6)$$

$$\sqrt{n} \geq \frac{(z_\alpha - z_{p_1}) \cdot \sigma}{\mu_1 - \mu_0}$$

$$n \geq \left[\frac{(z_\alpha - z_{p_1}) \cdot \sigma}{\mu_1 - \mu_0}\right]^2.$$

Since we must use an integer value for n, we should always round our answer up. For H_a: $\mu < \mu_0$, we can show that

$$n \geq \left[\frac{(z_\alpha + z_{p_1}) \cdot \sigma}{\mu_1 - \mu_0}\right]^2.$$

We leave the derivation of this result as a homework exercise.

For H_a: $\mu \neq \mu_0$, we can obtain a first approximation for the required sample size by using the formula associated with the particular tail that contains the specific alternative value of interest. Since there is a positive, but usually very small, probability that we could reject H_0 in the other tail, the actual sample size required could be slightly smaller. We can find the precise sample size through an iterative process. Let n^* be the calculated sample size based on the tail that contains the specific alternative value of interest. We use equation (4.5) to find the actual power for this sample size. We then let $n^* = n^* - 1$ and recalculate the power. We continue to subtract 1 until the actual power is less than the desired value. The required sample size is the previous n^*.

| Example 4.6 | **Filling Milk Cartons—Continued** |

Consider the testing procedure we developed in Example 4.4. Since we wish to detect whether the process overfills the cartons, these are the appropriate hypotheses:

$$H_0: \quad \mu = 260$$

$$H_a: \quad \mu > 260.$$

Suppose we need to determine the probability that our test rejects the null hypothesis if the true mean shifts to 261.5 g when we use five cartons per sample. In this case,

$$\gamma = \frac{\mu_1 - \mu_0}{\sigma/\sqrt{n}}$$

$$= \frac{261.5 - 260}{1.65/\sqrt{5}}$$

$$= 2.03.$$

From the previous example, we reject the null hypothesis if our test statistic exceeds $z_\alpha = 2.326$. By applying equation (4.3), we obtain the power by

$$\text{power} = P(Z > z_\alpha - \gamma)$$

$$= P(Z > 2.326 - 2.03)$$

$$= 0.3836.$$

As a result, we reject the null hypothesis about 38% of the time when the true mean amount is 261.5 g.

Suppose management insists that we increase our sample size so that the testing procedure rejects the null hypothesis at least 80% of the time when the true mean amount is 261.5 g. To find the appropriate sample size, we first must find $z_{p_1} = z_{.80}$. By the appropriate use of Table 1 in the appendix, we find $z_{.80} = -0.84$. We now can apply equation (4.7), which yields

$$n \geq \left[\frac{(z_\alpha - z_{p_1}) \cdot \sigma}{\mu_1 - \mu_0} \right]^2$$

$$\geq \left[\frac{(2.33 - (-0.84)) \cdot 1.65}{261.5 - 260} \right]^2$$

$$\geq 12.2.$$

Since we must use an integer sample size, we require a minimum of 13 cartons a day in order to meet management's request.

⫶ *Computer Exercises*

4.5 Use statistical software to illustrate Type I and Type II error rates for the hypotheses

$$H_0: \quad \mu = 6$$
$$H_a: \quad \mu > 6$$

when sampling is from a normal distribution with a variance known to be 12. In this situation, the appropriate test statistic is

$$Z = \frac{\bar{y} - 6}{\sqrt{12/n}}.$$

Suppose we wish to use a .05 significance level for our test. We thus should reject the null hypothesis whenever $Z > 1.645$.

a. Use a statistical software package to generate 1000 random samples, each of size 4, from a normal distribution with a mean of 6 and a variance of 12 that corresponds to the null hypothesis. Calculate the sample mean and Z for each sample. Count the number of samples that yield values of $Z > 1.645$—that is, that reject the null hypothesis. Comment on your results.

b. Use a statistical software package to generate 1000 random samples, each of size 4, from a normal distribution with a mean of 8 and a variance of 12. Calculate the sample mean and Z for each sample. Count the number of samples that yield values of $Z > 1.645$—that is, that reject the null hypothesis. Comment on your results.

c. Use a statistical software package to generate 1000 random samples, each of size 4, from a normal distribution with a mean of 10 and a variance of 12. Calculate the sample mean and Z for each sample. Count the number of samples that yield values of $Z > 1.645$—that is, that reject the null hypothesis. Comment on your results.

d. Use a statistical software package to generate 1000 random samples, each of size 16, from a normal distribution with a mean of 6 and a variance of 12 that corresponds to the null hypothesis. Calculate the sample mean and Z for each sample. Count the number of samples that yield values of $Z > 1.645$—that is, that reject the null hypothesis. Comment on your results.

e. Use a statistical software package to generate 1000 random samples, each of size 16, from a normal distribution with a mean of 8 and a variance of 12. Calculate the sample mean and Z for each sample. Count the number of samples that yield values of $Z > 1.645$—that is, that reject the null hypothesis. Comment on your results.

f. Use a statistical software package to generate 1000 random samples, each of size 16, from a normal distribution with a mean of 10 and a variance of 12. Calculate the sample mean and Z for each sample. Count the number of samples that yield values of $Z > 1.645$—that is, that reject the null hypothesis. Comment on your results.

4.6 Use statistical software to illustrate Type I and Type II error rates for the hypotheses

$$H_0: \quad \mu = 6$$

$$H_a: \quad \mu > 6$$

when sampling is from a uniform distribution with a variance known to be 12. In this situation, the appropriate test statistic is

$$Z = \frac{\bar{y} - 6}{\sqrt{12/n}}.$$

Suppose we wish to use a 0.05 significance level for our test. We thus should reject the null hypothesis whenever $Z > 1.645$.

a. Use a statistical software package to generate 1000 random samples, each of size 4, from a uniform distribution over the interval 0 to 12 that has a mean of 6 and a variance of 12 that corresponds to the null hypothesis. Calculate the sample mean and Z for each sample. Count the number of samples that yield values of $Z > 1.645$—that is, that reject the null hypothesis. Comment on your results.

b. Use a statistical software package to generate 1000 random samples, each of size 4, from a uniform distribution over the interval 2 to 14 that has a mean of 8 and a variance of 12. Calculate the sample mean and Z for each sample. Count the number of samples that yield values of $Z > 1.645$—that is, that reject the null hypothesis. Comment on your results.

c. Use a statistical software package to generate 1000 random samples, each of size 4, from a uniform distribution over the interval 4 to 16 that has a mean of 10 and a variance of 12. Calculate the sample mean and Z for each sample. Count the number of samples that yield values of $Z > 1.645$—that is, that reject the null hypothesis. Comment on your results.

d. Use a statistical software package to generate 1000 random samples, each of size 16, from a uniform distribution over the interval 0 to 12 that has a mean of 6 and a variance of 12 that corresponds to the null hypothesis. Calculate the sample mean and Z for each sample. Count the number of samples that yield values of $Z > 1.645$—that is, that reject the null hypothesis. Comment on your results.

e. Use a statistical software package to generate 1000 random samples, each of size 16, from a uniform distribution over the interval 2 to 14 that has a mean of 8 and a variance of 12. Calculate the sample mean and Z for each sample. Count the number of samples that yield values of $Z > 1.645$—that is, that reject the null hypothesis. Comment on your results.

f. Use a statistical software package to generate 1000 random samples, each of size 16, from a uniform distribution over the interval 4 to 16 that has a mean of 10 and a variance of 12. Calculate the sample mean and Z for each sample. Count the number of samples that yield values of $Z > 1.645$—that is, that reject the null hypothesis. Comment on your results.

⋗ Exercises

4.10   Kane (1986) discusses the concentricity of an engine oil seal groove. Historically, the standard deviation for the concentricity is 0.7. The target average concentricity is 5.6. To monitor this process, he periodically takes a random sample of three measurements.

a. A recent sample yielded a sample mean of 5.8. Conduct a hypothesis test to determine whether the true mean concentricity has changed. Use a 0.05 significance level.

b. Find the p-value associated with the test in part **a**.

c. In Exercise 4.1, you constructed a 95% confidence interval for this situation. Use this interval to determine whether the true mean concentricity has changed. Discuss the relationship of the 95% confidence interval and the corresponding hypothesis test.

d. Find the power of this test to detect a change in the true mean concentricity to 5.8.

e. Find the sample size required to achieve an approximate power of 0.9 when the true mean concentricity is 5.8.

f. What did you assume to do these analyses? How can you check these assumptions?

4.11   Runger and Pignatiello (1991) consider a plastic injection molding process for a part with a target critical width dimension of 100 and a historic standard deviation of 8. Periodically, clogs form in one of the feeder lines, causing the mean width to change. As a result, the operator periodically takes random samples of size four.

a. A recent sample yielded a sample mean of 101.4. Conduct a hypothesis test to determine whether the true mean width has increased. Use a 0.01 significance level.

b. Find the p-value associated with the test in part **a**.

c. In Exercise 4.2, you constructed a 99% confidence interval for this situation. Use this interval to determine whether the true mean width has changed. Discuss the relationship of the 99% confidence interval and the corresponding hypothesis test.

d. Find the power of this test to detect a change in the true mean width to 102.

e. Find the sample size required to achieve a power of 0.8 when the true mean width is 102.

f. What did you assume to do these analyses? How can you check these assumptions?

4.12 Yaschchin (1995) discusses a process for the chemical etching of silicon wafers used in integrated circuits. This company wishes to detect an increase in the thickness of the silicon oxide layers because thicker layers require longer etching times. Process specifications state a target value of 1 micron for the true mean thickness. Historically, the layer thickness have a standard deviation of 0.06 micron.

a. A recent random sample of four wafers yielded a sample mean of 1.134. Conduct a hypothesis test to determine whether the true mean thickness has increased. Use a significance level of 0.05.

b. Find the *p*-value associated with the test in part **a**.

c. In Exercise 4.3, you constructed a 95% confidence interval for this situation. Use this interval to determine whether the true mean thickness has changed. Discuss the relationship of the 95% confidence interval and the corresponding hypothesis test.

d. Find the power of this test to detect a change in the true mean thickness to 1.01.

e. Find the sample size required to achieve a power of 0.85 when the true mean thickness is 1.01.

f. What did you assume to do these analyses? How can you check these assumptions?

4.13 The modulus of rupture (MOR) for a particular grade of pencil lead is known to have a standard deviation of 250 psi. Process standards call for a target value of 6500 psi for the true mean MOR. For each batch, an inspector tests a random sample of 16 leads. Management wishes to detect any change in the true mean MOR.

a. A recent random sample yielded a sample mean of 6490. Conduct a hypothesis test to determine whether the true mean MOR has changed from the target. Use a 0.10 significance level.

b. Find the *p*-value associated with the test in part **a**.

c. In Exercise 4.4, you constructed a 90% confidence interval for this situation. Use this interval to determine whether the true mean MOR has changed. Discuss the relationship of the 90% confidence interval and the corresponding hypothesis test.

d. Find the power of this test to detect a change in the true mean MOR to 6400.

e. Find the sample size required to achieve an approximate power of 0.85 when the true mean MOR is 6400.

f. What did you assume to do these analyses? How can you check these assumptions?

4.14 The yields from an ethanol–water distillation column have a standard deviation of 1%. Process specifications call for a target yield of 93%. Management wishes to detect any decrease in the true mean yield.

a. A random sample of eight recent batches produced the following yields. Conduct a hypothesis test to determine whether the true mean yield has decreased. Use a 0.01 significance level.

0.90	0.93	0.95	0.86
0.90	0.87	0.93	0.92

b. Find the p-value associated with the test in part **a**.
c. In Exercise 4.5, you constructed a 99% confidence interval for this situation. Use this interval to determine whether the true mean yield has changed. Discuss the relationship of the 99% confidence interval and the corresponding hypothesis test.
d. Find the power of this test to detect a change in the true mean yield to 92.5%.
e. Find the sample size required to achieve a power of 0.95 when the true mean yield is 92.5%.
f. What did you assume to do these analyses? How can you check these assumptions?

4.15 The ocean swell produces spectacular eruptions of water through a hole in the cliff at Kiama, Australia, known as the Blowhole. Historically, the time between eruptions has been 60 seconds with a standard deviation of 10 seconds.

a. A random sample of the times at which 10 successive eruptions occurred was measured in seconds. Conduct a hypothesis test to determine whether the true mean time between eruptions has increased. Use a 0.01 significance level.

60	55	77	56	68
89	73	61	69	61

b. Find the p-value associated with the hypothesis test in part **a**.
c. In Exercise 4.7, you constructed a 99% confidence interval for this situation. Use this interval to determine whether the true mean time between eruptions has changed. Discuss the relationship of the 99% confidence interval and the corresponding hypothesis test.
d. Find the power of this test to detect a change in the true mean time between eruptions to 65 seconds.
e. Find the sample size required to achieve a power of 0.95 when the true mean time between eruptions is 65 seconds.
f. What did you assume to do these analyses? How can you check these assumptions?

4.16 Data on the concentration of polychlorinated biphenyl (PCB) residues in a series of lake trout from Cayuga Lake, NY, can be found in Bates and Watts (1988). Each whole fish was mechanically chopped, ground, and thoroughly mixed, and 5-gram samples taken. The samples were treated and PCB residues in parts per million (ppm) were estimated using column chromatography. Historically, the PCB residues have a standard deviation of 0.7 ppm. An important question of interest is whether the PCB residues exceed 1.4 ppm.

a. A subset of the random sample of the PCB residues was measured in ppm. Conduct a hypothesis test to determine whether the true mean PCB residues have increased. Use a 0.10 significance level.

0.6	1.6	0.5	1.2	2.0
1.3	2.5	2.2	2.4	1.2

b. Find the *p*-value associated with the hypothesis test in part **a**.

c. In Exercise 4.8, you constructed a 90% confidence interval for this situation. Use this interval to determine whether the true mean PCB residues have changed. Discuss the relationship of the 90% confidence interval and the corresponding hypothesis test.

d. Find the power of this test to detect a change in the true mean PCB residues to 1.475 ppm.

e. Find the sample size required to achieve a power of 0.80 when the true mean PCB residues are 1.475 ppm.

f. What did you assume to do these analyses? How can you check these assumptions?

4.17 A study was done on the propagation of an ultrasonic stress wave through a substance. In particular, attenuation (the decrease in amplitude of the stress wave measured in neper/cm) in fiber-glass-reinforced polyester composites was measured. The target attenutation is 2.3 neper/cm. Attenuation values are known to have a standard deviation of 0.25.

a. A random sample of attenuations produced the following values. Conduct a hypothesis test to determine whether the true mean attenuation has changed. Use a 0.05 significance level.

2.6	2.5	2.2	2.0	2.1	2.3	2.6

b. Find the *p*-value associated with the hypothesis test in part **a**.

c. In Exercise 4.9, you constructed a 95% confidence interval for this situation. Use this interval to determine whether the true mean attenuation has changed. Discuss the relationship of the 95% confidence interval and the corresponding hypothesis test.

d. Find the power of this test to detect a change in the true mean attenuation to 2.1 neper/cm.

e. Find the sample size required to achieve a power of 0.90 when the true mean PCB residues are 2.1 neper/cm.

f. What did you assume to do these analyses? How can you check these assumptions?

4.18   Consider the hypotheses

$$H_0: \quad \mu = \mu_0$$
$$H_a: \quad \mu < \mu_0$$

when the population variance is known. Derive the sample size required to achieve a power of p_0 with a significance level of α.

4.19   Consider the hypotheses

$$H_0: \quad \mu = \mu_0$$
$$H_a: \quad \mu > \mu_0$$

when the population variance is known. Derive the corresponding one-sided confidence interval. Discuss what this interval means. What are its relative advantages and disadvantages?

▷ 4.3   Inference for a Single Mean

One-Sided t-Tests

First, we consider the one-sided alternative when the variance is unknown, which is best illustrated through an example. The data are real, but for proprietary reasons they have been transformed.

Example 4.7 | **Porosities of Battery Plates**

Nickel–hydrogen (Ni–H) batteries use a nickel plate as the anode. A critical quality characteristic is the plate's porosity, which controls the interface of the anode with the potassium hydroxide electrolyte solution. For this particular battery cell, the manufacturer has set a target porosity of 80% as measured by a standard test. The sintering process, whereby the plates are "fired" at high temperature, essentially controls the plate's porosity. The production people have expressed concerns that the plate is being overfired and thus is not sufficiently porous. They plan to take a random sample of ten plates and test their porosities. For the purposes of statistical analysis, use a 0.05 significance level.

1. **State hypotheses:**   These are the two possible one-sided alternatives:

$$H_0: \quad \mu = \mu_0 \qquad H_0: \quad \mu = \mu_0$$
$$H_a: \quad \mu < \mu_0 \qquad H_a: \quad \mu > \mu_0$$

Figure 4.11 | Critical Region for H_a: $\mu < \mu_0$

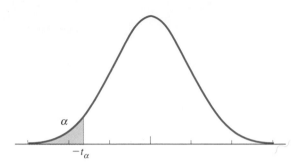

In this situation, we are concerned only with showing that the true mean porosity is less than the target. Consequently, our hypotheses are

$$H_0: \quad \mu = 80$$

$$H_a: \quad \mu < 80.$$

2. **State the test statistic:** Our test statistic is of the form

$$t = \frac{\overline{y} - \mu_0}{s/\sqrt{n}}.$$

The degrees of freedom for this statistic are $n - 1$.

3. **State the critical region:** The critical region depends on the specific alternative hypothesis. Consider H_a: $\mu < \mu_0$. Figure 4.11 illustrates that we reject H_0 if $t < -t_{n-1,\alpha}$, where $t_{n-1,\alpha}$ is the appropriate value from Table 2 in the appendix. Now consider H_a: $\mu > \mu_0$. Figure 4.12 illustrates that we reject

Figure 4.12 | Critical Region for H_a: $\mu > \mu_0$

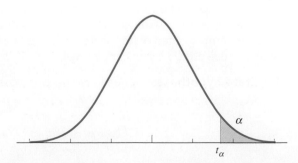

H_0 if $t > t_{n-1,\alpha}$, where $t_{n-1,\alpha}$ is the appropriate value from Table 2 of the appendix. In this case, $H_a : \mu < 80$, so we reject H_0 if

$$t < -t_{n-1,\alpha}$$
$$< -t_{9,0.05}$$
$$< -1.833.$$

4. **Conduct the experiment and find t:** The random sample yielded the following results:

79.1	79.5	79.3	79.3	78.8
79.0	79.2	79.7	79.0	79.2

We first need to find the sample mean, the sample variance, and the sample standard deviation:

$$\bar{y} = \frac{1}{n} \sum_{i=1}^{n} y_i = \frac{792.1}{10} = 79.21$$

$$s^2 = \frac{n \sum_{i=1}^{n} y_i^2 - \left(\sum_{i=1}^{n} y_i\right)^2}{n \cdot (n-1)}$$

$$= \frac{10(62{,}742.85) - (792.1)^2}{10(9)}$$

$$= 0.06767$$

$$s = \sqrt{s^2} = \sqrt{0.06767} = 0.26.$$

Our test statistic is

$$t = \frac{\bar{y} - \mu_0}{s/\sqrt{n}}$$

$$= \frac{79.21 - 80}{0.26/\sqrt{10}}$$

$$= -9.60.$$

5. **Reach conclusions and state them in English:** Since -9.60 is less than -1.833, we may reject the null hypothesis. Alternatively, the p-value is 0.000, which is less than $\alpha = 0.05$. We thus have evidence to suggest that the true mean porosity is less than 80%, which supports the contention that the sintering process is overfiring the plates.

   The hypothesis test tells us only that $\mu = 80$ is not plausible in light of the data. What are reasonable values for the true mean porosities? To answer this question, we must construct an appropriate confidence interval. By equation (4.1), the form of our $(1 - \alpha) \cdot 100\%$ confidence interval is

$$\overline{y} \pm t_{n-1, \alpha/2} \frac{s}{\sqrt{n}}.$$

In this case, a 95% confidence interval is

$$79.21 \pm t_{9, .025} \frac{0.26}{\sqrt{10}} = 79.21 \pm 2.262(0.0822)$$
$$= 79.21 \pm 0.19$$
$$= (79.02, 79.40).$$

As a result, we believe that the plausible values for the true mean porosity are between 79.02 and 79.40%. We need to adjust the process to increase the porosity, on average, approximately 0.8%.

Using Confidence Intervals as an Alternative Analysis—One-Sided Case   Many engineers and statisticians prefer to use confidence intervals rather than hypothesis tests. In this approach, we construct the appropriate confidence interval using α at the stated significance level. If the interval contains the hypothesized value for the population mean, then we simply conclude that it is plausible, which is equivalent to failing to reject the null hypothesis. On the other hand, if, as in our example, the confidence interval does not contain the hypothesized value, then we conclude that the hypothesized value is not plausible and that the plausible values reside in the calculated interval.

We have noted that using confidence intervals in this way is equivalent to performing a two-sided hypothesis test. In general, when we reject the null hypothesis for a one-sided test, the two-sided confidence interval does not contain the hypothesized value. Occasionally, however, the two approaches will reach different conclusions. In such situations, some analysts construct a one-sided confidence interval, which we leave as a homework exercise.

Checking Assumptions   In statistical analyses, we should check our assumptions whenever possible. With a sample size of ten, we need to be sampling from a distribution that is roughly normal—one that is single peaked, symmetric, and with tails that die rapidly. To check that this is the case, we should use a stem-and-leaf display. Figure 4.13, which gives this display, indicates a single-peaked pattern that is reasonably symmetric and has tails that seem to die relatively quickly. The center of this plot is between 79.2 and 79.3, and the range is from 78.8 to 79.7. Figure 4.14 shows the normal probability plot for these data. The plot appears reasonably close to a straight line, which supports the contention that the data come from a well-behaved distribution. On the whole, there appear to be no major problems with using a *t*-statistic in this situation.

| | **Figure 4.13** | The Stem-and-Leaf Display for the Ni–H Plate Porosities |

Stem	Leaves	Number	Depth
78.●:	8	1	1
79.*:	001	3	4
79.t:	2233	4	
79.f:	5	1	2
79.s:	7	1	1

The Two-Sided t-Test

Next, we consider the two-sided alternative, which again is best illustrated through an example.

| **Example 4.8** | **Grinding of Silicon Wafers for Integrated Circuits** |

Roes and Does (1995) present data on the grinding of silicon wafers used in integrated circuits. Philips Semiconductors grinds wafers in batches of 31. For a particular product, Philips has a target thickness of 244 μm. To monitor this process, it samples five wafers from each batch. Philips wants to ensure product consistency, so it seeks to discover whether the wafers are either too thick or too thin. We shall treat these five wafers as a random sample. For the purposes of statistical analysis, we shall use a 0.01 significance level.

| **Figure 4.14** | The Normal Probability Plot for the Plate Porosities |

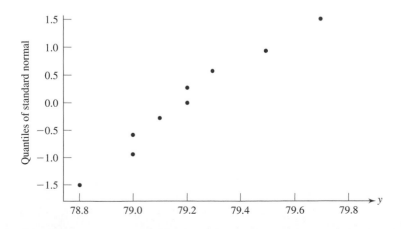

1. **State hypotheses:** For the two-sided alternative, the hypotheses are

$$H_0: \quad \mu = \mu_0$$
$$H_a: \quad \mu \neq \mu_0.$$

In this particular case,

$$H_0: \quad \mu = 244$$
$$H_a: \quad \mu \neq 244.$$

2. **State the test statistic:** Again, we shall use

$$t = \frac{\bar{y} - \mu_0}{s/\sqrt{n}}.$$

3. **State the critical region:** Consider $H_a: \mu \neq \mu_0$. Figure 4.15 illustrates that we reject H_0 if $|t| > t_{n-1,\alpha/2}$, where $t_{n-1,\alpha/2}$ is the appropriate value from Table 2 in the appendix. In this case, we shall reject the null hypothesis if

$$|t| > t_{n-1,\alpha/2}$$
$$> t_{4,.005}$$
$$> 4.604.$$

4. **Conduct the experiment and find t:** The random sample yielded these results: 240, 243, 250, 253, and 248. We first need to find the sample mean, the sample variance, and the sample standard deviation.

$$\bar{y} = \frac{1}{n} \sum_{i=1}^{n} y_i = \frac{1234}{5} = 246.8 \qquad s^2 = 27.7 \qquad s = 5.263.$$

Figure 4.15 | Critical Region for $H_a: \mu \neq \mu_0$

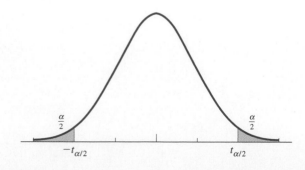

Our test statistic is

$$t = \frac{\overline{y} - \mu_0}{s/\sqrt{n}}$$

$$= \frac{246.8 - 244}{5.263/\sqrt{5}}$$

$$= 1.19.$$

5. **Reach conclusions and state them in English:** Since $|1.19|$ is not greater than 4.604, we cannot reject the null hypothesis. Using statistical software, we obtain a p-value of 0.299. Again, we cannot reject the null hypothesis. We thus do not have sufficient evidence to suggest that the true mean thickness is different from 244 μm, so we do not have sufficient evidence to adjust the grinding process. Within the hypothesis testing framework, we recognize that 244 μm is a perfectly reasonable value for the true mean thickness and stop at this point. We have no real reason to construct a confidence interval if we take this approach seriously.

Using Confidence Intervals as an Alternative Analysis—Two-Sided Case   Since we conducted a two-sided test, we can use our standard confidence interval as the basis for an alternative analysis, as we illustrate in the next example.

Example 4.9 | **Grinding of Silicon Wafers for Integrated Circuits—Revisited**

A 99% confidence interval for the true mean thickness is

$$\overline{y} \pm t_{n-1,\alpha/2} \frac{s}{\sqrt{n}} = \overline{y} \pm t_{4,.005} \frac{s}{\sqrt{n}}$$

$$= 246.8 \pm 4.604 \frac{5.263}{\sqrt{5}}$$

$$= 246.8 \pm 10.8$$

$$= (236.0, 257.6).$$

As a result, plausible values for the true mean thickness range from 236.0 to 257.6 μm. The nominal mean thickness, 244 μm, is squarely within this interval. Thus, it is perfectly plausible, and we have no real basis for rejecting it as a reasonable value. In other words, we have no real evidence to suggest that it cannot be the true mean thickness, which is exactly the same conclusion we reached in the hypothesis test.

Unlike the hypothesis testing approach, the confidence interval allows us to clearly state the range of plausible values for the true mean thickness. In this case, we see that 244 μm is quite plausible, but we also see that the range of plausible values is quite wide, which conveys a sense of the precision of our procedure.

Checking Assumptions   The sample size of five really makes checking our assumptions problematic. A stem-and-leaf display makes no sense with only five observations. Figure 4.16 gives the normal probability plot for these data. With such a small sample size, we should conclude that problems exist only if we see gross departures from the straight line. In this case, the data appear to follow a straight line, which supports the contention that the data do come from a roughly normal distribution. The derivation of the t-statistic required that the data themselves follow a normal distribution. However, it is well known that this statistic is robust (insensitive) to mild departures from normality. As the sample size increases, the t-statistics can tolerate even greater departures from normality.

Prediction Intervals

Whereas the confidence interval focuses on the true mean of the response, the prediction interval considers the prediction of individual responses. Such an interval must consider both the variability in the estimation of the true mean value of the response and the variability of the individual responses around this mean. A $(1 - \alpha) \cdot 100\%$ prediction interval is

$$\overline{y} \pm t_{n-1,\alpha/2}s \cdot \sqrt{1 + \frac{1}{n}}.$$

The constant 1 in this expression takes care of the variability associated with the individual responses. This interval assumes that the values predicted are independent of the data used to construct the interval itself.

Figure 4.16 | The Normal Probability Plot for the Silicon Wafer Thicknesses

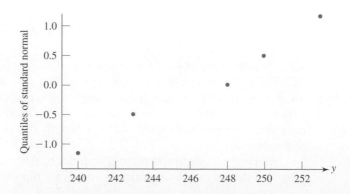

Prediction intervals make sense only to the extent that the sample on which we based the interval represents the population or process of interest. If the population or process changes over time, then our prediction interval can become useless.

Example 4.10 | **Porosities of Battery Plates—Revisited**

We can construct a 95% prediction interval for these porosities:

$$\bar{y} \pm t_{n-1,\alpha/2}s \cdot \sqrt{1 + \frac{1}{n}} = 79.21 \pm t_{9,.025}(0.26) \cdot \sqrt{1 + \frac{1}{10}}$$

$$= 79.21 \pm 2.262 \cdot (0.2727)$$

$$= 79.21 \pm 0.62$$

$$= (78.59, 79.83).$$

Unless this process changes (a big assumption), we expect approximately 95% of the data to fall within the interval (78.59, 79.53). Note that even this interval does not contain the originally hypothesized value of 80 for the true mean porosity.

⋗ Exercises

4.20 Consider the problem in Example 4.6. Derive the corresponding one-sided confidence interval and reanalyze these data. Discuss what this interval means. What are its relative advantages and disadvantages?

4.21 Yashchin (1992) studied the thicknesses of metal wires produced in a chip-manufacturing process. Ideally, these wires should have a target thickness of 8 microns. These are the sample data:

8.4	8.0	7.8	8.0	7.9	7.7	8.0	7.9	8.2	7.9
7.9	8.2	7.9	7.8	7.9	7.9	8.0	8.0	7.6	8.2
8.1	8.1	8.0	8.0	8.3	7.8	8.2	8.3	8.0	8.0
7.8	7.9	8.4	7.7	8.0	7.9	8.0	7.7	7.7	7.8
7.8	8.2	7.7	8.3	7.8	8.3	7.8	8.0	8.2	7.8

a. Conduct the most appropriate hypothesis test using a 0.05 significance level.
b. Construct a 95% confidence interval for the true mean thickness.
c. Construct a 95% prediction interval for the thicknesses.

      d. What did you assume to do the analyses? Can you evaluate how well these data meet the assumptions? If yes, determine how comfortable you are with them. If not, explain why.

      e. Discuss your conclusions about the chip-manufacturing process based on this analysis.

4.22 Montgomery (2004, p. 248) reports results for a process that manufactures high-voltage supplies with a nominal output of 350 V. The production people are concerned that the process is beginning to produce power supplies with a true mean output voltage somewhat greater than the nominal value. The voltages for the last four power supplies tested are 351.4, 351.5, 351.2, and 351.6.

      a. Conduct the most appropriate hypothesis test using a 0.10 significance level.

      b. Construct a 90% confidence interval for the true mean voltage.

      c. Construct a 90% prediction interval for the voltages.

      d. What did you assume to do the analyses? Can you evaluate how well these data meet the assumptions? If yes, determine how comfortable you are with them. If not, explain why.

      e. Discuss your conclusions about this manufacturing process based on this analysis.

4.23 Farnum (1994, p. 195) discusses a chrome plating process. Small electric currents are run through a chemical plate that contains nickel, resulting in a thin plating of the metal on the part. Since the bath loses nickel as the plating proceeds, the operators periodically add more nickel to the bath. Operating standards call for a nickel concentration of 4.5 oz/gal. The process runs three shifts per day. The following are the bath concentrations at the beginning of the day for five days. Assume that these concentrations form a random sample: 4.8, 4.5, 4.4, 4.2, and 4.4.

      a. Conduct the most appropriate hypothesis test using a 0.05 significance level.

      b. Construct a 95% confidence interval for the true mean concentration.

      c. Construct a 95% prediction interval for the concentrations.

      d. What did you assume to do the analyses? Can you evaluate how well these data meet the assumptions? If yes, determine how comfortable you are with them. If not, explain why.

      e. Discuss your conclusions about the chrome plating process based on this analysis.

4.24 DeVor, Chang, and Sutherland (1992, pp. 406–407) discuss the production of polyol, which is reacted with isocynate in a foam molding process. Variations in the moisture content of polyol cause problems in controlling the reaction with isocynate. Production has set a target moisture content of 2.125%. The following data represent 27 moisture analyses over a four-month period:

2.29	2.22	1.94	1.90	2.15	2.02	2.15	2.09	2.18
2.00	2.06	2.02	2.15	2.17	2.17	1.90	1.72	1.75
2.12	2.06	2.00	1.98	1.98	2.02	2.14	2.10	2.05

a. Conduct the most appropriate hypothesis test using a 0.01 significance level.
b. Construct a 99% confidence interval for the true mean moisture content.
c. Construct a 99% prediction interval for the moisture contents.
d. What did you assume to do the analyses? Can you evaluate how well these data meet the assumptions? If yes, determine how comfortable you are with them? If not, explain why.
e. Discuss your conclusions about the foam molding process based on this analysis.

4.25 Weaver (1990) examined a galvanized coating process for large pipes. Standards call for an average coating weight of 200 lb per pipe. These data are the coating weights for a random sample of 30 pipes:

216	202	208	208	212	202	193	208	206	206
206	213	204	204	204	218	204	198	207	218
204	212	212	205	203	196	216	200	215	202

a. Conduct the most appropriate hypothesis test using a 0.01 significance level.
b. Construct a 99% confidence interval for the true mean coating weight.
c. Construct a 99% prediction interval for the coating weights.
d. What did you assume to do the analyses? Can you evaluate how well these data meet the assumptions? If yes, determine how comfortable you are with them. If not, explain why.
e. Discuss your conclusions about the coating process based on this analysis.

4.26 Holmes and Mergen (1992) studied a batch operation at a chemical plant where an important quality characteristic was the product viscosity, which had a target value of 14.90. Production personnel use a viscosity measurement for each 12-hour batch to monitor this process. These are the viscosities for the past ten batches:

| 13.3 | 14.5 | 15.3 | 15.3 | 14.3 |
| 14.8 | 15.2 | 14.9 | 14.6 | 14.1 |

a. Conduct the most appropriate hypothesis test using a 0.10 significance level.
b. Construct a 90% confidence interval for the true mean viscosity.
c. Construct a 90% prediction interval for the viscosities.
d. What did you assume to do the analyses? Can you evaluate how well these data meet the assumptions? If yes, determine how comfortable you are with them. If not, explain why.
e. Discuss your conclusions about this batch operation based on this analysis.

4.27 McNeese and Klein (1991) looked at the average particle size of a product with a specification of 70–130 microns and a target of 100 microns. Production personnel measure the particle size distribution using a set of screening sieves. They test one sample a day to monitor this process. The average particle sizes for the past 25 days are listed here:

99.6	92.1	103.8	95.3	101.6
102.8	100.9	100.5	102.7	96.9
101.5	96.7	96.8	97.8	104.7
103.2	97.5	98.3	105.8	100.6
102.3	93.8	102.7	94.9	94.9

a. Conduct the most appropriate hypothesis test using a 0.05 significance level.
b. Construct a 95% confidence interval for the true mean particle size.
c. Construct a 95% prediction interval for the particle sizes.
d. What did you assume to do the analyses? Can you evaluate how well these data meet the assumptions? If yes, determine how comfortable you are with them. If not, explain why.
e. Discuss your conclusions about this product based on this analysis.

4.28 Calik, Vural and Özdamar (1997) reported on shake flask cultures, as well as oxygen transfer kinetics studied in laboratory-scale bioreactors. Suppose the oxygen uptake rate has a nominal value of 4 kmol. It is thought that increasing the agitation rate will increase the oxygen uptake rate. After increasing the agitation rate, a random sample of five produced the following oxygen uptake rates:

7.992	4.800	3.910	5.280	8.160

a. Conduct the most appropriate hypothesis test using a 0.05 significance level.
b. Construct a 95% confidence interval for the true mean oxygen uptake rate.
c. Construct a 95% prediction interval for oxygen uptake rates.
d. What did you assume to do the analyses? Can you evaluate how well these data meet the assumptions? If yes, determine how comfortable you are with them. If not, explain why.
e. Discuss your conclusions about the oxygen uptake rate based on this analysis.

4.29 Eight measurements of bulk resistivity of silicon wafers were made at NIST. The wafers were doped with phosphorous by neutron transmutation doping in order to have nominal resistivities of 196 ohm.cm.

196.3052	196.1240	196.1890	195.9884
196.2005	196.0052	195.8763	196.2090

   a. Conduct the most appropriate hypothesis test using a 0.10 significance level.
   b. Construct a 90% confidence interval for the true mean resistivity.
   c. Construct a 90% prediction interval for resistivities.
   d. What did you assume to do the analyses? Can you evaluate how well these data meet the assumptions? If yes, determine how comfortable you are with them. If not, explain why.
   e. Discuss your conclusions about the resistivity based on this analysis.

4.30 Rajniak and Yang (1994) studied a hysteresis-dependent adsorption process with water vapor-silica gel at $25°$ C and relative pressure at 360. Cyclic adsorption–desorption processes are widely employed in separation and purification of gases. Ideally, the adsorptions should be stable at .250. Twelve adsorptions measured in g/g were taken:

0.203	0.222	0.254	0.277	0.195	0.270
0.250	0.276	0.252	0.215	0.256	0.254

   a. Conduct the most appropriate hypothesis test using a 0.01 significance level.
   b. Construct a 99% confidence interval for the true mean adsorption.
   c. Construct a 99% prediction interval for adsorptions.
   d. What did you assume to do the analyses? Can you evaluate how well these data meet the assumptions? If yes, determine how comfortable you are with them. If not, explain why.
   e. Discuss your conclusions about the adsorption based on this analysis.

⋗ 4.4 Inference for Proportions

In Chapter 3, we introduced the binomial distribution and described its rather wide applicability to engineering problems. In many situations, we need to address interesting questions about proportions. The normal approximation to the binomial provides a firm basis for estimating and testing the true proportion of "successes" in a population, p.

> **VOICE OF EXPERIENCE**
>
> *Whenever possible, use continuous data over categorical data.*

Suppose we take a random sample of size n from a binomial population with parameter p. Let Y be the number of successes observed in the sample. An appropriate and commonsense estimate of p is the *sample proportion* given by

$$\hat{p} = \frac{Y}{n},$$

which is the number of successes in the sample divided by the sample size. Let $q = 1 - p$. By the normal approximation to the binomial, if $np \geq 5$ and $nq \geq 5$

(preferably both $\geq$ 10), then the distribution of Y may be well approximated by a normal distribution with

$$E(Y) = np \quad \text{and,} \quad \text{var}(Y) = npq.$$

As a result, it can be shown that

$$E(\hat{p}) = E\left(\frac{Y}{n}\right) = p$$

$$\text{var}(\hat{p}) = \text{var}\left(\frac{Y}{n}\right) = \frac{pq}{n}.$$

Many engineering applications based on proportions require quite large sample sizes. Binomial data classify each observation as either a success or a failure. In the process, we often lose a great deal of information. For example, in Section 3.8 we considered the packaging of 50-pound bags of graphite. A bag is considered a success if the packaged weight is between 48 and 52 pounds; otherwise, it is considered a failure. In essence, we have thrown away the actual weight by simply calling it a success or a failure. Bags that are exactly 50 pounds are much more of a success than a bag weighing 51.9 pounds. A bag weighing 52.1 pounds is a failure, but it is only trivially worse than a bag weighing 51.99 pounds, which is considered a success. Whenever possible, it is better to use the actual raw data rather than to classify the observations as successes or failures.

Hypothesis Tests for p

Once again, we illustrate this technique through an example.

Example 4.11 | **Breaking Strengths of Carbon Fibers**

Padgett and Spurrier (1990) analyze the breaking strengths of carbon fibers used in fibrous composite materials (see Exercise 2.4). These fibers measure 50 mm in length and 7–8 microns in diameter. Specifications state that fibers with a breaking strength of less than 1.2 GPa (gigapascals) are defective. Suppose that historically, this process has produced 10% nonconforming. How can we develop an appropriate *monitoring* procedure?

Consider a sequence of hypothesis tests where we collect a random sample of n fibers each shift and test their breaking strengths. We then classify each fiber as conforming or nonconforming. Our hypothesis test focuses on the true proportion of fibers that are nonconforming.

1. **State hypotheses:** Let p_0 be the nominal value of p for our test. In general, the possible sets of hypotheses are

$$H_0: \quad p = p_0 \qquad H_0: \quad p = p_0 \qquad H_0: \quad p = p_0$$
$$H_a: \quad p < p_0 \qquad H_a: \quad p > p_0 \qquad H_a: \quad p \neq p_0$$

   For most monitoring situations, we wish to detect either an increase or a decrease in the proportion nonconforming. Why? In the short term, we must ensure that the true proportion of nonconforming fibers is not excessive. Thus, we must detect when the proportion has increased. In the long term, we would like never to produce nonconforming fibers. Thus, we must detect when the proportion decreases so we can learn why. Since we really are interested in both $p > 0.10$ and $p < 0.10$, our hypotheses are

$$H_0: \quad p = 0.10$$
$$H_a: \quad p \neq 0.10.$$

2. **State the test statistic:** Under the null hypothesis, $p = p_0$. Let $q_0 = 1 - p_0$. We thus know the standard error. As a result, our test statistic is

$$Z = \frac{\hat{p} - p_0}{\sqrt{(p_0 \cdot q_0)/n}}.$$

   How large should n be and how do we determine its size? We base our decision for n assuming that H_0 is true. From the normal approximation to the binomial, we need to choose n such that these conditions are met:

$$np_0 \geq 5$$
$$nq_0 \geq 5.$$

   Preferably, both np_0 and nq_0 should be greater than or equal to 10. If we use this more stringent requirement, then

$$n \cdot (0.10) \geq 10$$
$$n \geq \frac{10}{0.1}$$
$$n \geq 100.$$

3. **State the critical region:** Again, the critical regions depend on the alternative hypotheses. Consider $H_a: p < p_0$. Figure 4.17 illustrates that we reject H_0 if $Z < -z_\alpha$. Now consider $H_a: p > p_0$. Figure 4.18 illustrates that we reject H_0 if $Z > z_\alpha$. Finally, consider $H_a: p \neq p_0$. Figure 4.19 illustrates that we reject H_0 if $|Z| > z_{\alpha/2}$. In our particular case, we need to determine an

Figure 4.17 | Critical Region for H_a: $p < p_0$

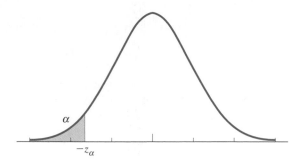

appropriate significance level. So far, our choices have been 0.10, 0.05, and 0.01. Since we conduct this test every shift, we really are more concerned about a Type I error, rejecting the null hypothesis when it is true, than a Type II error, failing to detect a change in the true proportion of nonconforming fibers. Why? If we make a Type I error, we must search for the cause of a change when none is present. Since we are conducting this test so frequently, we run the risk of constantly searching for problems that do not exist. On the other hand, because we are conducting this test so frequently, if we do not pick up a change in the true proportion on any given sample, we should pick it up on a subsequent sample. Thus, we shall use $\alpha = 0.01$. Since H_a: $p \neq 0.10$, we reject the null hypothesis if

$$|Z| > z_{\alpha/2}$$
$$> z_{0.005}$$
$$> 2.576.$$

Actually, in industry, the critical region would be to reject H_0 if $|Z| > 3.0$, which is equivalent to using $\alpha = 0.0027$.

Figure 4.18 | Critical Region for H_a: $p > p_0$

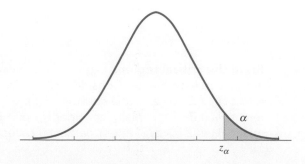

Figure 4.19 | Critical Region for H_a: $p \neq p_0$

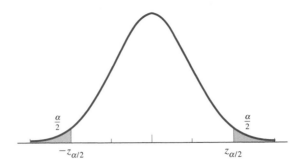

4. **Conduct the experiment and find Z:** Treat the data given in Exercise 2.4 as the most recent sample from this process. Out of the 100 fibers tested, six have breaking strengths less than 1.2 GPa and are nonconforming. Thus,

$$\hat{p} = \frac{6}{100} = 0.06$$

$$Z = \frac{\hat{p} - p_0}{\sqrt{(p_0 q_0)/n}}$$

$$= \frac{0.06 - 0.10}{\sqrt{(0.10)(0.90)/100}}$$

$$= -1.33.$$

5. **Reach conclusions and state them in English:** Since $|-1.33| < 2.576$ (or p-value of 0.1866), we do not have sufficient evidence to reject the null hypothesis. As a result, we do not have sufficient evidence to conclude that the proportion of defectives has changed. Thus, we do not have sufficient evidence to suggest that we should search out the presence of any changes to our process.

Using Confidence Intervals as an Alternative Analysis

By a process similar to the one we used to derive equation (4.1), we can establish that an appropriate $(1 - \alpha)100\%$ confidence interval for p is given by

$$\hat{p} \pm z_{\alpha/2} \sqrt{\frac{pq}{n}}.$$

This interval depends on p, however, which is the very entity we are trying to estimate! As a result, we use $\hat{p}$ and $\hat{q} = 1 - \hat{p}$ to estimate the appropriate standard error. Because we use $\hat{p}$ to estimate the standard error, our confidence interval is not exactly equivalent to the two-sided hypothesis test. Even though we use an estimated standard error, there is no theoretical basis for using a t-statistic for our interval. Instead, the traditional $(1 - \alpha) \cdot 100\%$ confidence interval for p is

$$\hat{p} \pm z_{\alpha/2} \sqrt{\frac{\hat{p}\hat{q}}{n}}.$$

We can use this result to determine the sample size required to produce an interval that is approximately $\pm B$. Let p_0 be a reasonable initial guess for p, and let $q_0 = 1 - p_0$. Following the same logic we used to derive equation (4.2), we obtain

$$n \geq p_0 \cdot q_0 \left[\frac{z_{\alpha/2}}{B} \right]^2.$$

If we truly have no idea of the general order of magnitude for p_0, we traditionally use a value of 0.5 because this value maximizes the resulting sample size and thus represents a worst case situation.

Example 4.12 | **Breaking Strengths of Carbon Fibers—Revisited**

A 99% confidence interval for the true proportion of nonconforming fibers is

$$\hat{p} \pm z_{\alpha/2} \sqrt{\frac{\hat{p}\,\hat{q}}{n}} = 0.06 \pm 2.576 \sqrt{\frac{(0.06)(0.94)}{100}}$$

$$= 0.06 \pm 0.06$$

$$= (0.00, 0.12).$$

VOICE OF EXPERIENCE

Tests on proportion often require large sample sizes.

This interval suggests that the plausible value for the true proportion of nonconforming fibers is somewhere between 0% and 12%. Once again, the nominal value of 10% is a plausible value, which is equivalent to failing to reject the null hypothesis.

Finding the Sample Size   Suppose management wants a 99% confidence interval, which estimates this proportion to ± 0.02. If we can assume that the nominal value of 10% is in the appropriate neighborhood of the true value, then we can find the appropriate sample size by

$$n \geq p_0 \cdot q_0 \left[\frac{z_{\alpha/2}}{B} \right]^2$$

$$\geq (0.10) \cdot (0.90) \left[\frac{2.576}{0.02} \right]^2$$

$$\geq 1493.$$

As a result, we need a sample size of roughly 1500 to estimate the true proportion of nonconforming fibers to the precision requested by management.

Exercises

4.31 Airplanes approaching the runway for landing are required to stay within the localizer (a certain distance left and right of the runway). When an airplane deviates from the localizer, it is sometimes referred to as an exceedence. Consider one airline at a medium-sized airport. Historically, the airline has experienced 12% exceedence. In an effort to improve the exceedence, pilots went through a new training program. After the program, a random sample of 250 landings at the airport found 22 in exceedence.

    a. Conduct the most appropriate hypothesis test using a 0.05 significance level.

    b. Construct a 95% confidence interval for the true proportion of landings that are in exceedence.

4.32 A manufacturing company produces water filters for home refrigerators. The process has typically produced about 4% defective. A recently designed experiment has led to changing the seal to reduce defects. With the process running using the new seal, a random sample of 300 filters yielded 7 defects.

    a. Conduct the most appropriate hypothesis test using a 0.10 significance level.

    b. Construct a 90% confidence interval for the true proportion of defective water filters.

4.33 DeVor, Chang, and Sutherland (1992, pp. 164–165, 258) discuss a cylinder boring process for an engine block. Specifications require that these bores be 3.5199 ± 0.0004 in. Management is concerned that the true proportion of cylinder bores outside the specifications is excessive. Current practice is willing to tolerate up to 10% outside the specifications. Out of a random sample of 165, 36 were outside the specifications.

    a. Conduct the most appropriate hypothesis test using a 0.01 significance level.

    b. Construct a 99% confidence interval for the true proportion of bores outside the specifications.

4.34 Marcucci (1985) looked at nonconforming brick from a brick manufacturing process. Typically, 5% of the brick produced is not suitable for all purposes. Management monitors this process by periodically collecting random samples and classifying the bricks as conforming or nonconforming. A recent sample of 214 bricks yielded 18 nonconforming.

    a. Conduct the most appropriate hypothesis test using a 0.01 significance level.

    b. Construct a 99% confidence interval for the true proportion of nonconforming bricks.

4.35 DeVor, Chang, and Sutherland (1992, pp. 209–210) examined a process for manufacturing electrical resistors that have a nominal resistance of 100 ohms with a specification of ±2 ohms. Suppose management has expressed a concern that the true proportion of resistors with resistances outside the specifications has increased from the historical level of 10%. A random sample of 180 resistors yielded 46 with resistances outside the specifications.

a. Conduct the most appropriate hypothesis test using a 0.05 significance level.
b. Construct a 95% confidence interval for the true proportion of resistors outside the specifications.

4.36 Operators make the nickel plates for nickel–hydrogen batteries by carefully pouring nickel powder into a frame. An important characteristic of the plate at this point in the process is its initial net weight. Production management imposes a strict weight standard on this product. Historically, only 35% of the plates made meet the specifications. The material for the rejected plates is immediately reused, so the real loss as the result of failing to meet the standard is labor. A new operator, whom management suspects is not as skillful as her more experienced colleagues, made 191 attempts in one day of work. Of these 191, 38 met the specifications.

a. Conduct the most appropriate hypothesis test using a 0.05 significance level.
b. Construct a 95% confidence interval for the true proportion of attempts that fail to meet the specifications.

4.37 An automobile manufacturer gives a 5-year/60,000-mile warranty on its drive train. Historically, 7% of this manufacturer's automobiles have required service under this warranty. Recently, a design team proposed an improvement that should extend the drive train's life. A random sample of 200 cars underwent 60,000 miles of road testing; the drive train failed for 12.

a. Conduct the most appropriate hypothesis test using a 0.05 significance level.
b. Construct a 95% confidence interval for the true proportion of automobiles with drive trains that fail.

4.38 Historically, 10% of the homes in Florida have radon levels higher than recommended by the Environmental Protection Agency. Radon is a weakly radioactive gas known to contribute to health problems. A city in north central Florida has hired an environmental consulting group to determine whether it has a greater than normal problem with this gas. A random sample of 200 homes indicated that 25 had radon levels exceeding EPA recommendations.

a. Conduct the most appropriate hypothesis test using a 0.05 significance level.
b. Construct a 95% confidence interval for the true proportion of homes with excessive levels of radon.

4.5 Inference for Two Independent Samples

Until now we have been solely interested in a single parameter, either μ or p, of a single population. We thus have used only one sample as the basis for our inferences. More typically, we are interested in the characteristics of two or more distinct populations. For example, do two different levels of paint viscosity produce different mean coating thicknesses? Do the mean amounts of ground beef delivered by a particular packaging process change from day to day? Do different brands of humidifier have the same mean output moisture rates? The basic way to address these questions is to collect random samples from both populations. The resulting methodology allows us to analyze the simplest factoral experiment where we have a single factor at two levels. Each level represents a different population.

Consider a ground beef packaging process. Suppose we wish to compare the true mean amount of ground beef delivered on one day to the amount delivered on the next. Let $y_{11}, y_{12}, \ldots, y_{1n_1}$ be a random sample of size n_1 taken from the first day. Let μ_1 and σ_1^2 be the population mean and variance for this population, respectively. Let $\overline{y}_1$ and s_1^2 be the sample mean and the sample variance calculated from this random sample. In a similar manner, let $y_{21}, y_{22}, \ldots, y_{2n_2}$ be a random sample of size n_2 taken from the second day. Let μ_2 and σ_2^2 be the population mean and variance for this population. Let $\overline{y}_2$ and s_2^2 be the sample mean and the sample variance calculated from this random sample.

We can address the relationship between the two means by looking at their *difference,* or $\mu_1 - \mu_2$. The possible outcomes are

- If $\mu_1 - \mu_2 > 0$, then the true mean amount delivered on the first day is larger than the true mean amount delivered on the second day; thus, $\mu_1 - \mu_2 > 0$ is equivalent to $\mu_1 > \mu_2$.
- If $\mu_1 - \mu_2 < 0$, then the true mean amount delivered on the first day is smaller than the true mean amount delivered on the second; thus, $\mu_1 - \mu_2 < 0$ is equivalent to $\mu_1 < \mu_2$.
- If $\mu_1 - \mu_2 \neq 0$, then the true mean amounts delivered are different; thus, $\mu_1 - \mu_2 \neq 0$ is equivalent to $\mu_1 \neq \mu_2$.

The real parameter of interest to us in this situation is $\mu_1 - \mu_2$.

Estimating the Difference and Its Variability

The most appropriate estimator of $\mu_1 - \mu_2$ is the commonsense one, the difference between the two sample means: $\overline{y}_1 - \overline{y}_2$. It is easy to show that

$$E(\overline{y}_1 - \overline{y}_2) = \mu_1 - \mu_2.$$

Thus, $\overline{y}_1 - \overline{y}_2$ is an unbiased estimator of $\mu_1 - \mu_2$. *If the two samples are independent of each other,* this estimator's standard error is

$$\sqrt{\frac{\sigma_1^2}{n_1} + \frac{\sigma_2^2}{n_2}}.$$

In general, we really do not know σ_1^2 and σ_2^2. In such cases, we require the additional assumption that

$$\sigma_1^2 = \sigma_2^2 = \sigma^2.$$

We thus are assuming that the true variances for both populations are the same and that the only difference between the two populations is the mean. Under this assumption, the standard error becomes

$$\sqrt{\frac{\sigma_1^2}{n_1} + \frac{\sigma_2^2}{n_2}} = \sqrt{\frac{\sigma^2}{n_1} + \frac{\sigma^2}{n_2}} = \sigma\sqrt{\frac{1}{n_1} + \frac{1}{n_2}}.$$

The common variance assumption may seem overly restrictive, but in general, it is fairly reasonable. It is well known that the procedure we are about to outline is extremely robust to this assumption. Most texts suggest that the common variance assumption is reasonable as long as the ratio of the larger variance to the smaller is less than four. This common variance assumption underlies the classical analysis of designed experiments.

How should we estimate σ^2? Under the common variance assumption, both s_1^2 and s_2^2 estimate σ^2. We would like to use as much of the information in the data as possible. It thus makes sense to use both s_1^2 and s_2^2. The best way to combine them is by a weighted average, where the weights are the respective degrees of freedom, which results in s_p^2, the "pooled" estimate of σ^2:

$$s_p^2 = \frac{(n_1 - 1) \cdot s_1^2 + (n_2 - 1) \cdot s_2^2}{n_1 + n_2 - 2}.$$

By the way we have defined s_p^2, it is always between s_1^2 and s_2^2. The degrees of freedom for this statistic are $n_1 + n_2 - 2$.

With this information, we can develop both hypothesis tests and confidence intervals.

Hypothesis Tests for the Difference

Once again, the best way to introduce this technique is through an example.

Example 4.13 | **Packaging of Ground Beef**

Maxcy and Lowry (1984) looked at a packaging process for ground beef over a series of days. An interesting question is whether the true mean amount delivered by this process changes from day to day. Ten packages are accurately weighed and randomly selected over the course of the day on two consecutive days. For the purposes of statistical analysis, we shall use a 0.05 significance level.

1. **State hypotheses:** Let δ_0 be a hypothesized difference—that is, a difference of particular interest. Often, but not always, $\delta_0 = 0$. Here are the three possible sets of hypotheses:

H_0: $\mu_1 - \mu_2 = \delta_0$ H_0: $\mu_1 - \mu_2 = \delta_0$ H_0: $\mu_1 - \mu_2 = \delta_0$

H_a: $\mu_1 - \mu_2 < \delta_0$ H_a: $\mu_1 - \mu_2 > \delta_0$ H_a: $\mu_1 - \mu_2 \neq \delta_0$

In our particular case,

$$H_0: \quad \mu_1 - \mu_2 = 0$$
$$H_a: \quad \mu_1 - \mu_2 \neq 0.$$

2. **State the test statistic:** The test statistic is

$$t = \frac{\bar{y}_1 - \bar{y}_2 - \delta_0}{s_p \cdot \sqrt{\dfrac{1}{n_1} + \dfrac{1}{n_2}}}.$$

In this case, since $\delta_0 = 0$, our test statistic is

$$t = \frac{\bar{y}_1 - \bar{y}_2}{s_p \cdot \sqrt{\dfrac{1}{n_1} + \dfrac{1}{n_2}}}.$$

The degrees of freedom for this statistic are $n_1 + n_2 - 2$.

3. **State the critical region:** These are the critical regions for this test:

- For $H_a: \mu_1 - \mu_2 < \delta_0$, we reject H_0 if $t < -t_{n_1+n_2-2,\alpha}$.
- For $H_a: \mu_1 - \mu_2 > \delta_0$, we reject H_0 if $t > t_{n_1+n_2-2,\alpha}$.
- For $H_a: \mu_1 - \mu_2 \neq \delta_0$, we reject H_0 if $|t| > t_{n_1+n_2-2,\alpha/2}$.

In our particular case, we reject the null hypothesis if

$$|t| > t_{n_1+n_2-2,\alpha/2}$$
$$> t_{18,0.025}$$
$$> 2.101.$$

4. **Conduct the experiment and find t:** Table 4.3 gives the weights (in grams) of the selected packages. For the first day, $\bar{y}_1 = 1391.50$ and $s_1^2 = 28.409$. For the second day, $\bar{y}_2 = 1398.65$ and $s_2^2 = 51.696$. We next need to calculate s_p^2:

$$\begin{aligned} s_p^2 &= \frac{(n_1 - 1) \cdot s_1^2 + (n_2 - 1) \cdot s_2^2}{n_1 + n_2 - 2} \\ &= \frac{9(28.409) + 9(51.696)}{18} \\ &= 40.053. \end{aligned}$$

Table 4.3	The Packaging of Ground Beef Data

First Day					Second Day				
1397.8	1394.8	1391.7	1400.0	1393.5	1410.0	1393.9	1405.9	1404.2	1387.3
1391.2	1384.0	1391.0	1385.7	1385.3	1398.5	1399.9	1392.5	1402.5	1391.8

Thus,

$$s_p = \sqrt{s_p^2} = \sqrt{40.053} = 6.33.$$

Our test statistic is

$$t = \frac{\overline{y}_1 - \overline{y}_2}{s_p \cdot \sqrt{\dfrac{1}{n_1} + \dfrac{1}{n_2}}}$$

$$= \frac{1391.50 - 1398.65}{6.33 \cdot \sqrt{\dfrac{1}{10} + \dfrac{1}{10}}}$$

$$= -2.526.$$

5. **Reach conclusions and state them in English:** Since $|-2.526| > 2.101$ (or since p-value $= 0.021$ is less than $\alpha = 0.05$), we may reject the null hypothesis. We do have evidence to suggest that the true mean amounts of ground beef are different on the two days. The industrial engineer assigned to this process should seek to identify the causes for the difference and eliminate them.

Since we reject the null hypothesis, we need to give a range of plausible values for the true difference. We can construct a $(1 - \alpha) \cdot 100\%$ confidence interval for $\mu_1 - \mu_2$ by

$$(\overline{y}_1 - \overline{y}_2) \pm t_{n_1+n_2-2,\alpha/2}\, s_p \cdot \sqrt{\frac{1}{n_1} + \frac{1}{n_2}}.$$

In this case, a 95% confidence interval for the true mean difference for these two days is

$$(\overline{y}_1 - \overline{y}_2) \pm t_{n_1+n_2-2,\alpha/2}\, s_p \cdot \sqrt{\frac{1}{n_1} + \frac{1}{n_2}}$$

$$= (1391.50 - 1398.65) \pm 2.101(6.33)\sqrt{\frac{1}{10} + \frac{1}{10}}$$

$$= -7.15 \pm 5.95$$

$$= (-13.1, -1.2).$$

Thus, the plausible values for the true difference in mean amounts range from -13.1 to -1.2 g.

Using Confidence Intervals As an Alternative Analysis

Many engineers and statisticians prefer this approach because it allows us to bypass completely the formal hypothesis test.

| Example 4.14 | **Packaging of Ground Beef—Revisited** |

From the last example, the 95% confidence interval for the true mean difference is $(-13.1, -1.2)$. Thus, the plausible values for the true difference in mean amounts range from -13.1 to -1.2 g. Since zero does not fall within this interval, we have no reason to believe that it is plausible. We thus have evidence to suggest that the true mean amount delivered on the second day is greater than the true mean delivered on the first.

Checking Assumptions

In order to do this analysis, we made three assumptions:

1. The distributions of the amounts of ground beef delivered for both days

> **VOICE OF EXPERIENCE**
>
> *The two-sample t-test is very robust both to normality and to constant variance.*

   are reasonably normal or mound-shaped (for sample sizes of ten, the amounts follow a distribution that is single peaked, roughly symmetric, and with tails that die rapidly). If this assumption is grossly violated, then alternative analyses (usually nonparametric methods) should be used.
2. The two random samples are independent.
3. The variances for each day are the same.

To check these assumptions, consider Figure 4.20, the side-by-side stem-and-leaf display, Figure 4.21, the parallel boxplots, Figure 4.22, the normal probability plot for the first day, and Figure 4.23, the normal probability plot for the second day. Figure 4.20 reveals that both data sets are single peaked, roughly symmetric, and with tails that die rapidly. The second day's amounts seem to be centered higher than the first day's, but the spreads for the amounts look similar. Figure 4.21 confirms these results. The median value for the second day is greater

| Figure 4.20 | The Side-by-Side Stem-and-Leaf Display Comparing the Amounts of Ground Beef Delivered on Two Days |

	First Day		Second Day			
Stem	Leaves	Number	Depth	Leaves	Number	Depth
138	455	3	3	7	1	1
139	111347	6		12389	5	6
140	0	1	1	45	2	3
141				0	1	1

Figure 4.21 | Parallel Boxplots Comparing the Amounts of Ground Beef Delivered on Two Days

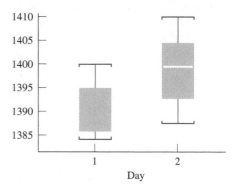

than the value for the first. The lengths of the boxes, which are the interquartile ranges, are similar and indicate similar variabilities. We see no apparent outliers. Figure 4.22 does not appear to be a perfectly straight line and does give a little concern that the data from the first day may not come from a very well-behaved distribution. Figure 4.23 looks much better in this regard. Since the data were taken on two completely different days, we have no particular reason to doubt the independence of the two samples. On the whole, we should feel reasonably comfortable about our assumptions in this case.

Figure 4.22 | The Normal Probability Plot for the Ground Beef Amounts on the First Day

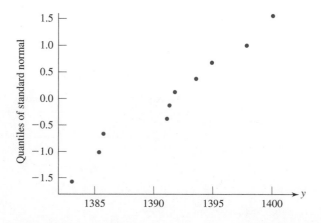

Figure 4.23 | The Normal Probability Plot for the Ground Beef Amounts on the Second Day

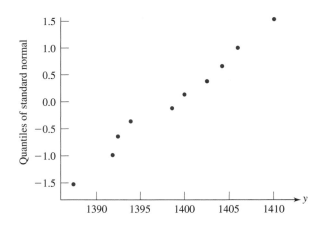

∷ *Exercises*

4.39 Eibl, Kess, and Pukelsheim (1992) studied the impact of viscosity on the observed coating thickness produced by a paint operation. For simplicity, they chose to study only two viscosities: "low" and "high." Up to a certain paint viscosity, higher viscosities cause thicker coatings. The engineers do not know whether they have hit that limit or not. They thus wish to test whether the higher viscosity paint leads to thicker coatings. Here are the coating thicknesses:

Low Viscosity							
1.09	1.12	0.83	0.88	1.62	1.49	1.48	1.59
0.88	1.29	1.04	1.31	1.83	1.65	1.71	1.76

High Viscosity							
1.46	1.51	1.59	1.40	0.74	0.98	0.79	0.83
2.05	2.17	2.36	2.12	1.51	1.46	1.42	1.40

a. Analyze these data using parallel boxplots.
b. Conduct the appropriate hypothesis test using a 0.05 significance level.
c. Construct a 95% confidence interval for the true difference in the mean coating thicknesses.
d. What did you assume to do the analyses? Can you evaluate how well these data meet the assumptions? If yes, determine how comfortable you are with them. If not, explain why.

e. Discuss the conclusions of the hypothesis test and confidence interval relative to the boxplot.

4.40 A manufacturer of aircraft (see Montgomery 1991, pp. 242–244) monitors the viscosity of primer paint. The viscosities for two different time periods are listed here:

Time Period 1					Time Period 2				
33.8	33.1	34.0	33.8	33.5	33.5	33.3	33.4	33.3	34.7
34.0	33.7	33.3	33.5	33.2	34.8	34.6	35.0	34.8	34.5
33.6	33.0	33.5	33.1	33.8	34.7	34.3	34.6	34.5	35.0

a. Analyze these data using parallel boxplots.
b. Conduct the appropriate hypothesis test using a 0.05 significance level.
c. Construct a 95% confidence interval for the true difference in the mean viscosities.
d. What did you assume to do the analyses? Can you evaluate how well these data meet the assumptions? If yes, determine how comfortable you are with them. If not, explain why.
e. Discuss the conclusions of the hypothesis test and confidence interval relative to the boxplot.

4.41 Galinsky and colleagues (1993) studied the impact of sensory modalities (either aural or visual) on people's ability to monitor a specific display for critical events to which they must respond. Such tasks are critical components of jobs like air traffic control, industrial quality control, robotic manufacturing operations, and nuclear power plant monitoring. One aspect of the study focused on the difference in response to aural and visual stimuli. In particular, they monitored the motor activity of the subject's dominant wrist as a measure of "restlessness" or "fidgeting." The greater the activity, the more restless the subject. Galinsky and her colleagues recorded these numbers of wrist movements over 10-minute periods of time:

Auditory					Visual				
418	236	281	416	578	386	517	617	870	892
329	197	397	677	698	416	574	782	838	885

a. Analyze these data using parallel boxplots.
b. Conduct the appropriate hypothesis test using a 0.01 significance level.
c. Construct a 99% confidence interval for the true difference in the mean numbers of wrist movements.
d. What did you assume to do the analyses? Can you evaluate how well these data meet the assumptions? If yes, determine how comfortable you are with them. If not, explain why.
e. Discuss the conclusions of the hypothesis test and confidence interval relative to the boxplot.

4.42 An independent consumer group tested radial tires from two major brands to determine whether there were any differences in the expected tread life. The data (in thousands of miles) are given here:

Brand 1						Brand 2				
50	54	52	47	61		57	61	47	52	53
56	51	51	48	56		57	56	53	67	58
53	43	58	52	48		62	56	56	62	57

a. Analyze these data using parallel boxplots.
b. Conduct the appropriate hypothesis test using a 0.05 significance level.
c. Construct a 95% confidence interval for the true difference in the mean tread lives.
d. What did you assume to do the analyses? Can you evaluate how well these data meet the assumptions? If yes, determine how comfortable you are with them. If not, explain why.
e. Discuss the conclusions of the hypothesis test and confidence interval relative to the boxplot.

4.43 Nelson (1989) compared two brands of ultrasonic humidifiers with respect to the rate at which they output moisture. The following data are the maximum outputs (in fluid ounces) per hour as measured in a chamber controlled at a temperature of 70°F and a relative humidity of 30%:

Brand 1				Brand 2			
14.0	14.3	12.2	15.1	12.1	13.6	11.9	11.2

a. Analyze these data using parallel boxplots.
b. Conduct the appropriate hypothesis test using a 0.10 significance level.
c. Construct a 90% confidence interval for the true difference in the mean viscosities.
d. What did you assume to do the analyses? Can you evaluate how well these data meet the assumptions? If yes, determine how comfortable you are with them. If not, explain why.
e. Discuss the conclusions of the hypothesis test and confidence interval relative to the boxplot.

4.44 The following data are the yields for the last 8 hours of production from two ethanol–water distillation columns:

Column 1							
70	74	73	72	72	73	72	73

Column 2							
71	74	72	71	72	70	72	72

a. Analyze these data using parallel boxplots.
b. Conduct the appropriate hypothesis test using a 0.10 significance level.
c. Construct a 90% confidence interval for the true difference in the mean yields.
d. What did you assume to do the analyses? Can you evaluate how well these data meet the assumptions? If yes, determine how comfortable you are with them. If not, explain why.
e. Discuss the conclusions of the hypothesis test and confidence interval relative to the boxplot.

4.45 Capurro, Brozzo and Cirillo (1997) studied two modern high-strength microalloyed structural steels. Thirty Charpy V notched specimens were machined from a 50-mm-thick plate of steel A. Similarly, a set of 30 specimens were machined from a piece of pipe of steel B. All specimens were tested in four-point bending tests at a temperature of 77 K and the values of fracture load recorded. The data are

Steel A

24700	31300	29400	31500	28700	31900	28400	27200	32800	30200
30200	32700	30900	42000	22800	36000	28000	27700	30500	28500
40400	31300	32600	34700	33100	23200	24300	35500	25300	24700

Steel B

34510	28730	36380	32060	25810	21900	28450	37510	29960	39730
21010	28880	35150	28780	35970	37960	28840	26400	24770	26580
34110	26750	23540	32300	34140	33140	28330	32220	22040	28610

a. Analyze these data using parallel boxplots.
b. Conduct the appropriate hypothesis test using a 0.05 significance level.
c. Construct a 95% confidence interval for the true difference in the mean fracture loads.
d. What did you assume to do the analyses? Can you evaluate how well these data meet the assumptions? If yes, determine how comfortable you are with them. If not, explain why.
e. Discuss the conclusions of the hypothesis test and confidence interval relative to the boxplot.

4.46 Penner and Watts (1991) investigated whether the time to drill holes in rock differs using "dry" or "wet" drilling. In dry drilling, compressed air is used to flush the cuttings, and in wet drilling water is used. Each method is used on 12 rocks. The drilling times (in 1/100 minutes) for the two methods are

Dry Drilling						Wet Drilling					
727	965	904	987	847	918	607	549	762	665	588	798
814	750	804	989	902	939	704	772	780	599	603	699

The plausible values for this difference range from −8.58 to −5.62, which again indicates that method 2 yields higher octane ratings than method 1. In particular, method 2 seems to yield ratings somewhere between 5.6 and 8.6 points higher.

Using Confidence Intervals as an Alternative Analysis

This approach allows us to bypass completely the formal hypothesis test. Confidence intervals are a simple, powerful, and direct tool if the two methods give different results on the average.

Example 4.16	**Testing Octane Blends—Revisited**

From the previous example, a 99% confidence interval for the true mean difference in the methods is (−8.58, −5.62). Since this interval does not contain zero, we may conclude that method 2 seems to yield ratings somewhere between 5.6 and 8.6 points higher than method 1.

Checking Assumptions

To perform this analysis, we assumed that the *differences* follow a distribution such that $\bar{d}/(s_d/\sqrt{n})$ follows a *t*-distribution with, in this particular case, 31 degrees of freedom. For so many degrees of freedom, we are mostly concerned with whether the distribution is single peaked and the tails die rapidly. We can tolerate a certain amount of skew in the data with so large a number of degrees of freedom. Figure 4.24 gives the stem-and-leaf display for these differences. The differences have been rounded to the nearest integer. The data show a single clear peak and the tails do die rapidly. Figure 4.25 gives the normal probability plot

Figure 4.24	The Stem-and-Leaf Display for the Octane Differences

Stem	Leaves	Number	Depth
−1t:	2	1	1
−1*:	0001111	7	8
−•:	8888888899	10	
−s:	666677	6	14
−f:	445	3	8
−t:	2223	4	5
−*:	1	1	1

Figure 4.25 | The Normal Probability Plot for the Octane Differences

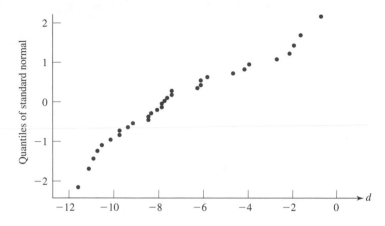

for these differences. This plot has a fairly straight line, which indicates that the data do come from a well-behaved distribution. We should feel very comfortable about our assumptions.

Paired Versus Independent t-Test

How we actually conduct an experiment determines whether we should use the paired t-test or the two-independent-samples t-test. Many experimental situations can easily lend themselves to either approach. A reasonable question is: When should an experimenter pursue a paired structure? Pairing works well when the sampling units available for the study differ widely among themselves. Then pairing allows us to remove the sampling unit to sampling unit variability. If the sampling units differ widely, removing this variability makes our estimate of the standard deviation much smaller. In turn, the denominator of our test statistic decreases, which makes the observed value of our test statistic larger in absolute value. As a result, we are more likely to reject the null hypothesis (we increase the power of our test). On the other hand, pairing the data also reduces the number of degrees of freedom available for our analysis. Decreasing the number of degrees of freedom makes the critical value for our test statistic slightly larger in absolute value, making it slightly more difficult to reject the null hypothesis (we slightly decrease the power of our test). A quick examination of the t-table reveals that the number of degrees of freedom available for analysis becomes critical for very small numbers of sampling units.

In general, we should obtain paired data whenever we know that the sampling units differ significantly from one another. The reduction in variability typically more than compensates for the slight increase in the critical value for the test. On the other hand, if we have no reason to believe that the sampling units differ—that is, that they are truly homogeneous—then the two-independent-samples structure makes a great deal of sense, particularly if very few sampling units are available for the test.

Exercises

4.48 An experiment was carried out on scale models in a wave tank to investigate how the choice of mooring method affected the bending stress produced in a device used to generate electricity from wave power at sea. The model system was subjected to the same sample of 18 sea states with each of the two mooring methods. The resulting data (root mean square bending moment in Newton-meters) are

Sea state	1	2	3	4	5	6	7	8	9
Method 1	2.23	2.55	7.99	4.09	9.62	1.59	8.98	0.82	10.83
Method 2	1.82	2.42	8.26	3.46	9.77	1.40	8.88	0.87	11.20

Sea state	10	11	12	13	14	15	16	17	18
Method 1	1.54	10.75	5.79	5.91	5.79	5.50	9.96	1.92	7.38
Method 2	1.33	10.32	5.87	6.44	5.87	5.30	9.82	1.69	7.41

a. Conduct the appropriate hypothesis test using a 0.10 significance level.
b. Construct a 90% confidence interval for the true mean difference in bending stress.
c. What did you assume to do the analyses? Can you evaluate how well these data meet the assumptions? If yes, determine how comfortable you are with them. If not, explain why.

4.49 Dunn (1989) studied the estimated lengths (to the nearest 0.1 inch) of 15 pieces of string as assessed by two raters. The two raters always overestimated the length of the string. The distances from the actual measured length for each rater are

String	1	2	3	4	5	6	7
Graham	1.3	0.9	1.5	0.5	1.7	0.8	1.0
Brian	1.5	1.0	1.3	0.9	0.5	0.5	1.6

String	8	9	10	11	12	13	14
Graham	0.4	1.5	0.3	0.3	1.2	1.6	0.3
Brian	0.8	1.4	0.4	0.5	0.5	1.3	0.9

a. Conduct the appropriate hypothesis test using a 0.05 significance level.
b. Construct a 95% confidence interval for the true mean difference in raters.
c. What did you assume to do the analyses? Can you evaluate how well these data meet the assumptions? If yes, determine how comfortable you are with them. If not, explain why.

4.50 The value of the turbine prediction coefficient was calculated for 13 pumps using two different prediction methods. Williams (1994) gives the data as

Pump	1	2	3	4	5	6	7
Childs	0.78	1.64	2.25	0.80	1.09	1.39	0.47
Stepanoff	0.60	1.82	2.25	0.80	0.17	0.91	0.47
Pump	8	9	10	11	12	13	
Childs	2.66	0.44	1.11	0.48	0.41	0.56	
Stepanoff	2.88	0.23	1.11	0.48	0.67	0.89	

a. Conduct the appropriate hypothesis test using a 0.01 significance level.
b. Construct a 99% confidence interval for the true mean difference in prediction methods.
c. What did you assume to do the analyses? Can you evaluate how well these data meet the assumptions? If yes, determine how comfortable you are with them. If not, explain why.

4.51 Grubbs (1983) presented data on the running times of 20 fuses. Two operators, acting independently, measured these times for the fuses:

Operator 1	4.85	4.93	4.75	4.77	4.67	4.87	4.67	4.94	4.85	4.75
Operator 2	5.09	5.04	4.95	5.02	4.90	5.05	4.90	5.15	5.08	4.98
Operator 1	4.83	4.92	4.74	4.99	4.88	4.95	4.95	4.93	4.92	4.89
Operator 2	5.04	5.12	4.95	5.23	5.07	5.23	5.16	5.11	5.11	5.08

a. Conduct the most appropriate hypothesis test using a 0.05 significance level.
b. Construct a 95% confidence interval for the true mean difference in the times.
c. What did you assume to do the analyses? Can you evaluate how well these data meet the assumptions? If yes, determine how comfortable you are with them. If not, explain why.

4.52 Nickel–hydrogen batteries use nickel plates as the anode. After the plates are sintered or fired in a high-temperature furnace, they are grouped into lots of 40 plates each and then placed into an "electrode deposition" (ED) bath where they are placed under an electrical load. This bath controls the electrical properties of the cell. An important characteristic of the nickel plates batteries is "stress growth." As the battery cell undergoes its charge–discharge cycle, the plates actually begin to expand due to the stress. One of the engineers believes that the more porous the plate, the greater the stress growth. He wants to conduct a test to confirm this belief. The engineer knows that the specific conditions of the ED bath have a major impact on stress growth. Since no two ED bath runs are identical, the engineer expects a lot of variability in the stress growth purely from the ED baths. To minimize the impact of the ED baths, he has set up each ED lot so that 20 plates have "low" porosity and 20 plates have "high" porosity. After the ED run, the engineer randomly selects five low-porosity plates to make a test battery cell and five high-porosity plates to form a second test cell. The data are the average percent increases in the plates' thicknesses after 200 charge–discharge cycles. The following values summarize the results from 16 different ED lots:

Low porosity	1.43	3.56	2.03	0.92	3.21	3.08	3.69	2.81
High porosity	3.00	9.41	3.81	1.81	4.42	2.19	1.02	2.81
Low porosity	2.34	2.39	0.95	2.01	1.98	1.59	1.04	1.66
High porosity	4.47	4.18	3.46	2.67	1.23	1.95	0.51	0.08

a. Conduct the most appropriate hypothesis test using a 0.05 significance level.
b. Construct a 95% confidence interval for the true mean difference in stress growth.
c. What did you assume to do the analyses? Can you evaluate how well these data meet the assumptions? If yes, determine how comfortable you are with them. If not, explain why.

4.53 Measuring the actual dimensions of a manufactured part is a classical problem in many different disciplines, especially mechanical and industrial engineering. A mechanical engineer must grapple with the thickness of nickel plates for a nickel–hydrogen battery. By the way the plate is made, he can consistently identify specific locations on each plate. Thus, location A on the first plate measured is the same as location A on the second plate. He believes that one specific location, A, of the plate is consistently thicker than another specific location, B. The actual thicknesses (in mm) of ten plates are listed here:

Location A	31.10	31.10	30.90	30.80	32.20	30.40	29.65	29.85	29.85	30.65
Location B	29.75	29.75	30.15	30.80	30.20	30.40	30.35	29.75	29.15	30.50

a. Conduct the most appropriate hypothesis test using a 0.05 significance level.
b. Construct a 95% confidence interval for the true mean difference in the thicknesses.

c. What did you assume to do the analyses? Can you evaluate how well these data meet the assumptions? If yes, determine how comfortable you are with them. If not, explain why.

4.54 Eck Industries, Inc. (Mee, 1990, also see Example 2.1), manufactures cast aluminum cylinder heads used for liquid-cooled aircraft engines. The wall thicknesses are critical, particularly in high-altitude applications. The company seeks to compare two methods for measuring these thicknesses: ultrasound (U), which is nondestructive, and sectioning (S) the heads, which obviously is destructive. Sectioning is more accurate, but ultrasound allows the company to test a part and then ship it. The company would like to see whether the methods give different measurements on the average. Measurements of 18 heads are listed here:

Method U	0.223	0.193	0.218	0.201	0.231	0.204	0.228	0.223	0.215
Method S	0.224	0.207	0.216	0.204	0.230	0.203	0.222	0.225	0.224
Method U	0.223	0.237	0.226	0.214	0.213	0.233	0.224	0.217	0.210
Method S	0.223	0.226	0.232	0.217	0.217	0.237	0.224	0.219	0.192

a. Conduct the most appropriate hypothesis test using a 0.05 significance level.
b. Construct a 95% confidence interval for the true mean difference in the thicknesses.
c. What did you assume to do the analyses? Can you evaluate how well these data meet the assumptions? If yes, determine how comfortable you are with them. If not, explain why.

4.55 Van Nuland (1992) daily compares two temperature instruments: one coupled to a process computer and the other used for visual control. Ideally, these two instruments should agree. The following data are the two temperatures over a five-day period:

Day	1	2	3	4	5
Temp. 1	84.6	84.5	84.4	84.6	84.3
Temp. 2	85.2	85.1	84.9	85.3	85.0

a. Conduct the most appropriate hypothesis test using a 0.05 significance level.
b. Construct a 95% confidence interval for the true mean difference in temperatures.
c. What did you assume to do the analyses? Can you evaluate how well these data meet the assumptions? If yes, determine how comfortable you are with them. If not, explain why.

4.56 A maintenance manager must test a new repair method that should increase the expected time between repairs. For each machine used in the study, she recorded the last time between failures prior to using the new method, which she called "Current," and the first time between failures after using the new method, which she called "New." These are the times (in hours):

Machine	1	2	3	4	5	6	7	8	9	10
Current	155	222	346	287	115	389	183	451	140	252
New	211	345	419	274	244	420	319	505	396	222

a. Conduct the most appropriate hypothesis test using a 0.05 significance level.
b. Construct a 95% confidence interval for the true mean difference.
c. What did you assume to do the analyses? Can you evaluate how well these data meet the assumptions? If yes, determine how comfortable you are with them. If not, explain why.

❧ **4.7 Inference for Variances**

A Single-Sample Variance

Since the sample variance, s^2, is a random variable, it also has a sampling distribution. If we take a random sample from a normal distribution we can show that

$$\frac{(n-1)s^2}{\sigma^2}$$

follows a χ^2 distribution with $n - 1$ degrees of freedom. This result depends quite heavily on the normality assumption. As a result, serious departures from normality should concern us. Figure 4.26 plots the pdf for a χ^2 distribution with four degrees of freedom. This distribution is quite skewed to the right.

Let $\chi^2_{n-1,\alpha}$ represent the value from the χ^2 table (Table 3 in the appendix) that corresponds to a right-hand tail area of α for a χ^2 distribution with $n - 1$ degrees of freedom. We then observe that

$$P\left(\chi^2_{n-1,1-\alpha/2} \leq \frac{(n-1)s^2}{\sigma^2} \leq \chi^2_{n-1,\alpha/2} \right) = 1 - \alpha.$$

Figure 4.27 illustrates this interval. With some algebra, we get

$$P\left(\frac{(n-1)s^2}{\chi^2_{n-1,\alpha/2}} \leq \sigma^2 \leq \frac{(n-1)s^2}{\chi^2_{n-1,1-\alpha/2}} \right) = 1 - \alpha.$$

Figure 4.26 | The pdf for a χ^2 Distribution with Four Degrees of Freedom

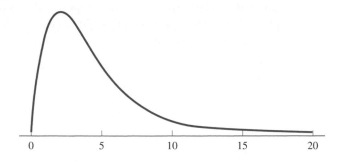

From this result, we get the $(1 - \alpha) \cdot 100\%$ confidence interval for σ^2:

$$\left(\frac{(n-1)s^2}{\chi^2_{n-1,\alpha/2}}, \frac{(n-1)s^2}{\chi^2_{n-1,1-\alpha/2}} \right).$$

We also can use the χ^2 distribution to develop an appropriate hypothesis test. Once again, we introduce this procedure through an example.

Example 4.17 | **Variability in Whiteness of Titanium Dioxide**

A paint manufacturer uses a large amount of titanium dioxide in its coatings. Titanium dioxide is the primary white pigment used in paints and coatings. People measure the "whiteness" of this pigment using a scale of 0–30, with 30 being essentially perfect white. Recently, this manufacturer switched vendors for its titanium dioxide. The new vendor claims that its titanium dioxide averages 25 on

Figure 4.27 |

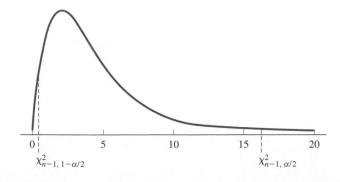

the whiteness scale, with a variance of 0.4. The paint manufacturer, based on previous experience, doubts that the new vendor really has such a small amount of variability in its product. The manufacturer plans to treat the next ten shipments as a random sample and to address this question by a formal hypothesis test with a .05 significance level.

1. **State hypotheses:** In general, these are the forms of the hypotheses:

$$H_0: \sigma^2 = \sigma_0^2 \qquad H_0: \sigma^2 = \sigma_0^2 \qquad H_0: \sigma^2 = \sigma_0^2$$
$$H_a: \sigma^2 < \sigma_0^2 \qquad H_a: \sigma^2 > \sigma_0^2 \qquad H_a: \sigma^2 \neq \sigma_0^2$$

   In this particular case, the appropriate set of hypotheses is

$$H_0: \sigma^2 = 0.4$$
$$H_a: \sigma^2 > 0.4.$$

2. **State the test statistic:** Our test statistic is

$$\chi^2 = \frac{(n-1)s^2}{\sigma_0^2}.$$

   The degrees of freedom for this statistic are $n - 1$.

3. **State the critical region:** These are the critical regions:

   - For $H_a: \sigma^2 < \sigma_0^2$, we reject H_0 if $\chi^2 < \chi_{n-1,1-\alpha}^2$.
   - For $H_a: \sigma^2 > \sigma_0^2$, we reject H_0 if $\chi^2 > \chi_{n-1,\alpha}^2$.
   - For $H_a: \sigma^2 \neq \sigma_0^2$, we reject H_0 if $\chi^2 < \chi_{n-1,1-\alpha/2}^2$ or $\chi^2 > \chi_{n-1,\alpha/2}^2$.

   In this case, we reject the null hypothesis if

$$\chi^2 > \chi_{n-1,\alpha/2}^2$$
$$> \chi_{9,0.05}^2$$
$$> 19.02.$$

4. **Conduct the experiment and find t:** These are the actual whiteness measurements for the ten shipments:

24	25	27	25	26
26	24	25	26	25

   For these data,

$$s^2 = 0.9.$$

Thus, our test statistic is

$$\chi^2 = \frac{(n-1)s^2}{\sigma_0^2}$$

$$= \frac{9(0.9)}{0.4}$$

$$= 20.25.$$

5. **Reach conclusions and state them in English:** Since $20.25 > 16.92$, we may reject the null hypothesis. We thus have sufficient evidence to conclude that the variability of the new vendor's product exceeds the stated claim.

   Since we reject the null hypothesis, we need to give a range of plausible values for the true variance. A $(1 - \alpha) \cdot 100\%$ confidence interval for σ^2 is given by

$$\left(\frac{(n-1)s^2}{\chi_{n-1,\alpha/2}^2}, \frac{(n-1)s^2}{\chi_{n-1,1-\alpha/2}^2} \right),$$

which yields an interval of (0.43, 3.00). Thus, the plausible values for the variance range from 0.43 to 3.00. We observe that the claimed value of 0.4 does not fall within this interval, which reinforces our conclusion from the hypothesis test.

As usual, we should check our assumptions. Both the confidence interval and the hypothesis test strongly depend on the assumption that individual whiteness measurements form a random sample from a normal distribution. Exercise 4.63 is to check the reasonableness of this assumption within the context of this example.

Once again, we can completely bypass the formal hypothesis test by constructing the appropriate confidence interval. In this particular situation, one could argue that a one-sided confidence interval is most appropriate. Again, Exercise 4.63 is to derive this interval and conduct the appropriate analysis.

The Ratio of Two-Sample Variances

VOICE OF EXPERIENCE

Please do not test for equal variances prior to the two-sample test since it is as much a test for normality as a test for equality of variance.

In some cases, we need to compare two variances. Consider two different random samples. Let s_1^2 and s_2^2 be the sample variances from the first and second random samples, respectively. Similarly, let σ_1^2 and σ_2^2 be the true population variances for the first and second samples. Finally, let n_1 and n_2 be the respective sample sizes. Once again, we must assume that the individual observations, the y's, for both random samples follow normal distributions.

Figure 4.28 | A Typical *F*-Distribution

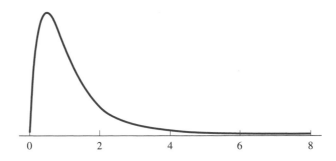

We can show that

$$\frac{s_1^2/\sigma_1^2}{s_2^2/\sigma_2^2} = \left(\frac{s_1^2}{s_2^2}\right)\left(\frac{\sigma_2^2}{\sigma_1^2}\right)$$

follows an *F*-distribution with $n_1 - 1$ numerator degrees of freedom and $n_2 - 1$ denominator degrees of freedom. In many cases, we assume that $\sigma_1^2 = \sigma_2^2$, in which case s_1^2/s_2^2 follows an appropriate *F*-distribution, which will prove useful in Chapter 6.

Figure 4.28 plots the pdf for a typical *F*-distribution. Like the χ^2 distribution, the *F*-distribution is heavily skewed to the right. As the denominator degrees of freedom increase, the *F*-distribution begins to converge to the χ^2 distribution with the same degrees of freedom as the numerator of the *F*-statistic. Table 4 in the appendix gives the critical values for the *F*-distribution.

We can use the *F*-distribution to develop an appropriate hypothesis test. We introduce this procedure through an example.

Example 4.18 | **Pellets from Plastic Suppliers**

A company that manufactures calculators is in the process of selecting from two plastics suppliers. Pellets from fifteen randomly selected batches of plastic from each supplier are pressed into wafers of uniform thickness. The breaking strength is recorded for each wafer. It has already been determined there is no statistical difference between the mean breaking strength of Supplier 1 and Supplier 2. One way to determine which supplier to go with is by cost. Assuming there is no cost difference, another option is to choose the supplier with the smaller variability (more consistent product). The variances for each supplier are $s_1^2 = 48.57$ and $s_2^2 = 6.93$. We shall use a 0.10 significance level.

1. **State hypotheses:**

$$H_0: \sigma_1^2 = \sigma_2^2.$$
$$H_a: \sigma_1^2 \neq \sigma_2^2.$$

2. **State the test statistic:** The test statistic is

$$F = (\text{larger sample variance})/(\text{smaller sample variance})$$

For our example, $F = s_1^2/s_2^2$. The degrees of freedom are $n_1 - 1$ numerator and $n_2 - 1$ denominator.

3. **State the critical region:**

$$\text{We reject } H_0 \text{ if } F > F_{n_1-1, \, n_2-1, \, \alpha/2}.$$

In this case, we reject the null hypothesis if

$$F > F_{n_1-1, \, n_2-1, \, \alpha/2}$$
$$> F_{14,14,0.10/2}$$
$$> F_{14,14,0.05}$$
$$> 2.48.$$

4. **Conduct the experiment and find F:**
   For these data, $s_1^2 = 48.57$ and $s_2^2 = 6.93$. Thus, our test statistic is

$$F = s_1^2/s_2^2 = 48.57/6.93 = 7.01.$$

Most statistical software packages will do this analysis and typically report p-values. From the output below, the p-value is 0.001.

Test for Equal Variances: Supplier 1, Supplier 2

```
          F-Test (normal distribution)
   Test statistic = 7.01,   p-value = 0.001
```

5. **Reach conclusions and state them in English:** Since 7.01 is greater than 2.48, we may reject the null hypothesis. We have sufficient evidence to conclude that the variability of Supplier 1 is greater than the variability of Supplier 2. We could reach the same conclusions using the p-value: Reject the null hypothesis since the p-value $= 0.001 < \alpha = 0.10$.

As usual, we should check our assumptions. The hypothesis test strongly depends on the assumption that the individual breaking strength measurements form a random sample from a normal distribution. Figure 4.29 shows that the normality assumption is valid for these data.

| Figure 4.29 | The Normal Probability Plot of the Breaking Strengths for Suppliers |

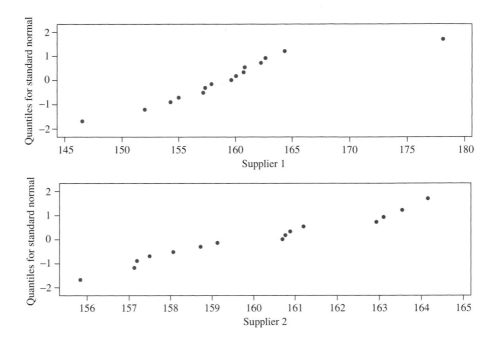

We must caution against testing the variances before conducting a two-sample independent t-test. One of George Box's (1953) more famous quotes is, "To make the preliminary test on variances is rather like putting to sea in a rowing boat to find out whether conditions are sufficiently calm for an ocean liner to leave port!" He points out the test on two variances is as much a test on normality. The two-sample independent t-test is more sensitive to violations of the constant variance assumption when the sample sizes are small. Unfortunately, the test on two variances has problems picking up a difference when the sample sizes are small. On the other hand, the two-sample independent t-test is very robust to the constant variance assumption when the sample sizes are large. However, the two-sample test on variances is likely to pick up very small differences in the variances when the sample sizes are large.

⁖ Exercises

4.57 Albin (1990) studied aluminum contamination in recycled PET plastic from a pilot plant operation at Rutgers University. She collected 26 samples and measured, in parts per million (ppm), the amount of aluminum contamination. The designer of this operation claims that the standard deviation for these concentrations is 60.

Consider these data:

291	222	125	79	145	119	244	118	182
63	30	140	101	102	87	183	60	191
119	511	120	172	70	30	90	115	

a. Test the claim about the variability using a 0.05 significance level.
b. Construct a 95% confidence interval for the true variance of the concentrations.
c. What did you assume to do the analyses? Can you evaluate how well these data meet the assumptions? If yes, determine how comfortable you are with them. If not, explain why.

4.58 DeVor, Chang, and Sutherland (1992, pp. 406–407) discuss the production of polyol, which is reacted with isocynate in a foam molding process. Variations in the moisture content of polyol cause problems in controlling the reaction with isocynate. Production claims that these moisture contents have a population variance of 0.02. The following data represent 27 moisture analyses over a four-month period:

2.29	2.22	1.94	1.90	2.15	2.02	2.15	2.09	2.18
2.00	2.06	2.02	2.15	2.17	2.17	1.90	1.72	1.75
2.12	2.06	2.00	1.98	1.98	2.02	2.14	2.10	2.05

a. Test this claim about the variability using a 0.05 significance level.
b. Construct a 95% confidence interval for the true variance of the moisture contents.
c. What did you assume to do the analyses? Can you evaluate how well these data meet the assumptions? If yes, determine how comfortable you are with them. If not, explain why.

4.59 McNeese and Klein (1991) looked at the average particle size of a product. Specifications suggest that the variance of the particle sizes is 144. Production personnel measured the particle size distribution using a set of screening sieves. They tested one sample a day and found these average particle sizes for the past 25 days:

99.6	92.1	103.8	95.3	101.6
102.8	100.9	100.5	102.7	96.9
101.5	96.7	96.8	97.8	104.7
103.2	97.5	98.3	105.8	100.6
102.3	93.8	102.7	94.9	94.9

a. Test the claim about the variability using a 0.05 significance level.
b. Construct a 95% confidence interval for the true variance of the particle sizes.
c. What did you assume to do the analyses? Can you evaluate how well these data meet the assumptions? If yes, determine how comfortable you are with them. If not, explain why.

4.60 Carbon fibers are usually subjected to an oxidative surface treatment that leads to an improvement in the interface strength when they are incorporated into a polymer mix. Baillie and Bader (1994) investigated the strength of single fibers embedded in resin using fragmentation tests. Ideally, the fragment strengths should have a variance of 1 GPa. Consider the following sample:

4.3	5.0	6.0	6.8	5.5	6.0	7.6	5.7	5.1	6.6

a. Test this criterion about the variability using a 0.01 significance level.
b. Construct a 99% confidence interval for the true variance of the fragment strengths.
c. What did you assume to do the analyses? Can you evaluate how well these data meet the assumptions? If yes, determine how comfortable you are with them. If not, explain why.

4.61 A study to characterize the physical behavior of steel-concrete beams under periodic load was carried out at the National Institute of Standards and Technology. The response variable is deflection (from rest point) of the steel-concrete beam. The variance should be no higher than 70. A sample of nine beams gave

−568	−568	−577	−578	−552	−576	−560	−568

a. Test this criterion about the variability using a 0.05 significance level.
b. Construct a 95% confidence interval for the true variance of the deflections.
c. What did you assume to do the analyses? Can you evaluate how well these data meet the assumptions? If yes, determine how comfortable you are with them. If not, explain why.

4.62 Moselhi and Hassanein (2003) analyzed a 957-m stretch of a recently constructed two-lane highway. A borehole test for silt was conducted. The goal of the study is to show that the variation in the thickness is less than 0.05 m. The data are

0.6	1.1	0.8	0.6	0.4	0.8
0.8	0.7	0.8	0.5	0.4	0.6

a. Test this criterion about the variability using a 0.10 significance level.

b. Construct a 90% confidence interval for the true variance of the thickness.

c. What did you assume to do the analyses? Can you evaluate how well these data meet the assumptions? If yes, determine how comfortable you are with them. If not, explain why.

4.63   Consider the Whiteness of Titanium Dioxide data given in Example 4.17.

a. Derive the appropriate one-sided confidence interval and conduct the appropriate analysis.

b. What did you assume to do the analysis? Evaluate how well the measurements meet these assumptions.

4.64   Ismail, Sobieh, and Abdel-Fattah (1996) investigated the production of cyclodextrin glucosyltransferase enzymes by bacterial cultures. Enzyme production was done in shaken and surface cultures. An important question is whether the two methods have equal variances. The protein content of the cultures in mg ml^{-1} is

Shaken	1.91	1.66	2.64	2.62	2.57	1.85
Surface	1.71	1.57	2.51	2.30	2.25	1.15

a. Conduct the appropriate hypothesis test using a 0.10 significance level.

b. What did you assume to do the analyses? Can you evaluate how well these data meet the assumptions? If yes, determine how comfortable you are with them. If not, explain why.

4.65   McGrath et al. (1995) describe materials prepared by selective functionalization of olefin-containing polymers to produce novel EPDM and polybutadiene polyols. The properties of the final urethane can be modified by the addition of other agents, such as 1,4-butanediol. One question of interest is how the variation is affected by increasing the ratio of 1,4-butanediol to ployol. The ultimate strength in psi is given:

Ratio-1:1	513	1415	619	1699
Ratio-1:2	1278	2528	758	2332

a. Conduct the appropriate hypothesis test using a 0.10 significance level.

b. What did you assume to do the analyses? Can you evaluate how well these data meet the assumptions? If yes, determine how comfortable you are with them. If not, explain why.

4.66   In Aktan and Khan (1992), spaghetti made from durum wheats with strong Vic gluten was dried at a high temperature of 80° C and a low temperature of 40° C. Of interest was whether the variability in the retention times was the same at the two drying temperatures. A sample of times yielded

40° C						80° C					
28.68	19.76	20.40	22.96	22.31	25.27	25.07	28.44	30.37	23.55	22.75	22.11
24.29	23.75	26.68	30.69	19.13	27.66	24.09	18.97	26.48	27.39	19.62	20.22

a. Conduct the appropriate hypothesis test using a 0.10 significance level.
b. What did you assume to do the analyses? Can you evaluate how well these data meet the assumptions? If yes, determine how comfortable you are with them. If not, explain why.

4.67 Tests of product quality using inspectors can lead to serious error problems. Benson and Ohta (1986) evaluated the performance of inspectors in a new company using novice and experienced inspectors. Each inspector classified 200 products as defective or nondefective. The number of errors was

Novice						Experienced					
25	35	26	40	46	20	31	15	25	19	28	17
45	31	33	29	21	49	19	18	24	10	20	21

a. Conduct the appropriate hypothesis test about the variability of the novice versus experienced inspectors using a 0.10 significance level.
b. What did you assume to do the analyses? Can you evaluate how well these data meet the assumptions? If yes, determine how comfortable you are with them. If not, explain why.

4.68 Kishimoto, Sakasai, and Ara (1996) looked at new methods to reconstruct plasma shape and plasma current distribution from magnetic measurements. A question of interest is whether the variation in error (calculated value − measured value) is the same for magnetic fields and flux fields using a combined method of genetic algorithms and neural networks. A sample of errors follows:

Magnetic Fields				Flux Fields			
−0.054	−0.019	−0.149	−0.088	0.003	−0.432	0.017	−0.207
−0.136	−0.155	−0.155	−0.125	−0.255	−0.109	0.153	−0.791
−0.162	−0.313	−0.027	−0.057	0.005	−0.984	0.564	−0.027

a. Conduct the appropriate hypothesis test about the variability in error between magnetic field and flux field using a 0.10 significance level.
b. What did you assume to do the analyses? Can you evaluate how well these data meet the assumptions? If yes, determine how comfortable you are with them. If not, explain why.

4.69   A diagnostic study was done on Ligustrum lucidum in relation to atmospheric pollutants in Cordoba City, Argentina, by Pignata et al. (1997). The study area receives regional pollutants and was categorized by traffic level. The level of pollutants was recorded for two traffic levels

Traffic=2	3.17	3.10	0.73	1.80	3.03	3.06
Traffic=3	1.20	3.17	2.87	2.10	2.13	2.60

a.   Conduct the appropriate hypothesis test about the pollutant variability of the traffic levels using a 0.10 significance level.

b.   What did you assume to do the analyses? Can you evaluate how well these data meet the assumptions? If yes, determine how comfortable you are with them. If not, explain why.

⋗ 4.8 Transformations and Nonparametric Analyses

In this chapter, we talked a great deal about checking our assumptions. A legitimate question is: What should we do if our assumptions are grossly violated? The *t*-test is quite robust to the normality assumption, but too often we encounter data that follow very poorly behaved distributions, especially when we deal with lifetime or reliability data.

VOICE OF EXPERIENCE

Nonparametric procedures are less powerful and more sensitive to the independence assumption.

Two common approaches for dealing with such data are using transformations and nonparametric analyses. When we transform the data, we replace them with a mathematical function of the data. In Chapter 6, we shall illustrate their use. Nonparametric analyses attempt to make the least restrictive assumptions possible about the parent distribution of the data. These procedures often replace the data with their ranks. Standard statistical software packages make it easy for the analyst to use either approach when the situation warrants.

Sometimes the data are not well behaved or do not follow a normal distribution. This occurs with reliability data and in manufacturing with only a one-sided specification limit. It may be possible in these settings to transform the data by a mathematical function such that the assumptions do hold for the transformed data. Basically, a transformation changes the scale of the data. Engineering data are often well modeled by differential equations. Thus, the log is a typical transformation since it is well founded in engineering theory. Other common transformations for engineering data include square root and inverse.

Example 4.19 | **Aluminum Contamination**

Albin (1990) studied aluminum contamination in recycled PET plastic from a pilot plant operation at Rutgers University. She collected 26 samples and measured, in

Table 4.5	The Aluminum Contamination Data

291	222	125	79	145	119	244	118	182
63	30	140	101	102	87	183	60	191
119	511	120	172	70	30	90	115	

parts per million (ppm), the amount of aluminum contamination. The maximum acceptable level of aluminum contamination, on average, is 160. The data are given in Table 4.5. Notice that there are several high-contamination observations, including one exceeding 500.

The output from a one-sample t-test of $H_0: \mu = 160$ versus $H_a: \mu < 160$ is shown in Table 4.6. Using a significance level of 5%, the conclusion is to fail to reject the null hypothesis; in other words, it appears that the contamination level is unacceptable. However, this conclusion is dependent on the assumption of normality, which may not hold in this example because of possible outliers on the high side. The normal probability shown in Figure 4.30 indicates that it may not be safe to assume that the aluminum contamination readings follow a normal distribution.

The data in this example can be modeled by a lognormal distribution, so it makes sense to try a log transformation. The hypothesis on the natural log scale would be $H_0: \mu = 5.075 = \ln(160)$ versus $H_a: \mu < 5.075$. The output for the test on the natural log scale is given in Table 4.7. Using a significance level of 5%, the conclusion is to reject the null hypothesis, in other words, it appears that the contamination level is acceptable. This reversal of the results will not always occur, but in this example it illustrates how serious violations of assumptions can lead to incorrect interpretations of the data.

A normal probability plot for the natural log of the contamination numbers is given in Figure 4.31. The assumption of normality is more reasonable for the transformed data. Brevity is not meant to belittle the importance of transformations; we will see uses in Chapter 6.

An alternative to transforming the data is to use a procedure that does not make any formal assumption about the distribution of the data. These procedures are called nonparametric. (We often prefer to use a transformation because it will result in a more powerful statistical analysis.) Nonparametric tests use the probability distributions of the sampled populations rather than specific parameters from these populations (such as the mean). The methods use the relative ranks of the

Table 4.6	MINITAB Output for the Aluminum Contamination Data

Variable	N	Mean	StDev	SE Mean	T	P
Contamination	26	142.654	98.204	19.259	-0.90	0.188

Figure 4.30 | The Normal Probability Plot for the Aluminum Contamination Data

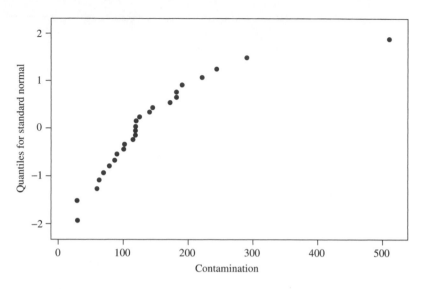

data instead of the actual observation values. In the case of the one-sample t-test, the alternative nonparametric test is called the **sign test.** Essentially, we count the number of observations that "support" the alternative hypothesis. If the alternative hypothesis is true, we expect significantly more observations to support the alternative (fewer than 160 in our example).

Computer packages can be used to count the observations and calculate a p-value, which can then be compared to a significance level. Table 4.8 shows the appropriate output from a sign test for the contamination data. Using a significance level of 5%, the conclusion is to reject the null hypothesis; in other words, it appears that the contamination level is acceptable. Note that the p-value for the nonparametric test, 0.0378, is greater than the p-value for the t-test on the transformed data, 0.011. This is because using the actual observation values (as is or transformed) is more powerful than using a nonparametric procedure.

When the data are skewed (for example, lifetimes of circuits), the mean will be drawn toward the long tail of the distribution and can misrepresent the center. In these cases, it is more accurate to use the median as the estimator of μ since it is less affected by outliers. Section 4.5 discussed the comparison of means from

Table 4.7 | MINITAB Output for the Transformed Aluminum Contamination Data

Variable	N	Mean	St Dev	SE Mean	T	P
Ln (contamination)	26	4.77286	0.63144	0.12384	−2.44	0.011

Figure 4.31 | The Normal Probability Plot for the Transformed Aluminum Contamination Data

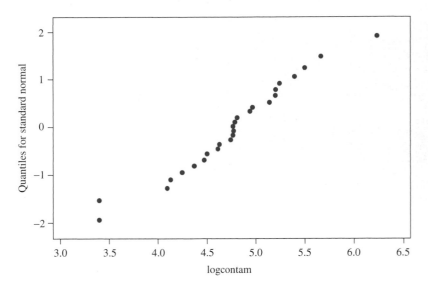

two distinct populations using a two-sample *t*-test. The analogous nonparametric procedure is called the **Wilcoxon rank sum test.** The null hypothesis is that the medians represented by η are equal, $H_0: \eta_1 = \eta_2$. The general idea is to treat the two samples as one big sample and rank the data from smallest to largest. Then, add up the ranks for each sample separately. If there is a difference between the two populations, then one of the samples should have most of the high ranks and thus a higher sum of ranks. The *p*-value can be used to determine whether the difference is statistically significant.

Consider a study to determine if a difference exists in the precipitation for two U.S. Geological Survey stations in Colorado. The data in Table 4.9, taken from Gastwirth and Mahmoud (1986), represent precipitation in inches. We note the data from Station 2 do not represent a normal distribution. The data have been ranked, and the ranks for the two stations are relatively balanced. Table 4.10 gives the *p*-value as 0.9616; therefore, the null hypothesis of equal precipitation between the stations cannot be rejected.

For paired data, the **signed rank test** can be applied to the differences. When data are not well behaved, transformations and nonparametric procedures can be useful tools for analyzing the data.

Table 4.8 | MINITAB Output for the Aluminum Contamination Data Using Nonparametrics

	N	Below	Equal	Above	P	Median
Contamination	26	18	0	8	0.0378	119.0

Table 4.9	The Precipitation Data

Station 1	Rank	Station 2	Rank
127.96	9	114.79	7
210.37	13	109.11	6
203.24	12	330.33	17
108.91	5	85.54	1
178.21	11	117.64	8
285.37	15	302.74	16
100.85	4	280.55	14
85.89	2	145.11	10
		95.36	3
	Sum = 71		Sum = 82

Table 4.10	MINITAB Output for the Precipitation Data Using the Wilcoxon Test

Wilcoxon Rank Sum Test and CI: Station 1, Station 2

```
            N   Median
Station 1   8   153.1
Station 2   9   117.6


95.1 Percent CI for ETA1-ETA2 is (-119.9, 88.5)
W = 71.0
Test of ETA1 = ETA2 vs ETA1 not = ETA2,  P-value = 0.9616
```

⠿ Exercises

4.70 Two types of electrical hazards are electrical shock and electrical arc/flash burn. Electrical shock requires that a person have contact with the exposed energized electrical conductor or be within the air flashover (breakdown) distance of the conductor. Curable burn distance is defined as the distance of a person from an arc source for a just curable burn—where the skin temperature stays less than 80°C. Jamil, Jones and McClung (1997) investigated whether the curable burn distance is less than 2.5 feet at 480 volts. A sample of ordered distances from their work is given as

1.14	1.31	1.70	1.86
1.86	1.99	2.15	4.16

a. Check the data for normality.
b. Use appropriate software to conduct the appropriate nonparametric hypothesis test using a 0.05 significance level.
c. The p-value for the t-test is 0.094. Would your conclusion be the same as in part **b**?
d. Try several common transformations for engineering data (square root, inverse, natural log) and determine if they are useful in transforming the data to normal.
e. If a successful transformation was found in part **d**, conduct the appropriate t-test on the transformed data using a 0.05 significance level. Does your conclusion differ from that in part **b**?

4.71 Lawless (1982) gives failure times in millions of revolutions from fatigue endurance tests for deep-groove ball bearings from a major supplier. A quality engineer believes the bearings are lasting less than 80 million revolutions, which contradicts the supplier's claim. An ordered subset of the data is given:

28.92	33.00	41.52	42.12	45.60	48.40	51.84	51.96	54.12	55.56	67.80
68.64	68.64	68.88	84.12	93.12	98.64	105.12	105.84	127.92	173.40	

a. Check the data for normality.
b. Use appropriate software to conduct the appropriate nonparametric hypothesis test using a 0.10 significance level.
c. The p-value for the t-test is 0.106. Would your conclusion be the same as in part **b**?
d. Try several common transformations for engineering data (square root, inverse, natural log) and determine if they are useful in transforming the data to normal.
e. If a successful transformation was found in part **d**, conduct the appropriate t-test on the transformed data using a 0.10 significance level. Does your conclusion differ from that in part **b**?

4.72 Watanabe and Terami (1989) conducted a study of excessive transitory migration of guppy populations; 40 adult female guppies were placed in the left compartment of an experimental aquarium tank which was divided in half by a glass plate. After the plate was removed, the number of fish passing through the slit from the left compartment to the right one, and vice versa, were monitored every minute for 30 minutes. If an equilibrium were reached, the researchers would expect the median number of fish remaining in the left compartment to be 20.

The data are

16	11	12	15	14	16	18	15	13	15
14	14	16	13	17	27	14	22	18	19
17	28	20	23	29	19	21	17	21	17

a. Check the data for normality.
b. Use appropriate software to conduct the appropriate nonparametric hypothesis test using a 0.05 significance level.
c. Try several common transformations for engineering data (square root, inverse, natural log) and determine if they are useful in transforming the data to normal.
d. If a successful transformation was found in part **c**, conduct the appropriate *t*-test on the transformed data using a 0.05 significance level. Does you conclusion differ from that in part **b**?

4.73   Consider a study on the properties of striated muscles in crayfish. The experiment was conducted to compare the biochemical properties of fast and slow muscles. The researchers excised fast-muscle fibers from 12 crayfish and tested for uptake of calcium (moles per milligram). They did the same for slow-muscle fibers from 12 different crayfish. The data are

Fast Muscle				Slow Muscle			
0.39366	0.36993	0.00065	0.49406	0.91170	0.02194	0.67700	0.79373
0.06434	0.23915	1.87403	0.14753	0.08673	1.07372	1.19230	0.11252
0.00114	1.22716	0.00577	1.45579	0.00009	0.02544	0.11024	0.00155

a. Check the data for normality.
b. Use appropriate software to conduct the appropriate nonparametric hypothesis test using a 0.10 significance level.

4.74   Two types of electrical hazards are electrical shock and electrical arc/flash burn. Electrical shock requires that a person have contact with the exposed energized electrical conductor or be within the air flashover (breakdown) distance of the conductor. Curable burn distance is defined as the distance of a person from an arc source for a just curable burn—where the skin temperature stays less than 80° C. Jamil, Jones, and McClung (1997) investigated whether the curable burn distance is different at 208 volts and 480 volts. A sample of ordered distances from their work is given as

208 volts				480 volts			
1.14	1.31	1.70	1.86	0.71	0.86	1.15	1.22
1.86	1.99	2.15	4.16	1.41	1.52	1.63	1.99

a. Check the data for normality.
b. Use appropriate software to conduct the appropriate nonparametric hypothesis test using a 0.05 significance level.

4.75   McCool (1980) gives the results of a life test on rolling contact fatigue of ceramic ball bearings. Twenty specimens were randomly split into two groups with ten tested at a stress of 0.87 and ten tested at a stress of 0.99. The ordered failure times are

Stress = 0.87					Stress = 0.99				
1.67	2.20	2.51	3.00	3.90	0.80	1.00	1.37	2.25	2.95
4.70	7.53	14.70	27.80	37.40	3.70	6.07	6.65	7.05	7.37

a. Check the data for normality.
b. Conduct the appropriate nonparametric hypothesis test using a 0.10 significance level.

4.76   A study was carried out on the calibration of ozone monitors. Consider two measurement devices used to measure the ozone concentration. Thirteen locations are chosen and measured by each device:

Device 1	338.8	118.1	888.0	228.1	668.5	9.2	0.3
Device 2	337.4	118.2	884.6	226.5	666.3	10.1	0.4

Device 1	0.1	10.8	119.6	449.2	339.3	668.4
Device 2	0.6	11.6	120.2	448.9	339.1	669.1

a. Check the differences for normality.
b. Use appropriate software to conduct the appropriate nonparametric hypothesis test using a 0.10 significance level.

4.77   Vrca et al. (1997) evaluated eleven prisoners of war in Croatia for neurological impairment after their release from a Serbian detention camp. All 11 released POWs received blows to the head and neck during imprisonment. Neurological impairment was assessed by measuring the amplitude of the visual evoked potential (VEP) in both eyes at two points in time: 157 days and 379 days after release from prison with higher VEP implying greater neurological impairment.

157 Days	2.46	4.09	3.93	4.51	4.99	4.42	1.14	4.30	7.56	7.07	7.98
379 Days	3.73	5.46	7.04	4.73	4.71	6.08	1.42	8.70	7.37	8.46	7.16

a. Check the differences for normality.
b. Conduct the appropriate *t*-test using a 0.05 significance level.
c. Use appropriate software to conduct the appropriate nonparametric hypothesis test using a 0.05 significance level.

4.78 Lin and Cole (1997) describe the development of the dynamic model for the Global Positioning System Block IIR space vehicle. The test-derived spacecraft and solar array assembly models were coupled using the component modal synthesis method to form the system dynamic model to be used for prediction. The frequencies for thirteen nodes were measured and also predicted from the dynamic model. The differences (predicted − measured) for the nodes are

0.0	0.6	0.2	4.2	−1.5	−1.6	−1.8
0.5	0.1	0.4	1.2	0.8	1.0	

a. Check the differences for normality.
b. Use appropriate software to conduct the appropriate nonparametric hypothesis test using a 0.05 significance level.

4.9 Case Study

Statistics plays a vital role in market field tests for new or revised products. Full field tests are extremely expensive and require proper planning. Part of the planning process includes a pretest, which is useful for these reasons:

- It determines whether the test instrument (usually a questionnaire) is appropriate.
- It provides an estimate of the experimental error in order to calculate an appropriate sample size.
- It confirms that the new or revised product really should be fully tested in the field.

This last point is important. Research and development (R&D) may feel that the new product is clearly superior, but they may lack objectivity. The pretest can serve as a reality check before marketing expends major resources on the full field test.

The Marketing Department of Faber-Castell had determined that Faber's basic ballpoint pen needed to be revised. R&D produced a lot of a new prototype ballpoint pen. Ten individuals *working at the production facility* were asked to evaluate the overall quality of the prototype compared with the current product and with several competitors' pens. For simplicity, we restrict our attention to the new prototype and Faber's leading competitor's pen. Since we expect significant differences in preferences from individual to individual, we should use a paired data approach. Thus, each individual rated the prototype and the competitor's pen on a scale of 1–10, with 1 being extremely poor and 10 being excellent. The actual ratings are given in Table 4.11.

Figure 4.32 shows the stem-and-leaf display of the results. This plot is not quite bell-shaped but close enough for us to proceed with the analysis. The center

Table 4.11	Ballpoint Pen Marketing Data

Individual	Competitor	Prototype	Difference
1	7	8	−1
2	6	7	−1
3	8	9	−1
4	10	8	2
5	2	9	−7
6	5	5	0
7	6	6	0
8	6	8	−2
9	4	10	−6
10	6	9	−3

appears to be just below zero, which indicates a possible preference for the prototype. Figure 4.33 gives the boxplot of the results. Again, the typical value seems to be slightly negative. We do not see any outliers. Figure 4.34 gives the normal probability plot of these differences. This plot, which is typical when dealing with *ordinal* data, suggests that the data may not come from a particularly well-behaved distribution. We may want to consider the equivalent nonparametric test. In this case, the appropriate nonparametric analysis test gives the same conclusion, so we shall concentrate on the standard paired *t*-test.

Since R&D really wishes to establish that the prototype is better than the competitor's pen, these are the appropriate hypotheses:

$$H_0: \quad \delta = 0$$
$$H_a: \quad \delta < 0.$$

Figure 4.32	The Stem-and-Leaf Display for the Pen Pretest Differences

Stem	Leaves	Number	Depth
−s:	67	2	2
−f:			
−t:	23	2	4
−*:	111	3	
*:	00	2	3
t:	2	1	1

Figure 4.33 | The Boxplot for the Pen Pretest Differences

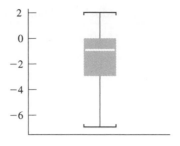

The test statistic is

$$t = \frac{\bar{d} - \delta_0}{s_d/\sqrt{n}}.$$

For a pretest, we should use a 0.05 significance level. Thus, we reject the null hypothesis if

$$t < -t_{n-1,\alpha}$$
$$< -t_{9,0.05}$$
$$< -1.833.$$

For these data, we have

$$\bar{d} = -1.9 \quad \text{and} \quad s_d = 2.77.$$

Figure 4.34 | The Normal Probability Plot for the Pen Pretest Differences

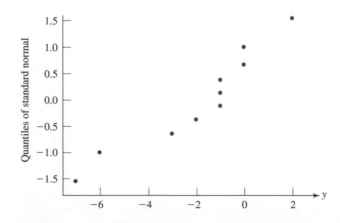

Thus, our test statistic is

$$t = \frac{\bar{d}}{s_d/\sqrt{n}}$$

$$= \frac{-1.9}{2.77/\sqrt{10}}$$

$$= -2.17.$$

We therefore have sufficient evidence to reject the null hypothesis. As a result, we have evidence to suggest that people really do prefer the prototype. Of course, one must remember that the people used in this study all worked at the production facility. Consequently, we may conclude only that people who work for Faber-Castell actually prefer the prototype. Nonetheless, the pretest has served as a reality check for R&D. They do have a reasonable basis for proceeding to the full field test where marketing can get a better sense of how the prototype should fare against the competition.

We can construct a 95% confidence interval for the true difference:

$$\bar{d} \pm t_{n-1,\alpha/2} \frac{s_d}{\sqrt{n}} = -1.9 \pm t_{9,0.025} \frac{2.77}{\sqrt{10}}$$

$$= -1.9 \pm 2.262(0.88)$$

$$= -1.9 \pm 1.98$$

$$= (-3.88, 0.08).$$

The plausible values for this difference range from -3.88 to 0.08, which seems to contradict the results of our hypothesis test. We must keep in mind that we conducted a *one-sided* hypothesis test; however, our confidence interval is *two-sided*. In this case, we have sufficient evidence to reject the null hypothesis in favor of the appropriate one-sided alternative but not enough evidence to reject in favor of the two-sided alternative.

⦂ 4.10 Ideas for Projects

1. Get data from a single section of a laboratory class where each student or several teams of students conduct experiments that all should achieve some target value. For example, in organic chemistry laboratory classes, many students perform the same organic reaction experiment and record their yields. Theoretically, the basic chemical reaction should produce a specific yield. Perform an appropriate analysis to test the class results relative to the theoretical value. Discuss whether the appropriate alternative hypothesis

should be one- or two-sided. Also, discuss differences and any interesting features. Other good sources are surveying classes and unit operations laboratories.

2. If data from two sections of the laboratory class are available, repeat the first project idea, testing whether there are any significant differences between the two sections. Discuss whether the data should be paired.

3. Compare the class performance on two different homework assignments or two different examinations. If there are significant differences, come up with a reasonable explanation. Discuss whether the data should be paired.

4. Collect data that will address some personal question of interest that requires the comparison of two groups. Perform the appropriate analysis. For example, you may wish to compare your typical daily expenses with those of a close friend. For the next two weeks, you and your friend should record all of your expenses. If you perform a two-group analysis, discuss whether the data should be paired.

⠋ *References*

1. Aktan, B., and Khan, K. (1992). Effect of high-temperature drying pasta on quality parameters and on solubility, gel electrophoresis, and reversed-phase high-performance liquid chromatography of protein components. *Cereal Chemistry, 69*, 288–295.

2. Albin, S. L. (1990). The lognormal distribution for modeling quality data when the mean is near zero. *Journal of Quality Technology, 22*, 105–110.

3. Anagonye, A., and Stephenson, D. A. (2002). Modeling cutting temperatures for turning inserts with various tool geometries and materials. *Journal of Manufacturing Science and Engineering, 124*, 544–552.

4. Baillie, C. A., and Bader, M. G. (1994). Strength studies of single carbon fibers in model composite fragmentation tests. *Composites 25*, 401–406.

5. Bates, D. M., and Watts, D. G. (1988). *Nonlinear regression analysis and its applications.* New York: John Wiley.

6. Benson, P. G., and Ohta, H. (1986). Classifying sensory inspectors with heterogenous inspection-error probabilities. *Journal of Quality Technology, 18*, 79–90.

7. Box, G. E. P. (1953). Non-normality and tests on variances. *Biomatrics, 40*, 318–335.

8. Calik, G., Vural, H., and Özdamar, T. H. (1997). Bioprocess parameters and oxygen transfer effects in the growth of Pseudomonas dacunhae for L-alanine production. *Chemical Engineering Journal, 65*, 109–116.

9. Capurro, M., Brozzo, P., and Cirillo, P. (1997). Statistical Aspects of cleavage resistance of high strength microalloyed structural steel. *The Institute of Materials*, 645–649.

10. DeVor, R. E., Chang, T., and Sutherland, J. W. (1992). *Statistical quality design and control: Contemporary concepts and methods.* New York: Macmillan.

11. Dunn, G. (1989). *Design and analysis of reliability data.* London: Edward Arnold.

12. Eibl, S., Kess, U., and Pukelsheim, F. (1992). Achieving a target value for a manufacturing process: A case study. *Journal of Quality Technology, 24*, 22–26.

13. Farnum, N. R. (1994). *Modern statistical quality control and improvement.* Pacific Grove, CA: Duxbury Press.

14. Galinsky, T. L., Rosa, R. R., Warm, J. S., and Dember, W. W. (1993). Psychophysical determinants of stress in sustained attention. *Human Factors, 35*, 603–614.

15. Gastwirth, J. L., and Mahmoud, H. (1986). An efficient robust nonparametric test for scale change data from a gamma distribution. *Technometrics, 28*, 83.

16. Grubbs, F. E. (1983). Grubbs' estimators (precision and accuracy of measurement). *Encyclopedia of statistical sciences*, Vol. 3 (S. Kotz and N. L. Johnson, Eds.). New York: John Wiley.

17. Holmes, D. S., and Mergen, A. E. (1992). Parabolic control limits for the exponentially weighted moving average control charts. *Quality Engineering, 4*, 487–495.

18. Ismail, A. S., Sobieh, U. I., and Abdel-Fattah, A. F. (1996). Biosynthesis of cyclodextrin glucosyltransferase and β-cyclodextrin by *Bacillus macerans* 314 and properties of the crude enzyme. *The Chemical Engineering Journal, 61*, 247–253.

19. Jamil, S., Jones, R. A., and McClung, L. B. (1997). Arc and flash burn hazards at various levels of an electrical system. *IEEE Transactions of Industry Applications, 33*, 359–366.

20. Kane, V. E. (1986). Process capability indices. *Journal of Quality Technology, 18*, 41–52.

21. Kishimoto, M., Sakasai, K., and Ara, K. (1996). Reconstruction of plasma current profile of Tokamaks using combinatorial optimization techniques. *IEEE Transactions on Plasma Science, 24*, 528–538.

22. Lawless, J. F. (1982). *Statistical models and methods for lifetime data*. New York: John Wiley.

23. Lin, C. S., and Cole, T. R. (1997). Dynamic model for global positioning system block IIR space vehicle. *Journal of Spacecraft and Rockets, 34*, 354–359.

24. Marcucci, M. (1985). Monitoring multinomial processes. *Journal of Quality Technology, 17*, 86–91.

25. Maxcy, R. B., and Lowry, S. R. (1984). Evaluating variability of filling operations. *Food Technology, 38* (12), 51–55.

26. McCool, J. L. (1980). Confidence limits for Weibull regression with censored data. *IEEE Transactions on Reliability, R-29*, 145–150.

27. McGrath et al. (1995). Novel polymeric alcohols by controlled catalytic polymer functionalization. *Journal of Applied Polymer Science, 56*, 533–543.

28. McNeese, W. H., and Klein, R. A. (1991). Measurement systems, sampling, and process capability. *Quality Engineering, 4*, 21–39.

29. Mee, R. W. (1990). An improved procedure for screening based on a correlated, normally distributed variable. *Technometrics, 32*, 331–337.

30. Montgomery, D. C. (1991). *Introduction to quality control*, 2nd ed. New York: John Wiley.

31. Moselhi, O., and Hassanein, A. (2003). Data acquisition and analysis for highway construction using geographic information systems. *Canadian Journal of Civil Engineering, 30*, 533–542.

32. Nelson, P. R. (1989). Multiple comparisons of means using simultaneous confidence intervals. *Journal of Quality Technology, 21*, 232–241.

33. The Open University (1984). *MDST242 Statistics in Society, Unit B5: Review*, 2nd ed. Milton Keynes: The Open University.

34. Padgett, W. J., and Spurrier, J. D. (1990). Shewhart-type charts for percentiles of strength distributions. *Journal of Quality Technology, 22*, 283–288.

35. Penner, R., and Watts, D. G. (1991). Missing information. *American Statistician, 45*, 4–9.

36. Pignata, M. L., Canas, M. S., Carreras, A., and Orellana, L. (1997). Exploring chemical variables in Ligustrum lucidum Ait. F. Tricolor (Rehd.) Rehd. in relation to air pollutants and environmental conditions. *Environmental Management, 21*, 793–801.

37. Rajniak, P., and Yang, R. T. (1994). Hysteresis-Dependent adsorption–desorption cycles: generalization for isothermal conditions. *AIChE Journal, 40*, 913–924.

38. Roes, K. C. B., and Does, R. J. M. M. (1995). Shewhart-type charts in nonstandard conditions. *Technometrics, 37*, 15–40 (with discussion).

39. Runger, G. C., and Pignatiello, J. J., Jr. (1991). Adaptive sampling for process control. *Journal of Quality Technology, 23*, 135–155.

40. Snee, R. D. (1981). Developing blending models for gasoline and other mixtures. *Technometrics, 23*, 119–130.

41. Van Nuland, Y. (1992). Maintaining calibration control with a control chart. *Quality Progress, 25* (3), 152.

42. Wasserman, G. S., and Wadsworth, H. M. (1989). A modified Beattie procedure for process modeling. *Technometrics, 31*, 415–421.

43. Watanabe, M., and Terami, H. (1989). Excessive transitory migration of guppy populations. II. Analysis of possible conspecific-following tendency. *Zoological Science, 6* (3), 573–578.

44. Weaver, W. R. (1990). The foreman's view of quality control. *Quality Engineering, 3*, 257–280.

45. Williams, A. A. (1994). The turbine performance of centrifugal pumps: a comparison of prediction methods. *Proceedings of the Institution of Mechanical Engineers Part A: Journal of Power and Energy.* London: Edward Arnold, 64.

46. Yashchin, E. (1992). Analysis of CUSUM and other Markov-type control schemes by using empirical distributions. *Technometrics, 34*, 54–63.

47. Yashchin, E. (1995). Estimating the current mean of a process subject to abrupt changes. *Technometrics, 37*, 311–323.

48. Vrca, A. et al. (1997). The use of visual evoked potentials to follow-up prisoners of war after release from detention camps. *Collegian Anthropologicum, 21*, p. 232.

Control Charts

> ## 5.1 Overview

Process Changes over Time

Industry has come to realize that to compete in a global market, it must produce *high-quality* goods and services as *efficiently* as possible. The active use of statistics plays an important role in the pursuit of these goals. In this chapter, we study how we can *monitor* an engineering process to ensure that important characteristics of interest remain as close to specified target values as possible. Here are examples from industry:

VOICE OF EXPERIENCE

All engineering process change over time.

- An industrial engineer needs to monitor the process for filling 8-oz cartons of milk.
- An electrical engineer needs to maintain the thickness of silicon wafers at 244 μm.
- A chemical engineer needs to keep the molecular weight of a polymer at a specified value.

Consider the manufacture of pen barrels. One important characteristic is the outside diameter at the point where the inside diameter of the cap forms a seal. Let y_i be the outside diameter of the ith barrel produced at some time t, and let $\mu(t)$ be the *process mean*, which is the true mean outside diameter at this time. An appropriate model is

$$y_i = \mu(t) + \epsilon_i,$$

where ϵ_i is random error associated with this barrel. For the moment, we shall assume that the *process variance*, σ^2, which is a measure of the variability among the random errors, is constant. Let μ_0 be the desired or "target" value. If $\mu(t) = \mu_0$, then we have achieved the desired diameter for the fit.

An injection molding process produces these pen barrels. Some of the factors associated with this process that control the outside diameter are the injection temperature, injection pressure, cooling temperature, and the formulation. A change in any of these factors leads to a change in the true mean outside diameter,

| Figure 5.1 | Typical Drift in a Process Mean over Time |

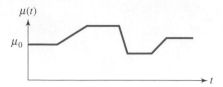

and these factors most assuredly do change over time. Figure 5.1 illustrates how the true mean outside diameter can drift over time. In this case, the injection molding process starts out producing barrels that have the desired mean outside diameter. Over time, the processing conditions change, which causes this mean to drift. As the mean shifts, the fit between the barrel and the cap deteriorates, becoming either too tight or too loose.

Dealing with Process Changes

Engineers must consider the most appropriate response to changes in the process mean. Five possible options are

1. Follow the philosophy *caveat emptor* ("let the buyer beware"), where management does little or nothing about these changes. Sometimes manufacturers find this strategy quite successful *in the short term* when demand far exceeds supply. In such a period, one can sell even poor-quality goods. Eventually, however, supply and demand equilibrate. Quality becomes a determining factor in purchasing decisions, and a reputation for poor quality lasts for a long time.
2. Inspect every pen barrel and "cull out" the unacceptable (100% inspection). This strategy requires a large amount of resources to detect problems and then to correct them. However, this strategy does minimize the chances of unacceptable product reaching the consumer. It is interesting that 100% inspection is rarely, if ever, 100% effective.
3. Produce large "lots" (batches of pen barrels) and use statistics to determine whether the entire lot is unacceptable (acceptance sampling). This strategy uses statistics as an *enumerative tool* to minimize the resources required to detect problems. The manufacturer accepts a certain risk that unacceptable product will slip through the system; however, this risk is well defined and determined to be reasonable. This strategy does not control the resources required to correct problems.
4. Constantly monitor the true mean diameter and adjust the process whenever it begins to drift away from the target condition (statistical process control). This strategy requires sampling from the process on a regular basis and uses statistics as an *analytic tool* to detect problems before

they can do serious damage to the overall quality of the product. Statistical process control seeks to make the product "right the first time." In so doing, we begin to reduce the resources required to correct problems.

5. Create processes that cannot produce unacceptable product. This strategy uses statistics to design efficient experiments within a continuous improvement philosophy to drive the process to an ideal state.

All manufacturing firms, at one time or another, have exercised one or more of these options. In areas of true competition, especially on a global scale, many manufacturers have learned that

- Engineers must detect when processes begin to drift away from the target quality as soon as possible by actively monitoring them through *control charts.*
- Engineers must design processes that minimize the chances of producing unacceptable product.

Processes In or Out of Control

Many practitioners define a process to be *in control* if it is stable, predictable, and repeatable. In some cases, we deal with processes that are easy to adjust back to their target values when they stray. In other applications, we may have an ideal value for our characteristic, but we cannot adjust the process well enough to hit this value precisely. We then assure ourselves that the process is set close enough to this ideal, and we let the process establish its own natural target. In either case, the appropriate goal of a sound monitoring scheme is to keep the process mean, $\mu(t)$, as close as possible to this target value, which we call μ_0. When the process mean is at this value, or $\mu(t) = \mu_0$, we say that the process is *in control,* and we mean that any deviation in the characteristic of interest from the target condition is purely due to random chance. Any attempt to change the process mean, which some people call *tinkering,* will actually introduce more variability into the process and thus do more harm than good.

In some cases, particularly in the early stages of process control, the target value, μ_0, may not be "ideal" for our characteristic of interest. The first step of any sound quality improvement program is to get the process in control—that is, stable around some known value. Later, we can use *experimental design* techniques to move the process mean to its ideal value.

Traditionally, statistical process control assumes that a process remains in control until acted upon by some outside force, called an *assignable cause.* This cause shifts $\mu(t)$ to some value $\mu_1 \neq \mu_0$, and then we say that the process is *out of control.* Figure 5.2 illustrates this shift. Statistical process control assumes that the process mean remains at the new value, μ_1, until someone takes appropriate corrective action. The purpose of statistical process control is to signal when a process shifts out of control. Someone then must investigate the cause of the shift and correct it.

| Figure 5.2 | The Form of a Process Shift Assumed by Standard Statistical Process Control |

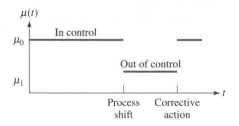

The Basic Idea of a Control Chart

Engineers use *control charts* to monitor processes. Some engineers and statisticians, most notably W. Edwards Deming, classify the control chart as a statistical method unto itself with little or no tie to other statistical methodologies except sampling. (For years, Deming was known more for his work in sampling than in statistical process control.)

Practitioners now talk about Phase I and Phase II control charts. Phase I occurs at the very beginning of the monitoring process and deals with estimating the appropriate control chart parameters. Phase II begins once these parameters are considered adequately estimated. In some sense, the Phase I control chart is an exploratory data analysis tool, similar in spirit to stem-and-leaf displays and boxplots. Phase II control charts are best viewed as a sequence of hypothesis tests. For the purposes of developing the control chart, we shall assume a phase II situation for three reasons:

1. It better reflects statistical thinking because it shows the explicit tie between the control chart and another fundamental statistical procedure.
2. It provides a formal basis for evaluating the properties of control charts, which we shall see in Section 5.8.
3. The popular CUSUM chart, which we shall present in Section 5.10, makes statistical sense only when viewed as a sequence of hypothesis tests.

Control charts take a random sample, usually called a *subgroup,* from a process on a regular basis, such as once every 30 minutes, once an hour, once every 2 hours, twice a shift (every 4 hours), once a shift (every 8 hours), or once a day. As soon as the data are collected, someone, usually the operator, tests whether the process is in or out of control. The next example illustrates the basic idea.

| Example 5.1 | **Filling Milk Cartons** |

Maxcy and Lowry (1984) report on a process for filling milk cartons that nominally contain 8 fluid ounces (fl oz). Since the specific gravity of milk is approximately 1.033, they can measure the volume by measuring the weight. In this case, 8 fl oz should weigh 245 g. U.S. Department of Agriculture regulations require this process to maintain a mean amount delivered of 260 g to ensure virtually no underfills.

An engineer must devise a monitoring strategy that detects when the process begins to move away from this target mean amount.

If we directly observed the true mean amount of milk delivered, we could easily monitor this process. Unfortunately, we never truly know this value. Consider the following approach:

1. Take random samples of five cartons per shift.
2. Use the sample mean amount of milk delivered, $\bar{y}$, to estimate μ.
3. Perform a hypothesis test each shift of the form

$$H_0: \quad \mu = \mu_0$$

$$H_a: \quad \mu \neq \mu_0.$$

Whenever we reject the null hypothesis, we conclude that the true mean amount of milk delivered has shifted away from the target value of 260 g, and we have evidence to suggest that the process is out of control. Someone then needs to investigate the cause of the shift and correct it. On the other hand, if we fail to reject the null hypothesis, then we conclude that we should take no corrective action *at this time*.

In this case, the standard deviation for the amounts delivered is 1.65 g. Assume that a sample of size five is sufficiently large to assume that the sample means follow a normal distribution by the Central Limit Theorem. Since we know the population standard deviation, our test statistic is

$$Z = \frac{\bar{y} - \mu_0}{\sigma / \sqrt{n}} \, .$$

We reject the null hypothesis and conclude that the mean amount of milk delivered has shifted if $|Z| > z_{\alpha/2}$.

Since production people conduct this test each shift, we would like to minimize the number of actual calculations and restate the critical region in terms of $\bar{y}$ rather than Z. Concluding that the mean has shifted if $|Z| > z_{\alpha/2}$ is equivalent to rejecting H_0 if either $Z > z_{\alpha/2}$ or $Z < -z_{\alpha/2}$. Thus, we act as if the process is in control if

$$-z_{\alpha/2} \leq Z \leq z_{\alpha/2}.$$

Since

$$Z = \frac{\bar{y} - \mu_0}{\sigma / \sqrt{n}},$$

we can rewrite this inequality in the following manner:

$$-z_{\alpha/2} \leq \frac{\bar{y} - \mu_0}{\sigma / \sqrt{n}} \leq z_{\alpha/2}$$

$$-z_{\alpha/2} \, \frac{\sigma}{\sqrt{n}} \leq \bar{y} - \mu_0 \leq z_{\alpha/2} \, \frac{\sigma}{\sqrt{n}}$$

$$\mu_0 - z_{\alpha/2} \, \frac{\sigma}{\sqrt{n}} \leq \bar{y} \leq \mu_0 + z_{\alpha/2} \, \frac{\sigma}{\sqrt{n}}.$$

As a result, we conclude that the process is out of control if either

$$\overline{y} > \mu_0 + z_{\alpha/2}\,\frac{\sigma}{\sqrt{n}} \quad \text{or} \quad \overline{y} < \mu_0 - z_{\alpha/2}\,\frac{\sigma}{\sqrt{n}}.$$

We call

- $\mu_0 + z_{\alpha/2}(\sigma/\sqrt{n})$ the *upper control limit (UCL)*.
- $\mu_0 - z_{\alpha/2}(\sigma/\sqrt{n})$ the *lower control limit (LCL)*.

False Alarm Rates   Within the standard hypothesis testing framework, α represents the chances of making a Type I error, where we reject the null hypothesis when it actually is true. Within the control chart setting, rejecting the null hypothesis means that we conclude that the process is out of control. Someone then must investigate the process to correct the problem. Concluding that the process is out of control, when in fact it is in control, is called a *false alarm* because we are asking someone to isolate a cause that really is not there. Within this context, α represents the false alarm rate. If we were to use the typical $\alpha = 0.05$, then the probability of a false alarm on the next sample from an in control process is 0.05. One out of every 20 samples, on average, from an in control process will be a false alarm. Many control charts require samples every hour, 24 hours per day. An $\alpha = 0.05$ would lead to more than one false alarm per day, on average, which is unacceptable. A better approach chooses a very small α and relies on frequent sampling to signal out-of-control conditions. Typically, we use $z_{\alpha/2} = 3$, which corresponds to $\alpha = 0.0027$.

> **VOICE OF EXPERIENCE**
>
> *Even if the process remains in control, a control chart will signal sooner or later.*

The true false alarm rate depends on how well the Central Limit Theorem applies to the data. If the sample size, n, is large enough for us to assume that the sample means follow a normal distribution, then α accurately reflects the true false alarm rate. If n is not sufficiently large, then the true false alarm rate tends to be different from α. By choosing $z_{\alpha/2} = 3$, we guarantee that α is "small" even if it is not exactly 0.0027.

The $\overline{X}$-Chart   Figure 5.3 illustrates a graphical method, called an $\overline{X}$*-chart* ($\overline{X}$ simply means a sample mean), that provides a simple basis for performing these tests over time. This chart plots each observed sample mean. When a sample mean falls between the upper and lower control limits, we conclude that the process *appears* to be in control. Otherwise, we conclude that the process is out of control. In this approach, $\mu_0 + 3(\sigma/\sqrt{n})$ is the upper control limit and $\mu_0 - 3(\sigma/\sqrt{n})$ is the lower control limit. The upper and lower control limits define for us the expected amount of variability in the sample means if the process is in control. We say that a control chart *signals* an out-of-control situation whenever a sample mean falls above the *UCL* or below the *LCL*. Such a signal indicates the possible presence of an assignable cause.

In the milk carton example, the target value is 260 g, the population standard deviation is 1.65 g, and we take a random sample of five cartons each shift.

Figure 5.3 | A Basic Illustration of a Control Chart

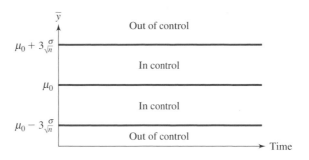

The control limits are calculated as shown here:

$$UCL = \mu_0 + 3\frac{\sigma}{\sqrt{n}}$$

$$= 260.0 + 3\frac{1.65}{\sqrt{5}}$$

$$= 260.0 + 2.21$$

$$= 262.21$$

$$LCL = \mu_0 - 3\frac{\sigma}{\sqrt{n}}$$

$$= 260.0 - 3\frac{1.65}{\sqrt{5}}$$

$$= 260.0 - 2.21$$

$$= 257.79.$$

Figure 5.4 is the control chart for the 20 shifts studied by Maxcy and Lowry. It signals several possible out-of-control shifts (shifts 1, 6, 8, 14, 15, 16, 17). Consequently, this milk delivery process appears to be unstable and requires management attention.

Sources of Variability and Stratification

A control chart provides a basis for keeping a specified process in control, which means that it is stable and predictable. The control limits define what constitutes in control, and the width of the control limits depends on the size of the process standard deviation, σ. The proper application of control charts requires us to think very carefully about what constitutes our process variability.

Figure 5.4 | The Milk Packaging Control Chart

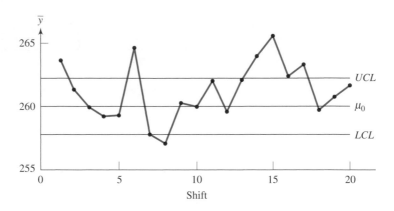

The statistical theory underlying the control chart assumes that the process variability is purely the result of random error. Unfortunately, in too many control chart applications, several well-defined sources produce much of the process variability. Failure to account for these sources can greatly influence the effectiveness of our monitoring procedure. On the other hand, identifying and removing these sources of variability can lead to significant quality improvement. The next two examples illustrate this point.

Example 5.2 | **Operators Making Nickel Plates**

Operators make the nickel plates for nickel–hydrogen batteries by carefully pouring nickel powder into a frame. An important characteristic of the plate at this point in the process is its initial net weight. Production management imposes a strict weight standard. The three operators on each shift require a great deal of skill in order to meet the specification. Historically, only 35% of the plates made meet the specifications. The material for the rejected plates is immediately reused, so the real loss as the result of failing to meet the standard is labor.

The production supervisor wishes to create a control chart to monitor the number of successful plates made each hour. She has two choices:

1. Run a single chart that lumps all of the operators together.
2. Run a separate chart for each operator.

Recently, this supervisor attended a team-building seminar. She feels that a single chart would build a better sense of teamwork, so she decides on that approach. Just in case she needs the information, however, she also keeps the individual production totals by hour.

Figure 5.5 is the stem-and-leaf display for the initial results when she lumped all of the operators together. For convenience, we have omitted the depth information. This display appears to have two peaks, which indicates a possible problem.

Figure 5.5 | The Stem-and-Leaf Display for the Number of Successful Attempts Per Hour

Stem	Leaves
0s	667
0•	888999
1*	001
1t	3
1f	44444455555
1s	66666666677777
1•	8888999

Figure 5.6 gives the side-by-side stem-and-leaf display when she separated the individual operators. This display shows that operator B makes fewer successful attempts per hour than operators A and C. With a little extra training, operator B's performance began to equal that of the other two.

In this example, the operators are an important source of variability. Running a control chart on their collective performance actually obscures a major problem. On the other hand, running separate control charts on each operator brings the problem to light and leads to overall improvement. We call breaking a population into subpopulations *stratification*. Often multiple peaks in a stem-and-leaf display indicate a need for stratification. However, we sometimes need to stratify our population even when the stem-and-leaf looks relatively well-behaved, as in the next example.

Figure 5.6 | The Side-by-Side Stem-and-Leaf Display by Operator

Stem	Operator A	Operator B	Operator C
0s		667	
0•		888999	
1*		001	
1t	3		
1f	444455		44555
1s	6666677		6666777
1•	889		8899

| Figure 5.7 | The Stem-and-Leaf Display for the Resistances |

Stem	Leaves
97	0
98	0000
99	0000000000
100	000000000000
101	0000000000
102	0000
103	0

| Example 5.3 | **Production Process for 100-Ohm Resistors** |

A major manufacturer of electrical resistors uses three separate production lines to make 100-ohm resistors that have a specification of 100 ± 2 ohms. The superintendent for these lines wants to create a control chart on the actual resistances. Since each production line is nominally the same as the others, he elects to run a single control chart for the entire area. However, just in case, he also keeps the identity of each specific production line. Figure 5.7 shows the stem-and-leaf display for the combined results. It indicates a very well-behaved distribution with a peak at 100 ohms. It also indicates some problems meeting the stated specifications.

Digging deeper, the superintendent decided to look at each production line individually. Figure 5.8 is the side-by-side stem-and-leaf display. We see that production line 2 appears right on target, fully meeting the specifications. Production line 1 tends to produce resistors with slightly less than 100 ohms, and production line 3 tends to produce resistors with slightly more than 100 ohms. Once the superintendent identified the problem, he quickly corrected both line 1 and line 3.

| Figure 5.8 | The Side-by-Side Stem-and-Leaf Display by Production Line |

Stem	Line 1	Line 2	Line 3
97	0		
98	000	0	
99	000000	000	0
100	000	000000	000
101	0	000	000000
102		0	000
103			0

In general, we should run control charts on the smallest subpopulation or subprocess that yields enough data to warrant a chart. In so doing, we can determine important sources of variability and often correct them. Throughout the rest of this chapter, we assume that we have appropriately stratified the population or process for the purposes of the control chart.

Rational Subgroups

Most random sampling schemes assume that we are dealing with a static population. For example, we may need a random sample of five milk cartons from a much larger batch of 200. The sampling scheme concentrates on how we should select from a group of objects, in this case milk cartons, already produced. Control charts, on the other hand, require us to sample from a dynamic process, which presents two rather subtle problems:

1. Possible correlations among the items in our sample or subgroup (auto-correlation)
2. Changes in the process during the collection of the specific sample or subgroup

We say that two variables are correlated if knowledge of one gives us some information about the other. For example, ambient temperature and ambient humidity tend to be correlated, especially in Florida. The humidity tends to be higher if the temperature is high. Autocorrelation in a production process refers to correlation among successive production units. For example, consider the temperature of a production furnace. Typically, the temperature controller records the furnace temperature every second. Knowing that the furnace temperature at some time t is too high probably suggests that the furnace temperature is too high one second later. We thus say the furnace temperatures are autocorrelated.

> **VOICE OF EXPERIENCE**
>
> *Proper choice of subgroups requires sound engineering insights into the specific process.*

If we suspect the first problem, we should not take our sample all at one time. If we suspect the second problem, then we do not want to take our sample over too long a period of time. How we actually collect the samples can significantly influence what we consider to be our process variability!

The next example illustrates the concept of autocorrelation and its impact on our sampling strategy.

Example 5.4 | **Extrusion of Ceramic Pipe**

Ceramic pipe is made by extruding a clay–water mixture through a die. Clay is a fairly abrasive material that wears away the die over time. In addition, the pipe tends to swell slightly immediately after extrusion. The actual amount of swelling depends to a certain extent on the moisture content of the clay–water mixture. Two successive sections of pipe tend to have very similar moisture contents and face very similar die conditions. If the first section is too thick, we have every reason to suspect that the second is also too thick. In other words, knowledge that the first section is too thick gives us some information about the probability that the second is too thick. We thus say that the two events are *correlated*. In this case, we expect the two sections to be positively correlated.

What is the real consequence of this positive correlation on our control chart? Suppose we take a sample of four consecutive sections each hour as our sample. Since these sections are positively correlated, the variability among these four sections is less than the true variability among all the sections produced over the hour. Using samples of four successive sections will seriously understate the true process variability. Our control chart will thus tend to signal more false alarms than it should—in some cases, significantly more.

The correlation among pipe sections depends on how close in time they were extruded. We expect two successive pipe sections to be highly correlated, but we should not expect much, if any, correlation between a specific section and the 100th after it. As a result, a better sampling scheme chooses a section every 15 minutes to form the sample of four per hour.

Figure 5.9 illustrates the problem of autocorrelation on sampling for control charts. The process variability between two successive items is virtually nil. However, the true range in the process variability is from the top of the peaks to the bottom of the troughs. Taking all of the items in the subgroup too close together in time causes us to understate the true process variability. When we face this particular situation, taking items at random over the entire sampling period is a much better idea.

On the other hand, many processes do not exhibit much, if any, autocorrelation. Successive observations are essentially independent of one another; that is, knowledge that one item is too large or too small gives us no information about the chances that the next observation is too large or too small. In these situations, we are much more concerned about a shift occurring during the sampling process that will then inflate our estimate of the process variability.

Figure 5.9 | The Drift in *y* for a Typical Positively Autocorrelated Process

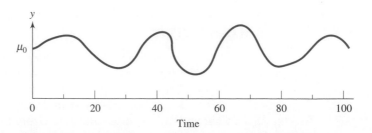

Rational subgrouping tries to balance these concerns. With a rational subgroup, we seek a sampling strategy that meets these objectives:

- Maximizes the differences between samples when assignable causes are truly present
- Minimizes the differences between samples when assignable causes are truly absent

Choosing an appropriate sampling strategy for control charts involves a careful balance of these competing concerns:

- The presence or absence of significant autocorrelation
- The chances that an assignable cause will occur during the collection of a specific subgroup
- The sample size requirements for at least nominally assuming the Central Limit Theorem (fortunately, this usually is not a major concern)
- The ability to detect shifts of real concern quickly, which we shall discuss in Sections 5.9 and 5.10.

Rational subgrouping does not always eliminate problems with autocorrelation. When such situations arise, engineers may use more advanced techniques, which are beyond the scope of this text. For the rest of this chapter, we shall assume that we are using an appropriate sampling scheme for obtaining our subgroups.

General Form of a Control Chart

We commonly classify control charts as either variables or attribute, depending on the data collected. *Variables charts* use data that are measured, at least theoretically, over a continuum, such as weights, lengths, or concentrations. We model such data by continuous random variables. *Attribute charts* use data that represent counts by classifications, such as the number of decorative bricks that fail to meet specifications or the number of accidents per month. We model attribute data by discrete random variables, usually either binomial or Poisson.

Both variables and attribute control charts use the Central Limit Theorem to define the control limits (we shall discuss one exception in Section 5.4). Let θ be the parameter associated with an important quality characteristic, let $\hat{\theta}$ be an appropriate estimator, and let $\sigma_{\hat{\theta}}$ be the standard error of this estimator. If we can assume the Central Limit Theorem, then

$$Z = \frac{\hat{\theta} - \theta}{\sigma_{\hat{\theta}}}$$

approximately follows a standard normal distribution.

Once again, we shall view a control chart as a sequence of hypothesis tests. Let θ_0 be the target value for this characteristic. If $\theta = \theta_0$, then the process is in control. On the other hand, if θ shifts in either direction away from this target value, then the process is out of control. Thus, the appropriate hypotheses are

$$H_0: \quad \theta = \theta_0$$
$$H_a: \quad \theta \neq \theta_0.$$

When we know the standard error, the appropriate test statistic is

$$Z = \frac{\hat{\theta} - \theta_0}{\sigma_{\hat{\theta}}}.$$

By the same logic as before, we can show that we should reject the null hypothesis and signal that the process is out of control if either

$$\hat{\theta} > \theta_0 + z_{\alpha/2}\sigma_{\hat{\theta}} \quad \text{or} \quad \hat{\theta} < \theta_0 - z_{\alpha/2}\sigma_{\hat{\theta}}.$$

Most control charts use *three standard error control limits:*

$$UCL = \theta_0 + 3\sigma_{\hat{\theta}}$$
$$LCL = \theta_0 - 3\sigma_{\hat{\theta}}.$$

If we may reasonably assume the Central Limit Theorem, then our actual false alarm rate with these limits is 0.0027. On the other hand, if the sample size does not justify the Central Limit Theorem, then three standard error limits should guarantee that the false alarm rate is small, although we may not be able to quantify exactly how small.

Phase I and Phase II Control Charts

It is rare that an engineer begins to monitor a process with sufficient information to assume that he/she knows the true target mean and the true process variance. In such a situation, the engineer begins to take rational subgroups on a regular basis. Phase I control charts refer to the process of estimating the control limits. Phase II control charts refer to a more mature monitoring process where the engineer has a great deal of faith that the target mean and true process variance are well estimated. Up to this point, we have described a Phase II control chart. It is under Phase II conditions that control charts are most purely a sequence of hypothesis tests. Under Phase I conditions, many practitioners view the control chart as an exploratory data analysis tool much like a boxplot.

In the first stage of a Phase I control chart, the engineer simply collects the information from the subgroups. At this point, he/she does not have enough information to establish appropriate control limits; hence, it is impossible to begin active control of the process. Typically, after twenty or so subgroups, the engineer assumes that all of these subgroups came from an in-control process (clearly, a questionable assumption!) and estimates the target mean and process variance accordingly. Often, at least one subgroup appears to be out of control. Ideally, the engineer should investigate what happened with that subgroup. Unfortunately, as with all retrospective investigations, too much time has passed, and the engineer has lost the opportunity to ascertain if the subgroup truly was the result of an out-of-control condition. Usually, the engineer drops that particular subgroup and recalculates the control limits. He/she continues this process until all of the out-of-control subgroups from the very initial period are dropped.

This first retrospective period leads to the first set of tentative control limits. The engineer then begins to use these limits as the basis for active control of the process. Good practice then updates the control limits every twenty or thirty subgroups until there are 100 or so assumed in-control subgroups used to calculate the control limits. At this point, the control limits are assumed to be well estimated, and the chart has matured to Phase II status.

⸬ Computer Exercises

5.1 Consider a process that is well modeled by a normal distribution with a variance of 4. Suppose the target value for the mean is 10 and that production monitors this process with an $\overline{X}$-chart using samples of size four. The resulting upper control limit is 13, and the lower control limit is 7.

   a. Generate 1000 random samples, each of size four, assuming that the process is in control. Count the number of times the procedure signals an out-of-control situation.

   b. Generate another 1000 random samples, each of size four, with the mean shifted to 11. Again, count the number of times the procedure signals an out-of-control situation.

   c. Generate another 1000 random samples, each of size four, with the mean shifted to 12. Again, count the number of times the procedure signals an out-of-control situation.

   d. Generate another 1000 random samples, each of size four, with the mean shifted to 13. Count the number of times the procedure signals an out-of-control situation.

   e. Comment on your results from parts a–d.

⸬ Exercises

5.1 Yashchin (1995) discusses a process for the chemical etching of silicon wafers used in integrated circuits. This process etches the layer of silicon dioxide until the layer of metal beneath is reached. This company monitors the thickness of the silicon dioxide layers because thicker layers require longer etching times. The layer has a target thickness of 1 micron and a historical standard deviation of 0.06 micron. The company uses subgroups of four wafers. The mean thicknesses for 40 subgroups follow. The data are in consecutive order, reading across the rows. The first observation is 1.006, the second is 1.037, and so on.

1.006	1.037	0.944	0.957	1.012	1.035	0.917	1.067
1.121	0.935	0.911	1.030	1.018	0.941	1.192	1.142
1.138	1.188	1.080	1.228	1.153	1.141	1.179	1.190
1.184	0.880	0.951	0.875	0.870	0.811	0.871	0.890
0.866	0.794	0.868	0.854	0.905	0.885	0.885	0.977

Calculate the appropriate control limits and plot the control chart. Comment on your results.

5.2   Kane (1986) discusses the concentricity of an engine oil seal groove. Concentricity measures the cross-sectional coaxial relationship of two cylindrical features. In this case, he studied the concentricity of an oil seal groove and a base cylinder in the interior of the groove. He measures the concentricity as a positive deviation using a dial indicator gauge. Historically, this process has produced an average concentricity of 5.6 with a standard deviation of 0.7. To monitor this process, he periodically takes a random sample of three measurements. The mean concentricities for 20 subgroups follow. The data are in consecutive order, reading across the rows. The first observation is 5.48, the second is 5.83, and so on.

| 5.48 | 5.83 | 6.69 | 6.04 | 5.64 | 5.23 | 4.89 | 5.72 | 5.39 | 6.19 |
| 5.78 | 5.76 | 5.82 | 5.63 | 5.73 | 5.05 | 4.62 | 5.74 | 6.60 | 6.81 |

Calculate the appropriate control limits and plot the control chart. Comment on your results.

5.3   Runger and Pignatiello (1991) consider a plastic injection molding process for a part that has a critical width dimension with a historical mean of 100 and a historical standard deviation of 8. Periodically, clogs form in one of the feeder lines, causing the mean width to change. As a result, the operator periodically takes random samples of size four. The mean widths for 30 subgroups follow. The data are in consecutive order, reading across the rows. The first observation is 93.77, the second is 105.09, and so on.

93.77	105.09	106.18	103.21	97.66
103.55	90.57	105.08	95.57	102.25
100.98	101.17	103.95	100.41	101.21
100.18	105.74	90.16	103.63	102.04
105.53	112.05	112.33	119.15	109.74
106.41	112.75	106.95	105.91	115.40

Calculate the appropriate control limits and plot the control chart. Comment on your results.

5.4 Porosity is one important quality characteristic of pencil lead, which is a ceramic material "fired" at high temperatures. The porosity measures the ultimate firing state of the material. In addition, the resulting pore structure permits the lead to absorb wax in the next production step. The wax smoothes the writing characteristics of the pencil. A particular pencil lead grade has a target porosity of 12.5. Historically, the porosities for this grade have had a standard deviation of 0.8. Production monitors the porosity of this grade by taking a random sample of size four from each lot of pencil lead. The mean porosities for 20 such lots follow. The data are in consecutive order, reading across the rows. The first observation is 12.88, the second is 12.68, and so on.

| 12.88 | 12.68 | 12.95 | 11.55 | 13.88 | 13.03 | 13.25 | 12.60 | 13.18 | 12.05 |
| 12.53 | 12.40 | 12.60 | 12.48 | 12.45 | 12.33 | 12.78 | 12.30 | 11.85 | 11.50 |

Calculate the appropriate control limits and plot the control chart. Comment on your results.

5.5 The 10-oz packaging line for a popular breakfast cereal has a target net weight of 10.5 oz. Historically, the process standard deviation is 0.1 oz. This company monitors the net weights by taking a random sample of five cereal boxes each hour. The mean net weights for 30 such subgroups follow. The data are in consecutive order, reading across the rows. The first observation is 10.56, the second is 10.56, and so on.

10.56	10.56	10.44	10.46	10.49	10.52	10.53	10.47	10.44	10.54
10.49	10.52	10.53	10.50	10.50	10.48	10.49	10.55	10.62	10.58
10.44	10.47	10.40	10.43	10.47	10.50	10.38	10.43	10.50	10.51

Calculate the appropriate control limits and plot the control chart. Comment on your results.

5.6 A major producer of a polymer commonly used in clothes must closely watch the molecular weight at a crucial stage in the process. The company has set a target value of 800. It measures the molecular weight four times each hour, at roughly 15-minute intervals. The operator uses the hourly average for control purposes. Historically, the molecular weights have had a standard deviation of 10. The average molecular weights for the past 24 hours follow. The data are in consecutive order, reading across the rows. The first observation is 795, the second is 786, and so on.

795	786	795	805	801	825	811	784
787	808	804	797	799	791	810	796
806	800	800	782	786	807	790	773

Calculate the appropriate control limits and plot the control chart. Comment on your results.

5.7   Spurrier and Thombs (1990) give data for 29 consecutive measurements on the inner diameter of a landing gear triunion. The target for the diameters is 98.00. Historically, the diameters have had a standard deviation of 4. Each mean listed below is the average of 4 values. The data are in consecutive order reading across rows. Thus, the first observation is 101.25, the second is 98.00, and so on.

101.25	98.00	95.50	97.25	93.50	95.25	93.00	85.25
96.25	97.50	94.50	103.75	100.50	109.50	97.25	102.00
99.00	102.00	101.25	101.25	98.75	99.75	100.25	92.50
93.75	97.50	95.60	95.75	99.75			

Calculate the appropriate control limits and plot the control chart. Comment on your results.

5.8   Venables (1989) reported a dew-retting process that involves softening flax stems by soaking them in water, thus enabling the separation of the linen fibers from the wooden material. Historically, the ret loss has been 17 percent with a standard deviation of 1.5. The process is monitored by taking a random sample of 3 observations and recording the ret loss as a percent. The mean for 18 subgroups follows. The data are in consecutive order reading across rows. Thus, the first observation is 16.90, the second is 16.57, and so on.

16.90	16.57	18.13	18.60	17.73	17.63	18.17	18.83	15.50
17.10	17.27	17.20	15.67	14.60	15.50	18.13	17.17	17.97

Calculate the appropriate control limits and plot the control chart. Comment on your results.

5.2 Specification Limits and Capability

The Need to Monitor Both the Mean and the Variability

We monitor processes to maintain an acceptable level of quality for our products. Typically with variables data, we define the quality of our products with *specification limits*. We can best illustrate this concept through an example.

Example 5.5 | **Ash Content of Graphite**

Graphite has many uses, from tires to brake linings to pencil lead. A critical quality characteristic of graphite is the ash content, which is the percent residual left

Figure 5.10	Distribution of Ash Contents When a Capable Process Is in Control

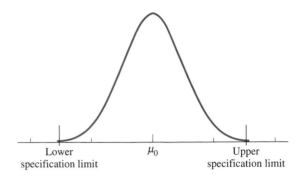

Lower
specification limit

μ_0

Upper
specification limit

after burning off the "volatiles" (mostly carbon). Lower ash contents mean more expensive grades of graphite. One particular grade of graphite has a specification of 16%–20%, with a target value of 18%. When the ash content exceeds 20%, the customer complains about the quality of the shipment. On the other hand, if the ash content is less than 16%, then the producer is "giving away" product.

The graphite producer grinds and blends four large batches of this grade of graphite each day. A technician takes a sample from each batch and performs the appropriate ash test. Historically, these results follow a normal distribution. Figure 5.10 illustrates the distribution of the ash contents for the batches when the process is in control. Virtually all of the product falls within the specification limits.

Figure 5.11 illustrates the distribution when the process mean ash content increases. Some of the batches now have unacceptably high ash contents due to the shift in the process mean. As a result, this producer should monitor this process mean closely to signal such a shift as soon as possible in order to minimize the number of batches with unacceptable ash contents.

Figure 5.11	Distribution of Ash Contents When the Process Mean Shifts Out of Control

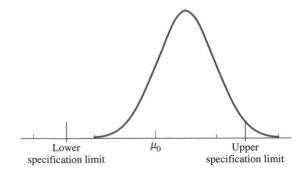

Lower
specification limit

μ_0

Upper
specification limit

Figure 5.12 | Distribution of Ash Contents When the Process Variance Increases

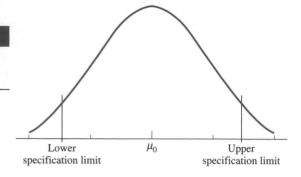

Lower specification limit μ_0 Upper specification limit

A shift in the mean is not the only way this process can produce unacceptable batches. Figure 5.12 illustrates the distribution of the ash contents when the process mean remains at the target value and the variability increases. The increase in variability causes some batches to have unacceptably high and other batches to have unacceptably low ash contents. As a result, this producer should monitor the process variability along with the process mean.

Over the long term, many manufacturers worry more about the process variability than the process mean. Figure 5.13 contrasts two different processes:

- A: the process mean is at the target.
- B: the process mean has shifted slightly from the target but the process variability is considerably smaller.

Even though the mean for process B is off target, it is less likely to produce unacceptable product than process A. The reduced variability for process B more than compensates for the shift in the process mean. In Chapters 7 and 8, we shall introduce experimental strategies for reducing process variability.

Many people confuse control limits and specification limits. Control limits come from sample data and are used to assess whether or not the process is stable over time. Specification limits define an acceptable range for individual products and are typically set by customers. Control limits can be thought of as the "voice of the process," while specification limits represent the "voice of the customer." It is very possible to have an out-of-control process with products within the specification limits and vice versa. Also, since specification limits are on individual observations, it makes no sense to add them to a control chart for the mean.

One common way to evaluate how well the process is behaving relative to the specification limits is to use capability indices. There are several measures

Figure 5.13 | Importance of Process Variability for Meeting Specifications

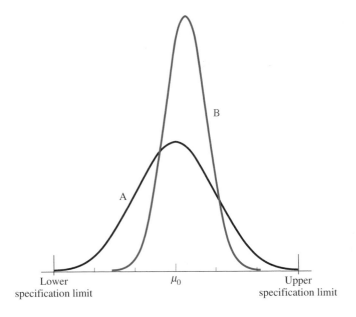

of process capability. Most of these measures assume the empirical rule where virtually all of the process falls with ±3 standard deviations. The simplest version of a capability index is

$$(USL - LSL)/6\sigma,$$

where good values are larger than 1. For more information about process capability, please refer to standard texts on statistical quality control such as Montgomery (2004).

⁖ 5.3 $\overline{X}$- and R-Charts

In Section 5.1, we developed the $\overline{X}$-chart from knowledge of both the target value for the mean and the process standard deviation. These are the appropriate control limits:

$$UCL = \mu_0 + 3\frac{\sigma}{\sqrt{n}}$$

$$LCL = \mu_0 - 3\frac{\sigma}{\sqrt{n}}.$$

The $\overline{X}$-chart allows us to monitor the *between-subgroup* variability. By between-subgroup variability, we mean the differences from subgroup to subgroup. The sample mean for each subgroup provides the appropriate basis for monitoring these differences. The control limits require that we know the *process* or *within-subgroup* variance and that this variance remains constant (stable) over time. If the within-subgroup variance is not stable, then our control limits for the $\overline{X}$-chart have no meaning. By within-subgroup variability, we mean the variability each subgroup exhibits within itself. The population variance controls this variability. Thus, by monitoring the within-subgroup variability, we actually monitor how stable σ^2 remains over time. Good rules of practice are

- The within-subgroup variability should be in control before we can actively monitor the between-subgroup variability.
- We should constantly monitor the within-subgroup variability to ensure that it remains stable.

Typically, engineers simultaneously run control charts for the subgroup mean and for some measure of the within-subgroup variability. In general, the within-subgroup variability tends to stay in control more than the between-subgroup variability.

The next example illustrates the most common way engineers simultaneously monitor the process mean and the process variance.

| Example 5.6 | **Grinding of Silicon Wafers** |

Roes and Does (1995) present data on the grinding of silicon wafers used in integrated circuits. Philips Semiconductors grinds wafers in batches of 31. For a particular product, Philips has a target thickness of 244 μm. To monitor this process, it samples five wafers from each batch.

The R-Chart   Engineers traditionally use the *R-chart* (for range) to monitor the process variance. Suppose we take random samples of size n on a regular basis. Roes and Does use $n = 5$. The *sample range* for the ith sample, R_i, is the largest of the n observations in the ith sample minus the smallest observation in the sample. More formally, let y_{ij} be the jth observation in the ith sample, let $y_{i(n)}$ be the largest observation in this sample, and let $y_{i(1)}$ be the smallest. The sample range is

$$R_i = y_{i(n)} - y_{i(1)}.$$

The sample range is an easy-to-calculate measure of the variability. Calculators did not exist when control charts were first developed, so ease of calculation was extremely important.

Engineers and statisticians have developed extensive tables for the properties of the sample range when the observations in the random sample come from a normal distribution. Let ρ_0 be the target value for the sample ranges. Thus, when the process is in control,

$$E[R_i] = \rho_0.$$

It can be shown that the standard error for the sample ranges, σ_R, is

$$\sigma_R = d_n^* \rho,$$

where d_n^* is a constant that depends on the sample size.*
    The three standard error control limits for the ranges are

$$UCL = \rho_0 + 3d_n^* \rho_0 = \rho_0(1 + 3d_n^*)$$
$$LCL = \rho_0 - 3d_n^* \rho_0 = \rho_0(1 - 3d_n^*).$$

To simplify the calculations, Table 5 of the appendix lists the values for the constants $D_3 = 1 - 3d_n^*$ and $D_4 = 1 + 3d_n^*$. Thus, the three standard error limits for the ranges are

$$UCL = D_4 \rho_0$$
$$LCL = D_3 \rho_0.$$

For sample sizes of six or fewer, D_3 is 0, in which case the lower control limit is 0. In our particular case, $n = 5$, so $D_3 = 0$ and $D_4 = 2.114$.
    When we start an R-chart, we rarely know ρ_0, so we need an appropriate estimate. Typically, we treat the first m subgroups as a *base period,* which provides the basis for estimating the required parameters for the control chart. Since we require reasonably precise estimates of these parameters, most authors suggest that m be at least 20 and preferably larger. Of course, the larger m is, the longer we must wait until we can estimate the control limits and the longer we must wait until we can actively monitor the process. If the process is in control during this base period, then a reasonable estimate of ρ_0 is $\overline{R}$ given by

$$\overline{R} = \frac{1}{m} \sum_{i=1}^{m} R_i.$$

The estimated control limits are

$$UCL = D_4 \overline{R}$$
$$LCL = D_3 \overline{R}.$$

Table 5.1 lists the thicknesses of 30 consecutive batches analyzed by Roes and Does. Since they did not have good prior information on the expected range for these data, consider the first 20 subgroups as a base period. We can estimate ρ_0 by

$$\overline{R} = \frac{1}{m} \sum_{i=1}^{m} R_i$$
$$= \frac{1}{20}(185)$$
$$= 9.25.$$

---

* $d_n^* = d_3/d_2$, where d_2 and d_3 are constants that depend on n. Table 5 of the appendix lists values for these constants.

Table 5.1 | The Sample Ranges for the Silicon Wafer Thickness Data

Batch	y_{i1}	y_{i2}	y_{i3}	y_{i4}	y_{i5}	R_i
1	240	243	250	253	248	13
2	238	242	245	251	247	13
3	239	242	246	250	248	11
4	235	237	246	249	246	14
5	240	241	246	247	249	9
6	240	243	244	248	245	8
7	240	243	244	249	246	9
8	245	250	250	247	248	5
9	238	240	245	248	246	10
10	240	242	246	249	248	9
11	240	243	246	250	248	10
12	241	245	243	247	245	6
13	247	245	255	250	249	10
14	237	239	243	247	246	10
15	242	244	245	248	245	6
16	237	239	242	247	245	10
17	242	244	246	251	248	9
18	243	245	247	252	249	9
19	243	245	248	251	250	8
20	244	246	246	250	246	6
21	241	239	244	250	246	11
22	242	245	248	251	249	9
23	242	245	248	243	246	6
24	241	244	245	249	247	8
25	236	239	241	246	242	10
26	243	246	247	252	247	9
27	241	243	245	248	246	7
28	239	240	242	243	244	5
29	239	240	250	252	250	13
30	241	243	249	255	253	14

Figure 5.14 | The R-Chart for the Silicon Wafer Thickness Data

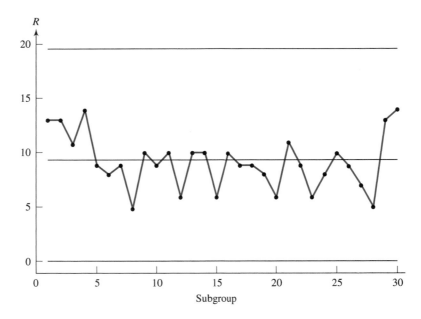

The upper control limit is

$$UCL = D_4\overline{R}$$
$$= 2.114\,(9.25)$$
$$= 19.55.$$

The lower control limit is

$$LCL = D_3\overline{R}.$$

Since $n \le 6$, $D_3 = 0$, and the lower control limit is 0. Figure 5.14 shows the R-chart for all 30 subgroups using these control limits. The process appears to be in control.

The $\overline{X}$-Chart Based on the Range   Since the R-chart indicated that the process variance is stable, we can construct the $\overline{X}$-chart to monitor the between-subgroup variability. We have already seen that if we know the process variance, then these are the appropriate control limits for this chart:

$$UCL = \mu_0 + 3\frac{\sigma}{\sqrt{n}}$$
$$LCL = \mu_0 - 3\frac{\sigma}{\sqrt{n}}.$$

Since in our case σ is not known, we use the R-chart as a basis for estimating these limits. It can be shown that

$$E[R_i] = d_2\sigma,$$

where d_2 is an appropriate constant that depends on n. Table 5 in the appendix gives this constant for various n's. An unbiased estimate of $3(\sigma/\sqrt{n})$ is

$$\frac{3}{d_2\sqrt{n}}\overline{R}.$$

For convenience, Table 5 of the appendix lists the values of $A_2 = 3/d_2\sqrt{n}$. As a result, when we do not know σ, the following are appropriate estimates of the control limits:

$$UCL = \mu_0 + A_2\overline{R}$$
$$LCL = \mu_0 - A_2\overline{R}.$$

For the grinding of silicon wafers example, $n = 5$, so $A_2 = 0.577$.

In many applications, we seek purely to get the subgroup means stable, in which case, we do not have a specific target value in mind. In other applications, we may have an ideal value for our characteristic, but we cannot adjust the process well enough to hit this value precisely. We then assure ourselves that the process is set close enough to this ideal, and we let the process establish its own natural target. In either of these two cases, we need to use the first m subgroups as a base period to estimate μ_0, the true target value for the subgroup means. Roes and Does decided to monitor this process relative to its own natural target rather than the nominal thickness of 244 μm. Frankly, most practitioners follow this strategy.

Let y_{ij} be the jth observation in the ith subgroup. The ith subgroup mean is

$$\overline{y}_i = \frac{1}{n}\sum_{j=1}^{n} y_{ij}.$$

If the process is in control over this base period, then an appropriate estimate of the target value, μ_0, is simply the overall mean, $\overline{\overline{y}}$, given by

$$\overline{\overline{y}} = \frac{1}{m}\sum_{i=1}^{m} \overline{y}_i.$$

The control limits for an $\overline{X}$-chart based on the sample range when we need to estimate the target value are

$$UCL = \overline{\overline{y}} + A_2\overline{R}$$
$$LCL = \overline{\overline{y}} - A_2\overline{R}.$$

Table 5.2 lists the data and the subgroup means for the grinding example. Consider the first 20 subgroups as the base period. The estimated target value is

$$\overline{\overline{y}} = \frac{1}{m} \sum_{i=1}^{m} \overline{y}_i$$

$$= \frac{1}{20}(4903.6)$$

$$= 245.18.$$

From the R-chart, $\overline{R} = 9.25$. The following are the resulting control limits:

$$UCL = \overline{\overline{y}} + A_2 \overline{R}$$

$$= 245.18 + 0.577\,(9.25)$$

$$= 245.18 + 5.34$$

$$= 250.52$$

$$LCL = \overline{\overline{y}} - A_2 \overline{R}$$

$$= 245.18 - 0.577\,(9.25)$$

$$= 245.18 - 5.34$$

$$= 239.84.$$

Figure 5.15 is the control chart with these limits. Again, we see an in-control process.

We should always check our assumptions when we first construct a control chart. The R-chart assumes

- The process is in control during the base period.
- The data follow a normal distribution.

The $\overline{X}$-chart based on the sample range assumes

- The R-chart is in control.
- The subgroup means are in control during the base period.
- The sample size is large enough to assume that the subgroup means follow a normal distribution by the Central Limit Theorem.

In the grinding of silicon wafers example, both the R- and the $\overline{X}$-charts appear in control. We next should look at Figure 5.16, the stem-and-leaf display of the data during the base period. This display appears roughly to follow a bell-shaped curve. As a result, we should feel reasonably comfortable about our control limits.

Table 5.2 | The Subgroup Means for the Silicon Wafer Thickness Data

Batch	y_{i1}	y_{i2}	y_{i3}	y_{i4}	y_{i5}	$\bar{y}_i$
1	240	243	250	253	248	246.8
2	238	242	245	251	247	244.6
3	239	242	246	250	248	245.0
4	235	237	246	249	246	242.6
5	240	241	246	247	249	244.6
6	240	243	244	248	245	244.0
7	240	243	244	249	246	244.4
8	245	250	250	247	248	248.0
9	238	240	245	248	246	243.4
10	240	242	246	249	248	245.0
11	240	243	246	250	248	245.4
12	241	245	243	247	245	244.2
13	247	245	255	250	249	249.2
14	237	239	243	247	246	242.4
15	242	244	245	248	245	244.8
16	237	239	242	247	245	242.0
17	242	244	246	251	248	246.2
18	243	245	247	252	249	247.2
19	243	245	248	251	250	247.4
20	244	246	246	250	246	246.4
21	241	239	244	250	246	244.0
22	242	245	248	251	249	247.0
23	242	245	248	243	246	244.8
24	241	244	245	249	247	245.2
25	236	239	241	246	242	240.8
26	243	246	247	252	247	247.0
27	241	243	245	248	246	244.6
28	239	240	242	243	244	241.6
29	239	240	250	252	250	246.2
30	241	243	249	255	253	248.2

Figure 5.15 | The $\overline{X}$-Chart Based on the Sample Range for the Silicon Wafer Thickness Data

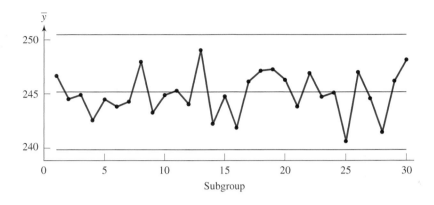

Figure 5.16 | Stem-and-Leaf Display for the Silicon Wafer Thickness Data

Stem	Leaves	Number	Depth
23f.:	5	1	1
23s.:	777	3	4
23•.:	88999	5	9
24*.:	000000011	9	18
24t.:	22222233333333	14	32
24f.:	44444555555555555	17	49
24s.:	666666666666677777777	21	
24•.:	8888888888999999	16	30
25*.:	00000000111	11	14
25t.:	23	2	3
25f.:	5	1	1

⁑ Exercises

5.9 Padgett and Spurrier (1990) analyze the breaking strengths of carbon fibers used in fibrous composite materials. These fibers measure 50 mm in length and 7–8 microns in diameter. Periodically, the manufacturer selects random samples of five fibers and tests their breaking stresses. Specifications require that 99% of the fibers must have a breaking stress of at least 1.2 GPa (gigapascals). Table 5.3 lists the breaking stresses (in GPa) from 20 such subgroups. Use all 20 subgroups as a base period.

Table 5.3 | Breaking Stress Data

Subgroup	y_{i1}	y_{i2}	y_{i3}	y_{i4}	y_{i5}
1	3.7	2.7	2.7	2.5	3.6
2	3.1	3.3	2.9	1.5	3.1
3	4.4	2.4	3.2	3.2	1.7
4	3.3	3.1	1.8	3.2	4.9
5	3.8	2.4	3.0	3.0	3.4
6	3.0	2.5	2.7	2.9	3.2
7	3.4	2.8	4.2	3.3	2.6
8	3.3	3.3	2.9	2.6	3.6
9	3.2	2.4	2.6	2.6	2.4
10	2.8	2.8	2.2	2.8	1.9
11	1.4	3.7	3.0	1.4	1.0
12	2.8	4.9	3.7	1.8	1.6
13	3.2	1.6	0.8	5.6	1.7
14	1.6	2.0	1.2	1.1	1.7
15	2.2	1.2	5.1	2.5	1.2
16	3.5	2.2	1.7	1.3	4.4
17	1.8	0.4	3.7	2.5	0.9
18	1.6	2.8	4.7	2.0	1.8
19	1.6	1.1	2.0	1.6	2.1
20	1.9	2.9	2.8	2.1	3.7

a. Calculate the appropriate control limits for an R-chart and plot the data. Comment on your results.

b. Calculate the appropriate control limits for an $\overline{X}$-chart based on R and plot the data. Comment on your results.

c. What did you assume to construct these control limits? Given the information in the base period, how comfortable are you with these assumptions?

5.10 A major manufacturer of writing instruments closely monitors the critical outside diameters a pen barrel. This company uses an injection molding process to make these barrels as well as the caps. To guarantee the fit of the cap to the barrel, the company must keep the critical outside diameter of the barrel as consistent as possible. Each hour, the operator takes a random sample of three barrels and measures the critical outside diameter. Table 5.4 gives the diameters for 25 such subgroups. Use the first 20 subgroups as a base period.

Table 5.4 | Outside Diameters of Pen Barrels

Subgroup	y_{i1}	y_{i2}	y_{i3}
1	0.379	0.376	0.379
2	0.378	0.377	0.378
3	0.378	0.378	0.378
4	0.378	0.377	0.377
5	0.378	0.378	0.378
6	0.378	0.378	0.377
7	0.379	0.379	0.379
8	0.379	0.378	0.377
9	0.378	0.378	0.377
10	0.377	0.377	0.378
11	0.381	0.379	0.377
12	0.379	0.380	0.379
13	0.378	0.378	0.379
14	0.377	0.380	0.378
15	0.379	0.378	0.380
16	0.379	0.381	0.379
17	0.379	0.381	0.379
18	0.379	0.378	0.379
19	0.378	0.379	0.377
20	0.379	0.379	0.378
21	0.380	0.378	0.379
22	0.378	0.381	0.380
23	0.379	0.380	0.380
24	0.378	0.379	0.379
25	0.377	0.377	0.377

a. Calculate the appropriate control limits for an R-chart and plot the data. Comment on your results.

b. Calculate the appropriate control limits for an $\overline{X}$-chart based on R and plot the data. Comment on your results.

c. What did you assume to construct these control limits? Given the information in the base period, how comfortable are you with these assumptions?

5.11 Snee (1983) examined the thicknesses of paint can ears. Periodically, the manufacturer took random samples of five cans each and measured the thickness of the ears. Table 5.5 gives the measurements (in units of 0.001 in.). Use the first 20 subgroups as a base period.

a. Calculate the appropriate control limits for an R-chart and plot the data. Comment on your results.

b. Calculate the appropriate control limits for an $\overline{X}$-chart based on R and plot the data. Comment on your results.

c. What did you assume to construct these control limits? Given the information in the base period, how comfortable are you with these assumptions?

5.12   A small manufacturer of brake linings closely monitors the incoming quality of a specific grade of graphite. This company receives the graphite in shipments of 20 pallets, with each pallet containing 40 fifty-pound bags of graphite. An inspector takes a small amount of graphite from four randomly selected bags from each pallet and performs a standard ash test, which burns off all the "volatiles" in the graphite. A low ash content indicates a high carbon content in the graphite. Table 5.6 lists the ash contents in the subgroups in the last shipment inspected. Use all 20 subgroups as a base period.

a. Calculate the appropriate control limits for an R-chart and plot the data. Comment on your results.

b. Calculate the appropriate control limits for an $\overline{X}$-chart based on R and plot the data. Comment on your results.

c. What did you assume to construct these control limits? Given the information in the base period, how comfortable are you with these assumptions?

5.13   A civil engineer who supervises a major interstate highway expansion recently started a control chart on the times required to fill dump trucks. Each day, he carefully times how long it takes to fill five randomly selected trucks. Table 5.7 gives the times for the first 25 days. Use the first 20 days as a base period.

a. Calculate the appropriate control limits for an R-chart and plot the data. Comment on your results.

b. Calculate the appropriate control limits for an $\overline{X}$-chart based on R and plot the data. Comment on your results.

c. What did you assume to construct these control limits? Given the information in the base period, how comfortable are you with these assumptions?

5.14   A chemical engineer supervises a new distillation column. Five times each shift, she determines the yield from the column. Table 5.8 lists the yields for 20 shifts. Use all 20 shifts as a base period.

a. Calculate the appropriate control limits for an R-chart and plot the data. Comment on your results.

b. Calculate the appropriate control limits for an $\overline{X}$-chart based on R and plot the data. Comment on your results.

c. What did you assume to construct these control limits? Given the information in the base period, how comfortable are you with these assumptions?

Table 5.5	The Thicknesses of Paint Can Ears				

Subgroup	y_{i1}	y_{i2}	y_{i3}	y_{i4}	y_{i5}
1	29	36	39	34	34
2	29	29	28	32	31
3	34	34	39	38	37
4	35	37	33	38	41
5	30	29	31	38	29
6	34	31	37	39	36
7	30	35	33	40	36
8	28	28	31	34	30
9	32	36	38	38	35
10	35	30	37	35	31
11	35	30	35	38	35
12	38	34	35	35	31
13	34	35	33	30	34
14	40	35	34	33	35
15	34	35	38	35	30
16	35	30	35	29	37
17	40	31	38	35	31
18	35	36	30	33	32
19	35	34	35	30	36
20	35	35	31	38	36
21	32	36	36	32	36
22	36	37	32	34	34
23	29	34	33	37	35
24	36	36	35	37	37
25	36	30	35	33	31
26	35	30	29	38	35
27	35	36	30	34	36
28	35	30	36	29	35
29	38	36	35	31	31
30	30	34	40	28	30

| Table 5.6 | Ash Contents of a Graphite Shipment |

Subgroup	y_{i1}	y_{i2}	y_{i3}	y_{i4}
1	19.2	19.5	19.3	19.3
2	19.0	18.7	18.9	18.3
3	19.0	18.5	18.4	18.6
4	19.1	19.0	19.0	18.9
5	18.5	18.4	18.3	18.4
6	18.9	18.7	18.7	18.6
7	19.8	19.4	19.3	19.3
8	19.3	19.5	19.2	19.2
9	19.6	19.6	20.2	19.3
10	18.8	19.2	19.1	18.8
11	18.8	19.2	18.6	18.7
12	19.9	20.1	20.4	20.0
13	19.2	19.2	19.0	19.1
14	20.6	20.1	20.0	20.2
15	20.2	19.9	19.7	19.7
16	20.0	19.6	19.6	19.6
17	19.9	19.8	19.7	19.8
18	20.1	19.8	19.8	19.7
19	20.0	19.9	19.9	20.6
20	20.1	20.0	19.8	19.9

5.15 In semiconductor processing, the basic experimental unit is a silicon wafer. Operations are performed on the wafer, but individual wafers can be grouped in multiple ways. The following is based on a case study performed by NIST of a lithography process. Three wafers are randomly selected from a cassette (typically a grouping of 25 wafers), and 10 cassettes are used in the study. Table 5.9 gives the line widths for the wafers.

a. Using the first 20 wafers as a base period, calculate the appropriate control limits for an R-chart and plot the data (wafer ranges). Comment on your results.

b. Using the first 20 wafers as a base period, calculate the appropriate control limits for an $\overline{X}$-chart based on R and plot the data (wafer means). Comment on your results.

Table 5.7	The Times to Fill Dump Trucks

Day	y_{i1}	y_{i2}	y_{i3}	y_{i4}	y_{i5}
1	18	21	19	19	21
2	19	25	18	17	23
3	18	17	23	18	22
4	19	18	19	21	19
5	21	19	22	21	22
6	19	25	10	17	16
7	15	21	21	19	22
8	19	16	18	23	22
9	20	16	13	19	18
10	19	25	16	19	16
11	15	17	23	17	15
12	18	23	22	22	15
13	21	21	26	24	20
14	17	22	17	22	17
15	19	17	20	25	17
16	16	24	12	19	24
17	15	14	22	19	17
18	17	19	23	19	16
19	15	24	18	20	19
20	18	20	25	24	22
21	22	14	16	21	19
22	23	20	20	19	19
23	15	21	16	14	22
24	17	18	15	19	19
25	17	18	16	19	19

c. What did you assume to construct these control limits? Given the information in the base period, how comfortable are you with these assumptions?

d. Using all 10 cassettes as a base period, calculate the appropriate control limits for an R-chart and plot the data (cassette ranges). Comment on your results.

e. Using all 10 cassettes as a base period, calculate the appropriate control limits for an $\overline{X}$-chart based on R and plot the data (cassette means). Comment on your results.

f. What did you assume to construct these control limits? Given the information in the base period, how comfortable are you with these assumptions?

| Table 5.8 | The Yields from a New Distillation Column |

Shift	y_{i1}	y_{i2}	y_{i3}	y_{i4}	y_{i5}
1	99.9	96.2	96.8	99.9	94.5
2	97.7	96.2	97.1	99.9	99.9
3	99.0	91.3	97.8	95.7	99.0
4	97.8	99.9	97.4	96.6	99.9
5	99.9	99.7	99.9	90.9	98.2
6	99.4	98.0	94.4	99.4	96.4
7	99.9	99.9	95.6	94.9	92.7
8	99.1	96.9	98.9	94.3	99.0
9	97.8	98.7	99.9	99.9	97.1
10	96.1	97.1	97.9	99.9	94.9
11	97.4	95.5	99.9	99.2	94.5
12	99.9	99.3	97.7	97.2	95.9
13	98.3	95.4	99.5	97.6	99.9
14	97.0	96.6	99.9	99.9	98.9
15	95.6	99.9	97.8	99.1	95.4
16	99.9	96.9	94.6	99.5	94.8
17	95.1	99.9	97.8	97.8	99.9
18	98.0	99.3	97.4	97.6	98.2
19	97.8	99.9	99.9	97.4	96.3
20	96.9	98.1	96.4	95.2	99.9

g. Compare the charts in a and b with those in d and e. Comment on their similarities and differences.

5.16 Maxcy and Lowry (1984) studied the performance of a milk-filling operation. Table 5.10 gives the gross weights of groups of five half-pint cartons of milk taken on 20 different days. Use all 20 subgroups as a base period.

a. Calculate the appropriate control limits for an R-chart and plot the data. Comment on your results.

b. Calculate the appropriate control limits for an $\overline{X}$-chart based on R and plot the data. Comment on your results.

c. What did you assume to construct these control limits? Given the information in the base period, how comfortable are you with these assumptions?

Table 5.9 | Line Widths for Silicon Wafers

Cassette	Wafer	y_{i1}	y_{i2}	y_{i3}
1	1	2.253	2.074	2.418
1	2	2.003	1.861	2.136
1	3	2.061	1.625	2.304
2	4	2.518	2.072	2.287
2	5	2.217	1.473	1.685
2	6	2.173	1.537	1.967
3	7	1.728	1.357	1.673
3	8	1.562	1.520	2.066
3	9	1.746	1.367	1.615
4	10	2.036	1.786	1.980
4	11	2.104	1.919	2.019
4	12	2.900	2.171	3.041
5	13	2.506	1.950	2.467
5	14	3.347	2.534	3.190
5	15	3.402	2.963	2.946
6	16	2.527	1.941	2.767
6	17	2.296	2.256	2.646
6	18	2.849	1.601	2.810
7	19	2.027	1.672	1.661
7	20	2.323	1.854	2.391
7	21	2.703	1.959	2.512
8	22	1.360	0.971	1.947
8	23	1.757	1.165	2.231
8	24	1.993	1.403	2.008
9	25	2.287	1.699	1.953
9	26	1.796	1.241	1.677
9	27	1.523	0.791	2.001
10	28	2.502	1.938	2.349
10	29	2.058	1.793	1.862
10	30	2.291	2.475	2.021

| Table 5.10 | Milk Gross Weights |

Subgroup	y_{i1}	y_{i2}	y_{i3}	y_{i4}	y_{i5}
1	263.5	264.7	262.6	263.9	264.2
2	259.4	262.7	257.6	264.7	263.0
3	260.5	259.9	260.1	260.7	259.1
4	260.2	259.1	260.4	257.2	259.4
5	260.4	258.0	260.4	260.8	257.2
6	264.8	265.1	263.8	265.8	254.9
7	255.0	262.3	260.8	256.2	254.9
8	254.6	259.5	257.4	255.0	258.9
9	260.1	260.1	260.9	259.8	260.7
10	258.2	261.3	260.7	260.5	259.5
11	263.0	263.4	260.6	262.0	261.2
12	259.5	260.7	259.4	259.2	259.7
13	262.6	262.2	262.0	261.8	262.0
14	262.9	266.1	265.3	262.7	263.3
15	263.9	266.2	266.3	266.8	265.0
16	262.9	262.8	265.1	261.0	260.9
17	264.0	263.0	265.1	262.9	262.0
18	259.6	258.1	260.6	263.8	256.8
19	261.8	259.4	261.2	261.4	260.2
20	262.2	259.7	264.5	260.7	261.1

⁙ 5.4 $\overline{X}$- and s^2-Charts

Engineers tend to use $\overline{X}$- and R-charts by tradition. A much less traditional method, but one strongly supported by statistical theory and by statistical thinking, uses $\overline{X}$- and s^2-charts. We illustrate these charts in the next example.

| Example 5.7 | **Grinding of Silicon Wafers—Revisited** |

The s^2-Chart   From a statistical perspective, the sample variance provides a much better measure of the within-subgroup variability, especially for moderate sample sizes. For very small sample sizes, the sample range is almost as efficient as the sample variance. As the sample size gets larger, however, the sample range becomes less efficient, and we really should prefer a procedure based on the sample variance. Most texts suggest that once the sample size gets to be seven to ten, we should use the sample variance. In addition, the three standard error control limits for the R-chart implicitly assume that the sample range follows a normal distribution, which is an especially bad assumption because the distribution of the sample range is quite skewed to the right for the small to moderate sample sizes commonly used to monitor processes. The increased use of computer software, especially spreadsheets, to perform the necessary calculations eliminates the computational advantage enjoyed by the sample range. Frankly, if we use the computer to generate our charts, then statistical theory suggests that we should never use the sample range.

Let σ_0^2 be the target value for the within-subgroup or process variance, and let s_i^2 be the sample variance for the ith subgroup. If the process is in control and if the data follow a normal distribution, then from Section 4.7 we know that

$$\frac{(n-1)s_i^2}{\sigma_0^2}$$

follows a χ^2 distribution with $n-1$ degrees of freedom. We can use this relationship to develop an appropriate monitoring scheme for the within-subgroup variability.

Consider a sequence of hypothesis tests of the form

$$H_0: \quad \sigma^2 = \sigma_0^2$$
$$H_a: \quad \sigma \neq \sigma_0^2.$$

The appropriate test statistic is

$$\chi^2 = \frac{(n-1)s_i^2}{\sigma_0^2}.$$

We reject the null hypothesis if either

$$\frac{(n-1)s_i^2}{\sigma_0^2} > \chi_{n-1,\alpha/2}^2 \quad \text{or} \quad \frac{(n-1)s_i^2}{\sigma_0^2} < \chi_{n-1,1-\alpha/2}^2.$$

For convenience, we need to restate the critical region in terms of s_i^2. Since both $n-1$ and σ_0^2 are positive, we can rewrite these inequalities as

$$s_i^2 > \frac{\chi_{n-1,\alpha/2}^2 \sigma_0^2}{n-1} \quad \text{and} \quad s_i^2 < \frac{\chi_{n-1,1-\alpha/2}^2 \sigma_0^2}{n-1}.$$

Three standard error control charts correspond to $\alpha = 0.0027$. Following this tradition, these are our control limits:

$$UCL = \frac{\chi_{n-1,0.00135}^2 \sigma_0^2}{n-1}$$

$$LCL = \frac{\chi_{n-1,0.99865}^2 \sigma_0^2}{n-1}.$$

For the grinding of silicon wafers example, $n = 5$. As a result, $\chi^2_{n-1,0.00135} = 17.800$ and $\chi^2_{n-1,0.99865} = 0.106$.

Often, when we start an s^2-chart, we do not know the process variance. In these situations, we treat the first m subgroups as a base period. If the process is in control, an appropriate estimate of the process variance is

$$\overline{s^2} = \frac{1}{m} \sum_{i=1}^{m} s_i^2,$$

which is an extension of the pooled estimate of the variance we used in the two-independent-samples t-test. Technically, $\overline{s^2}$ has $m \cdot (n-1)$ degrees of freedom. For most realistic base periods, the number of degrees of freedom is large enough to treat this estimate as the true value. The resulting control limits are

$$UCL = \frac{\chi^2_{n-1,0.00135}\overline{s^2}}{n-1}$$

$$LCL = \frac{\chi^2_{n-1,0.99865}\overline{s^2}}{n-1}.$$

Since Roes and Does did not have good prior information on the true process variance, consider the first 20 subgroups as a base period. Table 5.11 lists the data and the sample variances. We can estimate σ_0^2 by

$$\overline{s^2} = \frac{1}{m} \sum_{i=1}^{m} s_i^2$$

$$= \frac{1}{20}(298.9)$$

$$= 14.945.$$

The upper control limit is

$$UCL = \frac{\chi^2_{n-1,0.00135}\overline{s^2}}{n-1}$$

$$= \frac{\chi^2_{4,0.00135}\overline{s^2}}{n-1}$$

$$= \frac{17.800\,(14.945)}{4}$$

$$= 66.51.$$

Table 5.11 | The Data and Sample Variances for the Thickness of Silicon Wafer Data

Batch	y_{i1}	y_{i2}	y_{i3}	y_{i4}	y_{i5}	s_i^2
1	240	243	250	253	248	27.7
2	238	242	245	251	247	24.3
3	239	242	246	250	248	20.0
4	235	237	246	249	246	38.3
5	240	241	246	247	249	15.3
6	240	243	244	248	245	8.5
7	240	243	244	249	246	11.3
8	245	250	250	247	248	4.5
9	238	240	245	248	246	17.8
10	240	242	246	249	248	15.0
11	240	243	246	250	248	15.8
12	241	245	243	247	245	5.2
13	247	245	255	250	249	14.2
14	237	239	243	247	246	18.8
15	242	244	245	248	245	4.7
16	237	239	242	247	245	17.0
17	242	244	246	251	248	12.2
18	243	245	247	252	249	12.2
19	243	245	248	251	250	11.3
20	244	246	246	250	246	4.8
21	241	239	244	250	246	18.5
22	242	245	248	251	249	12.5
23	242	245	248	243	246	5.7
24	241	244	245	249	247	9.2
25	236	239	241	246	242	13.7
26	243	246	247	252	247	10.5
27	241	243	245	248	246	7.3
28	239	240	242	243	244	4.3
29	239	240	250	252	250	38.2
30	241	243	249	255	253	37.2

Figure 5.17 | The s^2-Chart for the Thicknesses of Silicon Wafers Data

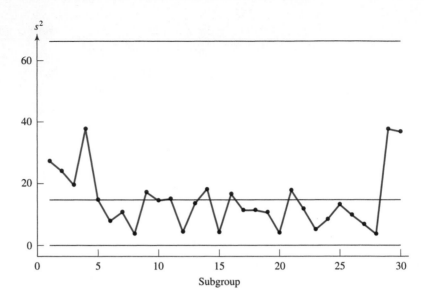

The lower control limit is

$$LCL = \frac{\chi^2_{n-1,0.99865}\overline{s^2}}{n-1}$$

$$= \frac{\chi^2_{4,0.99865}\overline{s^2}}{n-1}$$

$$= \frac{0.1058\,(14.945)}{4}$$

$$= 0.395.$$

Figure 5.17 shows the s^2-chart for all 30 subgroups using these control limits. Once again, the process appears to be in control.

The $\overline{X}$-Chart Based on s^2   Since the within-subgroup variability is stable, we can construct control limits to monitor the between-subgroup variability with the $\overline{X}$-chart. We have already seen that if we know the process variance, then the appropriate control limits for this chart are

$$UCL = \mu_0 + 3\frac{\sigma}{\sqrt{n}}$$

$$LCL = \mu_0 - 3\frac{\sigma}{\sqrt{n}}.$$

If σ is not known, we can substitute $\overline{s^2}$ for σ^2 as the basis for estimating these limits. Strictly speaking, we are conducting a sequence of t-tests; however, the degrees of freedom for these tests are $m \cdot (n-1)$ because we are using $\overline{s^2}$ to estimate σ^2. Since $m \cdot (n-1)$ is usually quite large, the appropriate critical value for the tests is essentially 3. The resulting control limits are

$$UCL = \mu_0 + 3\sqrt{\frac{\overline{s^2}}{n}}$$

$$LCL = \mu_0 - 3\sqrt{\frac{\overline{s^2}}{n}}.$$

If we need to estimate the target value for the subgroup means, the control limits are

$$UCL = \overline{\overline{y}} + 3\sqrt{\frac{\overline{s^2}}{n}}$$

$$LCL = \overline{\overline{y}} - 3\sqrt{\frac{\overline{s^2}}{n}}.$$

For the grinding of silicon wafers example, we have

$$n = 5$$
$$\overline{\overline{y}} = 245.18$$
$$\overline{s^2} = 14.945.$$

The resulting control limits are

$$UCL = \overline{\overline{y}} + 3\sqrt{\frac{\overline{s^2}}{n}}$$

$$= 245.18 + 3\sqrt{\frac{14.945}{5}}$$

$$= 245.18 + 5.19$$

$$= 250.37$$

$$LCL = \overline{\overline{y}} - 3\sqrt{\frac{\overline{s^2}}{n}}$$

$$= 245.18 - 3\sqrt{\frac{14.945}{5}}$$

$$= 245.18 - 5.19$$

$$= 239.99,$$

Figure 5.18 | The $\overline{X}$-Chart Based on s^2 for the Thicknesses of Silicon Wafer Data

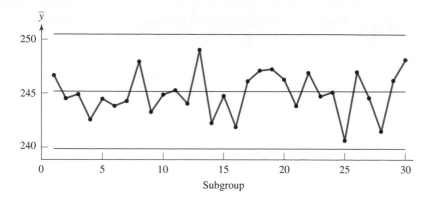

which are quite close to the limits based on the sample range. Figure 5.18 shows the resulting control chart. Once again, we see no evidence of any out-of-control conditions.

We should always check our assumptions when we first construct a control chart. The s^2-chart assumes

- The process is in control during the base period.
- The data follow a normal distribution.

Figure 5.19 | The Stem-and-Leaf Display for the Silicon Wafer Data

Stem	Leaves	Number	Depth
23f.:	5	1	1
23s.:	777	3	4
23•.:	88999	5	9
24*.:	000000011	9	18
24t.:	22222233333333	14	32
24f.:	44444555555555555	17	49
24s.:	666666666666677777777	21	
24•.:	8888888888999999	16	30
25*.:	00000000111	11	14
25t.:	23	2	3
25f.:	5	1	1

The $\overline{X}$-chart based on the sample variance assumes

- The s^2-chart is in control; that is, the process variance is constant.
- The subgroup means are in control during the base period.
- The sample size is large enough to assume that the subgroup means follow a normal distribution by the Central Limit Theorem.

In the grinding of silicon wafers example, both the s^2- and the $\overline{X}$-charts appear in control. We next should look at Figure 5.19, the stem-and-leaf display of the data during the base period. This display appears roughly to follow a bell-shaped curve. As a result, we should feel reasonably comfortable about our control limits.

⋗ Exercises

5.17 Consider the breaking strength data from Exercise **5.9**. Again, use all 20 subgroups as a base period.

a. Calculate the appropriate control limits for an s^2-chart and plot the data. Comment on your results.

b. Calculate the appropriate control limits for an $\overline{X}$-chart based on s^2 and plot the data. Comment on your results.

c. What did you assume to construct these control limits? Given the information in the base period, how comfortable are you with these assumptions?

5.18 Consider the outside diameters from Exercise **5.10**. Again, use the first 20 subgroups as a base period.

a. Calculate the appropriate control limits for an s^2-chart and plot the data. Comment on your results.

b. Calculate the appropriate control limits for an $\overline{X}$-chart based on s^2 and plot the data. Comment on your results.

c. What did you assume to construct these control limits? Given the information in the base period, how comfortable are you with these assumptions?

5.19 Consider the thicknesses of paint can ears from Exercise **5.11**. Again, use the first 20 subgroups as a base period.

a. Calculate the appropriate control limits for an s^2-chart and plot the data. Comment on your results.

b. Calculate the appropriate control limits for an $\overline{X}$-chart based on s^2 and plot the data. Comment on your results.

c. What did you assume to construct these control limits? Given the information in the base period, how comfortable are you with these assumptions?

5.20 Consider the ash content data from Exercise **5.12**. Again, use all 20 subgroups as a base period.

a. Calculate the appropriate control limits for an s^2-chart and plot the data. Comment on your results.

b. Calculate the appropriate control limits for an $\overline{X}$-chart based on s^2 and plot the data. Comment on your results.

c. What did you assume to construct these control limits? Given the information in the base period, how comfortable are you with these assumptions?

5.21 Consider the times to fill dump trucks from Exercise **5.13**. Again, use the first 20 days as a base period.

a. Calculate the appropriate control limits for an s^2-chart and plot the data. Comment on your results.

b. Calculate the appropriate control limits for an $\overline{X}$-chart based on s^2 and plot the data. Comment on your results.

c. What did you assume to construct these control limits? Given the information in the base period, how comfortable are you with these assumptions?

5.22 Consider the yield data from Exercise **5.14**. Again, use all 20 shifts as a base period.

a. Calculate the appropriate control limits for an s^2-chart and plot the data. Comment on your results.

b. Calculate the appropriate control limits for an $\overline{X}$-chart based on s^2 and plot the data. Comment on your results.

c. What did you assume to construct these control limits? Given the information in the base period, how comfortable are you with these assumptions?

5.23 Consider the wafer line widths from Exercise **5.15**.

a. Using the first 20 wafers as a base period, calculate the appropriate control limits for an s^2-chart and plot the data (wafer standard deviations). Comment on your results.

b. Using the first 20 wafers as a base period, calculate the appropriate control limits for an $\overline{X}$-chart based on s^2 and plot the data (wafer means). Comment on your results.

c. What did you assume to construct these control limits? Given the information in the base period, how comfortable are you with these assumptions?

d. Using all 10 cassettes as a base period, calculate the appropriate control limits for an s^2-chart and plot the data (cassette standard deviations). Comment on your results.

e. Using all 10 cassettes as a base period, calculate the appropriate control limits for an $\overline{X}$-chart based on s^2 and plot the data (cassette means). Comment on your results.

f. What did you assume to construct these control limits? Given the information in the base period, how comfortable are you with these assumptions?

g. Compare the charts in part **a** and **b** with those in part **d** and **e**. Comment on their similarities and differences.

5.24 Consider the gross weights from Exercise **5.16**. Again, use all 20 subgroups as a base period.

    a. Calculate the appropriate control limits for an s^2-chart and plot the data. Comment on your results.

    b. Calculate the appropriate control limits for an $\overline{X}$-chart based on s^2 and plot the data. Comment on your results.

    c. What did you assume to construct these control limits? Given the information in the base period, how comfortable are you with these assumptions?

⠖ 5.5 X-Chart

Some engineering processes do not produce data frequently enough to justify monitoring subgroup means. In such cases, we need to use each individual observation. An example illustrates this situation.

Example 5.8	**Viscosities from a Batch Chemical Process**

Holmes and Mergen (1992) studied a batch operation at a chemical plant where an important quality characteristic was the product viscosity. At the end of each 12-hour batch, an operator took a viscosity measurement. Since data from this process come infrequently, management required a monitoring procedure based on the individual viscosities. Such a procedure would allow the operator to determine whether the process is out of control at the end of each batch.

Let μ_0 be the target value for the process mean, and let σ^2 be the true process variance. If both are known and if we can assume that the viscosities follow a normal distribution, then the appropriate control limits are

$$UCL = \mu_0 + 3\sigma$$
$$LCL = \mu_0 - 3\sigma.$$

In this particular case, Holmes and Mergen knew neither μ_0 nor σ^2. In such a situation, we use the first m observations as a base period. Traditionally, engineers use the *moving range* to estimate the process variability. Let y_i be the ith viscosity observed, and let y_{i+1} be the next viscosity observed. The ith moving range, MR_i, is given by

$$MR_i = |y_{i+1} - y_i|.$$

Let $\overline{MR}$ be the average moving range over the base period. Since the moving range involves successive observations, we have $m - 1$ moving ranges over a base period of m observations. Thus,

$$\overline{MR} = \frac{1}{m-1} \sum_{i=1}^{m-1} MR_i.$$

If the target value is known and the process is in control over the base period, then the traditional estimates of the control limits are

$$UCL = \mu_0 + \frac{3 \cdot \overline{MR}}{d_2}$$

$$LCL = \mu_0 - \frac{3 \cdot \overline{MR}}{d_2},$$

where d_2 is an appropriate constant. Since we use successive observations to calculate the moving average, we are using subgroups of size 2 to estimate the variability. Thus, $d_2 = 1.128$, and the resulting control limits are

$$UCL = \mu_0 + 2.66 \cdot \overline{MR}$$
$$LCL = \mu_0 - 2.66 \cdot \overline{MR}.$$

If we do not know μ_0 and the process is in control over the base period, then we estimate it by

$$\overline{y} = \frac{1}{m} \sum_{i=1}^{m} y_i,$$

and the resulting control limits are

$$UCL = \overline{y} + 2.66 \cdot \overline{MR}$$
$$LCL = \overline{y} - 2.66 \cdot \overline{MR}.$$

Since Holmes and Mergen knew neither μ_0 nor σ^2, consider the first 40 observations as a base period. Table 5.12 lists the data and the moving ranges. In this particular case, since we use the first 40 observations as our base period, we have 39 moving ranges. The average moving range is

$$\overline{MR} = \frac{1}{m-1} \sum_{i=1}^{m-1} MR_i$$

$$= \frac{1}{39} (41.7)$$

$$= 1.07.$$

The estimate of the target value is

$$\overline{y} = \frac{1}{m} \sum_{i=1}^{m} y_i$$

$$= \frac{1}{40} (594.6)$$

$$= 14.87.$$

Table 5.12	The First 40 Viscosities and the Moving Ranges

Observation	Viscosity	MR_i	Observation	Viscosity	MR_i
1	13.3	1.2	21	14.9	1.2
2	14.5	0.8	22	13.7	1.5
3	15.3	0.0	23	15.2	0.7
4	15.3	1.0	24	14.5	0.8
5	14.3	0.5	25	15.3	0.3
6	14.8	0.4	26	15.6	0.2
7	15.2	0.3	27	15.8	2.5
8	14.9	0.3	28	13.3	0.8
9	14.6	0.5	29	14.1	1.3
10	14.1	0.2	30	15.4	0.2
11	14.3	1.8	31	15.2	0.0
12	16.1	3.0	32	15.2	0.7
13	13.1	2.4	33	15.9	0.6
14	15.5	2.9	34	16.5	2.5
15	12.6	2.0	35	14.0	1.1
16	14.6	0.3	36	15.1	1.9
17	14.3	1.1	37	17.0	2.1
18	15.4	0.2	38	14.9	0.1
19	15.2	1.6	39	14.8	0.8
20	16.8	1.9	40	14.0	

The control limits are

$$UCL = \overline{y} + 2.66 \cdot \overline{MR}$$
$$= 14.87 + 2.66\,(1.07)$$
$$= 14.87 + 2.85$$
$$= 17.72$$

$$LCL = \overline{y} - 2.66 \cdot \overline{MR}$$
$$= 14.87 - 2.66\,(1.07)$$
$$= 14.87 - 2.85$$
$$= 12.02.$$

Table 5.13 | The Next 100 Viscosities

Observation	Viscosity	Observation	Viscosity	Observation	Viscosity	Observation	Viscosity	Observation	Viscosity
41	15.8	61	16.0	81	15.7	101	14.8	121	15.6
42	13.7	62	14.9	82	13.0	102	15.6	122	15.7
43	15.1	63	13.6	83	13.9	103	14.5	123	16.4
44	13.4	64	15.3	84	16.2	104	14.9	124	14.5
45	14.1	65	14.3	85	13.8	105	16.0	125	14.9
46	14.8	66	15.6	86	16.5	106	15.0	126	14.6
47	14.3	67	16.1	87	14.2	107	14.7	127	15.5
48	14.3	68	13.9	88	14.9	108	15.1	128	14.7
49	16.4	69	15.2	89	14.7	109	15.4	129	15.0
50	16.9	70	14.4	90	15.0	110	16.0	130	13.8
51	14.2	71	14.0	91	14.4	111	18.6	131	14.0
52	16.9	72	14.4	92	14.4	112	16.0	132	15.8
53	14.9	73	13.7	93	15.4	113	15.9	133	14.8
54	15.2	74	13.8	94	16.3	114	14.5	134	15.8
55	14.4	75	15.6	95	15.0	115	15.1	135	16.7
56	15.2	76	14.5	96	15.7	116	14.2	136	16.4
57	14.6	77	12.8	97	14.5	117	17.6	137	15.3
58	16.4	78	16.1	98	15.5	118	13.5	138	15.7
59	14.2	79	16.6	99	14.4	119	15.3	139	15.0
60	15.7	80	15.6	100	14.4	120	15.0	140	16.8

Table 5.13 gives the next 100 observations from this process, and Figure 5.20 shows the control chart for all 140 observations. We see that observation 111 is a possible out-of-control situation that requires attention. The chart suggests that the viscosity for this batch has drifted higher than normal.

We should always check our assumptions when we first construct a control chart. The X-chart assumes

- The process is in control during the base period.
- The data follow a normal distribution.

In this example, the X-chart appears in control. We next should look at a stem-and-leaf display of the data during the base period, which is given in Figure 5.21 This display appears roughly to follow a bell-shaped curve. As a result, we should feel reasonably comfortable about our control limits.

| Figure 5.20 | The Control Chart for the Viscosity Data |

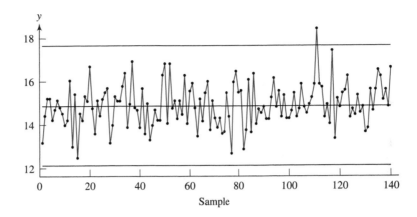

| Figure 5.21 | Stem-and-Leaf Display for the Viscosity Data |

Stem	Leaves	Number	Depth
12.:	6	1	1
13.:	1337	4	5
14.:	0011333556688999	16	
15.:	122222333445689	15	19
16.:	158	3	4
17.:	0	1	1

⫶ Exercises

5.25 Yashchin (1992) monitored the thicknesses of metal wires produced in a chip-manufacturing process. Ideally, these wires should have a target thickness of 8 microns. The thicknesses (in microns) follow. The data are in consecutive order, reading across the rows. The first observation is 8.4, the second is 8.0, and so on.

8.4	8.0	7.8	8.0	7.9	7.7	8.0	7.9	8.2	7.9
7.9	8.2	7.9	7.8	7.9	7.9	8.0	8.0	7.6	8.2
8.1	8.1	8.0	8.0	8.3	7.8	8.2	8.3	8.0	8.0
7.8	7.9	8.4	7.7	8.0	7.9	8.0	7.7	7.7	7.8
7.8	8.2	7.7	8.3	7.8	8.3	7.8	8.0	8.2	7.8

Use the first 30 observations as a base period.

a. Calculate the appropriate control limits and plot the control chart. Comment on your results.
b. What did you assume to construct these control limits? Given the information in the base period, how comfortable are you with these assumptions?

5.26 Cryer and Ryan (1990) monitor a chemical process where the quality characteristic is a color property. The data listed here are in consecutive order, reading across the rows. The first observation is 0.67, the second is 0.63, and so on.

0.67	0.63	0.76	0.66	0.69	0.71	0.72
0.71	0.72	0.72	0.83	0.87	0.76	0.79
0.74	0.81	0.76	0.77	0.68	0.68	0.74
0.68	0.69	0.75	0.80	0.81	0.86	0.86
0.79	0.78	0.77	0.77	0.80	0.76	0.67

Use the first 25 observations as a base period.

a. Calculate the appropriate control limits and plot the control chart. Comment on your results.
b. What did you assume to construct these control limits? Given the information in the base period, how comfortable are you with these assumptions?

5.27 Van Nuland (1992) daily compares two temperature instruments: one coupled to a process computer and the other used for visual control. Ideally, these two instruments should agree. As a result, he monitors the daily difference in the temperature readings (he is using a control chart based on *paired differences*). The temperature differences for 35 days follow. The data are in consecutive order, reading across the rows. The first observation is 0.3, the second observation is 0.0, and so on.

0.3	0.0	0.1	0.3	0.3
0.5	0.2	0.1	0.3	0.0
−0.1	0.5	0.4	0.1	0.1
−0.1	0.4	0.1	0.2	0.0
0.1	0.3	0.2	0.1	0.4
0.2	0.4	0.0	0.2	0.4
0.6	0.6	0.5	0.7	0.7

Use the first 30 observations as a base period.

a. Calculate the appropriate control limits and plot the control chart. Comment on your results.
b. What did you assume to construct these control limits? Given the information in the base period, how comfortable are you with these assumptions?

5.28 King (1992) monitors the net weights of a nominally 16-oz packaged product. An inspector collected a sample of 20 packages and accurately measured their net contents. The weights given here are in consecutive order, reading across the rows. The first observation is 16.4, the second observation is 16.4, and so on.

16.4	16.4	16.5	16.5	16.6	16.7	16.2	16.4	16.4	16.5
16.6	16.6	16.8	16.3	16.4	16.5	16.5	16.6	16.7	16.8

Use all the data as a base period.

a. Calculate the appropriate control limits and plot the control chart. Comment on your results.

b. What did you assume to construct these control limits? Given the information in the base period, how comfortable are you with these assumptions?

5.29 Roberts and Ling (1982) monitor the iron content of crushed blast-furnace slag. An inspector collected a sample of 30. The data are in consecutive order, reading across the rows. Thus, the first observation is 24, the second is 16, and so on.

24	16	24	18	18	10	14	16	18	20
21	20	21	15	16	15	17	19	16	15
15	13	24	22	21	24	15	20	20	25

Use all the data as a base period.

a. Calculate the appropriate control limits and plot the control chart. Comment on your results.

b. What did you assume to construct these control limits? Given the information in the base period, how comfortable are you with these assumptions?

5.30 Consider the semiconductor processing in Exercise **5.15**, which is based on a case study performed by NIST of a lithography process. The basic experimental unit is a silicon wafer. Three wafers are randomly selected from a cassette (typically a grouping of 25 wafers), and 10 cassettes are used in the study. The observations in y_{i2} are actually taken in the center of the wafer (the other y's are taken at different positions). We focus only on y_{i2} as shown in Table 5.14.

Use the first 20 wafers as a base period.

a. Calculate the appropriate control limits and plot the control chart. Comment on your results.

b. What did you assume to construct these control limits? Given the information in the base period, how comfortable are you with these assumptions?

Table 5.14	Line Width at One Position on Silicon Wafer

Cassette	Wafer	y_{i2}
1	1	2.074
1	2	1.861
1	3	1.625
2	4	2.072
2	5	1.473
2	6	1.537
3	7	1.357
3	8	1.520
3	9	1.367
4	10	1.786
4	11	1.919
4	12	2.171
5	13	1.950
5	14	2.534
5	15	2.963
6	16	1.941
6	17	2.256
6	18	1.601
7	19	1.672
7	20	1.854
7	21	1.959
8	22	0.971
8	23	1.165
8	24	1.403
9	25	1.699
9	26	1.241
9	27	0.791
10	28	1.938
10	29	1.793
10	30	2.475

❧ 5.6 *np*-Chart

We can develop a control chart to monitor the number of items that fail to meet specifications, which we typically model with a binomial distribution. The next example illustrates this technique.

Example 5.9 | **Nonconforming Bricks**

Marcucci (1985) looked at the number of nonconforming bricks from a manufacturing process. Management monitors this process by daily collecting random samples of 200 bricks and classifying them as conforming or nonconforming.

Suppose we take a random sample of size n each sampling period, where n remains constant over time. In our case, $n = 200$ and is the same for each sample. The *p*-chart, which we leave as Exercise 5.39, provides a basis for monitoring the number of nonconforming bricks when the sample size varies over time. Let p_0 be the target proportion for the number of nonconforming bricks, and let y_i be the number of nonconforming bricks in the ith subgroup. The expected value and variance of y_i are

> **VOICE OF EXPERIENCE**
>
> *For many engineering processes, the sample sizes for* np*-charts are quite large.*

$$E[y_i] = np_0$$

$$\mathrm{var}[y_i] = np_0q_0,$$

where $q_0 = 1 - p_0$. As long as the process is in control and we take the same size sample each time, the expected value and the variance for the number of nonconforming bricks remain constant.

We can now develop a monitoring scheme, called an *np-chart*, for the number of items that fail to meet the specifications. Consider a sequence of hypothesis tests of this form:

$$H_0: \quad np = np_0$$

$$H_a: \quad np \neq np_0.$$

If the smaller of np_0 and nq_0 is five or larger (preferably ten or larger because this condition will guarantee that the lower control limit is positive), the appropriate test statistic is

$$Z = \frac{y_i - np_0}{\sqrt{np_0q_0}}.$$

If the sample size remains constant each time, the appropriate control limits are

$$UCL = np_0 + 3 \cdot \sqrt{np_0q_0}$$

$$LCL = np_0 - 3 \cdot \sqrt{np_0q_0}.$$

With these control limits, operators need only count the number of nonconforming bricks in each subgroup. They then plot this count on the control chart. If the count falls within the control limits, they conclude that this process is in control and they do not need to take any action. On the other hand, if this count falls either above the upper control limit or below the lower control limit, then they conclude that the process is out of control and they must determine the cause of the problem and correct it.

Often, when we start a control chart, we do not know the true proportion of nonconforming items for the process. In such cases, we treat the first m subgroups as a base period. We can estimate the target proportion of items that fail to meet specifications by $\bar{p}$, which is the average proportion of items over the base period that fail to meet the specifications. If we take n items in each subgroup, then the total number of items inspected over the base period is mn. The total number of items that fail to meet the specifications is $\sum_{i=1}^{m} y_i$. The average proportion of items over the base period that fail to meet the specifications is

$$\bar{p} = \frac{1}{mn} \sum_{i=1}^{m} y_i.$$

If the process is in control over the base period, then the appropriate estimated control limits are

$$UCL = n\bar{p} + 3 \cdot \sqrt{n\bar{p}\bar{q}}$$
$$LCL = n\bar{p} - 3 \cdot \sqrt{n\bar{p}\bar{q}},$$

where $\bar{q} = 1 - \bar{p}$.

The numbers of nonconforming bricks are listed in Table 5.15. In this case, we have only 16 subgroups, which is small for a base period. Consequently, we shall use all 16 subgroups to estimate the control limits. The estimate of the proportion

Table 5.15 The Nonconforming Brick Data

Subgroup	y_i	Subgroup	y_i
1	9	9	13
2	8	10	31
3	12	11	18
4	8	12	15
5	16	13	15
6	9	14	16
7	11	15	10
8	12	16	9

of nonconforming bricks is

$$\bar{p} = \frac{1}{mn} \sum_{i=1}^{m} y_i$$

$$= \frac{1}{16 \cdot 200} (212)$$

$$= 0.06625.$$

The estimated control limits are

$$UCL = n\bar{p} + 3 \cdot \sqrt{n\bar{p}\bar{q}}$$
$$= 200 \cdot (0.06625) + 3\sqrt{200 \cdot (0.06625) \cdot (1.0 - 0.06625)}$$
$$= 13.25 + 10.55$$
$$= 23.80$$

$$LCL = n\bar{p} - 3 \cdot \sqrt{n\bar{p}\bar{q}}$$
$$= 200 \cdot (0.06625) - 3\sqrt{200 \cdot (0.06625) \cdot (1.0 - 0.06625)}$$
$$= 13.25 - 10.55$$
$$= 2.70.$$

| **Figure 5.22** | The *np*-Chart for the Nonconforming Bricks Data |

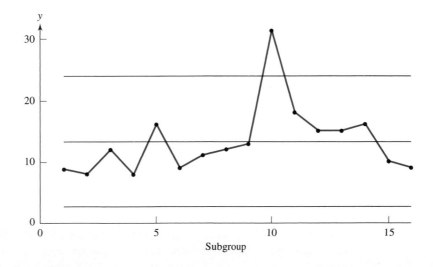

Figure 5.22 gives the control chart, which shows that sub-group 10 is a possible out-of-control situation that requires attention. The chart suggests that the number of nonconforming bricks from that day has drifted to a higher than normal level.

⫶ Exercises

5.31 Automobile hub caps typically are produced by a metal casting process that historically has been slow and expensive. A common problem facing many older casting processes is "flashing." In casting, liquid metal is shot into a mold and rapidly cooled. A flash commonly forms on the piece at the spot in the mold where the metal flows. A major automobile manufacturer closely monitors the incoming quality of the hub caps that come from a particular supplier by inspecting 200 hub caps from every incoming shipment. The following data are the number of hub caps with at least minor flashing in the last 40 shipments. The data are in consecutive order, reading across the rows. The first observation is 20, the second is 23, and so on.

20	23	20	15	25	24	27	18	17	20
26	15	20	21	15	18	12	25	16	25
24	27	21	19	14	23	17	20	19	20
23	18	25	22	17	20	22	24	15	11

Use the first 30 subgroups as a base period. Calculate the appropriate control limits and plot the control chart. Comment on your results.

5.32 A manufacturer of nickel–hydrogen batteries discovered a problem with "blisters" on its nickel plates. These blisters cause the resulting battery cell to short out prematurely. Each week, the manufacturer randomly selects 100 plates, constructs test cells, cycles these cells 50 times, and counts the number of plates that blister. The following data are the numbers of plates that blister in a 26-week period. The data are in consecutive order, reading across the rows. The first observation is 5, the second is 15, and so on.

5	15	7	11	3	12	7	11	12
7	12	8	10	8	6	4	5	7
9	9	8	0	11	11	8	7	

Use the first 20 subgroups as a base period. Calculate the appropriate control limits and plot the control chart. Comment on your results.

5.33 Felt-tip markers have shelf lives of approximately two years. Most manufacturers use accelerated life testing, whereby markers are placed in an oven at elevated temperatures for a given period of time, usually on the order of six weeks. The proportion that survive the elevated temperatures provides a good estimate of the proportion that should survive two years on a shelf. One major writing instrument company performs an accelerated life test on a random sample of 300 markers from each lot of markers. The following data are the numbers of markers that failed the accelerated life test in the last 40 lots. The data are in consecutive order, reading across the rows. The first observation is 14, the second is 16, and so on.

14	16	9	14	17	13	14	19	16	11
8	11	17	5	19	17	18	17	22	18
12	16	15	12	15	16	14	20	20	17
15	14	19	13	19	19	23	18	18	21

Use the first 30 subgroups as a base period. Calculate the appropriate control limits and plot the control chart. Comment on your results.

5.34 A major manufacturer of writing instruments uses high-speed equipment to assemble pencils. Each day, the operator randomly selects 10 gross of pencils $(10 \cdot 144, \text{ or } 1440)$ and classifies each as either OK or nonconforming. The following data are the numbers of nonconforming pencils on each of 30 days of production. The data are in consecutive order, reading across the rows. The first observation is 17, the second is 9, and so on.

17	9	15	17	12	12	17	15	15	10
14	9	15	10	15	17	21	14	18	10
19	21	20	26	19	21	18	20	15	20

Use the first 20 subgroups as a base period. Calculate the appropriate control limits and plot the control chart. Comment on your results.

5.35 A major semiconductor manufacturer tests 200 locations on a randomly selected wafer each hour from its new production process. The following data are the numbers of "dead" locations in the past 30 hours of production. The data are in consecutive order, reading across the rows. The first observation is 11, the second is 8, and so on.

11	8	11	7	11	10	1	11	15	8
11	15	9	11	15	11	7	4	14	10
8	9	5	5	12	7	9	14	15	11

Use the first 25 subgroups as a base period. Calculate the appropriate control limits and plot the control chart. Comment on your results.

5.36   A polymer chemist has developed a new adhesive. From each batch, an inspector makes 100 test specimens and subjects them to an accelerated life test. The following data are the numbers of test specimens that prematurely failed the life test in the last 40 batches. The data are in consecutive order, reading across the rows. The first observation is 24, the second is 16, and so on.

24	16	25	20	36	18	17	17	19	19
19	35	19	20	5	6	22	24	27	25
28	16	16	27	13	17	15	21	15	36
22	18	18	19	23	17	5	20	20	22

Use the first 30 subgroups as a base period. Calculate the appropriate control limits and plot the control chart. Comment on your results.

5.37   Airplanes approaching the runway for landing are required to stay within the localizer (a certain distance left and right of the runway). When an airplane deviates from the localizer, it is sometimes referred to as an exceedence. Each day, one airline randomly selects 200 flights and records the number in exceedence. The following data present the number of flights in exceedence on each of 30 days. The data are in consecutive order, reading across the rows. The first observation is 16, the second is 25, and so on.

16	25	23	16	15	23	24	15	23	13
21	19	23	28	30	19	18	15	18	10
24	22	20	27	21	26	30	25	26	30

Use the first 20 subgroups as a base period. Calculate the appropriate control limits and plot the control chart. Comment on your results.

5.38   A manufacturer is concerned about leaks from the gasket around their product. Each day they randomly select 200 products and record the number that leaked. The data are in consecutive order, reading across the rows. The first observation is 9, the second is 15, and so on.

9	15	11	12	12	8	9	9	12	13
14	8	7	7	16	11	9	13	19	5
10	9	11	10	15	16	16	11	15	15

Use the first 20 subgroups as a base period. Calculate the appropriate control limits and plot the control chart. Comment on your results.

Table 5.16	The Data for the Nonconforming Brick p-Chart

Subgroup	n_i	y_i	Subgroup	n_i	y_i
1	254	12	9	221	14
2	207	8	10	206	32
3	243	15	11	245	22
4	201	8	12	221	17
5	232	18	13	212	16
6	138	6	14	245	20
7	218	12	15	237	12
8	155	9	16	148	7

5.39 In some cases, we cannot get the same size sample each time. The proper control chart must adapt the control limits for the actual sample size used (called a *p-chart*). For example, the actual numbers of nonconforming bricks from Marcucci (1985) came from samples of differing sizes. Let n_i and let p_i be the actual sample size and the proportion of nonconforming bricks, respectively, for the ith subgroup. Use the hypothesis test for proportions described in Section 4.4 to develop an appropriate monitoring procedure based on the p_i's. Apply this procedure to the actual Marcucci data, which are given in Table 5.16.

5.7 c-Chart

We can develop a control chart to monitor small counts such as the number of incidents per period or the number of nonconformances per unit, where the size of the period or unit remains constant. We often can well model these counts by a Poisson distribution. The next example illustrates this technique.

Example 5.10	**Industrial Accident Data**

Lucas (1985) studied the number of accidents during a ten-year period at a major industrial facility. Historically, this company has strongly emphasized the importance of safety in its operations and has always striven to reduce the number of accidents over time. Corporate management expects each facility to closely monitor the accident rate.

Let c_i (for count) be the number of accidents in the ith calendar quarter (3-month period), and let λ be the expected rate of accidents. If we can model these counts by a Poisson distribution, then the expected value and variance for these counts are

$$E[c_i] = \lambda$$
$$\text{var}[c_i] = \lambda.$$

As long as the process is in control, the expected value and the variance for this count remain constant.

We can now develop a monitoring scheme, called a *c-chart*, for this count. Consider a sequence of hypothesis tests of the form

$$H_0: \quad \lambda = \lambda_0$$
$$H_a: \quad \lambda \neq \lambda_0.$$

If $\lambda \geq 5$ (preferably $\lambda \geq 10$ because that will guarantee that the lower control limit is positive), the appropriate test statistic is

$$Z = \frac{c_i - \lambda_0}{\sqrt{\lambda_0}}.$$

The resulting control limits are

$$UCL = \lambda_0 + 3\sqrt{\lambda_0}$$
$$LCL = \lambda_0 - 3\sqrt{\lambda_0}.$$

With these control limits, people need only count the number of incidents for each subgroup, and then plot this count on the control chart. If the count falls within the control limits, they conclude that this process is in control and they do not need to take any action. On the other hand, if this count falls either above the upper control limit or below the lower control limit, then they conclude that the process is out of control and they must determine the cause of the problem and correct it.

Often, when we start a control chart, we do not know the true expected count for the process. In such cases, we treat the first m subgroups as a base period. If the process is in control over the base period, then we can estimate the target count, λ_0, by $\bar{c}$, which is the average count over the base period and is given by

$$\bar{c} = \frac{1}{m} \sum_{i=1}^{m} c_i.$$

The appropriate estimated control limits for this situation are

$$UCL = \bar{c} + 3\sqrt{\bar{c}}$$
$$LCL = \bar{c} - 3\sqrt{\bar{c}}.$$

	The Accident Data by Quarter			
Table 5.17	Quarter	**Number of Accidents**	Quarter	**Number of Accidents**
	1	5	21	3
	2	5	22	4
	3	10	23	2
	4	8	24	0
	5	4	25	1
	6	5	26	3
	7	7	27	2
	8	3	28	2
	9	2	29	7
	10	8	30	7
	11	6	31	1
	12	9	32	4
	13	5	33	1
	14	6	34	2
	15	5	35	2
	16	10	36	1
	17	6	37	4
	18	3	38	4
	19	3	39	4
	20	10	40	4

Table 5.17 lists the numbers of accidents in each calendar quarter (3-month period) for our example. Lucas did not have a specific target for the accident rate at the beginning of the study. Consider the first 20 calendar quarters as the base period. An estimate of the in-control accident rate is

$$\bar{c} = \frac{1}{m} \sum_{i=1}^{m} c_i = \frac{1}{20} \cdot 120 = 6.0.$$

The resulting control limits are

$$UCL = \bar{c} + 3\sqrt{\bar{c}}$$
$$= 6.0 + 3\sqrt{6.0}$$
$$= 6.0 + 7.3$$
$$= 13.3$$

Figure 5.23 | The c-Chart for the Industrial Accident Data

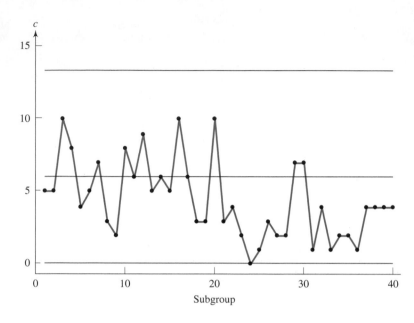

$$LCL = \bar{c} - 3\sqrt{\bar{c}}$$
$$= 6.0 - 3\sqrt{6.0}$$
$$= 6.0 - 7.3$$
$$= -1.3.$$

Since we cannot observe negative accidents, we use 0 for the lower control limit. Figure 5.23 shows the control chart. This process looks well in control over the base period. However, after the base period, only two observations are greater than $\bar{c}$. The other 18 observations all fall between $\bar{c}$ and the lower control limit, which is a clear sign that the accident rate has dropped at this facility. Such a reduction in the accident rate means that the facility's efforts to improve safety are working.

⫶ Exercises

5.40 Nelson (1987) considers a process in which an important quality characteristic is the number of flaws per length of wire. Routinely, inspectors examine 5000-m lengths of wire and count the number of flaws. The following data are the numbers of flaws in the last 30 sections inspected. The data are in consecutive order, reading across the rows. The first observation is 15, the second is 7, and so on.

15	7	13	13	5	8	15	10	10	7
14	16	15	14	21	10	15	15	13	24
22	18	18	14	8	11	6	10	1	3

Use the first 20 subgroups as a base period. Calculate the appropriate control limits and plot the control chart. Comment on your results.

5.41 The manufacture of silicon wafers used in integrated circuits requires the removal of contaminating particles of a certain size. Yashchin (1995) monitored a rinsing process for these wafers that rinses batches of 20 wafers with deionized water. The process then dries these wafers by spinning off the water droplets. Prior to loading the wafers into the rinser/dryer, production personnel count the number of contaminating particles. This count provides feedback on the cleanliness of the manufacturing environment. The following data are the counts per batch for 60 successive batches. The data are in consecutive order, reading across the rows. The first observation is 7, the second is 4, and so on.

7	4	9	9	2	10	3	6	6	5
5	7	5	7	3	4	8	4	5	5
9	8	8	8	13	10	6	10	11	3
11	13	9	11	13	15	6	10	11	12
12	2	4	7	2	4	7	6	7	4
6	4	6	6	8	5	6	9	3	6

Use the first 20 subgroups as a base period. Calculate the appropriate control limits and plot the control chart. Comment on your results.

5.42 A major automobile manufacturer inspects one car an hour for minor defects as the car rolls off the final assembly line. Virtually all of these defects are minor, usually cosmetic. The following data are the numbers of defects found on the last 40 cars inspected. The data are in consecutive order, reading across the rows. The first observation is 4, the second is 7, and so on.

4	7	5	6	9	5	10	5	6	9
7	5	6	2	4	8	9	7	7	6
8	9	8	6	7	9	5	8	6	15
1	6	5	6	6	4	10	3	7	3

Use the first 30 subgroups as a base period. Calculate the appropriate control limits and plot the control chart. Comment on your results.

5.43 A group of industrial engineering students created a control chart on the number of phone calls to the departmental office each hour. The following data are the numbers of calls during a 40-hour week. The data are in consecutive order, reading across the rows. The first observation is 4, the second is 14, and so on.

4	14	11	12	15	13	15	13
4	17	16	14	9	14	9	14
3	20	14	11	14	8	7	9
1	15	7	16	10	12	9	10
10	18	11	21	19	15	9	13

Use the first 30 subgroups as a base period. Calculate the appropriate control limits and plot the control chart. Comment on your results.

5.44 A long-distance carrier counts the number of calls that go through a critical station in a randomly selected minute every hour. The following data are the numbers of calls in the last 30 hours. The data are in consecutive order, reading across the rows. The first observation is 35, the second is 27, and so on.

35	27	33	44	46	31	35	29	39	68
64	39	38	31	37	16	19	17	23	33
31	41	28	26	49	30	44	44	32	33

Use the first 20 hours as a base period. Calculate the appropriate control limits and plot the control chart. Comment on your results.

5.45 An optical scanner manufacturer randomly selects one scanner an hour from its production line and tests it with a standard form. The company has designed this form to test the full capabilities of the scanner. As a result, it expects to see some errors. The following data are the numbers of errors in the past 20 hours of production. The data are in consecutive order, reading across the rows. The first observation is 14, the second is 14, and so on.

14	14	10	10	4	9	13	12	5	6
11	10	8	13	12	9	9	8	6	13

Use all 20 hours as a base period. Calculate the appropriate control limits and plot the control chart. Comment on your results.

5.46 Wrappers for candy bars are made on giant rolls, then shipped to the candy company for packaging. A manufacturer of wrappers monitors the process of the finished rolls by selecting a fixed sized cross-section of the roll and counting the number of imperfections. The following data are the number of imperfections in the last 30 sections inspected. The data are in consecutive order, reading across the rows. The first observation is 12, the second is 6, and so on.

12	6	3	10	5	7	7	5	7	7
10	8	9	9	5	8	5	8	3	8
5	8	5	7	7	7	12	6	11	7

Use the first 20 subgroups as a base period. Calculate the appropriate control limits and plot the control chart. Comment on your results.

5.47 A furniture manufacturer inspects one sofa a day for minor defects. Most of the defects are blemishes and loose stitching. The following data are the numbers of defects found on the last 30 sofas inspected. The data are in consecutive order, reading across the rows. The first observation is 4, the second is 0, and so on.

4	0	1	1	3	3	0	5	4	5
3	4	0	0	3	3	3	5	7	5
4	4	2	1	3	3	3	2	5	5

Use the first 20 subgroups as a base period. Calculate the appropriate control limits and plot the control chart. Comment on your results.

5.8 Average Run Lengths

The value of a control chart lies in its ability to detect assignable causes quickly with a minimum of false alarms. Too often, engineers act as if control charts never make mistakes; that is, whenever a control chart signals, there is an assignable cause, and when it fails to signal, no such cause is present. Unfortunately, like all statistical procedures, control charts do make mistakes. Control charts, sooner or later, must signal either due to a false alarm or due to an assignable cause. Statistical thinking allows us to quantify the consequences of these mistakes. We usually define the properties of control charts in terms of the run length.

Definition 5.1 | **Run Length**

The *run length*, N, is the number of samples taken before a control chart signals.

Typically, we use the *average run length, ARL,* defined by

$$ARL = E(N),$$

to describe the behavior of a specific control chart. To find the ARL, we first must know the distribution of N. Consider a control chart for μ when the process variance is known. Assume

- Each subgroup is independent of all other subgroups.
- The process mean is always at μ.

The second condition is an oversimplification. Essentially, it implies that we are looking at the properties of the procedure for a constant process mean. Under these assumptions, the run length, N, follows a *geometric* distribution.

The average run length makes sense only if we view the control chart as a sequence of hypothesis tests. Let $p(\mu)$ be the probability that the chart signals on any given sample when the process mean is μ. Thus, $p(\mu)$ represents the probability that a specific hypothesis test within the sequence rejects the null hypothesis. When the process is in control, $p(\mu_0) = \alpha$, where α is the false alarm rate. When the process is out of control, $p(\mu)$ represents the power of a specific hypothesis test when the process mean is exactly μ. Let $ARL(\mu)$ be the average run length when the process mean is μ. Since N follows a geometric distribution with parameter $p(\mu)$, we have

$$ARL(\mu) = E(N) = \frac{1}{p(\mu)}.$$

For an in-control process and three standard error limits, $\mu = \mu_0$ and $p(\mu_0) = 0.0027$, which is the false alarm rate. Thus,

$$ARL(\mu_0) = \frac{1}{0.0027} = 370.37.$$

For a standard $\overline{X}$-chart when we know the process variance, σ^2, we expect, on the average, to go 370.37 samples until a signal. For any other value of μ, we first must find the power of the hypothesis test for the specific value of the process mean, which is

$$p(\mu) = P\left(Z < -z_{\alpha/2} - \frac{\mu - \mu_0}{\sigma/\sqrt{n}}\right) + P\left(Z > z_{\alpha/2} - \frac{\mu - \mu_0}{\sigma/\sqrt{n}}\right)$$

by equation (4.5).

For example, consider a standard $\overline{X}$-chart when we know σ^2 that uses a sample size of four. Suppose from past experience we know that when the process goes out of control, the process mean shifts to $\mu_0 + \sigma$. On average, how long will the process operate in the out-of-control state before we see an out-of-control signal? To answer this question, we first must find the power of our basic hypothesis test to detect this shift on any given sample. When the process is out of control, $\mu = \mu_0 + \sigma$. The power is

$$p(\mu) = P\left(Z < -z_{\alpha/2} - \frac{\mu - \mu_0}{\sigma/\sqrt{n}}\right) + P\left(Z > z_{\alpha/2} - \frac{\mu - \mu_0}{\sigma/\sqrt{n}}\right)$$

$$= P\left(Z < -3 - \frac{\mu_0 + \sigma - \mu_0}{\sigma/\sqrt{4}}\right) + P\left(Z > 3 - \frac{\mu_0 + \sigma - \mu_0}{\sigma/\sqrt{4}}\right)$$

$$= P(Z < -5) + P(Z > 1)$$

$$= 0.1587. \tag{5.1}$$

The average run length is

$$ARL(\mu) = \frac{1}{p(\mu)} = \frac{1}{0.1587} = 6.3.$$

We see that this control chart averages 6.3 subgroups until it signals the presence of the assignable cause! Not all control charts pick up assignable causes quickly!

Table 5.18 summarizes the ARLs for various μ's for a standard $\overline{X}$-chart with known σ^2 when the sample size is four. Note these interesting features:

- For $\mu = \mu_0$, where the process is in control, the ARL is quite large, which means that we should not see many false alarms.
- For small shifts in μ, the ARLs can be quite large, which means that, with a sample size of four, this chart will not detect small shifts quickly.
- For large shifts in μ, the chart performs well, detecting the shift within one or two subgroups.

Table 5.19 considers the same control chart with $n = 9$. Increasing the sample size profoundly improves the ability of this control chart to detect small shifts. Naturally, we need to worry about rational subgroup issues as the sample size increases. Nonetheless, we can choose sufficiently large sample sizes for detecting shifts in the process mean rapidly for many engineering processes.

Table 5.18	Average Run Lengths When $n = 4$

μ	$p(\mu)$	$ARL(\mu)$
μ_0	0.0027	370.37
$\mu_0 + (0.5)\sigma$	0.0230	43.47
$\mu_0 + \sigma$	0.1587	6.30
$\mu_0 + (1.5)\sigma$	0.5000	2.00
$\mu_0 + 2\sigma$	0.8413	1.19
$\mu_0 + (2.5)\sigma$	0.9772	1.02

Table 5.19	Average Run Lengths When $n = 9$

μ	$p(\mu)$	$ARL(\mu)$
μ_0	0.0027	370.37
$\mu_0 + (0.5)\sigma$	0.0668	14.97
$\mu_0 + \sigma$	0.5000	2.00
$\mu_0 + (1.5)\sigma$	0.9332	1.07
$\mu_0 + 2\sigma$	≈ 0.99999	≈ 1.00
$\mu_0 + (2.5)\sigma$	≈ 1.00000	1.00

❧ *Exercises*

5.48   Confirm the average run lengths given in Table 5.18.

5.49   Confirm the average run lengths given in Table 5.19.

5.50   Construct a table similar to Table 5.18 when the sample size is two.

5.51   Construct a table similar to Table 5.18 when the sample size is five.

5.52   Construct a table similar to Table 5.18 when the sample size is 16.

❧ 5.9 Standard Control Charts with Runs Rules

From the preceding section, we know that standard control charts often have difficulty detecting small shifts quickly. Some engineers and statisticians suggest the use of one or more additional decision rules, often called *runs rules*.

Let θ_0 represent the target value for the parameter of interest, and let $\sigma_{\hat{\theta}}$ represent the appropriate standard error. The runs rules approach divides the chart into the following zones, as Figure 5.24 illustrates:

- Zone 1: $> \theta_0 + 3\sigma_{\hat{\theta}}$
- Zone 2: $> \theta_0 + 2\sigma_{\hat{\theta}}$ but $< \theta_0 + 3\sigma_{\hat{\theta}}$
- Zone 3: $> \theta_0 + \sigma_{\hat{\theta}}$ but $< \theta_0 + 2\sigma_{\hat{\theta}}$
- Zone 4: $> \theta_0$ but $< \theta_0 + \sigma_{\hat{\theta}}$
- Zone 5: $< \theta_0$ but $> \theta_0 - \sigma_{\hat{\theta}}$
- Zone 6: $< \theta_0 - \sigma_{\hat{\theta}}$ but $> \theta_0 - 2\sigma_{\hat{\theta}}$
- Zone 7: $< \theta_0 - 2\sigma_{\hat{\theta}}$ but $> \theta_0 - 3\sigma_{\hat{\theta}}$
- Zone 8: $< \theta_0 - 3\sigma_{\hat{\theta}}$

Figure 5.24 | The Zones Commonly Used for Runs Rules

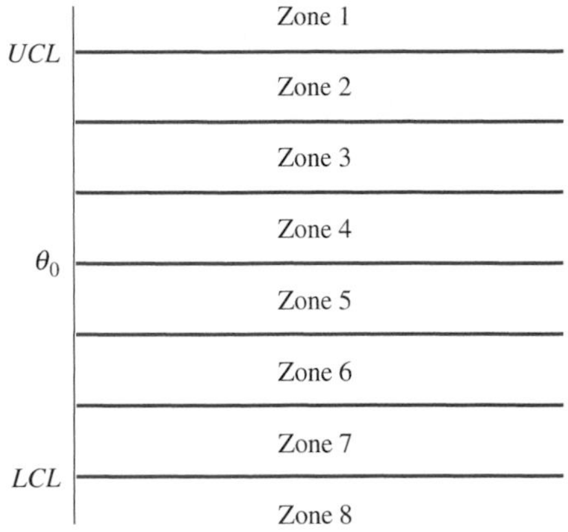

VOICE OF EXPERIENCE

Caution: *As we use more runs rules, we increase the chance of false alarms.*

The standard control chart signals whenever a point falls in either Zone 1 or Zone 8. The literature contains many additional decision rules, including these three common ones:

1. Two out of the last three points fall in Zone 2, or two out of the last three points fall in Zone 7.
2. Four out of the last five points fall in Zone 2 or 3, or four out of the last five points fall in Zone 6 or 7.
3. Eight points in a row fall in Zone 2, 3, or 4, or eight points in a row fall in Zone 5, 6, or 7.

How do these additional decision rules affect the properties of the chart? First, they reduce the average run length when the process is in control. Rule 3 by itself reduces the *ARL* to 128 from 370.37! Many practitioners use rule 3, but good statistical thinking should question its use. Second, the additional decision rules make the control chart more sensitive to small shifts. As a result, whenever rational subgrouping prevents us from using a large enough sample size, we should think very seriously about using these additional rules, especially rules 1 and 2. However, when we do use these additional rules, we must keep in mind the tradeoff between the increase in false alarms (the decrease in the *ARL* when in control) and the increased sensitivity to small shifts.

There are other runs rules that can be added to the charts besides zone rules. These patterns involve consecutive observations. For example, several points in a row, all increasing or decreasing may be an indication of a trend in the data. In manufacturing, an oscillating pattern typically is an indication of adjusting the process too often.

Example 5.11 | **Industrial Accident Data—Revisited**

When we originally analyzed these data, we considered the first 20 calendar quarters as the base period. The resulting estimate of the in-control accident rate was

$$\bar{c} = \frac{1}{m} \sum_{i=1}^{m} c_i$$

$$= \frac{1}{20} \cdot 120$$

$$= 6.0.$$

The resulting control limits were

$$UCL = 13.3$$
$$LCL = -1.3 = 0.$$

The appropriate warning limits are

$$\theta_0 + 2\sigma_{\hat{\theta}} = 6.0 + 2\sqrt{6.0} = 10.9$$
$$\theta_0 + \sigma_{\hat{\theta}} = 6.0 + \sqrt{6.0} = 8.4$$

Figure 5.25 | The c-Chart for the Industrial Accident Data with Runs Rules

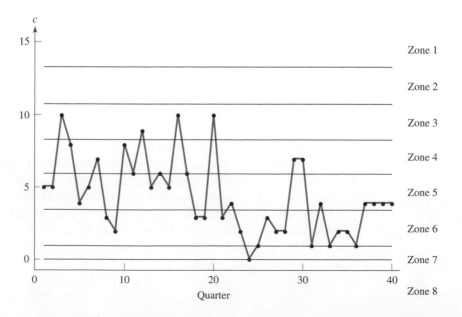

$$\theta_0 - \sigma_{\hat{\theta}} = 6.0 - \sqrt{6.0} = 3.6$$
$$\theta_0 - 2\sigma_{\hat{\theta}} = 6.0 - 2\sqrt{6.0} = 1.1.$$

Figure 5.25 shows the control chart with all eight zones. The standard decision rule does not indicate an assignable cause. Rule 1 does not help either. Rule 2, however, signals in quarter 25, indicating that the accident rate has actually decreased. As a result, the facility's safety efforts appear to be working.

Exercises

5.53   Repeat Exercise **5.3** using all three additional runs rules. Comment on your results.

5.54   Repeat Exercise **5.7** using all three additional runs rules. Comment on your results.

5.55   Repeat Exercise **5.12** using all three additional runs rules. Comment on your results.

5.56   Repeat Exercise **5.31** using runs rules 1 and 2 only. Comment on your results.

5.57   Repeat Exercise **5.34** using runs rules 1 and 2 only. Comment on your results.

5.58   Repeat Exercise **5.38** using runs rules 1 and 2 only. Comment on your results.

5.59   Repeat Exercise **5.41** using runs rules 1 and 2 only. Comment on your results.

5.60   Repeat Exercise **5.43** using all three additional runs rules. Comment on your results.

5.61   Repeat Exercise **5.46** using runs rules 1 and 2 only. Comment on your results.

5.10 CUSUM and EWMA Charts

Standard control charts have problems detecting small shifts because they use only the information contained in the current subgroup. A better approach for detecting small shifts uses the current plus at least some of the past subgroups, much like the runs rules use more than just the current subgroup. The basic idea is to accumulate information. Although no single subgroup indicates a problem, taken together the sequence of recent subgroups may signal a process shift.

Engineers and statisticians often use the cumulative sum (CUSUM) chart when they need to detect small shifts in a parameter of interest, especially when monitoring individual observations. The CUSUM chart attempts to use the *relevant* past subgroups. Of course, the trick is how we determine what subgroups constitute the relevant past. The CUSUM chart uses the *sequential probability ratio test* as the basis for determining how far back into the past to go in order to detect a trend. The sequential probability ratio test uses an important statistical concept best left for another course. Technically, the CUSUM chart is a sequence of sequential probability ratio tests designed for these hypotheses:

$$H_0: \quad \theta = \theta_0$$
$$H_a: \quad \theta = \theta_1,$$

where θ_0 and θ_1 are specific values. A subtle consequence of this approach is that CUSUM charts are inherently one-sided as opposed to the standard control charts we have described, which are all two-sided. To detect shifts in either direction, we must run two CUSUM procedures simultaneously. The resulting procedure signals a shift whenever one of the one-sided charts signals. The CUSUM chart tends to perform much better than the standard control chart for small shifts. The standard control chart actually tends to perform better for large shifts.

We present the CUSUM chart for individual, normally distributed observations, which is the usual case for this chart's use. Consider a sequence of individual observations from a normal distribution with mean μ and variance σ^2, where σ^2 is known. For the moment, consider the hypotheses

$$H_0: \quad \mu = \mu_0$$
$$H_a: \quad \mu = \mu_1,$$

where $\mu_1 > \mu_0$. Let S_i be the CUSUM statistic for the ith observation, where

$$S_i = \max[0, S_{i-1} + (z_i - d)].$$

In this expression, S_{i-1} is the value of the CUSUM statistic for the previous observation, $z_i = (y_i - \mu_0)/\sigma$, which is nothing more than standardizing the ith observation, y_i, and d is an appropriate constant, which depends on the size of the shift we wish to detect. We typically choose $S_0 = 0$ but not always. Whenever

$$S_{i-1} + (z_i - d) < 0,$$

the current information suggests that the process is more likely to be in control than out of control. Since we wish to detect a process shift as soon as possible, we reset the CUSUM statistic to 0. On the other hand, if $S_i > 0$, then the current information suggests that the process is more likely to be out of control: The larger S_i, the more likely the process is out of control. Once S_i exceeds some threshold, h, we conclude that the process is out of control.

These steps summarize the basic CUSUM chart:

1. Let $S_0 = 0$ be the initial value for the CUSUM statistic.
2. Signal a possible out of control state whenever $S_i > h$, where h is a suitably chosen bound.

For the case when $\mu_1 < \mu_0$, the CUSUM statistic is given by

$$S_i = \min[0, S_{i-1} + (z_i + d)],$$

and we signal an out-of-control state whenever $S_i < -h$.

The basic parameters of the CUSUM procedure are d and h. If we design the chart to detect a shift of one standard deviation ($\mu = \mu_0 \pm \sigma$), then the appropriate choice for d is 0.5. With this choice for d and $h = 5$, the in-control ARL is 465. In the absence of any other information, we typically use these values for d and h.

Example 5.12 | **Viscosities from a Cold Rolling Process**

Dodson (1995) studied an aluminum cold rolling process. The manufacturer discovered that it must control the coolant viscosity in order to produce aluminum that has an acceptable surface quality. Each day, a technician recorded this coolant's viscosity. Of particular concern to management was a drop in this viscosity. Table 5.20 lists the measurements on the first 20 days. For these data,

$$\bar{y} = 3.00$$

$$\overline{MR} = 0.0984.$$

Table 5.20 | The Coolant Viscosity Data

Day	Viscosity	Day	Viscosity
1	3.04	11	2.88
2	3.14	12	3.02
3	3.07	13	3.08
4	3.15	14	3.00
5	2.97	15	2.87
6	3.04	16	2.80
7	3.14	17	2.81
8	3.21	18	2.85
9	3.07	19	2.81
10	3.21	20	2.83

Figure 5.26 | The X-Chart for the Viscosity Data

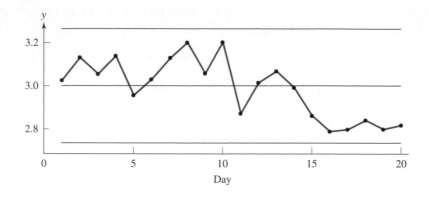

We use $\bar{y}$ to estimate μ_0. These are the control limits for the X-chart:

$$UCL = \bar{y} + 3\frac{\overline{MR}}{d_2} = 3 + 2.66 \cdot \overline{MR} = 3.262$$

$$LCL = \bar{y} - 3\frac{\overline{MR}}{d_2} = 3 - 2.66 \cdot \overline{MR} = 2.738.$$

Figure 5.26 gives the X-chart for these data and shows an in-control process.

Now, consider the CUSUM chart designed to detect a decrease in the viscosity. Once again, we use $\bar{y}$ to estimate μ_0. From our derivation of the control limits for the X-chart, we observe that an appropriate estimate of σ is

$$\hat{\sigma} = \frac{\overline{MR}}{d_2} = \frac{0.0984}{1.128} = 0.087.$$

The appropriate standardization of y_i is

$$z_i = \frac{y_i - \bar{y}}{\hat{\sigma}}.$$

Management decided that an appropriate CUSUM chart should use $S_0 = 0$, $d = 0.5$, and $h = 5$. The resulting CUSUM statistic is

$$S_i = \min\left[0, S_{i-1} + (z_i + d)\right]$$

$$= \min\left[0, S_{i-1} + \left(\frac{y_i - \bar{y}}{\hat{\sigma}} + 0.5\right)\right]$$

$$= \min\left[0, S_{i-1} + \left(\frac{y_i - 3.00}{0.087} + 0.5\right)\right].$$

We conclude that the process is out of control whenever $S_i < -5.0$.

| Table 5.21 | The CUSUM Statistics for the Viscosities |

Day	Viscosity	S_i	Day	Viscosity	S_i
1	3.04	0	11	2.88	−0.88
2	3.14	0	12	3.02	−0.15
3	3.07	0	13	3.08	0
4	3.15	0	14	3.00	0
5	2.97	0	15	2.87	−0.99
6	3.04	0	16	2.80	−2.79
7	3.14	0	17	2.81	−4.48
8	3.21	0	18	2.85	−5.70
9	3.07	0	19	2.81	−1.68
10	3.21	0	20	2.83	−3.14

Table 5.21 gives the data and the calculated values of the CUSUM statistic for the base period. The operator reset the CUSUM after the signal on day 18. Figure 5.27 shows the control chart for all 20 observations. Although the X-chart does not indicate any problems, the CUSUM chart signals on day 18 and looks like it will signal again shortly after day 20. In this example, we see how the CUSUM chart detects the small shift more quickly than the standard X-chart.

| Figure 5.27 | The CUSUM Chart for the Viscosity Data |

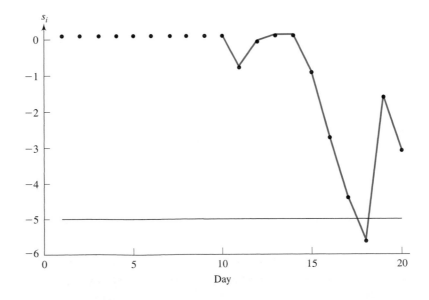

The exponentially weighted moving average (EWMA) control chart is another good alternative to the Shewhart control charts for detecting small shifts. The EWMA and CUSUM control charts perform similarly. We present the EWMA charts for individual, normally distributed observations. Consider a sequence of individual observations from a normal distribution with mean μ and variance σ^2, where σ^2 is known. Consider the hypotheses

$$H_0: \quad \mu = \mu_0$$

$$H_a: \quad \mu \neq \mu_0.$$

Let Z_i be the EWMA statistics for the ith observation, where

$$Z_i = \phi X_i + (1 - \phi) Z_{i-1}.$$

In this expression, X_i is the ith observation, Z_{i-1} is the value of the EWMA statistic for the previous observation, and $0 < \phi \leq 1$ is a weighting constant. (Even though traditionally λ is used for the weighting constant, we have used ϕ here because we have referred to the defect rate for Poisson distribution as λ.) We typically choose $Z_0 = \mu_0$, but sometimes we choose the average of a base period for Z_0.

The EWMA chart is constructed by plotting Z_i versus time i with appropriate control limits. To calculate the control limits for the EWMA chart, we need the Var(Z_i). Assume the observations, X_i are independent with variance σ^2; then the variance of Z_i is

$$\sigma_i{}^2 = \sigma^2 \phi / (2 - \phi)[1 - (1 - \phi)^{2i}].$$

The center line is typically the target value μ_0. Therefore, the control limits for the EWMA chart are as follows:

$$UCL = \mu_0 + k\sigma \sqrt{\frac{\phi}{(2 - \phi)}[1 - (1 - \phi)^{2i}]}$$

$$LCL = \mu_0 - k\sigma \sqrt{\frac{\phi}{(2 - \phi)}[1 - (1 - \phi)^{2i}]},$$

where k represents the width of the control limits. Typical values for ϕ are $0.05 \leq \phi \leq 0.25$ (see Crowder (1989) and Lucas and Saccucci (1990)). Also, $k = 3$, which corresponds to the usual 3 sigma limits, works reasonably well. Using $\phi = 0.1$ and $k = 2.7$ is approximately equivalent to using a CUSUM with $h = 5$ and $d = 0.5$.

Example 5.13 | Viscosities from a Cold Rolling Process—Revisited

In Example 5.12, we looked at a CUSUM control chart for the coolant viscosity in aluminum production. Table 5.20 lists the measurements on the first 20 days.

Day	Viscosity	Z_i	Day	Viscosity	Z_i
1	3.04	3.00355	11	2.88	3.05468
2	3.14	3.01720	12	3.02	3.05121
3	3.07	3.02248	13	3.08	3.05409
4	3.15	3.03523	14	3.00	3.04868
5	2.97	3.02871	15	2.87	3.03081
6	3.04	3.02983	16	2.80	3.00773
7	3.14	3.04085	17	2.81	2.98796
8	3.21	3.05777	18	2.85	2.97416
9	3.07	3.05899	19	2.81	2.95775
10	3.21	3.07409	20	2.83	2.94497

Table 5.22 | The EWMA Statistics for the Viscosities

Management decided to use an EWMA chart with $\phi = 0.1$ and $k = 2.7$. The resulting EWMA statistics is

$$Z_i = 0.1X_i + (1 - 0.1)Z_{i-1} = 0.1X_i + 0.9Z_{i-1}.$$

Figure 5.28 | The EWMA Chart for the Viscosity Data

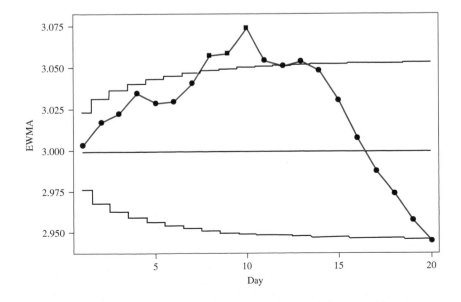

In this case, we will use the mean of the data as the target, $\mu_0 = \overline{X} = 2.9995$. Also, we will still use $\hat{\sigma} = .087$. This leads to the following control limits:

$$UCL = 2.9995 + 2.7(.087)\sqrt{\frac{0.1}{(2 - 0.1)}[1 - (1 - 0.1)^{2i}]}$$

$$LCL = 2.9995 - 2.7(.087)\sqrt{\frac{0.1}{(2 - 0.1)}[1 - (1 - 0.1)^{2i}]}.$$

Table 5.22 gives the data and the calculated values for the EWMA statistic. The EWMA chart, shown in Figure 5.28, does not signal a drop in viscosity, although the last data point is right at the lower control limit. These results are more in line with the Individuals control chart.

Exercises

5.62 Consider the thickness of metal wires data in Exercise **5.25**.
  a. Use a CUSUM chart with $d = 0.5$ and $h = 5.0$ to analyze the data. Comment on your results.
  b. Use an EWMA chart with $\phi = 0.1$ and $k = 2.7$ to analyze the data. Comment on your results.
  c. Compare all three charts.

5.63 Consider the color data in Exercise **5.26**.
  a. Use a CUSUM chart with $d = 0.5$ and $h = 5.0$ to analyze the data. Comment on your results.
  b. Use an EWMA chart with $\phi = 0.1$ and $k = 2.7$ to analyze the data. Comment on your results.
  c. Compare all three charts.

5.64 Consider the temperature differences data in Exercise **5.27**.
  a. Use a CUSUM chart with $d = 0.5$ and $h = 5.0$ to analyze the data. Comment on your results.
  b. Use an EWMA chart with $\phi = 0.1$ and $k = 2.7$ to analyze the data. Comment on your results.
  c. Compare all three charts.

5.65 Consider the weights data in Exercise **5.28**.
  a. Use a CUSUM chart with $d = 0.5$ and $h = 5.0$ to analyze the data. Comment on your results.
  b. Use an EWMA chart with $\phi = 0.1$ and $k = 2.7$ to analyze the data. Comment on your results.
  c. Compare all three charts.

5.11 Case Study

Pencil lead is a ceramic material that consists of clay and graphite. The clay provides the ceramic matrix that supports the graphite. Pencil lead formulation focuses on the ratio of clay to graphite: the more clay, the firmer the grade and the lighter the mark made by the pencil. One measure of the ratio of clay to graphite is the "ash" content. Inspectors take a sample of pencil lead from every lot and fire it at high temperature for a predetermined length of time (usually either 12 or 24 hours). This process burns off all of the "volatiles" (essentially the carbon), leaving the ash. The ash content is the weight of the remaining ash divided by the initial weight of the sample. We report the ash content as a percentage. A high ash content indicates too much clay in the formulation; a low ash content indicates too much graphite.

Faber routinely measures the ash content in each lot of pencil lead immediately after the firing step. Quality control uses subgroups of five lots as the basis for its control chart because this essentially represents one shift of production. Table 5.23 gives the ash content measurements for approximately a 2-month period. Faber used the first 30 subgroups as the base period.

The overall mean and the average sample variance for this base period were

$$\overline{\overline{y}} = 42.12$$
$$\overline{s^2} = 0.6922,$$

respectively. The control limits for the s^2-chart to monitor the within-subgroup variability were

$$UCL = 3.0803$$
$$LCL = 0.0183.$$

Figure 5.29 shows the resulting control chart *for the base period*. We see that subgroups 27 and 30 are out of control. Production investigated why and determined that the cause in each case was a rejected lot. In subgroup 27, the lot in question was too firm; in subgroup 30, the lot was too soft.

Since we could identify the assignable cause, we can drop these subgroups and recalculate the control limits. For the modified base period,

$$\overline{\overline{y}} = 42.09$$
$$\overline{s^2} = 0.4241.$$

The new control limits for the s^2-chart to monitor the within-subgroup variability were

$$UCL = 1.8874$$
$$LCL = 0.0112.$$

Table 5.23	The Pencil Lead Ash Data

Subgroup	y_{i1}	y_{i2}	y_{i3}	y_{i4}	y_{i5}	$\bar{y}$	s^2
1	43.0	42.5	42.2	43.5	42.6	42.76	0.253
2	42.5	42.5	42.2	42.3	42.6	42.42	0.027
3	40.2	42.0	41.4	42.1	42.0	41.54	0.638
4	41.5	40.3	41.7	41.2	42.3	41.40	0.540
5	42.4	41.7	42.5	42.5	43.2	42.46	0.283
6	40.7	41.0	41.4	42.0	42.0	41.42	0.342
7	42.2	42.0	42.0	42.5	43.7	42.48	0.507
8	42.6	43.7	41.3	42.7	42.7	42.60	0.730
9	42.5	42.1	42.2	42.3	40.3	41.88	0.802
10	42.6	42.6	40.5	42.1	42.4	42.04	0.783
11	41.0	42.3	42.2	42.3	40.9	41.74	0.523
12	41.0	40.3	40.5	40.1	40.5	40.48	0.112
13	40.7	40.8	40.5	41.5	42.3	41.16	0.548
14	41.9	41.8	41.7	41.6	41.4	41.68	0.037
15	41.3	41.7	42.8	42.4	41.5	41.94	0.403
16	41.7	42.7	41.0	40.4	42.3	41.62	0.877
17	40.4	42.4	42.9	42.1	42.7	42.10	0.995
18	42.0	42.8	42.3	42.6	42.5	42.44	0.093
19	42.5	42.6	42.0	42.0	42.6	42.34	0.098
20	42.9	41.8	42.4	42.4	42.5	42.40	0.155
21	42.6	42.2	42.3	41.1	41.2	41.88	0.467
22	42.3	42.3	42.3	41.6	41.6	42.02	0.147
23	42.1	43.0	42.1	42.2	41.6	42.20	0.255
24	41.8	41.8	42.2	43.3	41.9	42.20	0.405
25	41.9	43.6	43.1	43.0	43.2	42.96	0.403
26	43.0	43.7	42.2	42.7	43.0	42.92	0.297
27	41.7	45.0	43.4	40.0	43.6	42.74	3.718
28	43.2	41.7	41.5	43.0	42.9	42.46	0.633
29	43.4	43.3	43.8	42.9	41.9	43.06	0.523
30	38.3	42.9	43.8	43.1	43.5	42.32	5.172
31	41.3	41.5	41.5	41.9	41.4	41.52	0.052
32	41.3	41.7	41.7	41.9	41.4	41.60	0.060
33	41.7	41.7	41.7	42.0	41.8	41.78	0.017

Subgroup	y_{i1}	y_{i2}	y_{i3}	y_{i4}	y_{i5}	$\bar{y}$	s^2
34	41.7	41.7	41.6	41.3	41.3	41.52	0.042
35	41.8	41.7	41.7	42.4	41.6	41.84	0.103
36	41.3	42.0	41.5	41.1	41.6	41.50	0.115
37	40.1	38.8	40.9	41.0	39.4	40.04	0.903
38	40.4	41.5	41.0	40.7	39.6	40.64	0.503
39	41.6	40.7	41.6	39.4	40.0	40.66	0.948
40	40.3	41.3	41.3	41.4	40.6	40.98	0.247

Table 5.23   The Pencil Lead Ash Data (*Continued*)

Figure 5.30 gives the resulting control chart for the modified base period. We see no additional out of control subgroups.

Once we obtain a stable base period for the within-subgroup variability, we can proceed to the between-subgroup variability. The control limits, based on the modified base period, for the $\bar{X}$-chart to monitor the between-subgroup variability were

$$UCL = 42.97$$
$$LCL = 41.22.$$

Figure 5.29   The s^2-Chart for the Ash Content Data for the Initial Base Period

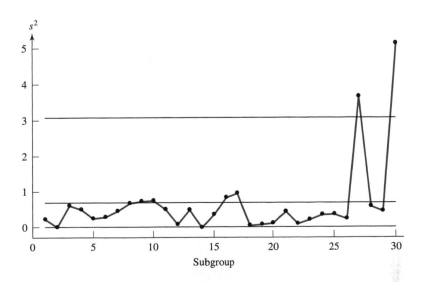

Figure 5.30 | The s^2-Chart for the Ash Content Data for the Modified Base Period

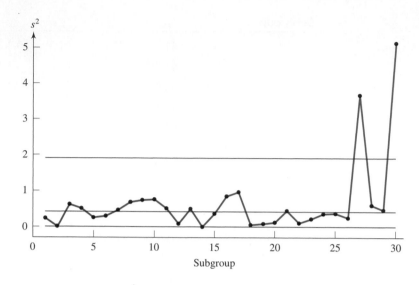

Figure 5.31 is the resulting control chart. We see that subgroups 12, 13, and 29 appear out of control. Production could find no problems with these specific subgroups. As a consequence, we have no basis for dropping them and recalculating the control limits.

Figure 5.32 shows the s^2-chart for the entire period, and Figure 5.33 gives the $\overline{X}$-chart. The s^2-chart indicates no new problems. The $\overline{X}$-chart, on the other hand, indicates that subgroups 37–40 have ash contents that are low. Production

Figure 5.31 | The $\overline{X}$-Chart for the Ash Content Data for the Modified Base Period

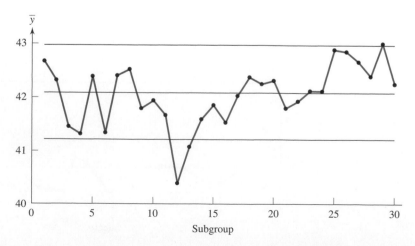

Figure 5.32	The s^2-Chart for the Ash Content Data for the Entire Period

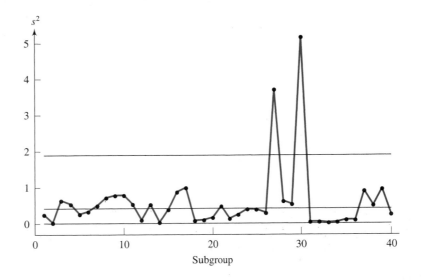

investigated and determined that these ash contents corresponded to a new shipment of one of the clays. Supervision quickly corrected the problem.

Finally, we need to confirm that the underlying assumptions for the control charts are reasonable. The s^2-chart assumes

- The process is in control during the base period.
- The data follow a normal distribution.

Figure 5.33	The $\overline{X}$-Chart for the Ash Content Data for the Entire Period

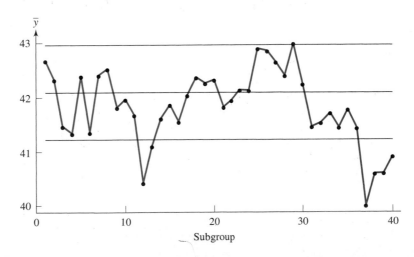

Figure 5.34 | The Stem-and-Leaf Display for the Ash Data

```
N = 140    Median = 42.2
Quartiles = 41.65, 42.6

Decimal point is at the colon

    7     7    40 : 1233344
   15     8    40 : 55557789
   27    12    41 : 000012233444
   49    22    41 : 5555666677777788889999
         42    42 : 00000000011111122222222233333333333444444
   49    30    42 : 5555555555566666666677777889999
   19    13    43 : 0000001222334
    6     6    43 : 567778
```

The $\overline{X}$-chart based on the sample variance assumes

- The s^2-chart is in control.
- The subgroup means are in control during the base period.
- The subgroup size is large enough to assume that the subgroup means follow a normal distribution by the Central Limit Theorem.

For the modified base period, the s^2-chart was in control, and we could not find any assignable causes for the $\overline{X}$-chart. We next should look at a stem-and-leaf display of the data during the base period, shown in Figure 5.34. This display appears roughly to follow a bell-shaped curve. As a result, we should feel reasonably comfortable about our control limits.

⁛ 5.12 Ideas for Projects

1. As a class, do Deming's Red Bead Experiment, which is a classic example of an np-chart. This experiment involves filling a bowl with a large number of red and white beads. Deming used 20% red. Make a paddle with 50 holes large enough to capture the beads. Use the paddle at least 20 times to sample from the bowl. Record the number of red beads in each sample. Use a control chart to analyze the results.
2. Weigh yourself daily for at least a month and record the results. Even better, weigh yourself two times a day for at least a month. Use a control chart to

analyze these data. If these weights are not in control, what are the assignable causes?

3. Record your daily expenditures on food or entertainment for a month. Use a control chart to analyze these data. If these expenditures are not in control, what are the assignable causes? Are there any periodic trends in the data (weekday versus weekend, for example)?

4. Many instructors use a catapult to teach basic statistical concepts. If you have access to one, fix the throwing conditions and launch the ball 30 times. Record the distances and use a control chart to analyze the results. If the process is not in control, what are the assignable causes?

5. Get data from a laboratory class where a large number of measurements are taken. For example, in organic chemistry laboratory classes, many students perform the same organic reaction experiment and record their yields. Ideally, the basic chemical reaction produces a specific yield. Use a control chart to analyze the consistency of the results. If the yields are not in control, what are the assignable causes? Other good sources are surveying classes and unit operations laboratories.

6. Use a control chart to analyze the average homework scores for the students in your class. If the scores are not in control, what are the assignable causes?

A generous reviewer suggested or inspired the following projects.

7. Count the number of cars stopped by a traffic light at a busy intersection for at least 20 cycles. If these counts are not in control, what are possible assignable causes?

8. Count the number of telephone calls made to the department office over 15-minute periods throughout the day. If these counts are not in control, what are possible assignable causes?

9. If you own an answering machine, count the number of messages you receive each day for a month. If these counts are not in control, what are possible assignable causes?

⠸ *References*

1. Crowder, S. V. (1989). Design of exponentially weighted moving average schemes. *Journal of Quality Technology, 21*, 155–162.

2. Cryer, J. D., and Ryan, T. P. (1990). The estimation of sigma for an X chart: $\overline{MR}/d_2$ or S/c_4. *Journal of Quality Technology, 22*, 187–192.

3. Dodson, B. (1995). Control charting dependent data: A case study. *Quality Engineering, 7*, 757–768.

4. Holmes, D. S., and Mergen, A. E. (1992). Parabolic control limits for the exponentially weighted moving average control charts. *Quality Engineering, 4*, 487–495.

5. Kane, V. E. (1986). Process capability indices. *Journal of Quality Technology, 18*, 41–52.

6. King, J. R. (1992). Tutorial comments on Lehrman manual [QE 4(1)]. *Quality Engineering, 5,* 107–122.

7. Lucas, J. M. (1985). Counted data CUSUM's. *Technometrics, 27,* 129–144.

8. Lucas, J. M., and Saccucci, M. S. (1990). Exponentially weighted moving average schemes: properties and enhancements. *Technometrics, 32,* 1–12.

9. Marcucci, M. (1985). Monitoring multinomial processes. *Journal of Quality Technology, 17,* 86–91.

10. Maxcy, R. B., and Lowry, S. R. (1984). Evaluating variability of filling operations. *Food Technology, 38*(12), 51–55.

11. Montgomery, D. C. (2004). *Introduction to statistical control,* 5th ed. New York: John Wiley.

12. Nelson, L. S. (1987). Comparison of Poisson means: The general case. *Journal of Quality Technology, 19,* 173–179.

13. Padgett, W. J., and Spurrier, J. D. (1990). Shewhart-type charts for percentiles of strength distributions. *Journal of Quality Technology, 22,* 283–288.

14. Roberts, H. V., and Ling, R. F. (1982). *Conversational statistics with IDA.* New York: Scientific Press/McGraw-Hill.

15. Roes, K. C. B., and Does, R. J. M. M. (1995). Shewhart-type chrts in nonstandard conditions. *Technometrics, 37,* 15–40 (with discussion).

16. Runger, G. C., and Pignatiello, J. J., Jr. (1991). Adaptive sampling for process control. *Journal of Quality Technology, 23,* 135–155.

17. Snee, R. D. (1983). Graphical analysis of process variation studies. *Journal of Quality Technology, 15,* 76–88.

18. Spurrier, J. D., and Thombs, L. A. (1990). Control charts for detecting cyclical behavior. *Technometrics, 32,* 163–170.

19. Van Nuland, Y. (1992). Maintaining calibration control with a control chart. *Quality Progress, 25*(3), 152.

20. Venables, W. N. (1989). Notes on elementary statistics. Department of Statistics, The University of Adelaide, Adelaide, Australia.

21. Yashchin, E. (1992). Analysis of CUSUM and other Markov-type control schemes by using empirical distributions. *Technometrics, 34,* 54–63.

22. Yashchin, E. (1995). Estimating the current mean of a process subject to abrupt changes. *Technometrics, 37,* 311–323.

6

Linear Regression Analysis

⋗ 6.1 Relationships Among Data

Engineers use *models,* which express the relationships among various engineering characteristics, to solve problems. These models allow engineers to predict a characteristic of interest, called the *response* or *dependent variable,* given the values of other characteristics, called the *independent variables.* We usually call these independent variables either the *regressors* or the *predictors.* For example, mechanical engineers routinely manipulate temperature (the regressor) to obtain a specific vapor pressure (the response) for steam. Chemical engineers manipulate the calcination temperature and stoichiometry (the regressors) of iron–cobalt hydroxides to maximize the surface area (the response) of a catalyst.

Engineers require data to estimate these models. In general, we obtain these data from either an *observational study* or a *designed experiment.* In observational studies, we merely observe the process, disturbing it only to the extent required to obtain data. We measure simultaneously, or as close to simultaneously as possible, the related engineering characteristics. In many observational studies, the distinction between the regressor and the response can be arbitrary. In a designed experiment, we actively manipulate the regressors, sometimes called the *factors* within this context, and then observe the resulting response. The next examples illustrate these techniques.

Example 6.1	**Relationship of Hardness and Young's Modulus for High-Density Penetrator Materials**

The military often uses high-density metals in munitions because these metals are better able to penetrate armor. Magness (1994) conducted an observational study of seven different high-density metals in which he measured the Rockwell hardness (the regressor) and the Young's modulus (the response) for a single specimen of each metal. The Young's modulus is the ratio of simple tension stress to the resulting strain parallel to the tension. Normally, we expect Young's modulus to

Table 6.1	The High-Density Penetrator Materials Data

Rockwell Hardness	Young's Modulus
41	310
41	340
44	380
40	317
43	413
15	62
40	119

increase as the Rockwell hardness increases. Table 6.1 gives the results for seven metals.

We will return to these data in Exercise 6.6.

Example 6.2	**Springs with Cracks**

Box and Bisgaard (1987) discuss a manufacturing operation for carbon-steel springs that have a severe problem with cracks. Basic metallurgy suggests that the cracking depends on these factors:

- The temperature of the steel before quenching
- The amount of carbon in the formulation
- The temperature of the quenching oil

The engineers use the percent of springs that do not exhibit cracking as their response. They seek to develop a model in these three factors to predict the cracking and to find conditions that will reduce or even eliminate the problem. The experimental results are listed in Table 6.2.

Engineers need data to build or to confirm models. Thus, after they collect the data, they must perform these tasks:

- Estimate the proposed model.
- Determine whether the hypothesized relationships truly exist.
- Determine the adequacy of the model.
- Determine the ranges for the regressors that allow reasonable prediction of the response.

| Table 6.2 | The Springs with Cracks Data |

Run	Steel Temp.	Percent Carbon	Oil Temp.	Percent Without Cracks
1	1450° F	0.50	70° F	67
2	1600° F	0.50	70° F	79
3	1450° F	0.70	70° F	61
4	1600° F	0.70	70° F	75
5	1450° F	0.50	120° F	59
6	1600° F	0.50	120° F	90
7	1450° F	0.70	120° F	52
8	1600° F	0.70	120° F	87

Initially, we shall restrict our attention to models that have only a single regressor and a response, which we call *simple linear regression*. Later in this chapter, we shall extend our analysis to models with two or more regressors, which we call *multiple linear regression*.

⑥ **6.2** Simple Linear Regression

Scatter Plots

The first step in analyzing the relationship between two characteristics of interest is to graph the data with the regressor on the horizontal or x-axis and the response on the vertical or y-axis. We call such a graph a *scatter plot*, and it allows us to visualize the relationship between the two characteristics. Seeing the nature of the relationship can help us to propose reasonable models.

| Example 6.3 | **Reflux Ratio for a Distillation Column** |

A chemical engineering professor used a pilot plant scale ethanol–water distillation column in the unit operations laboratory to illustrate the importance of the reflux ratio on the quality of the column's distillation. This column used a total condenser, which means all of the vapor from the top of the column is condensed to liquid. Let V represent the total vapor taken from the top of the column, let D represent the amount of the liquid that is taken off as final product, and let R represent the amount of liquid that is returned to the top of the column to "prime" it. By a basic material balance, we have

$$V = R + D.$$

Table 6.3 | The Reflux Ratio Data

Reflux Ratio	Concentration of Ethanol
20	0.446
30	0.601
40	0.786
50	0.928
60	0.950

The reflux ratio is R/D. For most separations of practical interest, this ratio is much greater than 1.

The professor divided the class into five teams. Only one team could use a specific piece of equipment each session. Over the semester, the professor had each team operate the column under exactly the same conditions except for the reflux ratio. After the column reached equilibrium, the students took a sample of the product and found the concentration of ethanol. Table 6.3 lists the results.

Figure 6.1 gives the scatter plot, which graphs each data pair of reflux ratio and concentration. We see that as the reflux ratio increases, so does the concentration of ethanol in the final product. This plot indicates that we can model the concentrations using a straight line in the reflux ratio. Although the data points display some curvature, a straight line may serve as a useful first approximation.

Figure 6.1 | The Scatter Plot for the Reflux Ratio Data

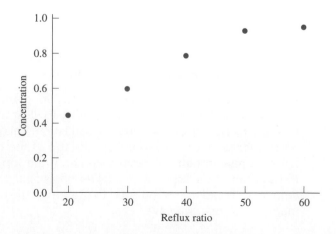

Least Squares Estimation

The Simple Linear Regression Model   Often, we can model the response as a straight line in the regressor. Let y_i be the ith response, and let x_i be the ith value for the regressor. The *simple linear regression model* is a linear relationship of the form

$$y_i = \beta_0 + \beta_1 x_i + \epsilon_i,$$

where β_0 is the y-intercept, β_1 is the slope, and ϵ_i is a random error. Usually, we assume that the random errors are independent with mean 0 and variance σ^2. Then the expected value of the response for any value of the regressor is

$$E[y] = \beta_0 + \beta_1 x_i.$$

This equation provides a basis for predicting the response given a specific value for the regressor.

The slope, β_1, represents the expected change in the response given a one-unit change in the regressor and determines the nature of the relationship between the response and the regressor. If $\beta_1 = 0$, then the response really does not depend on the regressor at all. The expected value of the response does not change as we change the values of the regressor, and we say that the response and the regressor are *uncorrelated*. If $\beta_1 < 0$, then the values of the response grow smaller as we increase the values of the regressor, and we say that the response and the regressor are *negatively correlated*. Conversely, if $\beta_1 > 0$, then the values of the response grow larger as we increase the values of the regressor, and we say that the response and the regressor are *positively correlated*. We should not confuse this sense of correlation with *causality*, which is necessary correlation and is an extremely strong claim about the nature of the relationship between the response and the regressor. Statistics can never establish the necessity of the relationship. Our models show only that the regressor and the response are related, not that one necessarily causes the other.

Our prediction equation depends on the y-intercept, β_0, and the slope, β_1, which in turn are parameters of the population and typically unknown. Traditionally, we estimate these parameters by the *method of least squares*, which minimizes in some meaningful way the errors in predicting the observed data.

A Measure of Overall Fit   The least squares estimates of β_0 and β_1 specifically minimize the *sum of the squares of the residuals*. Let $\hat{y}_i$ be the predicted value of the ith response. If b_0 is our estimate of the y-intercept, β_0, and if b_1 is our estimate of the slope, β_1, then the prediction equation is

$$\hat{y}_i = b_0 + b_1 x_i.$$

Figure 6.2 | Illustration of a Residual

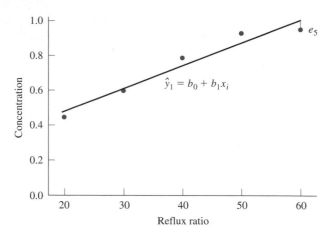

An appropriate measure of the quality of the fit of our model looks at the *residuals,* or the differences between the observed values for the response and the predicted values. We may view the residual as an estimate of the error for our prediction of the actual value. Let e_i be the ith residual defined by

$$e_i = y_i - \hat{y}_i.$$

Figure 6.2 illustrates the residual for the fifth pair of the reflux ratio data. The straight line represents the predicted values. The reflux ratio for the fifth pair is 60. The vertical distance from the straight line to the observed data value is the residual. In this case, since the actual concentration (y_5) is less than the value predicted by the straight line, the residual is negative. Values of e_i near zero, such as the one for a reflux ratio of 30, indicate a good fit.

In the spirit of the sample variance, an appropriate measure of the quality of the fit is the sum of squares for the residuals, SS_{res}, defined by

$$SS_{res} = \sum_{i=1}^{n} e_i^2$$
$$= \sum_{i=1}^{n} (y_i - \hat{y}_i)^2$$
$$= \sum_{i=1}^{n} [y_i - (b_0 + b_1 x_i)]^2.$$

Later, we shall see that SS_{res} provides a basis for estimating σ^2, which is the variance of the random errors.

Derivation of the Estimates   We call b_0 and b_1 the *least squares estimators* of β_0 and β_1 if they minimize SS_{res}. By minimizing SS_{res}, b_0 and b_1, in some sense, provide the best straight line equation to fit the data. From calculus, b_0 and b_1 minimize SS_{res} if they satisfy these equations:

$$\frac{\partial}{\partial b_0} \sum_{i=1}^{n} [y_i - (b_0 + b_1 x_i)]^2 = 0$$

$$\frac{\partial}{\partial b_1} \sum_{i=1}^{n} [y_i - (b_0 + b_1 x_i)]^2 = 0.$$

These derivatives result in the following "normal" equations:

$$n b_0 + b_1 \sum_{i=1}^{n} x_i = \sum_{i=1}^{n} y_i$$

$$b_0 \sum_{i=1}^{n} x_i + b_1 \sum_{i=1}^{n} x_i^2 = \sum_{i=1}^{n} x_i y_i.$$

Solving these two equations simultaneously, we get

$$b_0 = \bar{y} - b_1 \bar{x}$$

$$b_1 = \frac{SS_{xy}}{SS_{xx}},$$

where

$$\bar{x} = \frac{1}{n} \sum_{i=1}^{n} x_i$$

$$\bar{y} = \frac{1}{n} \sum_{i=1}^{n} y_i$$

$$SS_{xy} = \sum_{i=1}^{n} (x_i - \bar{x})(y_i - \bar{y})$$

$$= \sum_{i=1}^{n} x_i y_i - \frac{1}{n} \left(\sum_{i=1}^{n} x_i \right) \left(\sum_{i=1}^{n} y_i \right)$$

$$SS_{xx} = \sum_{i=1}^{n} (x_i - \bar{x})^2$$

$$= \sum_{i=1}^{n} x_i^2 - \frac{1}{n} \left(\sum_{i=1}^{n} x_i \right)^2.$$

Statistical software packages, spreadsheet programs, and many calculators find these estimates directly. Only rarely do engineers find these estimates by hand.

Table 6.4	Summary of Calculators for the Distillation Column Data

Reflux Ratio x_i	Concentration y_i	x_i^2	$x_i y_i$
20	0.446	400	8.92
30	0.601	900	18.03
40	0.786	1600	31.44
50	0.928	2500	46.40
60	0.950	3600	57.00
$\sum x_i = 200$	$\sum y_i = 3.711$	$\sum x_i^2 = 9000$	$\sum x_i y_i = 161.79$

Nonetheless, we can gain some insights into the nature of these estimates from these formulas.

Example 6.4	**Distillation Column Data—Estimating the Model**

From Table 6.4, we get

$$\bar{x} = \frac{1}{n} \sum_{i=1}^{n} x_i = \frac{1}{5}(200) = 40.0$$

$$\bar{y} = \frac{1}{n} \sum_{i=1}^{n} y_i = \frac{1}{5}(3.711) = 0.7422$$

$$SS_{xy} = \sum_{i=1}^{n} x_i y_i - \frac{1}{n} \left(\sum_{i=1}^{n} x_i \right) \left(\sum_{i=1}^{n} y_i \right)$$

$$= 161.79 - \frac{1}{5}(200)(3.711) = 13.35$$

$$SS_{xx} = \sum_{i=1}^{n} x_i^2 - \frac{1}{n} \left(\sum_{i=1}^{n} x_i \right)^2 = 9000 - \frac{1}{5}(200)^2 = 1000.$$

Our estimate of the slope is

$$b_1 = \frac{SS_{xy}}{SS_{xx}} = \frac{13.35}{1000} = 0.01335.$$

Figure 6.3 | The Plot of the Prediction Equation for the Reflux Ratio Data

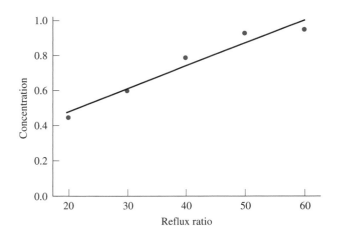

Our estimate of the *y*-intercept is

$$b_0 = \overline{y} - b_1\overline{x} = 0.7422 - 0.01335(40.0) = 0.2082.$$

The prediction equation is

$$\hat{y} = 0.2082 + 0.01335x.$$

Figure 6.3 shows how the estimated relationship fits the data. A straight line relationship provides a reasonable first approximation to the data, but the straight line does not perfectly fit the reflux ratio data. Rather, the plot indicates that the true relationship between the reflux ratio and concentration may be nonlinear. We may be able to generate a better model later.

Analysis of the Model

Tests and Intervals for the Slope   In many cases, the primary question we must address is whether the values of the response really depend on the specific values of the regressor. Trying to predict or to control the response by the regressor makes sense only if the two are related. Consider our model:

$$y_i = \beta_0 + \beta_1 x_i + \epsilon_i.$$

If $\beta_1 = 0$, then the response does not depend on the regressor. If $\beta_1 > 0$, then the response and the regressor have a positive relationship. Conversely, if $\beta_1 < 0$, then the response and the regressor have a negative relationship.

We have already seen that an appropriate estimator of β_1 is

$$b_1 = \frac{SS_{xy}}{SS_{xx}}.$$

Under our assumptions, we can show that

$$E(b_1) = \beta_1.$$

Thus, b_1 is an unbiased estimator of β_1. We can also show that the variance of b_1 is

$$\frac{\sigma^2}{SS_{xx}}.$$

Typically, we do not know σ^2; however, a reasonable estimate is called the mean squared residual, MS_{res}, which is

$$MS_{res} = \frac{SS_{res}}{df_{res}},$$

where df_{res} is the number of degrees of freedom associated with this estimate of the variance. In general,

$$df_{res} = (\text{number of observations}) - (\text{number of parameters estimated}).$$

In the simple linear regression case, we must estimate two parameters, β_0 and β_1, in order to calculate SS_{res}. If n is the total number of pairs of observations, then

$$df_{res} = n - 2$$
$$MS_{res} = \frac{SS_{res}}{n - 2}.$$

Under our assumptions, we can show that MS_{res} is an unbiased estimator of σ^2; that is,

$$E(MS_{res}) = \sigma^2.$$

Putting all this together, we get an appropriate estimate of the variance of b_1:

$$\widehat{\text{var}}(b_1) = \frac{MS_{res}}{SS_{xx}}.$$

If we can well model the random errors, the ϵ's, by a well-behaved distribution, then we can construct a $(1 - \alpha) \cdot 100\%$ confidence interval for β_1 by

$$b_1 \pm t_{n-2,\alpha/2} \cdot \sqrt{\frac{MS_{res}}{SS_{xx}}}.$$

The degrees of freedom are $n - 2$, which are the degrees of freedom associated with the estimate of the standard error.

Almost always we use computer software to perform the actual analysis. Let $\hat{\sigma}_{b_1}$ be the estimated standard error for b_1 given by

$$\hat{\sigma}_{b_1} = \sqrt{\frac{MS_{res}}{SS_{xx}}}.$$

Most statistical software packages print the estimated standard error for b_1 when they print out b_1. We can re-express the appropriate confidence interval by

$$b_1 \pm t_{n-2,\alpha/2} \cdot \hat{\sigma}_{b_1}.$$

In a similar manner, we can construct hypothesis tests for the slope.

| Example 6.5 | **Distillation Column Data—Testing the Slope** |

A reasonable question for these data considers the specific relationship between the reflux ratio and the concentration of ethanol in the final product. Unless otherwise stated, we shall use a 0.05 significance level throughout this chapter.

We first need to state our hypotheses. Let $\beta_{1,0}$ be a hypothesized value for β_1. In general, the hypotheses will have these forms:

$$H_0: \ \beta_1 = \beta_{1,0} \qquad H_0: \ \beta_1 = \beta_{1,0} \qquad H_0: \ \beta_1 = \beta_{1,0}$$
$$H_a: \ \beta_1 < \beta_{1,0} \qquad H_a: \ \beta_1 > \beta_{1,0} \qquad H_a: \ \beta_1 \neq \beta_{1,0}$$

Usually, $\beta_{1,0}$ will be zero, which corresponds to a null hypothesis of no relationship between the regressor and the response. In this particular case, a cursory knowledge of distillation suggests that as the reflux ratio increases, so does the concentration of ethanol in the final product. Thus, our hypotheses are

$$H_0: \ \beta_1 = 0$$
$$H_a: \ \beta_1 > 0.$$

If we can assume that the random errors follow a well-behaved distribution, then our test statistic is

$$t = \frac{b_1 - \beta_{1,0}}{\sqrt{MS_{res}/SS_{xx}}}.$$

Almost always, we use computer software to perform the actual analysis. Let $\hat{\sigma}_{b_1}$ be the estimated standard error for b_1 given by

$$\hat{\sigma}_{b_1} = \sqrt{\frac{MS_{res}}{SS_{xx}}}.$$

Most statistical software packages print the estimated standard error for b_1 when they print out b_1. We can re-express the appropriate test statistic by

$$t = \frac{b_1 - \beta_{1,0}}{\hat{\sigma}_{b_1}}.$$

The degrees of freedom for the test statistic are $n - 2$, which are the degrees of freedom associated with the estimate of the standard error.

In general, the critical regions are

- For H_a: $\beta_1 < \beta_{1,0}$, we reject H_0 if $t < -t_{n-2,\alpha}$.
- For H_a: $\beta_1 > \beta_{1,0}$, we reject H_0 if $t > t_{n-2,\alpha}$.
- For H_a: $\beta_1 \neq \beta_{1,0}$, we reject H_0 if $|t| > t_{n-2,\alpha/2}$.

In addition to the calculated value for the test statistic, computer software prints out the p-value. In Chapter 4, we learned that the p-value depends on the specific alternative hypothesis. Most software packages assume a two-sided alternative. When this is appropriate, we reject the null hypothesis whenever the given p-value is less than α.

In our case, we have a one-sided alternative, and we reject the null hypothesis if

$$t > t_{n-2,\alpha}$$
$$> t_{3,0.05}$$
$$> 2.353.$$

We have already found that $b_1 = 0.01335$. If we perform this analysis by hand, we next need to find SS_{res}. By definition,

$$SS_{res} = \sum_{i=1}^{n}(y_i - \hat{y}_i)^2.$$

An easier way to calculate SS_{res} uses SS_{total}, which is the overall variability in the data, given by

$$SS_{total} = \sum_{i=1}^{n}(y_i - \overline{y})^2$$
$$= \sum_{i=1}^{n} y_i^2 - \frac{1}{n}\left(\sum_{i=1}^{n} y_i\right)^2.$$

With some algebra, we can obtain the computational formula for SS_{res}:

$$SS_{res} = SS_{total} - b_1^2 SS_{xx}.$$

For the distillation column data, $\sum_{i=1}^{n} y_i^2 = 2.941597$. We have already found that $\sum_{i=1}^{n} y_i = 3.711$. Thus,

$$SS_{total} = \sum_{i=1}^{n} y_i^2 - \frac{1}{n} \left(\sum_{i=1}^{n} y_i \right)^2$$

$$= 2.941597 - \frac{1}{5}(3.711)^2$$

$$= 0.1872928.$$

We can find SS_{res} by

$$SS_{res} = SS_{total} - b_1^2 SS_{xx}$$

$$= 0.1872928 - (0.01335)^2(1000)$$

$$= 0.0090703.$$

We estimate the variance of the random errors by

$$MS_{res} = \frac{SS_{res}}{df_{res}}$$

$$= \frac{SS_{res}}{n-2}$$

$$= \frac{0.0090703}{3}$$

$$= 0.003023.$$

We now can evaluate our test statistic by

$$t = \frac{b_1}{\sqrt{MS_{res}/SS_{xx}}}$$

$$= \frac{0.01335}{\sqrt{0.003023/1000}}$$

$$= \frac{0.01335}{0.001739}$$

$$= 7.678.$$

Since $7.678 > 2.353$, we may reject the null hypothesis and conclude that the concentration of ethanol does increase as the reflux ratio increases.

Because we rejected the null hypothesis, we need to construct a range of plausible values for the slope. A 95% confidence interval for β_1 is

$$b_1 \pm t_{n-2,\alpha/2} \cdot \sqrt{\frac{MS_{res}}{SS_{xx}}} = 0.01335 \pm 3.183 \cdot \sqrt{\frac{0.003023}{1000}}$$

$$= 0.01335 \pm 0.00553$$

$$= (0.00782, 0.01888).$$

The plausible values for the slope range from 0.00782 to 0.01888, which again suggests a definite positive relationship between the reflux ratio and the ethanol concentration.

As usual, we should check our assumptions that the random errors are independent and follow a well-behaved distribution. Later in this chapter, we shall provide appropriate graphical analyses for doing this.

Using Confidence Intervals As an Alternative Analysis—One-Sided Case   In Chapter 4, we outlined how we can use confidence intervals to address questions of interest. We constructed the appropriate interval and checked to see whether the hypothesized value was plausible—that is, whether the hypothesized value fell within the calculated interval. In the process, we were able not only to conduct the test but also to determine whether the observed value was of practical significance. We can do the same analysis for slopes in simple linear regression.

Example 6.6 | **Distillation Column Data—Interval for the Slope**

Since our alternative hypothesis is greater than zero, we need only to compute a lower bound for the plausible values for the true slope. We take as our upper bound ∞. In this case, we do not need to split α between the two tails, so this lower bound is

$$b_1 - t_{n-2,\alpha} \cdot \sqrt{\frac{MS_{res}}{SS_{xx}}} = 0.01335 - 2.353 \cdot \sqrt{\frac{0.003023}{1000}}$$

$$= 0.01335 - 0.00409$$

$$= 0.00926.$$

The t-value used in the interval is the same as the t-value we used in our one-sided hypothesis test. Our interval is $(0.00926, \infty)$, which clearly does not include zero. This interval again suggests a definite positive relationship between the reflux ratio and the ethanol concentration.

Tests and Intervals for the Intercept   In many engineering problems, the y-intercept has little or no interpretation, and thus formal tests on β_0 make little sense. Nonetheless, we can easily extend the hypothesis test and confidence interval to the intercept whenever we have the least squares estimate and an appropriate estimate of its standard error.

We have already seen that an appropriate estimator of β_0 is

$$b_0 = \overline{y} - b_1 \overline{x}.$$

Under our assumptions, we can show that

$$E(b_0) = \beta_0.$$

Thus, b_0 is an unbiased estimator of β_0. We can also show that the variance of b_0 is

$$\sigma^2 \left(\frac{1}{n} + \frac{\overline{x}^2}{SS_{xx}} \right).$$

If our random errors are independent and follow a well-behaved distribution, then a $(1 - \alpha) \cdot 100\%$ confidence interval for the y-intercept, β_0, is

$$b_0 \pm t_{n-2,\alpha/2} \cdot \sqrt{MS_{res} \cdot \left(\frac{1}{n} + \frac{\overline{x}^2}{SS_{xx}} \right)}.$$

If we are interested in conducting a formal hypothesis test, then the test statistic is

$$\frac{b_0 - \beta_{0,0}}{\sqrt{MS_{res} \cdot \left(\frac{1}{n} + \frac{\overline{x}^2}{SS_{xx}} \right)}},$$

where $\beta_{0,0}$ is the hypothesized value for the y-intercept.

The Coefficient of Determination   We can partition the overall variability in the data, SS_{total}, into two components: SS_{reg}, which represents the variability explained by the model, and SS_{res}, which measures the variability left unexplained and usually attributed to error. We define the sum of squares due to the regression, SS_{reg}, by

> **VOICE OF EXPERIENCE**
>
> *Proper interpretation of R^2 requires subject area insights.*

$$SS_{reg} = \sum_{i=1}^{n} (\hat{y} - \overline{y})^2.$$

SS_{reg} measures the variability among the predicted values. With some algebra, we can show that

$$SS_{total} = SS_{reg} + SS_{res}.$$

If the true slope is zero, then the predicted values are all near the overall mean, $\overline{y}$, and the model does not fit the data well. Virtually all of the variability remains unexplained. As a result, SS_{reg} is relatively small and SS_{res} is relatively large. On the other hand, if the slope is pronouncedly positive or negative, then at least some of the predicted values are quite different from the overall mean. The model fits the data relatively well, and SS_{reg} is relatively large. The model thus accounts for much of the total variability, and SS_{res} is relatively small.

The *coefficient of determination*, R^2, uses the relative sizes of the variability explained by the regression model and the total variability to measure the overall adequacy of the model. It is defined by

$$R^2 = \frac{SS_{reg}}{SS_{total}} = \frac{SS_{total} - SS_{res}}{SS_{total}} = 1 - \frac{SS_{res}}{SS_{total}},$$

which guarantees that $0 \leq R^2 \leq 1$. We may interpret R^2 as the proportion of the total variability explained by the regression model.

For models that fit the data well, R^2 is near 1. Models that poorly fit the data have R^2 near 0. One problem with R^2 is what constitutes a "good" value. In many engineering situations, we see $R^2 > .90$, which means that our model explains more than 90% of the total variability. When physical phenomena are modeled, we may see R^2 virtually equal to 1. However, in very "noisy" systems, a good R^2 may be less than .5. Ultimately, we cannot determine what constitutes a good value for R^2 without some background on the specific application.

Many practitioners put too much focus on a high R^2. There are other diagnostics that should be used in conjunction with R^2 to assess the adequacy of the model. Also, a high R^2 could mean that the model is good for predicting the sample data but not necessarily good for predicting the population. On the other hand, if the purpose of the experiment is to identify important variables that relate to the response, a low R^2 may be fine. High values for R^2 are dependent on the experimental situation. For example, one would expect a high R^2 for engineering laboratory experiments where everything is controllable. However, R^2 values tend to be much lower in food tasting and agricultural experiments.

Some analysts prefer the *adjusted* R^2 defined by

$$R^2_{adj} = 1.0 - \frac{SS_{res}/df_{res}}{SS_{total}/(n-1)} = 1.0 - \frac{MS_{res}}{MS_{total}},$$

where $MS_{total} = SS_{total}/(n-1)$. R^2_{adj} corrects R^2 for the number of parameters in the model. We shall discuss R^2_{adj} in more detail in Section 6.3 when we discuss multiple linear regression. We tend to interpret R^2_{adj} similarly to R^2.

| Example 6.7 | **Distillation Column Data—R^2 and R^2_{adj}** |

We have already found that $SS_{total} = 0.1872928$. We next need to find SS_{reg}. By definition,

$$SS_{reg} = \sum_{i=1}^{n} (\hat{y} - \bar{y})^2.$$

However, a better formula computationally is

$$SS_{reg} = b_1^2 SS_{xx}.$$

Using the computational formula, we get

$$SS_{reg} = b_1^2 SS_{xx}$$
$$= (0.01335)^2 (1000)$$
$$= 0.1782225.$$

The resulting coefficient of determination is

$$R^2 = \frac{SS_{reg}}{SS_{total}} = \frac{0.1782225}{0.1872928} = 0.9516.$$

In this case, our model explains approximately 95% of the total variability. The adjusted R^2 is

$$R^2_{adj} = 1.0 - \frac{SS_{res}/df_{res}}{SS_{total}/(n-1)}$$
$$= 1.0 - \frac{0.0090703/3}{0.1872928/4}$$
$$= 0.9354.$$

In general, R^2 values greater than 0.9 are considered quite good. Thus, with $R^2 = 0.9516$ and $R^2_{adj} = 0.9354$, we appear to have an adequate model for predicting the ethanol concentrations given the reflux ratio.

The Overall F-Test We can also develop a formal testing procedure for the overall adequacy of the model. In general, the test considers the question whether there is some relationship between the response/regressors and the regression. In simple linear regression, this test focuses on the slope, β_1. Let MS_{reg} be the mean squared for the regression model, defined by

$$MS_{reg} = \frac{SS_{reg}}{df_{reg}},$$

where df_{reg} is the number of degrees of freedom for the model. In general,

$$df_{reg} = \text{number of parameters in the model} - 1$$
$$= \text{number of regressors}.$$

For the simple linear regression model, we have only two parameters in our model; thus, $df_{reg} = 1$. We can show that

$$E(MS_{reg}) = \sigma^2 + \beta_1^2 SS_{xx}.$$

We can view MS_{reg} as a standardized measure of the variance associated with our model.

Earlier, we stated that $E(MS_{res}) = \sigma^2$. If the model fits the data well and the true slope, β_1, is significantly different from zero, then MS_{reg} is large relative to MS_{res}. As a result, their ratio MS_{reg}/MS_{res}, is large. On the other hand, if the model does not fit the data, then this ratio is small, on the order of one or less. We can show that if the true slope is 0, then MS_{reg}/MS_{res} follows an F-distribution with 1 numerator degree of freedom and $n - 2$ denominator degrees of freedom. We can use this result to develop a test for the overall adequacy of the model. For simple linear regression, this test is exactly equivalent to the two-sided t-test for the slope. The next example outlines this testing procedure.

Example 6.8 | **Distillation Column Data—The Overall F-Test**

This procedure focuses purely on whether some relationship, either positive or negative, exists between the response and the regressor. Consequently, it is inherently a two-sided procedure. In general, this test evaluates the overall adequacy of the model. For simple linear regression, this test reduces to a two-sided test for the slope, in which case our hypotheses are

VOICE OF EXPERIENCE

The F statistic is still just a signal-to-noise ratio.

$$H_0: \quad \beta_1 = 0$$
$$H_a: \quad \beta_1 \neq 0.$$

We shall see in multiple regression that this test simultaneously evaluates all of the regression coefficients.

Strictly speaking, if the random errors follow a normal distribution, then our test statistic is

$$F = \frac{MS_{reg}}{MS_{res}}.$$

As long as the random errors follow a reasonably well-behaved distribution, we feel comfortable using this test statistic. For simple linear regression, we can show that this test statistic is the square of the t-statistic used to test the slope. In this situation, the degrees of freedom for the test statistic are 1 for the numerator and $n - 2$ for the denominator.

Since we have only one possible alternative hypothesis, we always reject H_0 if $F > F_{1,n-2,\alpha}$. If we use standard statistical software, we reject H_0 if the p-value associated with this test statistic is less than α. In our case, we reject the null hypothesis if

$$F > F_{1,n-2,\alpha}$$
$$> F_{1,3,0.05}$$
$$> 10.13.$$

If we perform this analysis by hand, we next need to find MS_{reg}. Earlier, we found that $SS_{reg} = 0.1782225$. We find MS_{reg} by

$$MS_{reg} = \frac{SS_{reg}}{df_{reg}}$$
$$= \frac{0.1782225}{1}$$
$$= 0.1782225.$$

We earlier found that $MS_{res} = 0.003023$. We now can evaluate our test statistic by

$$F = \frac{MS_{reg}}{MS_{res}}$$
$$= \frac{0.1782225}{0.003023}$$
$$= 58.956.$$

Apart from rounding errors, this value for the F-statistic is the square of the value for the t-statistic we used to test the slope originally.

We typically use the *analysis of variance* (ANOVA) table to summarize the calculations for this test. We call this an analysis of variance because we are testing the variance explained by the model relative to the variance left unexplained. In general, the ANOVA table has the form shown in Table 6.5. For our specific situation, the ANOVA table is given in Table 6.6.

Source refers to our partition of the total variability into two components: one for the regression model and the other for the residual or error. The degrees of

Table 6.5 A Generic ANOVA Table for Regression Analysis

Source	Degrees of Freedom	Sum of Squares	Mean Squares	F
Regression	df_{reg}	SS_{reg}	MS_{reg}	F
Residual	df_{res}	SS_{res}	MS_{res}	
Total	$n-1$	SS_{total}		

Table 6.6 | The ANOVA Table for Distillation Column Data

Source	Degrees of Freedom	Sum of Squares	Mean Squares	F
Regression	1	0.1782225	0.1782225	58.956
Residual	3	0.0090703	0.003023	
Total	4	0.1872928		

freedom for the model are

$$\text{number of parameters} - 1 = 2 - 1 = 1$$

for simple linear regression. The degrees of freedom for the residuals or the error are

$$\text{number of observations} - \text{number of parameters} = n - 2 = 3$$

for this particular situation. We obtain the mean squares by dividing the appropriate sum of squares by the corresponding degrees of freedom. We calculate the F-statistic by dividing the mean square for regression by the mean square for the residuals.

Since $58.956 > 10.13$, we may reject the null hypothesis and conclude that concentration of ethanol and the reflux ratio are related.

Reading a Computer-Generated Analysis

Engineers almost always use standard statistical software or spreadsheets to estimate and test models. However, we cannot fully understand these computer-generated analyses until we first know how the numbers were generated. From now on in this chapter, we shall rely purely on the computer to perform all of the calculations. The next example shows the relationship between the analysis of the distillation column data we developed by hand and the output of a standard statistical software package.

Example 6.9 | **Distillation Column Data—Using the Computer**

Table 6.7 shows the analysis generated by SAS software, a common statistical package. Other packages and most spreadsheets use a similar format. The analysis starts with the ANOVA table for the overall F-test. The F-statistic is the ratio of the mean square for the model and the mean square for error. Apart from rounding error, we see the same value for this statistic as we calculated by hand. The software also prints the associated p-value. Since the p-value $= 0.0046$, which is less than

| Table 6.7 | The Computer Analysis of the Distillation Column Data |

Analysis of Variance

		Sum of	Mean		
Source	DF	Squares	Square	F Value	Prob>F
Model	1	0.17822	0.17822	58.947	0.0046
Error	3	0.00907	0.00302		
C Total	4	0.18729			

Root MSE	0.05499	R-square	0.9516
Dep Mean	0.74220	Adj R-sq	0.9354
C.V.	7.40848		

Parameter Estimates

Variable	DF	Parameter Estimate	Standard Error	T for H0: Parameter=0	P > |T|
INTERCEPT	1	0.208200	0.07377113	2.822	0.0666
REFLUX	1	0.013350	0.00173880	7.678	0.0046

$\alpha = 0.05$, we are confident that some linear relationship exists between the ethanol concentration and the reflux ratio.

We next notice that R^2 is 0.9516 and R^2_{adj} is 0.9354, just as we calculated by hand. In general, R^2 values greater than 0.9 are considered quite good. As a result, this model appears to be adequate for predicting the ethanol concentration.

In the Parameter Estimates section of the analysis, we see the estimates of the intercept (INTERCEPT) and the slope associated with the reflux ratio (REFLUX). Beside each estimate is the appropriate estimated standard error. The value of the t-statistic is simply the ratio of the parameter estimate and the standard error. The software then gives the p-value testing whether the parameter is zero versus the alternative not equal to zero. The p-value for the test on the slope is exactly equal to the p-value for the overall F-test. We should not be surprised by this result because the two-sided test for the slope and the overall F-test are equivalent in simple linear regression.

In our original analysis, we conducted the one-sided test:

$$H_0: \quad \beta_1 = 0$$
$$H_a: \quad \beta_1 > 0.$$

In this case, because the alternative is $\beta_1 > 0$ and the estimate is greater than 0 (which supports the alternative), the appropriate p-value is one-half the value given by the printout. Thus, the appropriate p-value for the hypotheses of interest to us is 0.0023.

As usual, we should check our assumptions. Later in this chapter, we shall outline appropriate techniques.

Confidence and Prediction Intervals

Once we are satisfied with our model, we often use it for prediction. Suppose that we wish to estimate the expected response for some specific value for the regressor. Let x_0 be the specific value, not necessarily one used to estimate the model, and let $\hat{y}(x_0)$ be the resulting predicted value for the response from the estimated model. We have

$$\hat{y}(x_0) = b_0 + b_1 x_0.$$

A $(1-\alpha) \cdot 100\%$ confidence interval for the expected response when the regressor is set to x_0 is

$$\hat{y}(x_0) \pm t_{n-2,\alpha/2} \sqrt{MS_{res} \left(\frac{1}{n} + \frac{(x_0 - \overline{x})^2}{SS_{xx}} \right)}.$$

Most standard statistical software packages provide the option to calculate the confidence interval for all of the points used to estimate the model.

The width of this confidence interval depends on the distance of x_0 from $\overline{x}$. The width is a minimum when $x_0 = \overline{x}$. As x_0 gets farther from the center of the x's used to estimate the model, the confidence interval becomes wider. In some cases, we may wish to use our model for *extrapolation*, where we predict the response for an x_0 outside the range of the x's used to estimate the model. In such a case, the interval can become so wide that almost any value is plausible for the prediction. As a result, we generally do not recommend using the estimated model for extrapolation.

Usually, we are interested in the confidence interval for several values of the regressor. A $(1 - \alpha) \cdot 100\%$ *confidence band* for the expected values of the response is the plot of the $(1 - \alpha) \cdot 100\%$ confidence intervals for the values of the regressor over the region of interest.

Whereas the confidence interval focuses on the estimation of the expected value or the mean of the response, the prediction interval considers the estimation of individual responses. Such an interval must consider both the variability in the prediction of the expected value of the response and the variability of the individual responses around this expected value. A $(1 - \alpha) \cdot 100\%$ prediction interval is

$$\hat{y}(x_0) \pm t_{n-2,\alpha/2} \sqrt{MS_{res} \left(1 + \frac{1}{n} + \frac{(x_0 - \overline{x})^2}{SS_{xx}} \right)}.$$

The constant 1 in this expression takes care of the variability associated with the individual responses. Again, most standard software packages provide the option to calculate the prediction interval for every point used to estimate the model. The $(1 - \alpha) \cdot 100\%$ *prediction band* is the plot of the $(1 - \alpha) \cdot 100\%$ prediction intervals for the values of the regressor over the region of interest.

Table 6.8		The Confidence and Prediction Bands for the Distillation Column Data

Obs	Dep Var CONC	Predict Value	Lower95% Mean	Upper95% Mean	Lower95% Predict	Upper95% Predict
1	0.446	0.4752	0.3397	0.6107	0.2539	0.6965
2	0.601	0.6087	0.5129	0.7045	0.4092	0.8082
3	0.786	0.7422	0.6639	0.8205	0.5505	0.9339
4	0.928	0.8757	0.7799	0.9715	0.6762	1.0752
5	0.950	1.0092	0.8737	1.1447	0.7879	1.2305

Example 6.10	**Distillation Column Data—Confidence Bands**

Table 6.8 gives the confidence and prediction bands generated by SAS. Other statistical software packages and most spreadsheets use a similar format.

Figure 6.4 shows the plot of the original data, the estimated regression line, the 95% confidence band, and the 95% prediction band. We see that both the prediction and the confidence bands grow wider as the regressor moves away from the center of the region. As a result, we are less confident about our predictions when values of the regressor are distant from the center of the region used to estimate the model. In this case, all of the data fall within the upper and lower confidence bands, which does not always happen. We should expect, however, that about 95% of the data should fall between the upper and lower prediction bands.

Figure 6.4	The Prediction Equation, Confidence Bands, and Prediction Bands for the Distillation Data

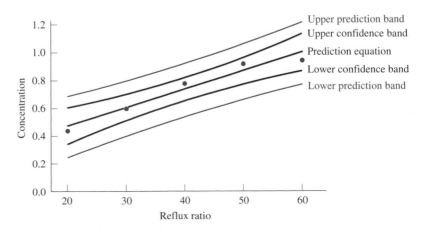

⫶ Exercises

6.1 Davidson (1993) studied the ozone levels in the South Coast Air Basin of California for the years 1976–1991. He believes that the number of days the ozone levels exceeded 0.20 ppm (the response) depends on the seasonal meteorological index, which is the seasonal average 850-millibar temperature (the regressor). Table 6.9 gives the data.

   a. Make a scatter plot of the data.
   b. Estimate the prediction equation.
   c. Perform a complete, appropriate analysis. Discuss your results and conclusions.
   d. Calculate and plot the 95% confidence and prediction bands.

6.2 Mandel (1984) looks at the relationship between the amount of nickel (the regressor) and the volume percent austenite in various steels. Table 6.10 gives the data.

   a. Make a scatter plot of the data.
   b. Estimate the prediction equation.
   c. Perform a complete, appropriate analysis. Discuss your results and conclusions.
   d. Calculate and plot the 95% confidence and prediction bands.

Table 6.9 | The Ozone Data

Year	Days	Index
1976	91	16.7
1977	105	17.1
1978	106	18.2
1979	108	18.1
1980	88	17.2
1981	91	18.2
1982	58	16.0
1983	82	17.2
1984	81	18.0
1985	65	17.2
1986	61	16.9
1987	48	17.1
1988	61	18.2
1989	43	17.3
1990	33	17.5
1991	36	16.6

Table 6.10 | The Steel Data

Amount of Nickel	Percent Austenite
0.608	2.11
0.634	1.95
0.651	2.27
0.658	1.95
0.675	2.05
0.677	2.09
0.702	2.54
0.710	2.51
0.730	2.33
0.750	2.26
0.772	2.47
0.802	2.80
0.819	2.95

6.3 Hsiue, Ma, and Tsai (1995) study the effect of the molar ratio of sebacic acid (the regressor) on the intrinsic viscosity of copolyesters (the response). Table 6.11 gives the data.

a. Make a scatter plot of the data.
b. Estimate the prediction equation.
c. Perform a complete, appropriate analysis. Discuss your results and conclusions.
d. Calculate and plot the 95% confidence and prediction bands.

Table 6.11 | The Sebacic Acid Data

Ratio	Viscosity
1.0	0.45
0.9	0.20
0.8	0.34
0.7	0.58
0.6	0.70
0.5	0.57
0.4	0.55
0.3	0.44

Table 6.12   The PET-LCP Data

Percent PET	Modulus
100	2.12
97.5	2.26
95	2.57
90	3.26
80	3.46
50	4.54
0	8.5

6.4   Mehta and Deopura (1995) studied the mechanical properties of spun PET-LCP blend fibers. They believe that the modulus (the response) depends on the percent of PET in the blend. Table 6.12 gives the data.

   a.   Make a scatter plot of the data.
   b.   Estimate the prediction equation.
   c.   Perform a complete, appropriate analysis. Discuss your results and conclusions.
   d.   Calculate and plot the 95% confidence and prediction bands.

6.5   Byers and Williams (1987) studied the impact of temperature (the regressor) on the viscosity (the response) of toluene–tetralin blends. Table 6.13 gives the data, which are for blends with a 0.4 molar fraction of toluene.

   a.   Estimate the prediction equation.
   b.   Perform a complete, appropriate analysis. Discuss your results and conclusions.
   c.   Calculate and plot the 95% confidence and prediction bands.

Table 6.13   The Viscosity Data

Temperature (°C)	Viscosity (mPa · s)
24.9	1.133
35.0	0.9772
44.9	0.8532
55.1	0.7550
65.2	0.6723
75.2	0.6021
85.2	0.5420
95.2	0.5074

Table 6.14	The Hardness Data

Rockwell Hardness	Young's Modulus
41	310
41	340
44	380
40	317
43	413
15	62
40	119

6.6 The military often uses high-density metals in munitions because these metals are better able to penetrate armor. Magness (1994) conducted an observational study of seven different high-density metals in which he measured the Rockwell hardness (the regressor) and the Young's modulus (the response) for a single specimen of each metal. The Young's modulus is the ratio of simple tension stress to the resulting strain parallel to the tension. Normally, we expect Young's modulus to increase as the Rockwell hardness increases. Table 6.14 gives the results.

   a. Estimate the prediction equation.

   b. Perform a complete, appropriate analysis. Discuss your results and conclusions.

   c. Calculate and plot the 95% confidence and prediction bands.

6.7 Anand and Ray (1995) study the relationship between the discharge of oil (the regressor) and the wear (the response) on the cylinder blocks for a specific type of hydraulic pump. Table 6.15 gives the data.

   a. Estimate the prediction equation.

   b. Perform a complete, appropriate analysis. Discuss your results and conclusions.

   c. Calculate and plot the 95% confidence and prediction bands.

6.8 Carroll and Spiegelman (1986) look at the relationship between the pressure in a tank (the response) and the volume of liquid (the regressor). Table 6.16 gives the data.

   a. Estimate the prediction equation.

   b. Perform a complete, appropriate analysis. Discuss your results and conclusions.

   c. Calculate and plot the 95% confidence and prediction bands.

6.9 A study is carried out to investigate the relationship between the average level of aflatoxin (parts per billion) in a mini-lot sample of peanuts (x) and the percentage of noncontaminated peanuts in the batch (y). The data from Draper and Smith (1981) are in Table 6.17.

Table 6.15 | The Wear Data

Discharge (mm)	Wear (L/min)
11	0.016
10	0.040
9	0.067
8	0.096
6	0.145
5	0.168
4	0.197
2	0.250

---

a. Make a scatterplot of the data.
b. Estimate the prediction equation.
c. Perform a complete, appropriate analysis. Discuss your results and conclusions.
d. Calculate and plot the 95% confidence and prediction bands.

6.10 Joglekar, Schuenemeyer and LaRiccia (1989) looked at the relationship between the percentage of hardwood in a batch of pulp (x) and the tensile strength

---

Table 6.16 | The Pressure Tank Data

Volume	Pressure	Volume	Pressure	Volume	Pressure
2084	4599	2842	6380	3789	8599
2084	4600	3030	6818	3789	8600
2273	5044	3031	6817	3979	9048
2273	5043	3031	6818	3979	9048
2273	5044	3221	7266	4167	9484
2463	5488	3221	7268	4168	9487
2463	5487	3409	7709	4168	9487
2651	5931	3410	7710	4358	9936
2652	5932	3600	8156	4358	9938
2652	5932	3600	8158	4546	10,377
2842	6380	3788	8597	4547	10,379

Table 6.17 | Noncontaminated Peanut Data

y	x	y	x	y	x
99.971	3.0	99.942	18.8	99.863	46.8
99.979	4.7	99.932	18.9	99.811	46.8
99.982	8.3	99.908	21.7	99.877	58.1
99.971	9.3	99.970	21.9	99.798	62.3
99.957	9.9	99.985	22.8	99.855	70.6
99.961	11.0	99.933	24.2	99.788	71.1
99.956	12.3	99.858	25.8	99.821	71.3
99.972	12.5	99.987	30.6	99.830	83.2
99.889	12.6	99.958	36.2	99.718	83.6
99.961	15.9	99.909	39.8	99.642	99.5
99.982	16.7	99.859	44.3	99.658	111.2
99.975	18.8				

of Kraft paper measured in psi made from the batch (y). The data are in Table 6.18.

a. Make a scatterplot of the data.
b. Estimate the prediction equation.
c. Perform a complete, appropriate analysis. Discuss your results and conclusions.
d. Calculate and plot the 95% confidence and prediction bands.

6.11 In a chemical process, batches of liquid are passed through a bed containing an ingredient that is absorbed by the liquid. In an attempt to relate the absorbed percentage of the ingredient (y) to the amount of liquid in the batch (x), Bissell (1992) gives the data in Table 6.19.

a. Make a scatterplot of the data.
b. Estimate the prediction equation.
c. Perform a complete, appropriate analysis. Discuss your results and conclusions.
d. Calculate and plot the 95% confidence and prediction bands.

6.12 A study considered the ability to predict cracking of latex paints on exposed wood surfaces based on accelerated cracking tests. Data on accelerated crack rating (x) and exposure cracking (y) are given below.

a. Make a scatterplot of the data.
b. Estimate the prediction equation.

Table 6.18	Paper Tensile Strength Data

y	x
6.3	1.0
11.1	1.5
20.0	2.0
24.0	3.0
26.1	4.0
30.0	4.5
33.8	5.0
34.0	5.5
38.1	6.0
39.9	6.5
42.0	7.0
46.1	8.0
53.1	9.0
52.0	10.0
52.5	11.0
48.0	12.0
42.8	13.0
27.8	14.0
21.9	15.0

c. Perform a complete, appropriate analysis. Discuss your results and conclusions.

d. Calculate and plot the 95% confidence and prediction bands.

x	y
2	2.86
2	4.00
3	5.65
3	4.57
4	5.28
4	6.26
5	5.45
5	6.42
6	8.71
6	8.33
7	8.26
7	9.24

Table 6.19	Absorbed Liquid Data

y	x
310	4.52
330	5.18
370	5.76
400	5.10
450	6.09
490	5.55
520	5.46
560	5.80
580	5.50
650	5.25
650	6.12
650	5.92
760	6.63
800	5.48
810	6.22
910	5.88
1020	6.99
1020	6.30
1160	6.86
1200	6.73
1230	6.38
1380	7.17
1460	7.23
1490	6.62

6.13 Williams and Lee (1985) investigated how the propagation of an ultrasonic stress wave through a substance depends on the properties of the substance. Data on fracture strength (x, as a percentage of ultimate tensile strength) and attenuation (y, in neper/cm, the decrease in amplitude of the stress wave) in fiberglass-reinforced polyester composites were read from a graph that appeared in the paper and is shown below.

a. Make a scatterplot of the data.
b. Estimate the prediction equation.
c. Perform a complete, appropriate analysis. Discuss your results and conclusions.
d. Calculate and plot the 95% confidence and prediction bands.

x	y
12	3.3
30	3.2
36	3.4
40	3.0
45	2.8
57	2.9
62	2.7
67	2.6
71	2.5
78	2.6
93	2.2
94	2.0
100	2.3
105	2.1

6.14 Fearn (1983) presents an experiment to calibrate a near-infrared reflectance instrument for the measurement of protein content of ground wheat samples. See Table 6.20. The aim of the experiment is to predict the protein content (y) from the reflectance of near infrared radiation at a wavelength of 2310 (x).

a. Make a scatterplot of the data.
b. Estimate the prediction equation.
c. Perform a complete, appropriate analysis. Discuss your results and conclusions.
d. Calculate and plot the 95% confidence and prediction bands.

6.3 Multiple Linear Regression

In most engineering problems, the response depends on several regressors. Multiple linear regression is a straightforward extension of simple linear regression to more than one regressor. Essentially, model estimation, model testing, and prediction all follow the same basic procedure.

Multiple Linear Regression Model

Often, we can well model engineering phenomena as a linear combination of several regressors. For example, Box and Bisgaard (1987) model the cracking of carbon-steel springs (the response) in terms of these factors:

- The temperature of the steel before quenching
- The amount of carbon in the formulation
- The temperature of the quenching oil

Table 6.20	Infrared Reflectance Instrument Data

Protein (%)	x
9.23	386
8.01	383
10.95	353
11.67	340
10.41	371
9.51	433
8.67	377
7.75	353
8.05	377
11.39	398
9.95	378
8.25	365
10.57	443
10.23	450
11.87	467
8.09	451
12.55	524
8.38	407
9.64	374
11.35	391
9.70	353
10.75	445
10.75	383
11.47	404

VOICE OF EXPERIENCE

In multiple linear regression, the regression coefficients, strictly speaking, are not slopes.

Let y_i be the percentage of springs that do not exhibit cracking for the ith batch produced. Let x_{i1} be the temperature of the steel for the ith batch, let x_{i2} be the amount of carbon in the ith batch, and let x_{i3} be the temperature of the quenching oil for the ith batch. A possible model for this situation is

$$y_i = \beta_0 + \beta_1 x_{i1} + \beta_2 x_{i2} + \beta_3 x_{i3} + \epsilon_i,$$

where

- β_0 is the y-intercept.
- β_1 is the coefficient associated with the steel temperature.

- β_2 is the coefficient associated with the amount of carbon.
- β_3 is the coefficient associated with the quenching oil temperature.
- ϵ_i is a random error.

Just as in simple linear regression, we usually assume that the random errors are independent with mean 0 and variance σ^2. Under these assumptions, the expected value of the response for the ith set of conditions for the regressors is

$$E[y] = \beta_0 + \beta_1 x_{i1} + \beta_2 x_{i2} + \beta_3 x_{i3}.$$

In the multiple regression setting, we must be very careful with our interpretation of the coefficients, the β's. The coefficient, β_j, represents the expected change in the response for a one-unit change in x_j *given that all the other regressors are held constant!* In the springs with cracks example, the engineers have complete control over all three regressors. As a result, they truly can change one regressor while holding the others constant. In many other engineering contexts, however, we really do not have this ability. In Chapter 1, we discussed an example involving a production process for a polymer filament. Supervisors kept a log of the actual temperatures and pressures for the reaction, which indicated that in the actual process, as the temperature increased, so did the pressure. In this situation, we could not make a one-unit change in temperature without making some kind of change in the pressure. We say that temperature and pressure are at least partially *collinear*. Collinearity is an important problem in regression analysis, which we shall introduce in Section 6.5.

Multiple linear regression means that the model is linear *in terms of the coefficients*. For example, engineers often use a Taylor series approximation to justify this type of polynomial model:

$$y_i = \beta_0 + \beta_1 x_{i1} + \beta_2 x_{i2} + \beta_{11} x_{i1}^2 + \beta_{22} x_{i2}^2 + \beta_{12} x_{i1} x_{i2} + \epsilon_i.$$

Since the model is linear in the coefficients, we could define new regressors $x_{i3} = x_{i1}^2$, $x_{i4} = x_{i2}^2$, and $x_{i5} = x_{i1} x_{i2}$ to produce an exactly equivalent model of the form

$$y_i = \beta_0 + \beta_1 x_{i1} + \beta_2 x_{i2} + \beta_3 x_{i3} + \beta_4 x_{i4} + \beta_5 x_{i5} + \epsilon_i.$$

Two other examples of multiple linear regression models are

$$y_i = \beta_0 + \beta_1 \log(x_{i1}) + \beta_2 \frac{1}{x_{i2}} + \epsilon_i$$

$$\log(y_i) = \beta_0 + \beta_1 x_{i1} + \beta_2 x_{i2} + \epsilon_i.$$

In each case, we could define either a new response or new regressors that produce exactly equivalent models that appear completely linear in the new terms.

Just like simple linear regression, we traditionally use the method of least squares to estimate the coefficients. Suppose our model has k regressors and is of the form

$$y_i = \beta_0 + \beta_1 x_{i1} + \cdots + \beta_k x_{ik} + \epsilon.$$

Let $b_0, b_1, \ldots, b_k$ be the corresponding estimates of the coefficients, and let $\hat{y}_i$ be the predicted value for the response for the ith set of conditions for the regressors. Thus,

$$\hat{y}_i = b_0 + b_1 x_{i1} + \cdots + b_k x_{ik}.$$

Once again, we define the sum of squares of the residuals by

$$SS_{res} = \sum_{i=1}^{n} (y_i - \hat{y}_i)^2.$$

We call $b_0, b_1, \ldots, b_k$ the least squares estimates if they minimize SS_{res}. From calculus, we find these estimates by taking the partial derivative of SS_{res} with respect to each parameter and setting them equal to zero, which produces the following $k + 1$ equations in $k + 1$ unknowns:

$$\frac{\partial}{\partial b_0} \sum_{i=1}^{n} [y_i - (b_0 + b_1 x_{i1} + b_2 x_{i2} + \cdots + b_k x_{ik})]^2 = 0$$

$$\frac{\partial}{\partial b_1} \sum_{i=1}^{n} [y_i - (b_0 + b_1 x_{i1} + b_2 x_{i2} + \cdots + b_k x_{ik})]^2 = 0$$

$$\vdots \qquad \vdots$$

$$\frac{\partial}{\partial b_k} \sum_{i=1}^{n} [y_i - (b_0 + b_1 x_{i1} + b_2 x_{i2} + \cdots + b_k x_{ik})]^2 = 0.$$

These derivatives result in the "normal" equations:

$$n b_0 + b_1 \sum_{i=1}^{n} x_{i1} + b_2 \sum_{i=1}^{n} x_{i2} + \cdots + b_k \sum_{i=1}^{n} x_{ik} = \sum_{i=1}^{n} y_i$$

$$b_0 \sum_{i=1}^{n} x_{i1} + b_1 \sum_{i=1}^{n} x_{i1}^2 + b_2 \sum_{i=1}^{n} x_{i1} x_{i2} + \cdots + b_k \sum_{i=1}^{n} x_{i1} x_{ik} = \sum_{i=1}^{n} x_{i1} y_i$$

$$b_0 \sum_{i=1}^{n} x_{i2} + b_1 \sum_{i=1}^{n} x_{i1} x_{i2} + b_2 \sum_{i=1}^{n} x_{i2}^2 + \cdots + b_k \sum_{i=1}^{n} x_{i2} x_{ik} = \sum_{i=1}^{n} x_{i2} y_i$$

$$\vdots$$

$$b_0 \sum_{i=1}^{n} x_{ik} + b_1 \sum_{i=1}^{n} x_{i1} x_{ik} + b_2 \sum_{i=1}^{n} x_{i2} x_{ik} + \cdots + b_k \sum_{i=1}^{n} x_{ik}^2 = \sum_{i=1}^{n} x_{ik} y_i.$$

Matrix Notation (Optional)   Matrix notation is convenient for some of our discussion in this chapter. Let **y** be the $n \times 1$ vector of responses

$$\mathbf{y} = (y_1, y_2, \ldots, y_n)'.$$

Let $\mathbf{X}$ be the $n \times (k+1)$ matrix defined by

$$\mathbf{X} = \begin{pmatrix} 1 & x_{11} & x_{12} & \ldots & x_{1k} \\ 1 & x_{21} & x_{22} & \ldots & x_{2k} \\ \vdots & \vdots & \vdots & & \vdots \\ 1 & x_{i1} & x_{i2} & \ldots & x_{ik} \\ \vdots & \vdots & \vdots & & \vdots \\ 1 & x_{n1} & x_{n2} & \ldots & x_{nk} \end{pmatrix}.$$

We sometimes call $\mathbf{X}$ the *model matrix*. Let $\mathbf{x}_i'$ be the ith row of $\mathbf{X}$. Thus,

$$\mathbf{x}_i' = (1, x_{i1}, x_{i2}, \ldots, x_{ik})$$

and represents the specific values for the regressors for the ith data point. We then can rewrite $\mathbf{X}$ as

$$\mathbf{X} = \begin{pmatrix} \mathbf{x}_1' \\ \mathbf{x}_2' \\ \vdots \\ \mathbf{x}_i' \\ \vdots \\ \mathbf{x}_n' \end{pmatrix}.$$

Finally, let $\mathbf{b}$ be the $(k+1) \times 1$ vector of estimated coefficients:

$$\mathbf{b} = (b_0, b_1, \ldots, b_k)'.$$

We now can rewrite the normal equations as

$$\mathbf{X}'\mathbf{X}\mathbf{b} = \mathbf{X}'\mathbf{y}.$$

If $\mathbf{X}'\mathbf{X}$ is nonsingular, then the least squares estimates of the coefficients are

$$\mathbf{b} = (\mathbf{X}'\mathbf{X})^{-1}\mathbf{X}'\mathbf{y}.$$

Standard statistical software packages and many spreadsheets use good algorithms for finding these estimates.

Model Testing and Prediction

Once we estimate the model, we need to address some basic questions such as

1. Is there a relationship between the response and at least one of the regressors?
2. What is the overall adequacy of the model?
3. If there is a relationship between the response and at least one of the regressors, which ones are related?

We can address each of these questions using straightforward extensions of techniques we developed in simple linear regression, as the next example illustrates.

Example 6.11 | **Stability of Catalyst**

Supercritical water oxidation has become a well-known treatment for conversions of aqueous wastes. Aki, Ding and Abraham (1996) studied the stability of chromium oxide in supercritical water oxidation. The effect of oxygen concentration (mol fr.), water concentration (mol/L) and residence time (seconds) on chromium effluent concentration was investigated. Table 6.21 gives the data based on their study.

R^2 and the Overall F-Test   The first question is whether there is any relationship between the response and at least one of the regressors. Our approach to this question looks at the amount of variability explained by the regression model (SS_{reg}) relative to the amount of variation left unexplained and presumed to be error (SS_{res}).

Just as in simple linear regression, we define the total variability by

$$SS_{total} = \sum_{i=1}^{n}(y_i - \overline{y})^2.$$

Once again, we can establish that

$$SS_{total} = SS_{reg} + SS_{res},$$

where

$$SS_{reg} = \sum_{i=1}^{n}(\hat{y}_i - \overline{y})^2$$

Table 6.21 | Chromium Effluent Concentration Data

Water Conc.	Oxygen Conc.	Time	Cr Conc.
5.66	0.025	5.40	49.95
6.38	0.017	5.64	47.59
3.43	0.069	5.23	38.97
3.88	0.054	5.81	46.16
5.06	0.030	6.04	48.93
4.39	0.036	6.56	51.68
6.20	0.014	6.43	48.70
3.73	0.052	6.44	46.21

is the variability explained by the model. If all of the coefficients associated with the regressors are truly zero, then the predicted values are all near the overall mean, and the model does not fit the data well. Virtually all of the variability remains unexplained. As a result, SS_{reg} is relatively small and SS_{res} is relatively large. On the other hand, if some of the coefficients associated with the regressors are not zero, then at least some of the predicted values differ from the overall mean. The model fits the data better, and SS_{reg} accounts for more of the total variability. If the model fits the data almost perfectly, then SS_{reg} is very large relative to SS_{res}.

We can use the coefficient of determination,

$$R^2 = \frac{SS_{reg}}{SS_{total}} = 1.0 - \frac{SS_{res}}{SS_{total}},$$

to evaluate the overall adequacy of the model. We interpret R^2 just as we did in simple linear regression. Ultimately, we cannot determine what constitutes a good value for R^2 without some background on the specific application.

One problem with R^2 is that it cannot decrease as we add more regressors to our model. As a result, many analysts prefer to use R^2_{adj}:

$$R^2_{adj} = 1.0 - \frac{SS_{res}/(n - k - 1)}{SS_{total}/(n - 1)}.$$

This measure "adjusts" R^2 for the degrees of freedom we use for the model. As k gets larger, $n - k - 1$ gets smaller. As we add terms to our model, R^2_{adj} will increase only if the new term reduces SS_{res} enough to compensate for the decrease in $n - k - 1$. Most standard statistical software packages routinely give both R^2 and R^2_{adj}.

We can also develop a formal testing procedure for the overall adequacy of the model. Let MS_{res} be the mean squared residual:

$$MS_{res} = \frac{SS_{res}}{df_{res}},$$

where df_{res} is the number of degrees of freedom for the model. In general,

$$df_{res} = \text{number of observations} - \text{number of parameters in the model}$$
$$= n - k - 1.$$

For the multiple linear regression models with k regressors, we have $k + 1$ parameters in our model to estimate; thus, $df_{res} = n - k - 1$. Under our model assumptions, we can show that

$$E(MS_{res}) = \sigma^2;$$

thus, MS_{res} is an unbiased estimator of σ^2.

Let MS_{reg} be the mean squared for the regression model defined by

$$MS_{reg} = \frac{SS_{reg}}{df_{reg}},$$

where df_{reg} is the number of degrees of freedom for the model. In general,

$$df_{reg} = \text{number of parameters in the model} - 1.$$

Since our models always have a y-intercept, we have $k + 1$ parameters in our model, and $df_{reg} = k$, the number of regressors. Under our model assumptions, we can show that

$$E(MS_{reg}) \geq \sigma^2,$$

and equals σ^2 if and only if the true coefficients associated with all the regressors are 0.

If some of the coefficients are significantly different from 0, then MS_{reg} is large relative to MS_{res}. As a result, MS_{reg}/MS_{res} is large. On the other hand, if the model does not fit the data, then this ratio is small, on the order of 1 or less. We can show that if the true value for all of the coefficients is 0, then MS_{reg}/MS_{res} follows an F-distribution with k numerator degrees of freedom and $n - k - 1$ denominator degrees of freedom.

We now can develop an appropriate hypothesis test to determine whether some relationship, either positive or negative, exists between the response and at least one of the regressors. If the response does not depend on at least one of the regressors, then all of the coefficients must be zero. In such a case, there seems to be little point in continuing the analysis. On the other hand, if the response depends on at least one regressor, then at least one of the coefficients must be different from zero. We then pursue follow-up analyses to determine which of the coefficients are nonzero.

Our hypotheses are

$$H_0: \quad \beta_1 = \beta_2 = \cdots = \beta_k = 0$$
$$H_a: \quad \text{At least one } \beta_j \neq 0 \qquad \text{for } j = 1, 2, \ldots, k.$$

Thus, this test jointly considers all of the coefficients.

Strictly speaking, if the random errors follow a normal distribution, then the test statistic is

$$F = \frac{MS_{reg}}{MS_{res}}.$$

As long as the random errors follow a reasonably well-behaved distribution, we feel comfortable using this test statistic. The degrees of freedom for the test statistic are k for the numerator and $n - k - 1$ for the denominator.

Since we have only one possible alternative hypothesis, we always reject H_0 if $F > F_{k,n-k-1,\alpha}$. If we use standard statistical software, we reject H_0 if the

Table 6.22	The Analysis of Variance Table for the Chromium Effluent Concentration Data

```
Analysis of Variance

Source           DF          SS          MS        F        P
Regression        3      98.298      32.766    29.25    0.004
Residual Errror   4       4.481       1.120
Total             7     102.779


S=1.05843       R-Sq = 95.6%        R-Sq(adj)= 92.4%
```

p-value associated with this test statistic is less than α. In our case, we reject the null hypothesis if

$$F > F_{k,n-k-1,\alpha}$$
$$> F_{3,4,0.05}$$
$$> 6.59.$$

We typically use an analysis of variance (ANOVA) table to summarize the calculations for this test. We call this an analysis of variance because we are testing the variance explained by the model relative to the variance left unexplained. Table 6.22 gives the analysis based on the MINITAB statistical software package. Source refers to our partition of the total variability into two components: one for the regression model and the other for the residual or error. Since we have three regressors, we have three degrees of freedom for the model. The degrees of freedom for the residuals or the error are

$$\text{number of observations} - \text{number of parameters} = n - k - 1 = 4$$

for this particular situation. We obtain the mean squares by dividing the appropriate sum of squares by the corresponding degrees of freedom. We calculate the F-statistic by dividing the mean square for regression by the mean square for the residuals. In this case, we obtain an F-value of 29.28 and a p-value of 0.004.

Since the p-value is less than $\alpha = .05$, we may reject the null hypothesis and conclude that the effluent concentration does depend on oxygen concentration, the water concentration, or the residence time. Furthermore, we observe that $R^2 = 0.956$, which tends to indicate a reasonable candidate model. Of course, we need to conduct follow-up analyses to determine whether we can develop a better model.

Tests on the Coefficients   Once we determine that the effluent depends on at least one of the regressors, we need to determine which specific ones. Let $\beta_{j,0}$ be a hypothesized value for β_j for $j = 1, 2, \ldots, k$. In general, these are the hypotheses:

$$H_0: \quad \beta_j = \beta_{j,0} \qquad H_0: \quad \beta_j = \beta_{j,0} \qquad H_0: \quad \beta_j = \beta_{j,0}$$
$$H_a: \quad \beta_j < \beta_{j,0} \qquad H_a: \quad \beta_j > \beta_{j,0} \qquad H_a: \quad \beta_j \neq \beta_{j,0}$$

Usually, $\beta_{j,0}$ will be zero, which corresponds to a null hypothesis of no relationship between the jth regressor and the response. It is important to note that these hypotheses test the relationship *given that the other regressors are held constant*. We thus must use some care in interpreting the results of these tests.

In this particular case, the engineers wish to determine which of the regressors appear to influence the effluent. As a result, these are the appropriate hypotheses:

$$H_0: \quad \beta_j = 0 \quad \text{for } j = 1, 2, \ldots, k$$
$$H_a: \quad \beta_j \neq 0 \quad \text{for } j = 1, 2, \ldots, k.$$

We next need a test statistic. Let $\hat{\sigma}_{b_j}$ be the estimated standard error for b_j. If we can assume that the random errors follow a well-behaved distribution, then our test statistic is

$$t = \frac{b_j - \beta_{j,0}}{\hat{\sigma}_{b_j}}.$$

The degrees of freedom for the test statistic are $n - k - 1$.

Our critical regions are listed next:

- For $H_a: \beta_j < \beta_{j,0}$, we reject H_0 if $t < -t_{n-k-1,\alpha}$.
- For $H_a: \beta_j > \beta_{j,0}$, we reject H_0 if $t > t_{n-k-1,\alpha}$.
- For $H_a: \beta_j \neq \beta_{j,0}$, we reject H_0 if $|t| > t_{n-k-1,\alpha/2}$.

As previously stated on page 382, in addition to the calculated value for the test statistic, computer software prints out the p-value. In Chapter 4, we learned that the p-value depends on the specific alternative hypothesis. Most software packages assume a two-sided alternative. When this is appropriate, we reject the null hypothesis whenever the given p-value is less than α.

In our case, we reject the null hypothesis if

$$|t| > t_{n-k-1,\alpha/2}$$
$$> t_{4,0.025}$$
$$> 2.777.$$

Table 6.23 gives the analysis based on the MINITAB statistical software package. We see the least squares estimates of the intercept (Constant), the coefficient associated with the water concentration (water conc.), the coefficient associated with the amount of oxygen concentration (oxygen conc.), and the coefficient associated with the residence time (time). Next to each estimate is the appropriate estimated

Table 6.23	The Tests on the Individual Coefficients for the Chromium Effluent Concentration Data

Predictor	Coef	SE Coef	T	P
Constant	154.41	25.14	6.14	0.004
Water Conc.	-11.610	2.376	-4.89	0.008
Oxygen Conc.	-827.5	145.1	-5.70	0.005
Time	-3.400	1.544	-2.20	0.092

standard error. The value of the t-statistic is simply the ratio of the parameter estimate and the standard error. The software then gives the p-value testing whether the parameter is zero versus the alternative not equal to 0.

We see that both the coefficients for the water concentration and the oxygen concentration have p-values less than $\alpha = 0.05$. As a result, we have evidence that both water and oxygen concentrations influence the chromium effluent concentration, given the other regressor in the model. Furthermore, since both coefficients are negative, we see lower chromium effluent concentration as we increase the water or oxygen concentration.

Once we conclude that the coefficient for the water concentration is not zero, many analysts suggest giving a range of plausible values for it. In general, we need to be very careful about confidence intervals for regression coefficients in multiple linear regression. We must keep in mind that these coefficients represent the estimated per-unit change associated with a specific regressor *when all of the other regressors are held constant*. These coefficients represent the estimated impact of the specific regressor once all of the other regressors are in the model. Thus, these coefficients represent the estimated impact above and beyond the impact of all the other regressors. Improperly applied, confidence intervals on the individual coefficients tend to give these coefficients more weight than they deserve. Many experts in regression analysis prefer joint confidence intervals for several coefficients of interest over confidence intervals for individual coefficients. Unfortunately, we must leave joint confidence intervals for more advanced texts.

A 95% confidence interval for the true value of this coefficient is

$$b_1 \pm t_{n-k-1,\alpha/2} \cdot \hat{\sigma}_{b_1} = -11.610 \pm 2.777(2.376)$$
$$= -11.610 \pm 6.598$$
$$= (-18.208, -5.012).$$

Thus, the plausible values for this coefficient are between -18.208 and -5.012, which confirms the negative relationship between the water concentration and the response.

Confidence and Prediction Intervals   Most statistical software packages provide the option to generate the 95% confidence and prediction bounds for the

Table 6.24	The Confidence and Prediction Intervals

Obs	Cr Conc.	Fit	SE Fit	95% CI	95% PI
1	49.9598	49.656	0.848	(47.301, 52.012)	(45.890, 53.423)
2	47.5871	47.095	0.758	(44.990, 49.200)	(43.480, 50.710)
3	38.9687	39.708	0.906	(37.192, 42.224)	(35.839, 43.577)
4	46.1603	44.926	0.512	(43.506, 46.347)	(41.662, 48.190)
5	48.9325	50.307	0.522	(48.773, 51.841)	(46.992, 53.622)
6	51.6833	51.353	0.756	(49.254, 53.452)	(47.742, 54.964)
7	48.7010	48.979	0.859	(46.596, 51.363)	(45.195, 52.763)
8	46.2116	46.179	0.701	(44.234, 48.125)	(42.655, 49.704)

data used to generate the model. Table 6.24 gives the confidence and prediction intervals generated by the MINITAB statistical software package for the chromium effluent concentration data.

How to come up with the final model in regression is an important topic. We will only discuss this briefly. In many regression problems, model reduction is done one term at a time beginning with the largest p-value for the highest-order term, because factors are correlated (see Section 6.5). Then, in general, terms are removed one at a time until all remaining terms are significant. However, in many planned or designed experiments, model reduction is done in one step by removing all terms that are insignificant because the factors are orthogonal or uncorrelated. We will talk more about planned experiments in the next two chapters.

Exercises

6.15 Chang and Shivpuri (1994) study the effect of the furnace temperature (x_1) and die close time (x_2) on the temperature difference (y) on the die surface in a die casting process. Ideally, they would like to minimize this difference. Table 6.25 summarizes the data. Perform a complete analysis. Discuss your results and conclusions.

6.16 Tracy, Young, and Mason (1992) studied the impact of temperature (x_1) and concentration (x_2) on the percentage of impurities (y) for a chemical process. Table 6.26 gives the data. Perform a complete analysis. Discuss your results and conclusions.

6.17 Lawson (1982) conducted a designed experiment to optimize the yield (the response) for a caros acid process. For proprietary reasons, he could not discuss the three specific regressors, x_1, x_2, and x_3. Table 6.27 lists the data in their actual run order. Perform a complete analysis. Discuss your results and conclusions.

Table 6.25	The Die Casting Data

Temp.	Time	y
1250	6	80
1300	7	95
1350	6	101
1250	7	85
1300	6	92
1250	8	87
1300	8	96
1350	7	106
1350	8	108

6.18 Said and colleagues (1994) studied the effect of the mole contents of cobalt (x_1) and calcination temperature (x_2) on the surface area of an iron–cobalt hydroxide catalyst. Table 6.28 gives the data. Perform as complete an analysis as possible. Discuss your results and conclusions.

Table 6.26	The Impurities in a Chemical Process Data

Temp.	Conc.	Percent Impurities
85.8	42.3	14.9
83.8	43.4	16.9
84.5	42.7	17.4
86.3	43.6	16.9
85.2	43.2	16.9
83.8	43.7	16.7
86.1	43.3	17.1
85.9	43.4	16.9
85.7	43.3	16.7
86.3	42.6	16.9
83.5	44.0	16.7
85.8	42.8	17.1
85.9	43.1	17.6
84.2	43.5	16.9

Table 6.27 | The Caros Acid Data

Run	x_1	x_2	x_3	y
1	−1	1	−1	77
2	1	1	1	92
3	0	0	0	81
4	−1	−1	1	86
5	1	−1	−1	67
6	0	0	0	82
7	0	0	−1	72
8	−1	0	0	84
9	0	1	0	81
10	0	0	1	87
11	0	0	0	82
12	0	−1	0	74
13	0	0	0	82
14	1	0	0	78
15	1	1	−1	80
16	−1	−1	−1	68
17	−1	1	1	92
18	0	0	0	81
19	0	0	0	80
20	1	−1	1	86
21	0	−1	0	76
22	0	0	−1	71
23	−1	0	0	86
24	0	0	0	82
25	0	0	0	81
26	0	0	1	86
27	0	1	0	82
28	1	0	0	79

6.19 Wauchope and McDowell (1984) studied the effect of the amount of extractable iron, the amount of extractable aluminum, and the pH of soils on the soils' adsorption of phosphate. Table 6.29 gives the data. Perform a complete analysis. Discuss your results and conclusions.

Table 6.28 | The Catalyst Data

Cobalt Contents	Temp.	Surface Area
0.6	200	90.6
0.6	250	82.7
0.6	400	58.7
0.6	500	43.2
0.6	600	25.0
1.0	200	127.1
1.0	250	112.3
1.0	400	19.6
1.0	500	17.8
1.0	600	09.1
2.6	200	53.1
2.6	250	52.0
2.6	400	43.4
2.6	500	42.4
2.6	600	31.6
2.8	200	40.9
2.8	250	37.9
2.8	400	27.5
2.8	500	27.3
2.8	600	19.0

6.20   Liu, Kan, and Chen (1993) studied the relationship among the superficial fluid velocity (x_1), the liquid viscosity (x_2), and the mesh size of the openings (x_3) on a dimensionless factor (y) used to describe pressure drops in a screen-plate bubble column. Table 6.30 summarizes the data. Perform as complete an analysis as possible. Discuss your results and conclusions.

6.21   A study was carried out to relate specific power, proportional to power per unit weight (x_1), flight range factor (x_2), payload as a fraction of gross weight of aircraft (x_3), and sustained load factor (x_4) to the first flight date, in months after January 1940 (y). The data from Cook and Weisberg (1982) are in Table 6.31. Perform a complete analysis. Discuss your results and conclusions.

6.22   Perch and Bridgewater (1980) investigated the relationship among the percent of fixed carbon (x_1), the percent ash (x_2), and the percent sulfur (x_3) on the coking heat (y) in BTU per pound. Table 6.32 summarizes the data. Perform a complete analysis. Discuss your results and conclusions.

Table 6.29 | The Soil Adsorption Data

Extractable Iron	Extractable Aluminum	pH	Adsorption Index
61	13	7.7	4
175	21	7.7	18
111	24	6.8	14
124	23	7.3	18
130	64	5.1	26
173	38	5.7	26
169	33	5.8	21
169	61	5.2	30
160	39	6.3	28
244	71	5.7	36
257	112	4.4	65
333	88	4.5	62
199	54	6.2	40

Table 6.30 | The Pressure Drop Data

Fluid Velocity	Liquid Viscosity	Mesh Size	y
2.14	10	0.34	28.9
4.14	10	0.34	26.1
8.15	10	0.34	22.8
2.14	2.63	0.34	24.2
4.14	2.63	0.34	15.7
8.15	2.63	0.34	18.3
5.60	1.25	0.34	18.1
4.30	2.63	0.34	19.1
4.30	2.63	0.34	15.4
5.60	10.1	0.25	12.0
5.60	10.1	0.34	19.8
4.30	10.1	0.34	18.6
2.40	10.1	0.34	13.2
5.60	10.1	0.55	22.8

(Continued)

| Table 6.30 | The Pressure Drop Data (*Continued*) |

Fluid Velocity	Liquid Viscosity	Mesh Size	y
2.14	112	0.34	41.8
4.14	112	0.34	48.6
5.60	10.1	0.25	19.2
5.60	10.1	0.25	18.4
5.60	10.1	0.25	15.0

| Table 6.31 | First Flight Data |

y	x_1	x_2	x_3	x_4
82	1.468	3.30	0.166	0.10
89	1.605	3.64	0.154	0.10
101	2.168	4.87	0.177	2.90
107	2.054	4.72	0.275	1.10
115	2.467	4.11	0.298	1.00
122	1.294	3.75	0.150	0.90
127	2.183	3.97	0.000	2.40
137	2.426	4.65	0.117	1.80
147	2.607	3.84	0.155	2.30
166	4.567	4.92	0.138	3.20
174	4.588	3.82	0.249	3.50
175	3.618	4.32	0.143	2.80
177	5.855	4.53	0.172	2.50
184	2.898	4.48	0.178	3.00
187	3.880	5.39	0.101	3.00
189	0.455	4.99	0.008	2.64
194	8.088	4.50	0.251	2.70
197	6.502	5.20	0.366	2.90
201	6.081	5.65	0.106	2.90
204	7.105	5.40	0.089	3.20
255	8.548	4.20	0.222	2.90
328	6.321	6.45	0.187	2.00

Table 6.32 | Coking Heat Data

x_1	x_2	x_3	y
83.2	11.2	0.61	625
78.9	5.1	0.60	680
76.1	5.3	1.65	680
72.2	8.1	1.06	710
73.2	7.0	1.02	710
73.8	6.0	0.75	685
70.6	8.6	0.74	705
68.4	9.8	1.09	685
70.5	6.7	0.76	680
63.2	11.8	1.85	700
55.8	10.4	0.71	720
56.3	8.8	1.70	705
57.1	5.3	0.93	730
55.6	6.6	0.90	715
54.7	6.5	1.54	705
53.4	4.5	1.10	730
60.4	9.9	1.08	725
60.8	8.1	1.41	710
61.9	6.8	1.03	710
61.8	6.8	0.99	700
61.9	6.6	0.90	715
61.1	6.4	0.91	710
59.0	7.6	1.36	740
59.3	7.0	1.31	730
56.6	7.6	1.07	730

6.23 The three basic structural elements of a data processing system are files, flows, and processes. Files are collections of permanent records in the system, flows are data interfaces between the system and the environment, and processes are functionally defined logical manipulations of the data. An investigation of the cost of developing software as related to files, flows, and processes was conducted. The data are in Table 6.33. Perform a complete analysis. Discuss your results and conclusions.

| Table 6.33 | Cost of Developing Software Data |

Files	Flows	Processes	Cost
4	44	18	22.6
2	33	15	15.0
20	80	80	78.1
6	24	21	28.0
6	227	50	80.5
3	20	18	24.5
4	41	13	20.5
16	187	137	147.6
4	19	15	4.2
6	50	21	48.2
5	48	17	20.5

⯮ 6.4 Residual Analysis

We estimate models based on their ability to explain the observed data. The residuals measure how well the model predicts the data used in the model's estimation.

VOICE OF EXPERIENCE

The residuals are fundamental for checking our assumptions.

In a sense, the residuals represent the failure of the model to predict the given data. As a result, the residuals provide a wealth of information on the quality of the analysis. Large residuals indicate either an outlier or a poor model (due to *model misspecification* or some violation of the underlying assumptions). Every measure of the adequacy of the chosen model we have developed depends on

$$SS_{res} = \sum_{i=1}^{n}(y_i - \hat{y}_i)^2.$$

We cannot appreciate how well our model estimates the data without a thorough analysis of the individual residuals. In this section, we emphasize some simple graphical techniques for this analysis. First, we outline methods for identifying systematic model misspecification. We then develop methods for identifying possible outliers and for checking the assumptions underlying our estimation and testing procedures. We conclude this section by showing how transformations often can improve our models.

Figure 6.5	An "Ideal" Residual Pattern

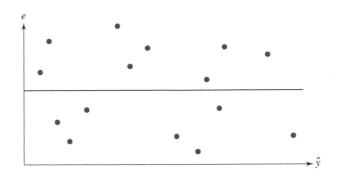

Checking for Model Misspecification

Model misspecification often reveals itself by a systematic pattern in the residuals. Two plots often prove useful for identifying this problem:

1. The residuals against the predicted values
2. The residuals against the regressors

In the case of simple linear regression, the two plots are essentially the same. If our model is reasonable, we expect the residual plots to look like a random pattern, as in Figure 6.5. Two common problems are systematic curvature, as in Figure 6.6, and a funnel shape, as in Figure 6.7. We generally can correct these problems either by adding appropriate terms—for example, an x^2 term—or by a transformation of the response. It is important to note that by transforming the response, we affect the distributional assumptions on our random errors. In many cases, the changes are nontrivial. We leave a more detailed description of this issue to a more advanced course.

Figure 6.6	Systematic Curvature in a Residual Plot

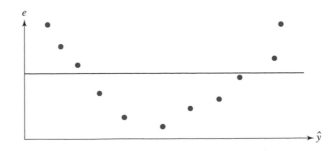

Figure 6.7	A Funnel Pattern in a Residual Plot

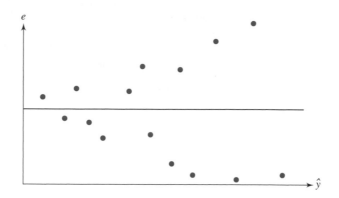

Example 6.12	**Vapor Pressure of Water**

Chemical and mechanical engineers often need to know the vapor pressure of water for specific temperatures. One approach requires the engineer to always refer to the infamous steam tables. Another approach seeks to use a simple model to predict the vapor pressure given the temperature. Thus, temperature is the regressor, and vapor pressure is the response. Table 6.34 lists the vapor pressures of water for various temperatures from 10° C to 60° C. Figure 6.8 is the scatter plot of these data. We see that as the temperature increases, so does the vapor pressure. This plot indicates that we may be able to model the vapor pressures using a straight line in the temperatures. Although the data display some curvature, a straight line may serve as a useful first approximation. This curvature provides us with an excellent opportunity to illustrate how we can use a residual plot to identify model misspecification. Later in this section, we shall use an appropriate transformation to improve this model.

Table 6.34	The Vapor Pressure of Water Data

Temp. (°C)	Vapor Pressure (mm Hg)
10	9.2
20	17.5
30	31.8
40	55.3
50	92.5
60	149.4

Figure 6.8	The Scatter Plot of the Vapor Pressure Data

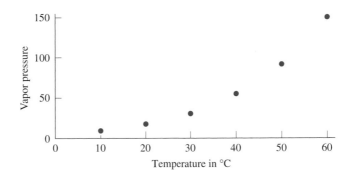

Table 6.35 gives the analysis of the simple linear regression model based on the SAS statistical software package. In this case, because we have only one regressor, the overall F-test and the test on temperature are equivalent. Since they both yield a p-value of 0.0036, we may safely conclude that the vapor pressure of water does depend on the temperature. The estimated coefficient is positive, so the vapor pressure does increase as the temperature increases.

We next notice that R^2 is 0.9038 and R^2_{adj} is 0.8798. This model may be adequate for predicting the vapor pressures by the temperatures; however, since we are dealing with physical phenomena, we may be able to generate a better model.

Table 6.35	The Computer-Generated Analysis of the Vapor Pressure Data

```
                     Analysis of Variance
                    Sum of           Mean
Source     DF      Squares         Square    F Value    Prob > F
Model       1   12879.28929    12879.28929    37.591     0.0036
Error       4    1370.45905      342.61476
C Total     5   14249.74833
        Root MSE      18.50986      R-square       0.9038
        Dep Mean      59.28333      Adj R-sq       0.8798
        C.V.          31.22270

                   Parameter Estimates
                  Parameter    Standard    T for H0:
Variable   DF      Estimate       Error   Parameter=0   P > |T|
INTERCEPT   1    -35.666667  17.23173798    -2.070       0.1073
T           1      2.712857   0.44247018     6.131       0.0036
```

Figure 6.9 | The Residuals Versus Predicted Values Plot for the Vapor Pressure Data

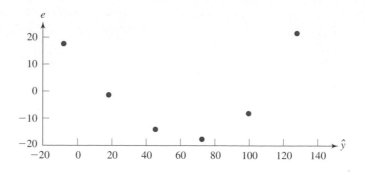

Figure 6.9 is a plot of the residuals against the predicted values. We see that the residuals for low temperatures are positive, for moderate temperatures they are negative, and for high temperatures they are positive again. The pattern appears quadratic. Figure 6.10 plots the residuals against the temperatures and reveals the same pattern. The pattern suggests that a better model should include a quadratic term for temperature or some other method to account for the curvature in the data. At the end of this section, we shall develop a model that deals with this curvature.

Outliers, Leverage, and Influence

We have already observed that outliers are data values that appear to be distinctly different from the rest of the data. In the regression setting, data may appear to be distinctly different in terms of the response or in terms of the regressors.

Figure 6.10 | The Residuals Versus Temperature Plot for the Vapor Pressure Data

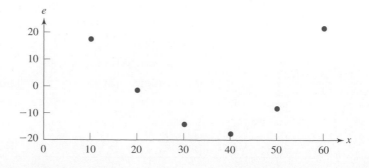

Thus, there are three possibilities for the different points:

1. Outliers, which are data points where the observed response does not appear to follow the pattern established by the rest of the data
2. Leverage points, which are data points that are distant from the other data points in terms of the *regressors*
3. Influential points, which try to combine the concepts of both leverage points and outliers

Outliers are extreme data values in terms of the *y*-direction. Leverage points are extreme data values in terms of the *x*'s. Influential points are extreme in a combined sense. The next two examples illustrate these concepts.

| Example 6.13 | **Surface Tension of Water-Based Coatings** |

The laboratory manager for a paint manufacturer asked a summer intern to make a series of water-based coatings by changing only the amount of the surfactant. The manager wanted the intern to see exactly what the surfactant does to the surface tension of the coating. Table 6.36 summarizes the data. Figure 6.11 gives the scatter plot of the data along with the estimated prediction equation.

The data value with an amount of 0.15 appears to be an outlier because it does not follow the general trend of the value in the *y*-direction; that is, the response for this amount looks different from the rest of the data. The laboratory manager rechecked this solution and determined that the intern had misread the instrument. The data value with an amount of 0.50 is a leverage point because it is quite distant from the rest of the data in terms of the amounts. In this case, however, the observed surface tension appears to be consistent with the trend established by the rest of the data. As a result, we have no reason to believe that it is overly influential.

| Table 6.36 | The Initial Surfactant Data |

Amount	Surface Tension
0.05	70
0.10	64
0.15	65
0.20	54
0.25	50
0.50	26

Figure 6.11 | The Scatter Plot and the Prediction Equation for the Initial Surfactant Data

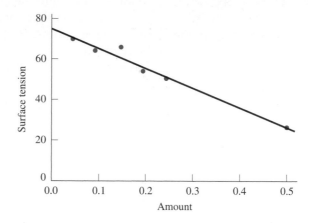

Example 6.14 | **Surface Tension of Water-Based Coatings—Revisited**

The laboratory manager later asked the intern to repeat the same experiment using a different surfactant. Table 6.37 gives the results, and Figure 6.12 shows the scatter plot along with the estimated prediction equation. In this case, the data point with an amount of 0.50 appears to be highly influential because its response does not seem to follow the same trend as the rest of the data and it is extreme relative to the other amounts. When we use least squares to estimate our model, influential points tend to draw the line to themselves, as Figure 6.12 illustrates. Many engineers' instincts would tell them to drop this influential point and to reanalyze the data. A better policy investigates this point to learn why it is so different. In general, there are three reasons a point is so influential:

VOICE OF EXPERIENCE

Outliers are not necessarily invalid data.

1. Someone made a recording error, which happens often and is easy to correct.
2. Someone made a fundamental error collecting the observation, which usually requires us to repeat the experiment at that data value to confirm.

Table 6.37 | The Second Set of Surfactant Data

Amount	Surface Tension
0.05	70
0.10	60
0.15	49
0.20	41
0.25	30
0.50	46

Figure 6.12 | The Scatter Plot and the Prediction Equation for the Second Set of Surfactant Data

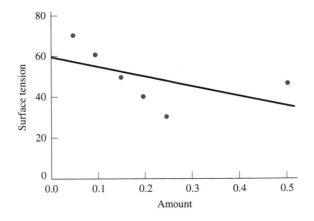

3. The data point is perfectly valid, in which case the model cannot account for the behavior.

Too often both statisticians and engineers become slaves to their models, even when they should not. George Box, a famous statistician who has worked for years with engineers, has often said, "All models are wrong. Some models are useful." Sometimes we throw out perfectly good data when we should be throwing out questionable models. Figure 6.13 is a typical plot of the surface tension of a water-based coating as we add surfactant. Although a straight-line model works well for part of the curve, it cannot work well over the entire range of interest. The laboratory manager wanted the intern to see this important aspect about surfactants firsthand. In fact, most water-based coatings formulations use amounts

Figure 6.13 | A Typical Surface Tension Versus Surfactant Amount Plot

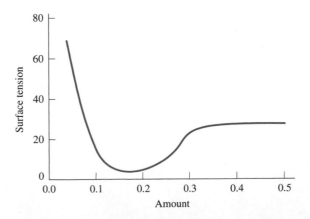

of surfactants out in the flat portion of the curve. Such a strategy makes the coating insensitive to the actual amount of surfactant added.

___

Studentized Residuals

Our regression model assumes that the random errors, the ϵ's, are independent with mean 0 and variance σ^2. It is natural to think that our residuals, the e's, are also independent with mean 0 and variance σ^2. In actuality, they are not. To appreciate why, we need to develop some notation.

Let $\hat{\mathbf{y}}$ be the $n \times 1$ vector of predicted values; thus,

$$\hat{\mathbf{y}} = (\hat{y}_1, \hat{y}_2, \ldots, \hat{y}_n)'.$$

In matrix notation,

$$\hat{\mathbf{y}} = \mathbf{Xb} = \mathbf{X}(\mathbf{X}'\mathbf{X})^{-1}\mathbf{X}'\mathbf{y}.$$

Let $\mathbf{e}$ be the $n \times 1$ vector of residuals; thus,

$$
\begin{aligned}
\mathbf{e} &= \mathbf{y} - \hat{\mathbf{y}} \\
&= \mathbf{y} - \mathbf{X}(\mathbf{X}'\mathbf{X})^{-1}\mathbf{X}'\mathbf{y} \\
&= [\mathbf{I} - \mathbf{X}(\mathbf{X}'\mathbf{X})^{-1}\mathbf{X}']\mathbf{y} \\
&= [\mathbf{I} - \mathbf{H}]\mathbf{y}
\end{aligned}
$$

where $\mathbf{H} = \mathbf{X}(\mathbf{X}'\mathbf{X})^{-1}\mathbf{X}'$. We often call $\mathbf{H}$ the *hat matrix*.

We call the ith diagonal element of $\mathbf{H}$ the ith *hat diagonal* and denote it by h_{ii}. The hat diagonal represents the distance the point $\mathbf{x}_i$ is from the center of all the $\mathbf{x}$'s. For a model that contains an intercept term, we can show that

$$\frac{1}{n} \leq h_{ii} \leq 1.$$

The ith hat diagonal is $1/n$ if $\mathbf{x}_i$ is in the exact center of the space defined by $\mathbf{X}$. For simple linear regression, $h_{ii} = 1/n$ if $x_i = \bar{x}$.

Recall that a leverage point is one that is distant from the overall center of the data in terms of the $\mathbf{x}$'s. The hat diagonals provide a measure of this distance. Most regression texts suggest that ith data point is a leverage point if

$$h_{ii} > \frac{2(k+1)}{n}.$$

Virtually all statistical software packages calculate the hat diagonals as standard practice.

We can show that for the ith residual, e_i,

$$E(e_i) = 0$$

$$\text{var}(e_i) = (1.0 - h_{ii})\sigma^2.$$

As a result, the variance of each residual depends on how far away the data point is from the overall center of the data in terms of the $\mathbf{x}$'s: the farther the data point is from the center, the smaller the variance for the corresponding residual.

The dependence of the variance for the ith residual on the data point's distance from the overall center of the data can make interpreting residual plots difficult. The usual approach to this problem is to standardize the residuals appropriately. We can show that

$$\frac{e_i}{\sigma \sqrt{1.0 - h_{ii}}}$$

follows a standard normal distribution. Unfortunately, we never know σ. At first glance, a reasonable approach uses the internally studentized statistic

$$\frac{e_i}{\sqrt{MS_{res}(1.0 - h_{ii})}},$$

which many people would presume follows a t-distribution with $n - k - 1$ degrees of freedom; however, in reality, it does not for quite technical reasons. The best approach uses the R-*student* statistic, r_i, defined by

$$r_i = \frac{e_i}{s_{(i)} \sqrt{1.0 - h_{ii}}},$$

where $s_{(i)}$ is an appropriate estimate of σ that does not use the ith residual. Some people call r_i the externally studentized residual. We can show that r_i truly follows a t-distribution with $n - k - 2$ degrees of freedom. Virtually all statistical software packages calculate either the internally or the externally standardized residuals as standard practice.

Many analysts use R-student as a measure, among many, of influence. Studentized residuals remove the scale of the data by transforming the residuals into t-statistics. This means that the magnitude of the studentized residuals has the same interpretation in every data set. For this reason, we recommend using the studentized residuals for graphically checking assumptions. A very rough rule of thumb suggests that $|r_i| > 3$ implies that the ith data point is either highly influential or an outlier and that $|r_i| > 2$ implies that we should investigate this point further. There are better techniques for formally determining whether a data value is an outlier or highly influential. Unfortunately, we must leave this discussion for a course that specializes in regression analysis.

| Example 6.15 | **Surface Tension of Water-Based Coatings—Continued** |

Tables 6.38 and 6.39 give the hat diagonals and the R-students for the two surface tension data sets. Both data sets use the same amounts, the x's, so we expect to see the same hat diagonals for these two data sets. Since we have one regressor,

Table 6.38	The Hat Diagonals and R-Students for the Initial Surfactant Data

Amount	Surface Tension	h_{ii}	r_i
0.05	70	0.3639	−0.2273
0.10	64	0.2590	−0.6737
0.15	65	0.1934	7.2393
0.20	54	0.1672	−0.6725
0.25	50	0.1803	−0.2718
0.50	26	0.8361	0.0559

$k = 1$, our cutoff value is

$$\frac{2(k+1)}{n} = \frac{2(2)}{6} = 0.6666.$$

The data point with an amount of 0.50 is a leverage point in both data sets. For the initial data set, the data value with an amount of 0.15 is an outlier because it has a large, in absolute value, R-student but no real leverage. For the second data set, the data value with an amount of 0.50 is an influential point because it also has a large, in absolute value, R-student and it has leverage.

Checking Assumptions

Least squares estimation of the model assumes that the random errors

1. Have an expected value of zero.
2. Have constant variance.
3. Are independent.

Table 6.39	The Hat Diagonals and R-Students for the Second Set of Surfactant Data

Amount	Surface Tension	h_{ii}	r_i
0.05	70	0.3639	1.3610
0.10	60	0.2590	0.4268
0.15	49	0.1934	−0.2308
0.20	41	0.1672	−0.6716
0.25	30	0.1803	−1.8122
0.50	46	0.8361	33.2225

Our testing procedures require the additional assumption that the residuals follow a well-behaved distribution. We can show that if we use least squares to estimate the model, then the sum of the residuals is zero. As a result, the sample mean of the residuals, which is our best estimate of the expected value, is always zero. Some useful graphical techniques for checking the other assumptions are

- A plot of the studentized residuals against the predicted values, which checks the constant variance assumption
- A plot of the studentized residuals against the regressors, which also checks the constant variance assumption
- A plot of the studentized residuals in time order, which checks the independence assumption
- A stem-and-leaf display of the residuals or of the studentized residuals, which checks the well-behaved distribution assumption
- A normal probability plot of the studentized residuals, which also checks the well-behaved distribution assumption

Some statistical software packages use the nonstudentized residuals in these plots. When possible, we should use the externally studentized residuals to check these assumptions, particularly the constant variance assumption, because we know that the nonstudentized residuals do not have the same variance.

Statisticians tend to focus the most attention on the constant variance assumption. In many engineering problems, the variability increases with the predicted value. The resulting residual plot displays a distinct funnel effect, which we illustrate in the next example. In other cases, the variability appears to depend on the specific values of at least one of the regressors. In either case, we sometimes can use an appropriate *transformation* of either the response or the regressors to correct these problems.

Too often, people lose the time order of the data, which can be rather unfortunate. Whenever possible, we should plot the residuals in their time order to ensure that no systematic biases occurred during the data collection. Such biases can result from important factors for which we did not collect data.

The stem-and-leaf display and the normal probability plot are methods for checking the shape of the distribution of the residuals. We discussed the stem-and-leaf display at length in Chapter 2. Most statistical software packages generate normal probability plots, which we introduced in Chapter 3, as a routine feature. If the data truly come from a normal distribution, then the resulting plot should look like a straight line. Deviations are evidence of nonnormality: The bigger the deviation, the bigger the problem. Problems with either of these displays may suggest that we consider an appropriate transformation of the response.

Example 6.16 | **Popcorn Data**

A popular student project in this course looks at making popcorn. The following data represent an attempt to find the optimal combination of burner setting (x_1), amount of oil (x_2), and "popping" time (x_3) to minimize the number of inedible kernels (y) when the corn is popped over a stove top. To minimize extraneous variability, the experimenter uses the same burner, the same pan, the same bag of popcorn, and the same initial amount for each run. In addition, the experimenter makes only one batch per day to ensure no residual heat effects from the previous batch. An inedible kernel is one that either does not pop or is burned. The same person evaluates each batch. From previous experience, the experimenter believes that an appropriate initial model is

$$y_i = \beta_0 + \beta_1 x_{i1} + \beta_2 x_{i2} + \beta_3 x_{i3} + \beta_{11} x_{i1}^2 + \beta_{22} x_{i2}^2 + \beta_{33} x_{i3}^2$$
$$+ \beta_{12} x_{i1} x_{i2} + \beta_{13} x_{i1} x_{i3} + \beta_{23} x_{i2} x_{i3} + \epsilon_i.$$

The experimenter used a *Box–Behnken design*, which is a commonly used experimental plan for fitting this model. Table 6.40 gives the results.

Table 6.41 summarizes the analysis based on the SAS statistical software package. The overall F-test indicates a marginal model because the p-value is greater than $\alpha = 0.05$. The R^2 value of 0.8548 indicates that the model accounts for more than 85% of the observed variability. The adjusted R^2_{adj} of 0.5934 indicates that the model is possibly overspecified. The individual t-tests indicate that only the coefficients associated with x_1, the burner setting, x_3, the popping time, and $x_1 x_3$, the interaction of burner setting and time, appear significant (have p-values less than 0.05). None of the terms involving x_2, the amount of oil, appear to be important.

Table 6.40	The Popcorn Data

Temp.	Oil	Time	y
7	4	90	24
5	3	105	28
7	3	105	40
7	2	90	42
6	4	105	11
6	3	90	16
5	3	75	126
6	2	105	34
5	4	90	32
6	2	75	32
5	2	90	34
7	3	75	17
6	3	90	30
6	3	90	17
6	4	75	50

| Table 6.41 | The Computer-Generated Analysis of the Popcorn Data |

Analysis of Variance

Source	DF	Sum of Squares	Mean Square	F Value	Prob>F
Model	9	8868.98333	985.44259	3.270	0.1026
Error	5	1506.75000	301.35000		
Total	14	10375.73333			

Root MSE	17.35944	R-square	0.8548
Dep Mean	35.53333	Adj R-sq	0.5934
C.V.	48.85395		

Parameter Estimates

Variable	DF	Parameter Estimate	Standard Error	T for H0: Parameter=0	P > |T|
INTERCEPT	1	2122.125000	619.70895134	3.424	0.0187
X1	1	-379.625000	123.20919710	-3.081	0.0274
X2	1	109.375000	91.65222992	1.193	0.2863
X3	1	-23.183333	8.21394647	-2.822	0.0370
X11	1	16.500000	9.03413665	1.826	0.1274
X22	1	-4.500000	9.03413665	-0.498	0.6395
X33	1	0.067778	0.04015172	1.688	0.1522
X12	1	-4.000000	8.67971774	-0.461	0.6643
X13	1	2.016667	0.57864785	3.485	0.0176
X23	1	-0.683333	0.57864785	-1.181	0.2907

Table 6.42 gives the residuals for this analysis. Figure 6.14 gives the stem-and-leaf display, which indicates that the residuals do follow a well-behaved distribution. We see a single peak, the data appear roughly symmetric, and the tails die rapidly. Figure 6.15 shows the boxplot and does not indicate any potential outliers.

Figure 6.16 gives the normal probability plot of the studentized results. We really do not see a straight line; instead, the pattern appears more S-like. Usually such a pattern suggests that we try some kind of transformation on the responses and reanalyze the data. Since both the stem-and-leaf display and the boxplot look reasonably good, we may or may not want to use a transformation.

Figure 6.17 is a plot the studentized residuals against the predicted values and suggests that the variability may increase with the predicted value (a funnel effect). Such a pattern may justify using a square root or log transformation of the response and reanalyzing the data.

Figure 6.18 is a plot of the studentized residuals against the burner settings. We see that the residuals appear to be less variable for a setting of 6 than for the

Table 6.42	The Residuals for the Popcorn Data

Obs	X1	X2	X3	Y	Predict	Residual
1	7	4	90	24	13.75	10.25
2	5	3	105	28	20.63	7.37
3	7	3	105	40	56.88	-16.88
4	7	2	90	42	28.00	14.00
5	6	4	105	11	4.38	6.62
6	6	3	90	16	21.00	-5.00
7	5	3	75	126	109.12	16.88
8	6	2	105	34	31.13	2.87
9	5	4	90	32	46.00	-14.00
10	6	2	75	32	38.63	-6.63
11	5	2	90	34	44.25	-10.25
12	7	3	75	17	22.38	-7.38
13	6	3	90	30	21.00	9.00
14	6	3	90	17	21.00	-4.00
15	6	4	75	50	52.88	-2.88

settings of 5 and 7. If the true effect of the burner setting is quadratic, then this pattern is consistent with the variability being related to the predicted value of the response. Figure 6.19 is a plot of the studentized residuals against the amount of oil. Frankly, we do not see any patterns. Figure 6.20 is a plot of the studentized residuals against the popping times. The residuals for 75 seconds appear to be slightly negative, and the residuals for 105 seconds appear slightly positive. This pattern may suggest some systematic departure from the model that warrants further analysis.

Figure 6.14	The Stem-and-Leaf Display for the Popcorn Residuals

```
N = 15    Median = -2.88
Quartiles = -7.38, 9
Decimal point is 1 place to the right of the colon
    3    3   -1 : 740
         5   -0 : 77543
    7    4    0 : 3779
    3    3    1 : 047
```

Figure 6.15 | The Boxplot for the Popcorn Residuals

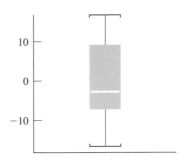

Figure 6.21 plots the studentized residuals in their time order. We see no pattern, so we should feel reasonably comfortable that there were no systematic effects over the course of the experiment and that the independence assumption holds.

Transformations

Engineers use transformations of the response for two basic reasons:

1. To correct problems with the underlying assumptions
2. To change the natural metric of the problem in accordance with engineering theory

Typically, we use some kind of transformation of the response when we encounter problems with the constant variance assumption or when the residuals appear to follow a distinctly nonnormal distribution. In other cases, engineering theory suggests an appropriate transformation of the data. For example, theory from physical chemistry suggests that the vapor pressure is an exponential function of the inverse of the temperature. We shall see that this insight provides a much better way to model our vapor pressure data.

Figure 6.16 | The Normal Probability Plot of the Studentized Residuals for the Popcorn Data

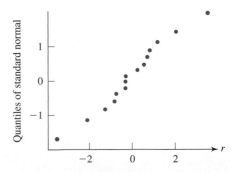

| Figure 6.17 | The Studentized Residuals Versus the Predicted Values for the Popcorn Data |

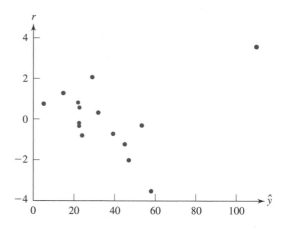

Two common transformations of the response are square root and log. Both transformations can help to make the variance more consistent over the region of interest and make the residuals appear more normal. We often use the square root transformation when we deal with count data. We often use the log transformation when we believe that the variance depends on the mean or when the response represents a time until some event.

| Example 6.17 | **Popcorn Data—Revisited** |

Table 6.43 gives the analysis for the square root transformation of the data based on the SAS statistical software package. The overall F-test appears to provide

| Figure 6.18 | The Studentized Residuals Versus Burner Settings for the Popcorn Data |

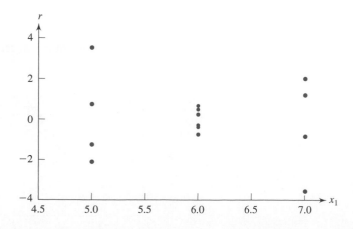

Figure 6.19 | The Studentized Residuals Versus Amounts of Oil for the Popcorn Data

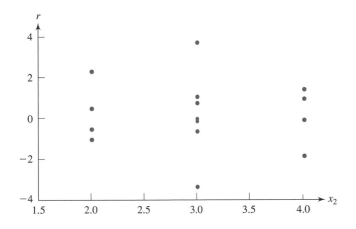

a little more evidence that at least one of the coefficients is nonzero, although it still is not significant relative to a 0.05 significance level. The R^2 of 0.8779 is reasonably good. We cannot directly compare it to the R^2 for the untransformed data, however, because the transformation changes the the units for our response (the transformation changes the *metric* of our problem). None of the terms involving x_2 appears important, which suggests that we drop all such terms from the model. In Chapter 7, we shall explain why we reanalyze data in this manner from a well-planned experiment.

Table 6.44 gives the analysis when we drop all terms involving x_2. Now the overall F-test suggests that at least one of the regressors does influence the number of inedible kernels. The burner setting (x_1), the popping time (x_3), and their interaction are all significant. The two pure quadratic terms appear to be less important.

Figure 6.20 | The Studentized Residuals Versus Popping Times for the Popcorn Data

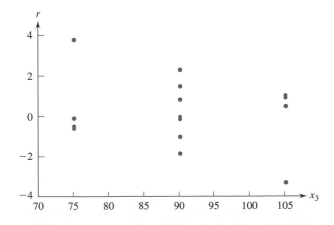

Figure 6.21 | The Time Plot of the Studentized Residuals for the Popcorn Data

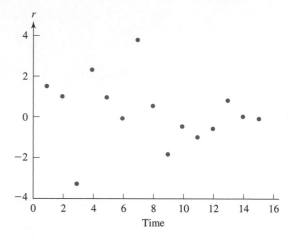

Table 6.43 | The Computer-Generated Analysis for the Square Root Transformation of the Popcorn Data

Analysis of Variance

Source	DF	Sum of Squares	Mean Square	F Value	Prob>F
Model	9	42.00055	4.66673	3.994	0.0708
Error	5	5.84157	1.16831		
C Total	14	47.84212			

Root MSE	1.08089	R-square	0.8779	
Dep Mean	5.68717	Adj R-sq	0.6581	
C.V.	19.00571			

Parameter Estimates

Variable	DF	Parameter Estimate	Standard Error	T for H0: Parameter=0	P > |T|
INTERCEPT	1	143.996737	38.58620960	3.732	0.0135
X1	1	-26.652326	7.67162697	-3.474	0.0178
X2	1	7.913474	5.70673079	1.387	0.2242
X3	1	-1.461835	0.51144180	-2.858	0.0355
X11	1	1.227799	0.56251098	2.183	0.0808
X22	1	-0.044361	0.56251098	-0.079	0.9402
X33	1	0.004355	0.00250005	1.742	0.1420
X12	1	-0.351916	0.54044307	-0.651	0.5437
X13	1	0.135582	0.03602954	3.763	0.0131
X23	1	-0.065476	0.03602954	-1.817	0.1289

Table 6.44	The Computer-Generated Analysis of the Popcorn Data When All Terms Involving x_2 Are Dropped

```
                        Analysis of Variance
                        Sum of          Mean
Source       DF         Squares        Square    F Value    Prob>F
Model        5          36.61997       7.32399    5.874     0.0110
Error        9          11.22215       1.24691
C Total      14         47.84212

        Root MSE          1.11665    R-square        0.7654
        Dep Mean          5.68717    Adj R-sq        0.6351
        C.V.             19.63455
                        Parameter Estimates
                        Parameter    Standard    T for H0:
Variable     DF         Estimate       Error    Parameter=0   P > |T|
INTERCEPT    1          167.556304  36.33435279    4.612      0.0013
X1           1          -27.749022   7.72784449   -3.591      0.0058
X3           1          -1.660992    0.51518963   -3.224      0.0104
X11          1          1.231211     0.57940084    2.125      0.0625
X33          1          0.004370     0.00257511    1.697      0.1239
X13          1          0.135582     0.03722164    3.643      0.0054
```

Many analysts prefer to use the smallest model that explains the data well. In that spirit, we would drop the pure quadratic terms. In this particular case, the experimenter's goal was to minimize the number of inedible kernels. The pure quadratic terms play an important role in finding optimum conditions, as we shall see. In such a situation, an analyst may justify including some marginal terms in the model, which is what this particular experimenter did.

We next need to check the residuals. Figure 6.22 shows the stem-and-leaf display. We see a pattern slightly skewed to the right for these residuals, though probably not enough to cause much concern about the results of our tests. Figure 6.23 gives the boxplot, which does not indicate any problems with outliers. Figure 6.24 is the normal probability plot for the studentized residuals. It indicates a more severe S-like pattern. As a result, the transformation may have worsened the shape. In this case, the transformation did not improve the distributional assumptions for our analysis.

Figure 6.25 is a plot of the residuals against the predicted values. Prior to the transformation, we saw some evidence that the variability increased with the predicted value. We do not see any apparent pattern to the residuals now. Thus, this plot indicates that the transformation has eliminated this problem.

Figure 6.22 | The Stem-and-Leaf Display for the Residuals After the Transformation of the Popcorn Data

```
N = 15    Median = -0.38
Quartiles = -0.74, 0.78
Decimal point is at the colon
     1     1   -1 : 3
           8   -0 : 98777541
     6     3    0 : 778
     3     3    1 : 035
```

Figure 6.26 is a plot of the residuals versus the burner settings. Prior to the transformation, these residuals appeared to be more variable at the settings of 5 and 7 than at 6. We do not see any apparent patterns in the residuals in this plot, which again suggests that the transformation has eliminated this problem. Figure 6.27 is a plot of the residuals against the popping times. Figure 6.28 plots the residuals in their time order. Again, we do not see any apparent pattern, which suggests that no systematic biases were present.

On the whole, the square root transformation appears to provide a better basis for modeling these data. We can conclude that a reasonable model for the data is

$$\sqrt{y} = 167.6 - 27.7x_1 - 1.66x_3 + 1.23x_1^2 + 0.0044x_3^2 + 0.14x_1x_3 + \epsilon.$$

We can use this model to find the burner setting and popping time that minimize the number of inedible kernels. We first take the partial derivatives with respect to x_1 and x_3 and set them equal to zero, which yields these equations:

$$2.46x_1 + 0.14x_3 = 27.7$$
$$0.0088x_3 + 0.14x_1 = 1.66.$$

Figure 6.23 | The Boxplot for the Residuals After the Transformation of the Popcorn Data

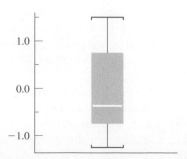

Figure 6.24 | The Normal Probability Plot of the Studentized Residuals After the Square Root Transformation

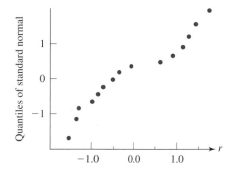

Solving then simultaneously, we find that the minimum number of inedible kernels occurs with a burner setting of 5.5 and a popping time of essentially 105 seconds. In Chapter 8, we shall outline experimental strategies for optimizing an engineering process along these lines.

Sometimes engineering theory suggests important transformations, as in the next example.

Example 6.18 | **Vapor Pressure of Water—Revisited**

Physical chemistry suggests that the vapor pressure should follow an exponential relationship to the inverse of the temperature. Specifically, let p_v be the vapor

Figure 6.25 | The Residuals Versus the Predicted Values After the Transformation of the Popcorn Data

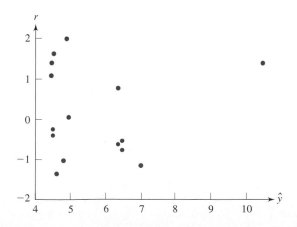

Figure 6.26 | The Residuals Versus Burner Settings After the Transformation of the Popcorn Data

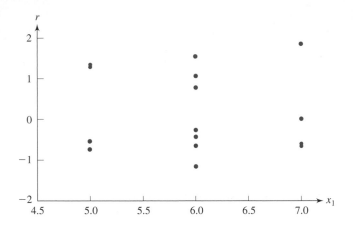

pressure, and let T be the temperature. The Clausius–Clapeyron equation states that

$$\ln(p_v) \propto -\frac{1}{T}.$$

Let y_i be the natural log of the ith vapor pressure, and let x_i be the inverse of the ith temperature. The Clausius–Clapeyron equation suggests that a reasonable model for the vapor pressures over a wide range of temperatures is

$$y_i = \beta_0 + \beta_1 x_i + \epsilon_i.$$

Figure 6.27 | The Residuals Versus Popping Times After the Transformation of the Popcorn Data

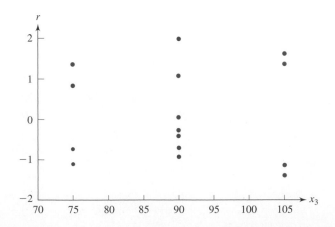

Figure 6.28 | The Time Plot of the Residuals After the Transformation of the Popcorn Data

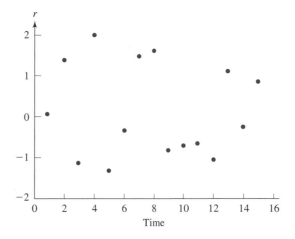

Table 6.45 lists the vapor pressures of water from 0° C to 100° C. Since we need the inverse of the temperatures, we must convert from degrees Celsius to degrees Kelvin by adding 273 to each temperature.

If y_i is the natural log of the ith vapor pressure and if x_i is the inverse of the ith temperature, then we can restate these data as in Table 6.46. Table 6.47 gives the analysis from the SAS statistical software package. The estimated model provides an excellent fit to the data. All of the tests have p-values of 0.0001, and both R^2 and

Table 6.45 | The Vapor Pressures of Water from 0° C to 100° C

Temp. (K)	Vapor Pressure (mm Hg)
273	4.6
283	9.2
293	17.5
303	31.8
313	55.3
323	92.5
333	149.4
343	233.7
353	355.1
363	525.8
373	760.0

| Table 6.46 | The Transformed Values for Temperature and Vapor Pressure |

x_i	y_i
0.00366	1.526
0.00353	2.219
0.00341	2.862
0.00330	3.459
0.00319	4.013
0.00310	4.527
0.00300	5.007
0.00291	5.454
0.00283	5.872
0.00275	6.264
0.00268	6.633

R^2_{adj} are 0.9999. Figure 6.29 shows the scatter plot for the transformed data, which form an excellent straight line. Figure 6.30 converts the prediction equation back to the original metric and shows a nonlinear curve even though we used a linear technique. Again, the prediction equation does an excellent job of explaining the data.

| Table 6.47 | The Computer Generated Analysis of the Transformed Vapor Pressure Data |

```
                    Analysis of Variance
                   Sum of        Mean
Source      DF     Squares      Square     F Value   Prob>F
Model       1      28.51104     28.51104   66715.469  0.0001
Error       9       0.00385      0.00043
C Total    10      28.51489

     Root MSE        0.02067     R-square        0.9999
     Dep Mean        4.34893     Adj R-sq        0.9999
     C.V.            0.47535

                    Parameter Estimates
                  Parameter     Standard    T for H0:
Variable   DF     Estimate        Error   Parameter=0   P > |T|
INTERCEPT   1     20.607379    0.06325352    325.790     0.0001
TEMP_INV    1  -5200.761791   20.13509535   -258.293     0.0001
```

Figure 6.29	The Scatter Plot for the Transformed Vapor Pressure Data

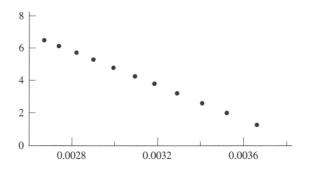

Figure 6.30	The Prediction Equation Transformed Back to the Original Scatter Plot for the Vapor Pressure Data

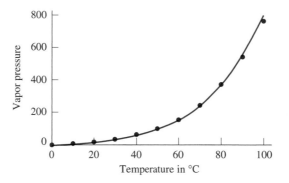

⫶ Exercises

6.24 Perform a thorough residual analysis of the ozone data given in Exercise **6.1**. If you feel a transformation is warranted, transform the data and reanalyze the data. Discuss your results and conclusions. Note that these data appear in time order.

6.25 Perform a thorough residual analysis of the steel data given in Exercise **6.2**. If you feel a transformation is warranted, transform the data and reanalyze the data. Discuss your results and conclusions. Note that these data appear in time order.

6.26 Perform a thorough residual analysis of the PET-LCP data given in Exercise **6.4**.

6.27 Perform a thorough residual analysis of the percent of noncontaminated peanuts data given in Exercise **6.9**. If you feel a transformation is warranted, transform the data and reanalyze the data. Discuss your results and conclusions.

6.28   Perform a thorough residual analysis of the tensile strength of Kraft paper data given in Exercise **6.10**. If you feel a transformation is warranted, transform the data and reanalyze the data. Discuss your results and conclusions.

6.29   Perform a thorough residual analysis of the springs with cracks data given in Example **6.2**. If you feel a transformation is warranted, transform the data and reanalyze the data. Discuss your results and conclusions.

6.30   Perform a thorough residual analysis of the caros acid data given in Exercise **6.17**. If you feel a transformation is warranted, transform the data and reanalyze the data. Discuss your results and conclusions. Note that these data appear in time order.

6.31   Perform a thorough residual analysis of the catalyst data given in Exercise **6.18**. If you feel a transformation is warranted, transform the data and reanalyze the data. Discuss your results and conclusions.

6.32   Perform a thorough residual analysis of the soil adsorption data given in Exercise **6.19**. If you feel a transformation is warranted, transform the data and reanalyze the data. Discuss your results and conclusions.

6.33   Perform a thorough residual analysis of the first flight data given in Exercise **6.21**. If you feel a transformation is warranted, transform the data and reanalyze the data. Discuss your results and conclusions.

6.34   Perform a thorough residual analysis of the coking heat data given in Exercise **6.22**. If you feel a transformation is warranted, transform the data and reanalyze the data. Discuss your results and conclusions.

6.35   From basic principles of physical chemistry, the viscosity is an exponential function of the temperature. Use appropriate transformations of the viscosity data given in Exercise **6.5** to perform a thorough residual analysis.

6.5 Collinearity Diagnostics

The Problem of Collinearity

Often, when we use an observational study to collect data, the individual regressors are actually related to one another. For many of these data sets, the relationships are minor enough so they do not present any real problems for the analysis. Occasionally, though, especially if we use an observational study on a normally operating process, the regressors are highly related, and we encounter significant problems with our analysis as a result. In some cases, the overall F-test will indicate that at least one of the individual coefficients is important but none of the t-tests for the individual coefficients is even close to significant. Why can this occur? Recall that the t-tests for the individual coefficients actually test the significance of the specific coefficient *given that all of the other regressors are in*

Figure 6.31 | A Data Set with a Collinearity Problem—The Picket Fence

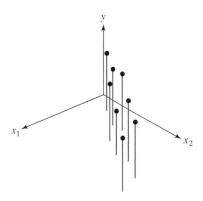

the model. If the regressors are highly related, then having any one of them in the model can essentially mean that the real contribution of all of the others is also present. We can avoid this problem entirely if we use a designed experiment where we manipulate the regressors in such a way as to ensure that they are unrelated to one another.

Figures 6.31–6.34 help to illustrate the problem of collinearity. Figure 6.31 shows the classic "picket fence" analogy for a data set with a collinearity problem. The *x*-values determine the location of the pickets, and the response values at these *x*'s determine the height of each picket. The *x*'s provide the basic support for estimating our model. When we estimate our regression model, we actually are trying to balance a plane on top of the *x*'s. In Figure 6.31, the *x*'s fall almost perfectly along a straight line. They offer little, if any, real support for balancing a plane on top of them. Some statisticians make the analogy to trying to balance a plane on top of a picket fence. Such a plane will not be stable. The plane's orientation can change dramatically with even a slight change in one of the responses.

Figure 6.32 | A Data Set with Orthogonal Regressors

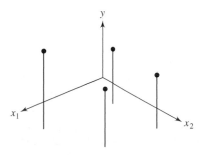

Figure 6.33	A Data Set with a Collinearity Problem—A Two-Dimensional View with Only the x's

Figure 6.32, on the other hand, illustrates how orthogonal regressors provide a more stable basis for balancing the plane. Orthogonal regressors are like table legs, which provide fairly rigid support for our model. Designed experiments generally produce either orthogonal or nearly orthogonal regressors, depending on the nature of the model to be estimated.

Figure 6.33 also helps us to see the source of the problem. We think we have two regressors, x_1 and x_2, which implies that we have two dimensions of interest. However, the two regressors tend to form a straight line with each other, so essentially we have only one dimension to support our model. Figure 6.34 shows how orthogonal regressors truly form a two-dimensional basis for estimating our model. In general, collinearity problems occur when we have k regressors, so we believe we have a k-dimensional basis for supporting our model, but because of relationships among these regressors, we essentially have something less than a k-dimensional basis available.

Five distinct consequences of collinearity are

1. The estimates of the coefficients are not stable.
2. The variances of the estimated coefficients are inflated.
3. The sign of the estimated coefficient is the opposite of what we expect.
4. Prediction away from the "picket fence" (the actual range of the observed data) is poor.
5. The power of the individual t-tests is greatly reduced.

Three common reasons for problems with collinearity are

1. The data collection method (sometimes called *sample* collinearity)
2. Inherent constraints on the population or process (sometimes called *population* collinearity)
3. The assumed model, particularly if it is overspecified

Sample collinearity is generally the result of a poor data collection method. Observational studies often have collinearity problems for this reason. Population collinearity is a more subtle issue. Montgomery and Peck (1992, p. 307) discuss an example in which an electrical utility investigated the effect of family income and house size on residential electricity consumption. For this particular region of the country, family income and size of house are highly related: the higher

| Figure 6.34 | A Data Set with Orthogonal Regressors—A Two-Dimensional View with Only the x's |

the income, the larger the house. Extremely few people with very low family incomes owned large houses. Conversely, very few wealthy families lived in extremely small houses. The nature of this relationship is not due to any artifact of the sampling scheme; rather, it is inherent to the population under study. Hence, this is an example of population collinearity. Finally, including too many terms in our model, especially when the data come from an observational study, increases the likelihood that at least two of the regressors are highly related. Reducing our model—that is, dropping some of the model terms—usually corrects the problem.

In engineering situations, we encounter problems with collinearity when we try to observe a process under fairly tight control. In such circumstances, we do not allow our regressors enough "room to roam"; that is, we do not allow them to cover the real region of interest. Designed experiments, on the other hand, force the regressors to cover the region of interest, thus avoiding the problem with collinearity.

More mathematically, collinearity depends on the columns of our model matrix, $\mathbf{X}$. If we can express at least one of the columns of $\mathbf{X}$ as a linear combination of the other columns, then we have perfect collinearity. Technically, the least squares estimates do not exist. We say that we have a problem with collinearity when we can almost express at least one of the columns of $\mathbf{X}$ as a linear combination of the other columns. From a mathematical perspective, we must look at the $\mathbf{X}'\mathbf{X}$ matrix to identify problems with collinearity.

| Example 6.19 | **Molecular Weight of a Polymer** |

An engineer assigned to a polymer process used an observational study to examine the current production process. She believed that the reaction temperature (x_1), the reaction pressure (x_2), the flow rate (x_3), and the amount of catalyst (x_4) control the molecular weight of this polymer. Table 6.48 gives her data. Table 6.49 is the statistical analysis generated by SAS. We see that the overall F-test is significant, but none of the individual coefficients is. We also note a relatively low R^2 of 0.5051, which is not uncommon for an observational study. The inconsistency between the overall F-test and the individual t-tests for the coefficients suggests a possible problem with collinearity.

| Table 6.48 | The Polymer Molecular Weight Data |

Temp.	Pressure	Flow Rate	Catalyst	Molecular Weight
258	49	107.5	2.3	831
248	51	104.1	2.5	823
256	53	107.6	2.5	846
243	45	101.2	2.5	803
254	53	106.9	2.5	848
247	48	103.4	2.5	821
242	45	103.0	2.4	815
240	46	101.5	2.5	802
249	48	104.6	2.4	830
258	55	110.0	2.8	831
259	54	108.9	2.4	820
251	49	104.8	2.2	812
258	53	108.0	2.2	839
250	50	105.1	2.4	804
257	53	109.2	2.5	843
248	45	102.9	2.2	833
246	47	102.9	2.2	808
249	50	105.1	2.2	826
241	47	100.8	2.2	809
247	48	104.1	2.5	808

Common Diagnostics

The literature suggests many different collinearity diagnostics, including these three common ones:

1. The correlation matrix for the estimated coefficients
2. Variance inflation factors
3. Condition numbers

All three of these measures depend upon $\mathbf{X'X}$.

The correlation matrix for the estimated coefficients is nothing more than the $\mathbf{X'X}$ matrix where we have scaled everything such that the diagonal elements are 1. The off-diagonal elements represent the correlations between each pair of regressors. These correlations must be between -1 and 1, inclusive. A correlation of 0 indicates that the two regressors are uncorrelated. A correlation

| Table 6.49 | The Computer-Generated Analysis of the Polymer Molecular Weight Data |

```
                     Analysis of Variance
                     Sum of          Mean
  Source     DF      Squares        Square     F Value     Prob>F
  Model       4    2090.37929     522.59482      3.827      0.0245
  Error      15    2048.42071     136.56138
  C Total    19    4138.80000

        Root MSE         11.68595      R-square          0.5051
        Dep Mean        822.60000      Adj R-sq          0.3731
        C.V.              1.42061

                     Parameter Estimates
                     Parameter     Standard     T for H0:
  Variable   DF      Estimate        Error    Parameter=0   P > |T|
  INTERCEPT   1    411.549722   167.53389350      2.457      0.0267
  TEMP        1      1.190355     1.90114540      0.626      0.5406
  PRES        1      0.138173     1.97394634      0.070      0.9451
  FLOW        1      1.078073     4.82740584      0.223      0.8263
  CAT         1     -2.803591    21.29209378     -0.132      0.8970
```

of -1 indicates that the two regressors are perfectly negatively correlated. Similarly, a correlation of 1 indicates that the two regressors are perfectly positively correlated. Values of the correlation near 1 or -1 indicate a strong relationship between the two regressors. Values around 0.5 or -0.5 indicate a moderate relationship.

The variance inflation factors are actually the diagonal elements of the inverse of the correlation matrix. Whereas the correlation matrix allows us to look at pairs of regressors, the variance inflation factors allow us to look at the joint relationships among a specified regressor and all the other regressors. Most texts suggest that variance inflation factors of 10 or more indicate a strong problem with collinearity. Values between 5 and 10 indicate a moderate problem.

The condition number also allows us to look at the joint relationships among the regressors. The matrix $\mathbf{X'X}$ must have $k + 1$ eigenvalues, which in turn must all be positive. The condition number is the ratio of the largest eigenvalue over each eigenvalue. Most texts suggest that any condition number greater than 1000 is evidence of a collinearity problem. To identify where the problem occurs, we look at the corresponding variance proportion, which is the proportion of the variance for a specified estimated coefficient that we can attribute to a specific eigenvalue. If an eigenvalue has a condition number greater than 1000, we then look at the variance proportions associated with it. Those regressors with variance proportions near 1 are affected by the collinearity problem.

SAS actually uses the singular values rather than the eigenvalues to compute its condition number. As a result, a large condition number in SAS is over 30. We still use the variance proportions as before.

The common ways we correct for collinearity are

- More data collection
- Subset models, where we drop regressors from the model
- Biased regression methods

The basic cause of the collinearity problem sometimes boils down to the fact that our data do not cover the region of interest in the regressors. We can correct this problem by collecting data in those areas we missed before. Subset models assume that because the regressors are highly related, we really do not need all of them in the model. The trick then becomes to find the most appropriate subset model. In other cases, we think that all of our regressors are important and should appear in our model. In these cases, we use such biased regression techniques as ridge regression or principal components regression. The details for all of these corrective measures are better left to a separate course on regression that can go into sufficient detail.

Example 6.20 | **Molecular Weight of a Polymer—Revisited**

Table 6.50 gives the collinearity diagnostics for these data. The correlation matrix information reveals that the temperature and the flow rate, with a correlation of -0.8849, are highly related to each other. In addition, the following pairs are at least moderately correlated:

- Temperature and catalyst, with a correlation of 0.5392
- Pressure and flow rate, with a correlation of -0.4175
- Flow rate and catalyst, with a correlation of -0.4657

The variance inflation factors indicate severe problems (values greater than 10) for temperature and flow rate, and a moderate problem (a value between 5 and 10) for pressure. The condition numbers also support these conclusions. We see four quite large condition numbers. Since we are dealing with SAS output, any condition number greater than 30 indicates a potential problem. The variance proportions indicate that temperature, pressure, and flow rate are the most affected. Taken together, these diagnostics suggest that these regressors are highly related, which explains why the overall F-test indicates a significant model but none of the t-tests for the individual coefficients is significant. As a result, we probably do not need all of these regressors in the model. The analyst used a subset model approach called *all possible regressions*, which suggested that the only regressor needed in the model is temperature.

Table 6.50	The Collinearity Diagnostics for the Polymer Molecular Weight Data

```
                         Correlation of Estimates
   CORRB        INTERCEP    TEMP      PRES      FLOW      CAT

   INTERCEP      1.0000   -0.3241    0.7211   -0.1319   -0.3314

   TEMP         -0.3241    1.0000    0.0031   -0.8849    0.5392

   PRES          0.7211    0.0031    1.0000   -0.4175   -0.1591

   FLOW         -0.1319   -0.8849   -0.4175    1.0000   -0.4657

   CAT          -0.3315    0.5392   -0.1591   -0.4657    1.0000
```

```
                        Variance
   Variable   DF     Inflation

   INTERCEP    1     0.00000000

   TEMP        1    18.86956294

   PRES        1     5.56244881

   FLOW        1    24.72544588

   CAT         1     1.62503058
```

```
              Condition Numbers and Variance Proportions
         Eigen-      Cond. Var Prop Var Prop Var Prop Var Prop Var Prop
   Number  value     Number INTERCEP  TEMP    PRES     FLOW     CAT
      1   4.99468   1.00000  0.0000  0.0000  0.0000  0.0000  0.0001
      2   0.00311  40.09564  0.0016  0.0007  0.0161  0.0004  0.6664
      3   0.00209  48.91789  0.0399  0.0004  0.2107  0.0001  0.0054
      4   .000112 211.36786  0.9513  0.0640  0.7197  0.0500  0.0606
      5   .000015 578.37005  0.0073  0.9349  0.0534  0.9495  0.2675
```

⊱ Exercises

6.36   Check the data given in Exercise **6.15** for collinearity problems. Perform thorough analyses on appropriate subset models.

6.37   Check the data given in Exercise **6.16** for collinearity problems. Perform thorough analyses on appropriate subset models.

6.38   Check the data given in Exercise **6.17** for collinearity problems. Perform thorough analyses on appropriate subset models.

6.39   Check the data given in Exercise **6.18** for collinearity problems. Perform thorough analyses on appropriate subset models.

6.40 Check the data given in Exercise **6.19** for collinearity problems. Perform thorough analyses on appropriate subset models.

6.41 Check the data given in Exercise **6.20** for collinearity problems. Perform thorough analyses on appropriate subset models.

6.42 Check the data given in Exercise **6.21** for collinearity problems. Perform thorough analyses on appropriate subset models.

6.43 Check the data given in Exercise **6.22** for collinearity problems. Perform thorough analyses on appropriate subset models.

6.6 Case Study

Pencil lead is a mixture of clay and graphite. The clay provides a ceramic matrix to support the writing medium, the graphite. Like all ceramics, pencil lead is "fired" or heated to extremely high temperatures in a furnace. The heating process initially drives out any residual moisture in the pencil lead mixture. Later, the heating process breaks down the water of hydration in the clay. The water of hydration actually creates a pore structure in the ceramic as the water of hydration exits. Once all of the water of hydration exits, the pores begin to close.

After the firing stage, the pencil lead is placed in hot wax for a period of time. The residual pore structure in the pencil lead allows the lead to absorb the wax into the ceramic matrix. The wax can actually add strength to the resulting pencil lead. More importantly, though, the wax provides lubrication, which makes for smoother writing quality.

An important quality characteristic of pencil lead is the transverse strength. The transverse strength is the force required to snap a pencil lead into two pieces under a two-point loading. Basic clay chemistry suggests that the transverse strength depends upon the amount of clay, the final porosity, and the outside diameter of the pencil lead. An extended study was conducted to confirm this hypothesis. It was impossible to determine the actual amount of clay used in a particular pencil lead. As a result, Faber-Castell used the residual ash content of the pencil lead. The ash content was the amount of solid material remaining after all combustibles (for example, graphite) are burned away. Table 6.51 summarizes a subset of the data.

A multiple linear regression model involving the ash content (Ash), the porosity (Porosity) and the outside diameter (Outer Diam.) is fit to strength. Table 6.52 gives the initial analysis. The overall F-test indicates that at least one of the coefficients is nonzero, p-value $= 0.000$. Table 6.52 also indicates Ash is not significant with a p-value $= 0.428$. Therefore, the term for ash content will be removed. Table 6.53 gives the final model, which includes only porosity and outside diameter. The R^2 of 0.666 is not overly high but is sufficient, because the purpose of the data collection was just to identify the relationship between the input variables and the response, strength.

Table 6.51			Strength of Pencil Lead Data						

Obs.	Ash	Por-osity	Outer Diam.	Strength	Obs.	Ash	Por-osity	Outer Diam.	Strength
1	42.2	12.9	0.087	1.25	34	43.0	13.6	0.088	1.30
2	43.8	13.7	0.090	1.20	35	42.7	13.9	0.090	1.35
3	42.1	15.6	0.087	0.85	36	41.6	14.1	0.090	1.30
4	42.0	13.3	0.086	1.15	37	42.4	15.6	0.083	0.60
5	45.0	12.2	0.088	1.55	38	42.9	14.2	0.086	0.90
6	42.5	14.3	0.085	1.00	39	43.5	15.4	0.088	0.75
7	41.9	13.1	0.085	1.35	40	43.3	15.2	0.086	0.75
8	42.4	13.8	0.086	1.05	41	41.4	15.5	0.093	1.00
9	41.9	14.4	0.085	1.15	42	42.7	14.3	0.087	0.75
10	42.1	15.5	0.086	0.90	43	42.8	14.1	0.085	1.00
11	42.2	12.7	0.087	1.30	44	42.0	13.6	0.084	1.10
12	43.4	13.4	0.089	1.20	45	42.2	13.4	0.092	1.30
13	41.9	12.9	0.084	1.35	46	42.4	14.1	0.091	1.30
14	42.4	15.5	0.084	1.00	47	42.8	16.1	0.087	0.60
15	43.3	13.8	0.085	1.10	48	42.5	14.9	0.086	0.95
16	42.2	13.3	0.087	1.25	49	42.3	15.5	0.091	1.15
17	40.0	14.0	0.088	1.20	50	43.3	15.9	0.089	0.65
18	42.7	16.0	0.090	0.95	51	40.4	13.0	0.084	1.35
19	41.9	12.7	0.084	1.05	52	41.7	14.2	0.083	1.20
20	42.5	16.0	0.087	0.85	53	42.3	12.8	0.083	1.25
21	41.1	14.3	0.086	1.25	54	42.9	14.3	0.090	1.15
22	41.9	15.2	0.087	0.85	55	40.3	14.2	0.084	0.95
23	42.5	14.0	0.085	0.80	56	41.3	15.0	0.088	0.80
24	41.4	16.0	0.085	0.60	57	42.0	12.5	0.086	1.20
25	41.0	14.4	0.085	0.70	58	42.2	15.6	0.084	0.60
26	38.9	15.8	0.086	0.60	59	42.5	13.7	0.088	1.05
27	42.5	14.9	0.087	0.90	60	42.8	17.8	0.088	0.40
28	42.2	12.5	0.083	1.05	61	40.5	14.2	0.084	0.65
29	41.6	15.0	0.087	1.10	62	40.1	14.3	0.085	0.95
30	42.6	15.4	0.091	1.20	63	42.9	17.9	0.085	0.55
31	42.7	13.4	0.083	0.70	64	41.7	12.9	0.092	1.45
32	44.1	15.8	0.088	0.60	65	41.0	15.5	0.092	1.20
33	43.2	13.3	0.083	1.25	66	42.5	13.9	0.093	1.50

(Continued)

Table 6.51 | Strength of Pencil Lead Data (*Continued*)

Obs.	Ash	Por-osity	Outer Diam.	Strength	Obs.	Ash	Por-osity	Outer Diam.	Strength
67	43.0	14.7	0.084	0.85	74	42.1	14.1	0.087	1.20
68	43.4	12.6	0.086	1.00	75	42.7	13.2	0.091	1.40
69	41.1	16.3	0.088	1.10	76	43.8	14.9	0.084	0.70
70	42.7	13.0	0.085	1.20	77	42.3	15.7	0.087	0.95
71	42.1	13.0	0.084	0.90	78	42.3	15.5	0.087	0.65
72	43.7	13.5	0.087	1.15	79	42.6	13.6	0.084	1.05
73	39.5	16.0	0.088	0.65	80	42.5	14.0	0.088	1.10

We need to check the residuals. Figure 6.35 is the normal probability plot for the studentized residuals. The residuals fall on a straight line, indicating the distributional assumption is fine. Figure 6.36 is the plot of the residuals against the predicted values. We do not see any apparent pattern to the residuals. Figure 6.37 plots the residuals in their time order. Again, we do not see any apparent pattern, which suggests that no systematic biases were present. The assumptions do not appear to be violated.

Table 6.52 | Initial Model

```
Analysis of Variance


Source           DF      SS        MS       F        P
Regression        3    3.5984    1.1995   51.25    0.000
Residual Error   76    1.7786    0.0234
Total            79    5.3770


Predictor       Coef      SE Coef        T        P
Constant      0.1206      0.9106      0.13     0.895
Ash          -0.01370     0.01719    -0.80     0.428
Porosity     -0.16477     0.01443   -11.42     0.000
Out Diam     44.287       6.766       6.55     0.000


S = 0.152980    R-Sq = 66.9%       R-Sq(adj) = 65.6%
```

Table 6.53 | Final Model with Only Porosity and Outside Diameter

Predictor	Coef	SE Coef	T	P
Constant	-0.4328	0.5878	-0.74	0.464
Porosity	-0.16338	0.01429	-11.43	0.000
Out Diam	43.765	6.719	6.51	0.000

S = 0.152617   R-Sq = 66.6%   R-Sq(adj) = 65.8%

Our analysis indicates that porosity and the outside diameter impact the lead's final strength, while ash content does not. It is interesting to note that the ash content was not important. A basic understanding of clay chemistry would suggest that the amount of clay, as measured by the ash, is an important driver of strength. It is crucial to note that these data come from an observational study, which we discussed in Chapter 1. Faber-Castell tries to maintain strict control on each batch's formulation. These efforts kept the ash content within a sufficiently narrow band that we could not see this effect. Often in observational studies, regressors that the analyst strongly believe to be important turn out to be insignificant *over the ranges encountered in the study.*

The porosity of the lead is a clear measure of the impact of the firing process in the furnace. Clearly, Faber-Castell is not able to control the firing process as

Figure 6.35 | The Normal Probability Plot of the Studentized Residuals for the Pencil Lead Strength Data

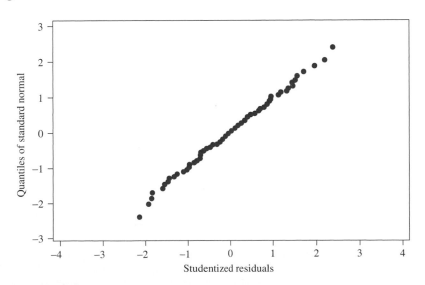

Figure 6.36 | The Residuals Versus the Predicted Values for the Pencil Lead Strength Data

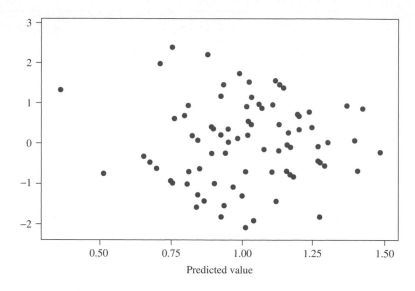

well as the formulation process. Our regression analysis strongly suggests that Faber-Castell should use control charts to monitor the porosity and to identify when the firing conditions change due to assignable causes.

Finally, it should be no surprise that the outside diameter of the lead impacts the final strength: the larger the diameter, the more material. Pencil lead, as a

Figure 6.37 | The Time Plot of the Residuals for the Pencil Lead Strength Data

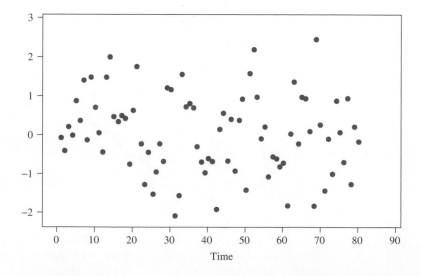

ceramic material, is quite abrasive. Pencil lead is extruded through a ceramic die. Over time, the pencil lead wears away the die, causing larger outside diameters. As a result, Faber-Castell closely monitors the outside diameters produced by each die. Unfortunately, an appropriate monitoring procedure for this situation is beyond the scope of this book.

⋗ 6.7 Ideas for Projects

1. Determine how well homework scores and class attendance predict test scores. Do a thorough residual analysis to confirm the underlying assumptions and to check for any interesting results. Comment on the nature of these relationships.
2. Almost all engineering laboratories focus on relationships among engineering characteristics. Use regression analysis followed by a thorough residual analysis to develop good models for these data.
3. Many instructors use a catapult to teach basic statistical concepts. If you have access to one, conduct a 2^3 factorial experiment (see Chapter 2) using the distance that the ball is thrown as the response. Use regression analysis followed by a thorough residual analysis to construct a good model. Place a box a fixed distance from the catapult, and use the model to try to throw the ball into the box.
4. If you are a team sports enthusiast, use regression analysis to model the performance of your favorite team. Determine which regressors appear most important for predicting success.

⋗ References

1. AKi, S. N., Ding, Z., and Abraham, M. A. (1996). Catalytic super critical water oxidation: stability of Cr_2O_3 catalyst. *AICHE Journal, 42,* 1995–2004.
2. Anand, K. N., and Ray, S. (1995). Determining wear limits of critical components for improving service quality of hydraulic pumps. *Quality Engineering, 8,* 249–254.
3. Bissell, A. F. (1992). Lines through the origin—is NO INT the answer. *Journal of Applied Statistics, 19,* 193–210.
4. Box, G. E. P., and Bisgaard, S. (1987). The scientific context of quality improvement. *Quality Progress, 22(6),* 54–61.
5. Byers, C. H., and Williams, D. F. (1987). Viscosities of binary and ternary mixtures of polyaromatic hydrocarbons. *Journal of Chemical and Engineering Data, 32,* 349–354.
6. Carroll, R. J., and Spiegelman, C. H. (1986). The effects of ignoring small measurement errors in precision instrument calibration. *Journal of Quality Technology, 18,* 170–173.

7. Chang, S. I., and Shivpuri, R. (1994). A multiple-objective decision-making approach for assessing simultaneous improvement in die life and casting quality in a die casting process. *Quality Engineering, 7,* 371–383.

8. Cook, R. D., and Weisberg, S. (1982). *Residuals and influences in regression.* New York: Chapman and Hall.

9. Davidson, A. (1993). Update on ozone trends in California's South Coast Air Basin. *Air and Waste, 43,* 226.

10. Draper, N. R., and Smith, H. (1981). *Applied regression analysis,* 2nd ed. New York: John Wiley.

11. Fearn, T. (1983). A misuse of ridge regression in the calibration of a near infrared reflectance instrument. *Applied Statistics, 32,* 73–79.

12. Hsiue, L-T., Ma, C-C. M., and Tsai, H-B. (1995). Separation and characterizations of thermotropic copolyesters of p-hydroxybenzoic acid, sebacic acid, and hydroquinone. *Journal of Applied Polymer Science, 56,* 471–476.

13. Joglekar, G., Schuenemeyer, J. H., and LaRiccia, V. (1989). Lack-of-fit testing when replicates are not available. *American Statistician, 43,* 135–143.

14. Lawson. J. S. (1982). Applications of robust regression in designed industrial experiments. *Journal of Quality Technology, 14,* 19–33.

15. Liu, C. H., Kan, M., and Chen, B. H. (1993). A correlation of two-phase pressure drops in screen-plate bubble column. *Canadian Journal of Chemical Engineering, 71,* 460–463.

16. Magness, L. S., Jr. (1994). High strain rate deformation behaviors of kinetic energy penetrator materials during ballistic impact. *Mechanics of Materials, 17,* 147–154.

17. Mandel, J. (1984). Fitting straight lines when both variables are subject to error. *Journal of Quality Technology, 16,* 1–14.

18. Mehta, S., and Deopura, B. L. (1995). Morphological and mechanical properties of PET-LCP blend fibers. *Journal of Applied Polymer Science, 56,* 169–175.

19. Montgomery, D. C., and Peck, E. A. (1992). *Introduction to linear regression analysis,* 2nd ed. New York: Wiley.

20. Perch and Bridgewafer (1980). *Iron and Steel Engineering, 57,* 47–50.

21. Said, A., Hassan, E., El-Awad, A., El-Salaam, K., and El-Wahab, M. (1994). Structural changes and surface properties of $CO_xFe_{3-x}O_4$ spinels. *Journal of Chemical Technology and Biotechnology, 60,* 161–170.

22. Tracy, N. D., Young, J. C., and Mason, R. L. (1992). Multivariate control charts for individual observations. *Journal of Quality Technology, 24,* 88–95.

23. Wauchope, R. D., and McDowell, L. L. (1984). Adsorption of phosphate, arsenate, methanearsonate, and cacodylate by lake and stream sediments: Comparisons with soils. *Journal of Environmental Quality, 13,* 499–504.

24. Williams, J. H., and Lee, S. S. (1985). Promising quantitative nondestructive evaluation techniques for composite materials. *Materials Evaluation, 43,* 561–565.

7

Introduction to 2^k Factorial-Based Experiments

In general, the best way to study an engineering process is to design an appropriate experiment. The next two chapters outline an important approach to experimentation called *response surface methodology* (RSM) originally espoused by Box and Wilson (1951), a statistician and a chemist. This approach uses a sequential philosophy of experimentation, whose ultimate goal is to optimize a process. RSM recognizes that many industrial situations allow us to get experimental results very quickly. For example, we can run one small experimental plan this week, analyze it next week, and then run a follow-up experiment based on what we learned the week after. RSM plans each experiment to support an appropriate regression model. In the early stages, RSM uses the fewest possible experimental runs to save resources for later in the optimization process. As we become more confident that we know where the true optimum conditions for our process lie, we begin to use more experimental runs. RSM recognizes that in industrial experimentation, each run is expensive in terms of either time or money. As a result, RSM seeks to use resources as efficiently as possible.

This chapter begins by outlining the simplest experimental design called the 2^k factorial (Sections 7.1 and 7.2). In many cases, the 2^k factorial design requires more resources than we can afford. As a result, we next describe a simple method for generating fractional factorial designs that cuts down on the number of experimental runs while still preserving the basic factorial structure (Section 7.3).

⇒ 7.1 The 2^2 Factorial Design

Consider a situation where we must determine the relationship between reaction temperature and pressure on the strength of polymer fibers. Our best data collection strategy uses a designed experiment where we systematically manipulate the reaction temperature and pressure and then observe our response, the strength. In this experiment, we call reaction temperature and pressure the *factors*. In a designed experiment, we manipulate the factors according to a well-defined strategy, called the *experimental design*. This strategy must ensure that we can separate out the *effects* due to each factor. The specified values of the factors used in the experiment are called the *levels*. Typically, we use a small number of levels for each factor, such as two or three. For example, we may use a "high" or $+1$ and a "low" or -1 level for both reaction temperature and pressure. We thus would use two levels for each factor. A *treatment combination* is a specific, distinct combination of the levels of each factor. Each time we carry out a treatment combination is an experimental *run* or *setting*. The experimental *design* or *plan* consists of a series of runs.

The experimental design and analysis basically are the same for categorical and continuous factors when we use only two levels for each factor. If all of the factors are continuous and we need more than two levels, however, we generally use either a standard two-level design (see Sections 7.1–7.3) plus center runs (see Section 7.2) or a response surface design (see Chapter 8). If all of the factors are categorical and we need more than two levels, we generally use classical factorial designs. We shall not discuss experimental designs and analysis for categorical factors with more than two levels in this text, but the interested student may read about them in Montgomery (2004). It remains an open research question when we need both categorical and continuous factors with more than two levels each. A full discussion of this problem is beyond the scope of this text. A reasonable approach is to run a different response surface design for each combination of the categorical factors.

Engineers often run two-level factorial experiments. If we have k factors, the full factorial design consists of every possible combination of the two levels for the k factors, and we have a total of 2^k distinct treatment combinations. We begin with the simplest case, where we have only two factors.

The Basic Design

| Example 7.1 | **The Carbon Monoxide Concentration from Burning Pine Wood** |

Rao and Saxena (1993) studied the effect of moisture and furnace temperature on the flue gases produced when pine wood is burned. They studied the composition of the flue gases that resulted. We concentrate on their results for carbon monoxide because it is an important pollutant. The researchers were concerned with a moisture content of 0%, which represents "bone dry" wood, and with a moisture content of 22.2%. They used furnace temperatures of $1100°$ K and $1500°$ K.

VOICE OF EXPERIENCE

The actual high and low levels used in an experiment require sound engineering insights.

The 2^2 factorial design uses as the treatment combinations every possible combination of two different factors, each at two levels. Since each factor has only two levels, we can name one the "low" or -1 level and the other the "high" or $+1$ level. In general, statisticians prefer to talk about designs in terms of the *design variables*. Let x_1 be the design variable associated with the moisture content.

Figure 7.1	The Basic 2^2 Factorial Design in Design Variables

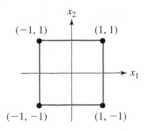

Thus,

$$x_1 = \begin{cases} -1 & \text{if the moisture content is } 0\% \\ 1 & \text{if the moisture content is } 22.2\%. \end{cases}$$

In a similar manner, let x_2 be the design variable associated with the furnace temperature. Thus,

$$x_2 = \begin{cases} -1 & \text{if the temperature is } 1100K \\ 1 & \text{if the temperature is } 1500K. \end{cases}$$

The basic 2^2 factorial in design variables is simply all of the possible combinations of the two levels:

x_1	x_2
−1	−1
1	−1
−1	1
1	1

Figure 7.1 shows that this design forms a square in terms of the design variables. This design contains $2^2 = 4$ distinct treatment combinations—hence, the name. All 2^2 factorial designs in the design variables reduce to this particular form. Rao and Saxena replicated this base design to get an estimate of the experimental error. Table 7.1 summarizes the actual design, and Table 7.2 gives the corresponding 2^2 factorial design in the natural units.

The Model and Calculating Effects

For the situation in Example 7.1, the largest model this design allows us to fit is

$$y_i = \beta_0 + \beta_1 x_{i1} + \beta_2 x_{i2} + \beta_{12} x_{i1} x_{i2} + \epsilon_i,$$

| Table 7.1 | The Pine Wood Experimental Data in the Design Units |

		Concentration of CO	
x_1	x_2	Replication 1	Replication 2
−1	−1	20.3	20.4
1	−1	13.6	14.8
−1	1	15.0	15.1
1	1	9.7	10.7

where

- y_i is the carbon dioxide concentration for the ith test run.
- β_0 is the y-intercept, which in this case is the overall mean response because we have centered the x's around zero.
- β_1 is the regression coefficient associated with the moisture content.
- β_2 is the regression coefficient associated with the temperature.
- β_{12} is the *interaction* coefficient for moisture and temperature.
- ϵ_i is a random error.

The interaction term allows the model to adjust the impact of the moisture content for the specific level of temperature used, and vice versa. We shall discuss this concept of interaction in more detail in the next example.

Traditionally, we estimate the *effects* of moisture, temperature, and their interaction. We can best illustrate what we mean by effects and their calculation through an example. We shall see that by dividing the appropriate effect by 2, we obtain the estimate of the regression coefficient associated with the corresponding design variable. Because of this relationship, we are more interested in the calculation of effects and interactions conceptually than computationally.

| Table 7.2 | The Pine Wood Experimental Data in the Natural Units |

Moisture (%)	Temp. (K)	Concentration of CO	
		Replication 1	Replication 2
0	1100	20.3	20.4
22.2	1100	13.6	14.8
0	1500	15.0	15.1
22.2	1500	9.7	10.7

| Example 7.2 | **The Pine Wood Experiment—Estimation of the Effects** |

We first need to introduce some notation that allows us to "name" the specific responses. Let x_1 be the design variable associated with factor A, and let x_2 be the design variable associated with factor B. For the moment, let factor A be the moisture content, and let factor B be the temperature. Thus,

$$x_1 = \begin{cases} -1 & \text{if factor A is at its low level} \\ 1 & \text{if factor A is at its high value} \end{cases}$$

$$x_2 = \begin{cases} -1 & \text{if factor B is at its low level} \\ 1 & \text{if factor B is at its high value.} \end{cases}$$

Let a denote the *average* of the responses when A is at its high level and B is at its low level. Similarly, let b denote the average of the responses when B is at its high level and A is at its low level. This convention gives us that ab is the average of the responses when both A and B are at their high levels. By convention, we let (1) represent the average of the responses when both factors are at their low levels. We can summarize the results of a 2^2 factorial experiment as shown here:

x_1	x_2	
−1	−1	(1)
1	−1	a
−1	1	b
1	1	ab

For the pine wood data, we have these values:

x_1	x_2		
−1	−1	(1)	20.35
1	−1	a	14.2
−1	1	b	15.05
1	1	ab	10.2

For factorial designs, we can estimate two types of effects:

1. The main effects associated with the factors
2. The interactions between factors

In general, if a factor has a positive main effect, we mean that the response, on the average, increases as the factor goes from its low level to its high level. Consider the main effect due to moisture content. A reasonable definition of this effect is

$$\begin{aligned} \text{effect of A} &= \text{effect of moisture} \\ &= \text{average response at the high level} \\ &\quad - \text{average response at the low level.} \end{aligned}$$

In terms of our conventions, this main effect becomes

$$\begin{aligned}
\text{effect of A} &= \frac{a + ab}{2} - \frac{(1) + b}{2} \\
&= \frac{14.2 + 10.2}{2} - \frac{20.35 + 15.05}{2} \\
&= 12.2 - 17.7 \\
&= -5.5.
\end{aligned}$$

In a similar manner, the main effect due to temperature is

$$\begin{aligned}
\text{effect of B} &= \text{effect of temperature} \\
&= \frac{b + ab}{2} - \frac{(1) + a}{2} \\
&= \frac{15.05 + 10.2}{2} - \frac{20.35 + 14.2}{2} \\
&= 12.625 - 17.275 \\
&= -4.65.
\end{aligned}$$

We see that the main effects for moisture and for temperature are about the same size, which indicates that they have a similar influence on the carbon monoxide concentration. In addition, we see that both moisture and temperature have a negative effect: The higher we go on either moisture or temperature, the lower the carbon monoxide concentration.

Concept of Interaction

If the factors A and B interact, then the effect of A depends on the specific level used of B. We can see what we mean by an interaction most clearly by an *interaction plot*, which simply plots the means for one factor given the levels of the other factor. Two quick numerical examples illustrate this concept.

VOICE OF EXPERIENCE

An interaction means that the effect of one factor depends on the specific level used of another factor.

Suppose the following table summarizes the results from a 2^2 factorial experiment:

x_1	x_2		y
−1	−1	(1)	2
1	−1	a	5
−1	1	b	4
1	1	ab	7

Figure 7.2 | An Interaction Plot When No Interaction Is Present

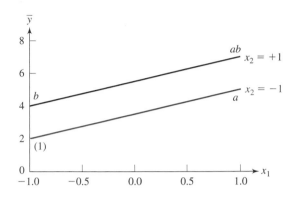

Figure 7.2 shows the resulting interaction plot. In this case, both factors A and B have positive effects. We also see that the effect of factor A is the same for both levels of B:

- The average response increases three units when we go from the low level of A ($x_1 = -1$) to the high level ($x_1 = +1$) for B at its low level ($x_2 = -1$).
- The average response increases three units when we go from the low level of A ($x_1 = -1$) to the high level ($x_1 = +1$) for B at its high level ($x_2 = +1$).

Thus, the two lines are parallel. Since the effect of one factor does not depend on the specific level used of the other, we say they have no interaction.

Suppose the following table summarizes the results from another 2^2 factorial experiment:

x_1	x_2		y
−1	−1	(1)	2
1	−1	a	2
−1	1	b	4
1	1	ab	7

Figure 7.3 shows the resulting interaction plot. There seems to be no effect due to factor A when B is at its low level, but there seems to be a positive effect due to A when B is at its high level. So, the effect of A does seem to depend on the level of B, and the lines are not parallel. Since the effect of one factor depends on the specific level used of the other, we say that the two factors interact.

Figure 7.4 gives the interaction plot for the pine wood experiment. The lines appear to be almost parallel, which suggests that the effect of moisture content does not depend on the level of temperature. Thus, we see no evidence for an interaction between moisture content and temperature. In addition, we can see the negative effects for both moisture content and temperature.

Figure 7.3 | An Interaction Plot When an Interaction Is Present

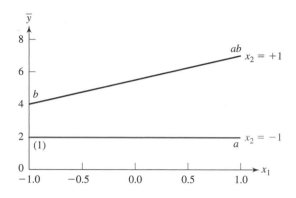

VOICE OF EXPERIENCE

The test for the presence of an interaction is a test for parallelism.

We can formally calculate an interaction effect from this concept of parallel lines. Consider the effect of factor B at the low level of A, which is given by

$$\Delta_1 = b - (1).$$

The effect of factor B at the high level of A is

$$\Delta_2 = ab - a.$$

If the two factors do not interact, then the effect of B does not depend on A. Thus, the effect of B is the same at the high and low levels of A. On the other hand, if A and B have a positive interaction, then we expect the response when both A and B are at their high levels to be larger than we expect from the main effects alone. Consequently, if A and B have a positive interaction, then we expect the effect of B at the

Figure 7.4 | The Interaction Plot for the Pine Wood Experiment

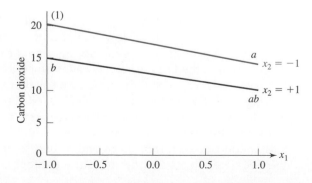

Figure 7.5 | Illustration of How to Calculate an Interaction

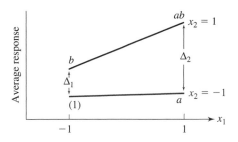

high level of A is larger than the effect at the low level of A, or $\Delta_2 > \Delta_1$. Figure 7.5 illustrates the situation when Δ_2 is greater than Δ_1. We define the interaction between A and B by

$$\text{interaction between A and B} = \frac{\Delta_2 - \Delta_1}{2}$$

$$= \frac{[ab - a] - [b - (1)]}{2}$$

$$= \frac{(1) + ab - a - b}{2}.$$

The denominator 2 ensures that the size of the interaction effect is consistent with the size of the main effects. We may view the definition of an interaction as the average difference in the effect of B over the levels of A. For the pine wood data,

$$\text{AB interaction} = \text{interaction between moisture and temperature}$$

$$= \frac{(1) + ab - a - b}{2}$$

$$= \frac{20.35 + 10.2 - 14.2 - 15.05}{2}$$

$$= 0.65.$$

We thus see a small positive interaction, which is much smaller in absolute value than either of the main effects. The relative size of this interaction suggests that the effect of moisture content really does not depend on the level of temperature used.

Table of Contrasts

We have seen how to use the definition of the main effects and the interaction to obtain these equations:

$$\text{effect of A} = \frac{a + ab}{2} - \frac{(1) + b}{2}$$

$$= \frac{-(1) + a - b + ab}{2}$$

$$\text{effect of B} = \frac{b + ab}{2} - \frac{(1) + a}{2}$$

$$= \frac{-(1) - a + b + ab}{2}$$

$$\text{interaction between A and B} = \frac{(1) + ab - a - b}{2}$$

$$= \frac{(1) - a - b + ab}{2}.$$

We can obtain the numerators for each of these expressions more easily by looking at our model

$$y_i = \beta_0 + \beta_1 x_{i1} + \beta_2 x_{i2} + \beta_{12} x_{i1} x_{i2} + \epsilon_i.$$

We see that the interaction really involves the *product* of the design variables associated with the two factors. This model also points out the importance of the overall mean, which in this case is the y-intercept β_0. We find the overall mean by averaging all the responses. Let I stand for the intercept. We produce the *table of contrasts*, which follows, by rewriting the table for our design to include the intercept and the interaction terms:

I	x_1	x_2	$x_1 x_2$	
1	−1	−1	1	(1)
1	1	−1	−1	a
1	−1	1	−1	b
1	1	1	1	ab

Consider the column for x_1. If we use the −1's and +1's in this column to combine the average response for each treatment combination, then we get

$$-(1) + a - b + ab$$

which is the numerator for the effect of A. In a similar manner, we can use the columns for x_2 and $x_1 x_2$ to get

$$-(1) - a + b + ab$$

$$(1) - a - b + ab,$$

which are the numerators for the main effect of B and the AB interaction, respectively. This table always tells us how to combine the average response for each treatment combination to form the numerator of our estimate of the effect.

For two-level factorial designs, the denominator for estimating effects and interactions will always be one-half of the number of distinct factorial treatment combinations. In the case of the 2^2, we have 4 different factorial treatment combinations, so our denominator is 2 for estimating the effects. We always use the total number of distinct treatment combinations in our denominator to estimate the intercept.

Traditionally, we have used the table of contrasts as the basis for estimating all of the effects. With the widespread use of statistical software and spreadsheets, we no longer need this table for computations. However, we shall see that it provides profound insights on how we can reduce the size of our two-level experiments under certain conditions.

Formal Analysis Using Regression

We can use the estimated effects to estimate the corresponding regression coefficients by recognizing that these coefficients are slopes. Slopes represent the expected change in the response when we increase one factor one unit while holding the other factors constant. Going from -1 to 1 in the design variable represents a movement of 2 units. Since the effect is the average change in the response by moving from the -1 to the $+1$ level of a factor, the appropriate estimate of the regression coefficient corresponding to a particular effect is

$$\text{estimated regression coefficient} = \frac{\text{estimated effect}}{2}.$$

This relationship holds due to the orthogonality of the 2^2 factorial design. Because of this relationship, we rarely calculate effects by hand. Instead, we use standard statistical software or even spreadsheets to obtain either the estimated effects directly or the estimated regression coefficients. Since the two are related by a constant multiple, we can show that the two analyses are algebraically equivalent.

Example 7.3	The Pine Wood Experiment—Regression Analysis

Table 7.3 gives the analysis for the pine wood experiment based on the SAS regression software. This analysis uses the design variables, the ± 1's, as the regressors.

The overall F-test indicates that the carbon monoxide concentration depends on at least one of the factors because the p-value (0.0003) associated with this test is much less than 0.05. The R^2 of 0.9884 indicates that our model accounts for more than 98% of the total variability, which is quite good for this kind of experiment. Since we used the design variables to generate this analysis, we see that the parameter estimate for each of the regression coefficients is exactly one-half the corresponding estimated effect that we computed by hand. We also

Table 7.3 | The Regression Analysis of the Pine Wood Experiment

```
                    Analysis of Variance
                 Sum of        Mean
   Source    DF   Squares      Square    F Value    Prob>F
   Model      3   104.59000    34.86333  113.377    0.0003
   Error      4     1.23000     0.30750
   C Total    7   105.82000

        Root MSE      0.55453     R-square      0.9884
        Dep Mean     14.95000     Adj R-sq      0.9797
        C.V.          3.70921

                   Parameter Estimates
                Parameter   Standard    T for H0:
   Variable  DF  Estimate     Error   Parameter=0  Prob > |T|
   INTERCEPT  1  14.950000  0.19605484   76.254     0.0001
   X1         1  -2.750000  0.19605484  -14.027     0.0001
   X2         1  -2.325000  0.19605484  -11.859     0.0003
   X12        1   0.325000  0.19605484    1.658     0.1727
```

see that both main effects are important because both of their *p*-values are much less than 0.05 (0.0001 for moisture and 0.0003 for temperature). In addition, the interaction between moisture and temperature appears unimportant, with a *p*-value of 0.1727. Ideally, we should perform a complete residual analysis within a regression context; however, for brevity, we leave that as an exercise for the student.

⁖ Exercises

7.1   Buckner, Cammenga, and Weber (1993) ran a two-factor experiment to look at the effects of pressure and the H_2/WF_6 ratio and determine their effect on the uniformity of the titanium nitride adhesion layer for a type of silicon wafer. We extracted a 2^2 factorial design from their larger experiment with these levels for these two factors:

Factor	Low Level	High Level
Pressure	15 Torr	70 Torr
H_2/WF_6 ratio	3	9

Table 7.4 gives the results. Estimate both main effects and the interaction.

Table 7.4	The Uniformity Experiment

Pressure	H_2/WF_6	Uniformity
15	3	8.6
70	3	3.4
15	9	6.9
70	9	5.1

7.2 Said and colleagues (1994) studied the effect of the calcination temperature and the mole fraction of cobalt on the surface area of an iron–cobalt hydroxide catalyst. From their work, we extracted the 2^2 factorial experiment summarized in Table 7.5. Estimate both main effects and the interaction.

7.3 Yi and Shrive (2003) studied the strength of hollow masonary walls with one course bond beams subjected to concentrated loads. The walls were modeled with different loading plate sizes and different loading locations along the wall. We extracted a 2^2 factorial design from their larger experiment, which is summarized in Table 7.6. Estimate both main effects and the interaction.

7.4 A chemical engineer used a replicated 2^2 factorial design to study the impact of inlet feed temperature and reflux ratio on the yield of gasoline from a distillation column. The yield was defined to be the concentration of gasoline in the product stream from the top of the column. Table 7.7 summarizes the experimental results.

a. Estimate the two main effects and the interaction.
b. Use a statistical software package to analyze these results.

7.5 An engineering statistics class ran a catapult experiment to develop a prediction equation for how far a catapult can throw a plastic ball. The class decided to manipulate two factors: how far back the operator draws the arm (angle), measured in degrees, and the height of the pin that supports the rubber band, measured in equally spaced locations. The class used a replicated 2^2 design. Table 7.8 summarizes the experimental results.

Table 7.5	The Calcination Data

Temp.	mole fraction	Surface Area
200	0.6	90.6
600	0.6	25.0
200	2.8	40.9
600	2.8	19.0

Table 7.6	The Hollow Masonary Walls Experiment

Loading Location	Loading Plate Length	Failure Load
400	160	342
1200	160	378
400	320	463
1200	320	517

    a.   Estimate the two main effects and the interaction.
    b.   Use a statistical software package to analyze these results.

7.6   Liu, Kan, and Chen (1993) conducted an experiment to determine the effect of gas velocity and fluid viscosity on a dimensional factor, K, related to the pressure drop across a screen plate used in bubble columns. Table 7.9 summarizes the results for a replicated 2^2 factorial design that we extracted from their data.

    a.   Estimate the two main effects and the interaction.
    b.   Use a statistical software package to analyze these results.

7.7   The fracture toughness of asphalt concrete increases at low temperature and then decreases at temperatures below a certain level. Some polymers are known to have the property of improving the temperature susceptibility of asphalt binder at low temperatures. Kim et al. (2003) evaluated the fracture toughness of some polymer-modified asphalt concretes. We extracted a 2^2 factorial design from their larger experiment, which is summarized in Table 7.10.

Table 7.7	The Distillation Column Experiment

Inlet Temp.	Reflux Ratio	Yield
550	4	88
600	4	90
550	8	95
600	8	97
550	4	87
600	4	91
550	8	94
600	8	98

Table 7.8	The Catapult Data

Angle	Height	Distance (inches)	
140	2	27	27
180	2	81	67
140	4	67	62
180	4	137	158

   a. Estimate both main effects and the interaction.
   b. Use a statistical software package to analyze these results.

7.8 A manufacturer of automotive accessories provides hardware to fasten the accessory to the car. Hardware is counted and packaged automatically. Specifically, bolts are dumped into a large metal dish, a plate that forms at the bottom of the dish rotates counterclockwise, and the bolts are forced along the narrow edge at the outside of the dish. The ledge spirals up to a point where the bolts are allowed to drop into a pan on a conveyor belt. As the bolt drops, it passes an electronic eye that counts it. The speed of rotation and the sensitivity of the electronic eye are varied to study their effects on the time it takes to count 20 bolts. The data are in Table 7.11.

   a. Estimate both main effects and the interaction.
   b. Use a statistical software package to analyze these results.

⋗ 7.2 The 2^k Factorial Design

The treatment combinations for the 2^k full factorial design consist of every possible combination of the two levels for the k factors. The analysis extends naturally from what we developed for the 2^2, as we illustrate for a three-factor case.

Table 7.9	The Pressure Drop Data

Velocity	Viscosity	K			
2.14	2.63	24.2	17.6	14.0	33.8
8.15	2.63	20.9	15.8	18.3	28.1
2.14	10	28.9	27.2	19.7	29.2
8.15	10	26.4	23.2	22.8	23.6

Table 7.10	The Asphalt Concrete Toughness Experiment

Mixture	Temp.	Toughness
Normal	−35	15.83
Polymer added	−35	13.69
Normal	−10	13.33
Polymer added	−10	15.62
Normal	−35	15.06
Polymer added	−35	13.83
Normal	−10	13.91
Polymer added	−10	15.86

The 2^3 Design

Example 7.4	**Springs with Cracks**

Box and Bisgaard (1987) discuss a manufacturing operation for carbon-steel springs that have a severe problem with cracks. Basic metallurgy suggests that the cracking depends on three factors:

- The temperature of the steel before quenching
- The amount of carbon in the formulation
- The temperature of the quenching oil

In this case, the engineers apply each treatment combination to an entire production lot of springs. Quality control inspects each spring for cracking. The engineers use the percent of springs that do not exhibit cracking as their response. They seek to determine the basic relationships among the three factors and the cracking. In the process, they hope to find conditions that will reduce or even eliminate the problem. They decided to pursue a two-level experiment involving these three factors. Table 7.12 lists the actual levels used.

Table 7.11	The Automotive Bolts Experiment

Speed	Sensitivity	Time		
2	6	22.56	18.44	18.01
6	6	26.84	28.25	34.40
2	10	12.97	17.13	14.93
6	10	68.52	71.22	71.59

Table 7.12	The Factors and Their Levels for the Springs with Cracks Experiment

Factor	Low Level	High Level
Steel temp.	1450° F	1600° F
Carbon	0.50%	0.70%
Oil temp.	70° F	120° F

The variables are defined as follows:

- x_1 is the design variable associated with the steel temperature before quenching.
- x_2 is the design variable associated with the carbon content.
- x_3 is the design variable associated with the quenching oil temperature.

The 2^3 factorial design uses every possible combination of the two levels for the three factors, which entails $2^3 = 8$ different treatment combinations. Table 7.13 gives this 2^3 design in the design variables. Figure 7.6 shows that this design forms a cube in terms of the design variables. Table 7.14 gives the design for this specific experiment in the natural units. The engineers ran the actual design in random order.

Hidden Replication For economic reasons, the engineers did not replicate the design, which complicates the analysis. Full 2^k factorial designs exhibit *hidden replication*. Often, when we run a 2^k factorial, at least one of the factors and all of the interactions that involve this factor really are not important. Of course, when we plan the experiment, we have little, if any, idea which of these factors will prove unimportant. For an unimportant factor, there essentially

Table 7.13	The 2^3 Design in the Design Variables

x_1	x_2	x_3
−1	−1	−1
1	−1	−1
−1	1	−1
1	1	−1
−1	−1	1
1	−1	1
−1	1	1
1	1	1

Figure 7.6	An Illustration of a 2^3 Factorial Design in the Design Variables

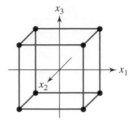

is no difference in the response when we go from the low to the high level. Including this factor really just provides replication. For the sake of argument, suppose that the oil quench temperature has no effect on the cracking of the springs. Then we can remove oil quench temperature as a factor. Our eight-run design then is shown in Table 7.15, a replicated 2^2 factorial design. The problem is how to identify any unimportant factors. Later in this section, we shall describe a procedure based on the normal probability plot.

Estimating Effects for the 2^3 Factorial   The 2^3 factorial design supports the estimation of this model:

$$y_i = \beta_0 + \beta_1 x_{i1} + \beta_2 x_{i2} + \beta_3 x_{i3}$$
$$+ \beta_{12} x_{i1} x_{i2} + \beta_{13} x_{i1} x_{i3} + \beta_{23} x_{i2} x_{i3} + \beta_{123} x_{i1} x_{i2} x_{i3} + \epsilon_i,$$

where

- y_i is the percent of springs that do not crack in the ith production lot.
- β_0 is the y-intercept, which is the overall average percent of springs that do not crack.

Table 7.14	The Springs with Cracks Experimental Data in the Natural Units

Steel Temp. (°F)	Percent Carbon	Oil Temp. (°F)
1450	0.50	70
1600	0.50	70
1450	0.70	70
1600	0.70	70
1450	0.50	120
1600	0.50	120
1450	0.70	120
1600	0.70	120

Table 7.15

x_1	x_2
-1	-1
1	-1
-1	1
1	1
-1	-1
1	-1
-1	1
1	1

- β_1, β_2, and β_3 are the regression coefficients or the slopes for the design variables.
- β_{12}, β_{13}, and β_{23} are the regression coefficients or the slopes for the two-factor interactions.
- β_{123} is the regression coefficient or the slope for the three-factor interaction.
- ϵ_i is the random error associated with the ith spring.

Table 7.16 is the table of contrasts including the observed responses for the springs with cracks experiment. The naming convention for the average of the responses at each treatment combination extends naturally from the 2^2 case. The letter a is the average of the responses when factor A is at its high level and all the other factors are at their low levels. In a similar manner, abc is the average of the responses when all three factors are at their high levels. In general, if the letter is present, the corresponding factor is at its high level. If a letter is absent, then its corresponding factor is at its low level.

Table 7.16 | The Table of Contrasts for the Springs with Cracks Experiment

I	x_1	x_2	x_3	$x_1 x_2$	$x_1 x_3$	$x_2 x_3$	$x_1 x_2 x_3$		
1	-1	-1	-1	1	1	1	-1	(1)	67
1	1	-1	-1	-1	-1	1	1	a	79
1	-1	1	-1	-1	1	-1	1	b	61
1	1	1	-1	1	-1	-1	-1	ab	75
1	-1	-1	1	1	-1	-1	1	c	59
1	1	-1	1	-1	1	-1	-1	ac	90
1	-1	1	1	-1	-1	1	-1	bc	52
1	1	1	1	1	1	1	1	abc	87

We can use the table of contrasts to estimate all the main effects and interactions. In each case, the denominator is one-half the number of factorial treatment combinations. Since we have $2^3 = 8$ treatment combinations, the denominator is 4 for each estimate. The columns of the table of contrasts tell us how to combine the average responses to form the denominators. Thus,

$$
\begin{aligned}
\text{effect of steel temperature (A)} \\
= \frac{-(1) + a - b + ab - c + ac - bc + abc}{4} \\
= \frac{-67 + 79 - 61 + 75 - 59 + 90 - 52 + 87}{4} \\
= 23.0.
\end{aligned}
$$

In a similar manner, we can obtain the following estimates of the other main effects:

$$
\text{effect of carbon (B)} = -5.0
$$

$$
\text{effect of oil temperature (C)} = 1.5.
$$

We can use the table of contrasts to find each two-factor interaction. For example, the estimate of the steel temperature by oil temperature (AC) interaction is given by

$$
\begin{aligned}
\text{AC interaction} &= \frac{(1) - a + b - ab - c + ac - bc + abc}{4} \\
&= \frac{67 - 79 + 61 - 75 - 59 + 90 - 52 + 87}{4} \\
&= 10.0.
\end{aligned}
$$

In a similar manner, we can obtain the following estimates for the other two-factor interactions; steel temperature by carbon (AB) and carbon by oil temperature (BC).

$$
\text{AB interaction} = 1.5
$$

$$
\text{BC interaction} = 0.0.
$$

Finally, we can estimate the three-factor interaction (ABC) by

$$
\begin{aligned}
\text{ABC interaction} &= \frac{-(1) + a + b - ab + c - ac - bc + abc}{4} \\
&= \frac{-67 + 79 + 61 - 75 + 59 - 90 - 52 + 87}{4} \\
&= 0.5.
\end{aligned}
$$

We can estimate the regression coefficients for the corresponding design variables by

$$\text{estimated regression coefficient} = \frac{\text{estimated effect}}{2}.$$

Because of this relationship, we rarely calculate effects by hand. Instead, we use standard statistical software or even spreadsheets to obtain either the estimated effects directly or the estimated regression coefficients. Since the two are related by a constant multiple, we can show that the two analyses are algebraically equivalent.

Table 7.17 gives the analysis based on the SAS regression software when we estimate the full model given by

$$y_i = \beta_0 + \beta_1 x_{i1} + \beta_2 x_{i2} + \beta_3 x_{i3}$$
$$+ \beta_{12} x_{i1} x_{i2} + \beta_{13} x_{i1} x_{i3} + \beta_{23} x_{i2} x_{i3} + \beta_{123} x_{i1} x_{i2} x_{i3} + \epsilon_i.$$

Table 7.17	The ANOVA Table for the Full Model

```
                        Analysis of Variance

                        Sum of          Mean
Source        DF        Squares        Square      F Value      Prob>F
Model          7     1317.50000     188.21429         .            .
Error          0        0.00000         .
C Total        7     1317.50000

        Root MSE          .            R-square       1.0000
        Dep Mean      71.25000         Adj R-sq          .
        C.V.              .

                        Parameter Estimates
                    Parameter   Standard    T for H0:
Variable     DF     Estimate     Error    Parameter=0   Prob > |T|
INTERCEPT     1     71.250000      .           .            .
X1            1     11.500000      .           .            .
X2            1     -2.500000      .           .            .
X3            1      0.750000      .           .            .
X12           1      0.750000      .           .            .
X13           1      5.000000      .           .            .
X23           1           0        .           .            .
X123          1      0.250000      .           .            .
```

Because our model has as many regression coefficients as we have observations, we are unable to estimate an error term. And without an appropriate error term, we are unable to perform any formal tests, which is a problem.

Many statistical software packages force the user to fit a hierarchical model. This means that if a model includes a higher-order term (such as AB), it must also contain all lower-order terms that compose it (in this case, A and B). Hierarchy promotes stability and consistency in the model. There are some engineering principles (such as Ohm's law) that do not follow model hierarchy, and sometimes better predictions can be obtained from nonhierarchical models.

Using Normal Probability Plots

The *normal probability plot,* which we introduced in Chapter 3, provides a powerful, informal, graphical method for determining important effects in a 2^k factorial design when we do not have an appropriate error term. By *important*, we mean that an effect is large in absolute value. The success of this procedure depends heavily on the assumption of *effect sparsity*. When we run an experiment, we rarely expect all of the effects to be large in absolute value, and we certainly do not expect all of the effects to be of equal absolute size. The normal probability plot often allows us to pick out the larger effects.

VOICE OF EXPERIENCE

Normal probability plots are a powerful but subjective tool for identifying important effects.

We may use a normal probability plot on either the effects or the estimated coefficients. For simplicity, this discussion will focus on the effects. In some sense, the normal probability plot parallels the global *F*-test from regression analysis. If the assumptions for our regression model are reasonable, then all of the estimated effects follow normal distributions. The global *F*-test has a null hypothesis that none of the effects is important, which means that all of the true effect sizes are zero. With factorial experiments, if all of the true effects are zero, then they all follow the same normal distribution. As a result, if all of the true effects are really zero, then a normal probability plot of them should form a straight line. Any estimated effect whose true size is not zero will tend to fall significantly away from this straight line. The basic idea of the normal probability plot is to let the estimated effects closest in absolute value to zero define a straight line. Any estimated effect that is significantly different from this line indicates an important effect.

Determining which estimated effects depart from the straight line requires some subjectivity. Several of my colleagues suggest the "fat pencil" rule, where the analyst places a pencil over the estimated effects closest in absolute value. We consider important any estimated main effects and interactions that are not covered by the pencil. Although this approach may seem somewhat ad hoc, in practice it works quite well as long as the assumption of effect sparsity holds.

Table 7.18 | The Estimated Effects from the Springs with Cracks Experiment

Source	Estimated Effect
Steel temp. (A)	23.0
Carbon (B)	−5.0
Oil temp. (C)	1.5
AB	1.5
AC	10.0
BC	0.0
ABC	0.5

Example 7.5 | **Springs with Cracks—Normal Probability Plot**

Table 7.18 lists the estimated effects we calculated by hand. We obtain the same conclusions if we use the estimated coefficients from any regression software package or spreadsheet. Figure 7.7 gives the normal probability plot for the springs with cracks data. Since we have seven effects, the plot has seven points. We see that the steel temperature (A) and the steel temperature by oil quench temperature interaction (AC) appear most important because these two terms do not lie on the straight line defined by the estimated effects smallest in absolute value. To generate the interaction plot, we need the average responses over the levels of carbon for these combinations:

- Steel temperature at its low level and quenching oil temperature at its low level ($x_1 = -1$ and $x_3 = -1$)

Figure 7.7 | The Normal Probability Plot of the Effects from the Springs with Cracks Experiment

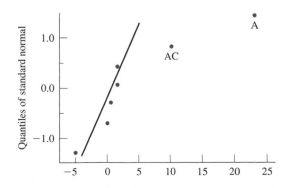

- Steel temperature at its high level and quenching oil temperature at its low level ($x_1 = +1$ and $x_3 = -1$)
- Steel temperature at its low level and quenching oil temperature at its high level ($x_1 = -1$ and $x_3 = +1$)
- Steel temperature at its high level and quenching oil temperature at its high level ($x_1 = +1$ and $x_3 = +1$)

These averages are

$$x_1 = -1 \text{ and } x_3 = -1: \quad \frac{(1) + b}{2} = 64.0$$

$$x_1 = +1 \text{ and } x_3 = -1: \quad \frac{a + ab}{2} = 77.0$$

$$x_1 = -1 \text{ and } x_3 = +1: \quad \frac{c + bc}{2} = 55.5$$

$$x_1 = +1 \text{ and } x_3 = +1: \quad \frac{ac + abc}{2} = 88.5.$$

Figure 7.8 gives the interaction plot, which indicates that the percent of springs without cracks increases as the steel temperature (x_1) increases. This makes a great deal of engineering sense because the heat treatment of steel changes the steel's crystalline structure. The higher temperature allows the steel to develop a stronger crystalline structure, which in turn reduces the cracking. Since the two lines are not parallel, we have evidence of an interaction. The rate of increase is much greater when the quenching oil temperature (x_3) is at 120° F than at 70° F. The lower oil temperature may induce some thermal stress, which results in some cracking. Because of this interaction, we must be careful in interpreting the main effects of

Figure 7.8   The Steel Temperature—Oil Quench Temperature Interaction Plot

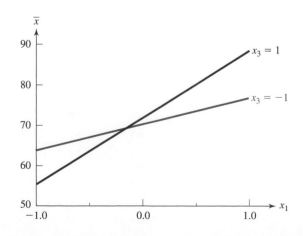

Table 7.19	The Analysis of Variance Table for the Reduced Springs with Cracks Experiment

```
                      Analysis of Variance
                      Sum of        Mean
   Source      DF    Squares       Square    F Value   Prob>F
   Model        3   1262.50000   420.83333    30.606    0.0032
   Error        4     55.00000    13.75000
   C Total      7   1317.50000

        Root MSE         3.70810     R-square       0.9583
        Dep Mean        71.25000     Adj R-sq       0.9269
        C.V.             5.20435

                      Parameter Estimates
                   Parameter    Standard    T for H0:
   Variable   DF    Estimate       Error   Parameter=0  Prob > |T|
   INTERCEPT   1  71.250000    1.31101106     54.347      0.0001
   X1          1  11.500000    1.31101106      8.772      0.0009
   X3          1   0.750000    1.31101106      0.572      0.5979
   X13         1   5.000000    1.31101106      3.814      0.0189
```

temperature and oil quench temperature, especially the latter. It is not completely true to say that the oil quench temperature has no effect on the percentage of springs without cracks. Instead, the effect of the oil quench temperature depends on what steel temperature we use. For the low level of steel temperature, the effect of the oil quench temperature is positive. The estimated main effect of the oil quench temperature is the average effect over the two levels of steel temperature. In this case, the negative effect at the low level of the steel temperature cancels out the positive effect at the high level of steel temperature.

Since the estimated main effect and all interactions involving the amount of carbon are so small, *at least over the range of carbon studied*, it does not seem to affect the percent cracking. We thus have reason to drop this factor from the analysis. The hidden replication property of the 2^3 design yields a replicated 2^2 design. Now we are able to perform a formal statistical analysis, given in Table 7.19. The overall F-test, with a p-value of 0.0032, indicates that the cracking depends on at least one of the factors. The R^2 indicates that the model explains more than 95% of the total variability, which is quite good. The tests on the individual factors confirm that the steel temperature main effect and the steel temperature by oil quench temperature interaction are important.

Center Runs

For true two-level designs, the basic design and analysis of the experiment are the same for categorical and continuous factors. When we have continuous factors, however, we often are interested in what happens to the response for settings in between those used in the experiment. Recall the steel springs with cracks example that has these factors and experimental ranges:

- Steel temperature from 1450° F to 1600° F
- Amount of carbon from 0.50% to 0.70%
- Quenching oil temperature from 70° F to 120° F

In this example, since all the factors are continuous, we at least conceivably could run the process at a steel temperature of 1500° F, an amount of carbon of 0.60%, and an oil quench temperature of 100° F. On the other hand, consider a factor such as supplier. We can arbitrarily designate one supplier as the −1 level and the other supplier as the +1 level. It really makes no sense to talk about the 0.25 level, however, because this factor is categorical.

When we run engineering experiments with all continuous factors, we often add *center runs* to the base design. As the name implies, center runs use as the settings the levels for the factors in the exact center of the experimental region. The exact center is the average of the low and high levels. In design variables, the center run has each design variable set to 0, which is the average of −1 and +1. In the steel springs with cracks example, these are the center runs in the natural units:

- The steel temperature at $\frac{1450 + 1600}{2} = 1525°$ F
- The amount of carbon at $\frac{0.50 + 0.70}{2} = 0.60\%$
- The quenching oil temperature at $\frac{70 + 120}{2} = 95°$ F.

Engineers use center runs for a variety of reasons including these three:

1. To serve as a "reality check"
2. To provide replication, which allows us to estimate an error term
3. To check for curvature

We typically center our experimental region around what we expect to be the best setting. The 2^k factorial design then moves an equal distance in all k dimensions in an effort to improve. Ideally, our experiment should include at least one experimental run at the point of expected best performance to serve as a reality check. Engineering economics often dictates minimal replication of the experiment. In such situations, cost prevents replicating every treatment combination. Rather than depending solely on the hidden replication property of 2^k factorial designs, many engineers replicate only a single treatment combination. The most logical candidate from both the engineering and statistical perspectives is the center. The concept of curvature requires some explanation. The strict first-order prediction equation

$$\hat{y}_i = b_0 + b_1 x_{i1} + \cdots + b_k x_{ik}$$

Figure 7.9 | An Illustration of the Plane Formed by a Strict First-Order Prediction Equation

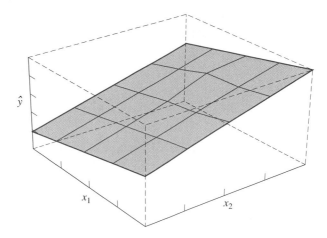

actually represents a plane through the experimental region, as illustrated in Figure 7.9. When we add the interaction terms to the prediction equation, we add a twist to this plane, as illustrated in Figure 7.10. Many engineering phenomena display true curvature, which requires pure quadratic terms, such as x_j^2, to model. Figure 7.11 illustrates the response surface generated by the second-order model

$$\hat{y}_i = b_0 + b_1 x_{i1} + b_2 x_{i2} + b_{11} x_{i1}^2 + b_{22} x_{i2}^2 + b_{12} x_{i1} x_{i2}.$$

Figure 7.10 | An Illustration of the Plane Formed by a First-Order Prediction Equation with Interactions

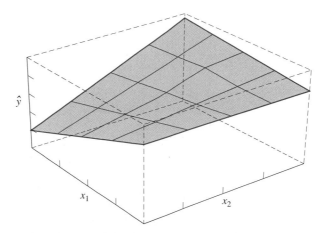

Figure 7.11 | An Illustration of the Plane Formed by a Second-Order Prediction Equation

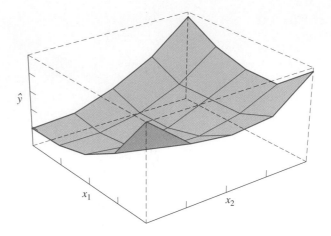

Center runs provide a method to test whether we need to include these pure quadratic or curvature terms in our model. The specifics of the analysis are beyond the scope of this text, but the interested student may read about them in Myers and Montgomery (2002, pp. 128–134).

Example 7.6 | **Concrete Strength**

The federal Highway Administration (FHWA) sponsored a project performed jointly by researchers from FHWA and the National Institute of Standards and Technology (NIST) Building Materials and Statistical Engineering Divisions. The purpose of this project was to investigate the use of statistical experimental design approaches to maximize the strength of a concrete mixture. Table 7.20 lists the factors used and their proposed levels.

The variables were defined as follows:

- x_1 is the design variable associated with the temperature (factor A).

Table 7.20 | The Factors and Their Levels for the Concrete Strength Experiment

Water	Coarse Agg.	Silica
0.3576	0.4071	0.0153
0.4329	0.4353	0.0247

Table 7.21	The Concrete Strength Experiment in Design Units		

x_1	x_2	x_3
−1	−1	−1
1	−1	−1
−1	1	−1
1	1	−1
−1	−1	1
1	−1	1
−1	1	1
1	1	1
0	0	0
0	0	0
0	0	0
0	0	0

- x_2 is the design variable associated with the coarse aggregates (factor B).
- x_3 is the design variable associated with the silica fume (factor C).

The researchers used a 2^3 factorial design with four center runs. Table 7.21 gives the design in design variables. The center runs use each of the factors at its exact middle value. In this case, these are the center runs:

- The water is set to $\frac{0.3576 + 0.4329}{2} = 0.39525$.
- Coarse aggregate is set to $\frac{0.4071 + 0.4353}{2} = 0.4212$.
- The silica fume is set to $\frac{0.0153 + 0.0247}{2} = 0.02$.

Table 7.22 gives the design in the natural units. The actual design was run in random order. The researchers used the center runs here mostly as check points.

The model for this experiment is the same as the model we used for the cracking of steel springs data. We shall extend our naming convention for the responses to let (0) represent the response at the center. Table 7.23 is the table of contrasts and the experimental results. We obtain our estimated effects in the same manner as before. The table of contrasts tells us how to combine the responses to form the numerator. The denominator is one-half the number of *factorial* treatment combinations. In this case, we have 8 factorial treatment combinations, so the denominator is 4. Since the table of contrasts indicates that the coefficient associated with the average response at the center is always zero except for the intercept, the center run really has no influence on the estimates of the main effects and interactions. Table 7.24 lists the estimated effects. Figure 7.12 gives the normal

| Table 7.22 | The Concrete Strength Experiment in Natural Units |

Water	Coarse Agg.	Silica
0.35760	0.4071	0.0153
0.43290	0.4071	0.0153
0.35760	0.4353	0.0153
0.43290	0.4353	0.0153
0.35760	0.4071	0.0247
0.43290	0.4071	0.0247
0.35760	0.4353	0.0247
0.43290	0.4353	0.0247
0.39525	0.4212	0.0200
0.39525	0.4212	0.0200
0.39525	0.4212	0.0200
0.39525	0.4212	0.0200

probability plot of the estimated main effects and the interactions. This plot indicates that only the interaction effect (AC) between water (A) and silica (C) appears to have an important effect on the strength because it is the only estimated effect that does not lie close to the straight line defined by the smallest estimated effects in absolute value. Apparently, the interaction between water and silica produces significant variations in the concrete strength.

| Table 7.23 | The Table of Contrasts and Experimental Results for the Concrete Strength Experiment |

I	x_1	x_2	x_3	$x_1 x_2$	$x_1 x_3$	$x_2 x_3$	$x_1 x_2 x_3$		
1	−1	−1	−1	+1	+1	+1	−1	(1)	58.27
1	1	−1	−1	−1	−1	+1	+1	a	55.06
1	−1	1	−1	−1	+1	−1	+1	b	58.73
1	1	1	−1	+1	−1	−1	−1	ab	52.55
1	−1	−1	1	+1	−1	−1	+1	c	54.88
1	1	−1	1	−1	+1	−1	−1	ac	58.07
1	−1	1	1	−1	−1	+1	−1	bc	56.60
1	1	1	1	+1	+1	+1	+1	abc	59.57
1	0	0	0	0	0	0	0	(0)	56.815

| Table 7.24 | The Estimated Effects for the Concrete Strength Experiment |

Source	Estimated Effect
Water	−0.8075
Coarse agg.	0.2925
Silica	1.1275
AB	−0.7975
AC	3.8875
BC	1.3175
ABC	0.6875

The engineers added the center runs to this design to serve as a reality check. Figure 7.13 plots the responses for each level of the water by silica interaction. The plot indicates no curvature in the response. Since the goal of the experiment is to maximize the concrete strength, the experimenters could choose either low water with low silica or high water with high silica. In the end, it was decided that the low settings would be used.

| Figure 7.12 | The Normal Probability Plot of the Standardized Effects for the Concrete Experiment |

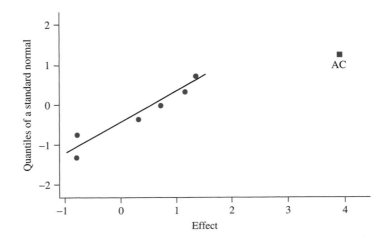

| Figure 7.13 | The Interaction Plot for Water and Silica |

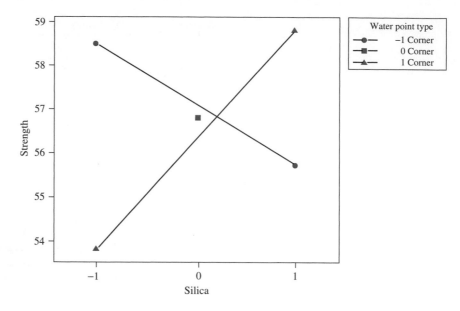

Exercises

7.9 Aceves-Mijares and colleagues (1995) consider a polysilicon deposition process commonly used in the manufacture of integrated circuits. Four factors of interest are pressure (A), temperature gradient (B), furnace temperature (C), and the location of the gas input (D). Suppose the engineers assigned to this process wish to conduct an appropriate two-level full factorial design using these levels:

Pressure	Temp. Gradient	Furnace Temp.	Gas Input
1 Torr	Flat	625° C	Bottom
3 Torr	Ramp	725° C	Top

  a. Give the appropriate factorial design in the design variables.
  b. Give the appropriate design in the natural units.
  c. Give the appropriate design in random order.

7.10 Consider the polymer filament example where the breaking strength of the filament depends on the amount of the catalyst (factor A), the polymerization temperature (factor B), and the polymerization pressure (factor C). Suppose the engineers assigned to this process need to conduct an appropriate two-level full factorial design using these levels:

Catalyst	Temp.	Pressure
1.5%	250° C	25 psig
3.0%	280° C	40 psig

a. Give the appropriate factorial design in the design variables.
b. Give the appropriate design in the natural units.
c. Give the appropriate design in random order.
d. Repeat parts a–c adding four center runs to the design.

7.11 Ul-Haq, White, and Adeleye (1995) investigated the freezing of aqueous solutions on cooled heat-conducting surfaces at atmospheric pressure to assess the freeze process efficiency. Five factors of interest are the cooling rate (A), ultrasonic intensity (B), presence of air in the freezing solution (C), initial soluble concentration (D), and freezing time (E). Suppose the engineers assigned to this process need to conduct an appropriate two-level full factorial design using these levels:

Cooling Rate	Ultrasonic Intensity	Presence of Air	Concentration	Freezing Time
−0.1	160	Yes	7.5	32.5
−0.2	200	No	12.5	57.5

a. Give the appropriate factorial design in design variables.
b. Give the appropriate design in natural units.
c. Give the appropriate design in random order.

7.12 Shina (1991) conducted an experiment to determine the impact of wave width (x_1), direction (x_2), flux (x_3), and angle (x_4) on the average number of short leads per batch produced by a wave-soldering process for printed circuit boards. Table 7.25 gives the experimental results.

a. Estimate all of the main effects and interactions. Use software as appropriate.
b. Perform a thorough analysis of these data and interpret your results.

7.13 Antony (2002) presents a catapult experiment to maximize distance, which can be easily taught to engineers and managers in organizations to train for design of experiments. The results of this experiment have been taken from a real live catapult experiment performed by a group of engineers in a company during the training program on DOE. The data are in Table 7.26.

a. Estimate all the main effects and interactions. Use software as appropriate.
b. Plot the estimated effects on a normal probability plot. Again, use software as appropriate.
c. Interpret your results.

7.14 A mechanical engineer studied the effect of tool speed (A), depth (B), and feed rate (C), on the life of a cutting tool. Table 7.27 lists the experimental results.

Table 7.25	The Results for the Wave-Soldering Experiment

| | | | | \multicolumn{2}{c}{y} |
x_1	x_2	x_3	x_4	Rep. 1	Rep. 2
−1	−1	−1	−1	6.00	10.00
1	−1	−1	−1	9.50	6.75
−1	1	−1	−1	20.00	17.25
1	1	−1	−1	26.70	10.30
−1	−1	1	−1	1.50	1.75
1	−1	1	−1	0.75	3.25
−1	1	1	−1	9.67	5.67
1	1	1	−1	7.30	9.00
−1	−1	−1	1	10.00	8.50
1	−1	−1	1	6.00	6.25
−1	1	−1	1	16.50	19.50
1	1	−1	1	19.30	17.70
−1	−1	1	1	0.25	4.25
1	−1	1	1	3.50	6.50
−1	1	1	1	2.00	3.75
1	1	1	1	6.00	8.70

a. Estimate all of the main effects and interactions. Use software as appropriate.
b. Plot the estimated effects on a normal probability plot. Again, use software as appropriate.
c. Interpret your results.

7.15 It is widely believed that spending more on high-quality batteries, using expensive gold-plated connectors, and storing batteries at low temperatures will improve overall battery life in remote-control cars. Wasiloff and Hargritt (1999) conducted an experiment to evaluate the validity of these practices. A simple electrical test circuit with an indicator light to detect the state of battery discharge was developed to permit testing. Table 7.28 lists the experimental results.

a. Estimate all the main effects and interactions. Use software as appropriate.
b. Plot the estimated effects on a normal probability plot. Again, use software as appropriate.
c. Interpret your results.

7.16 Eibl, Kess, and Pukelsheim (1992) used a 2^3 factorial design to study the impact of belt speed (factor A), tube width (factor B), and pump pressure (factor C) on the coating thickness for a painting operation. Table 7.29 gives the responses.

| Table 7.26 | | Catapult Experiment | | | |

Release Angle	Peg Height	Stop Position	Hook Position	Distance	
180	3	3	3	363	364
Full	3	3	3	401	406
180	4	3	3	416	460
Full	4	3	3	470	490
180	3	5	3	380	383
Full	3	5	3	437	440
180	4	5	3	474	477
Full	4	5	3	532	558
180	3	3	5	426	413
Full	3	3	5	474	494
180	4	3	5	480	502
Full	4	3	5	520	555
180	3	5	5	446	467
Full	3	5	5	512	550
180	4	5	5	480	485
Full	4	5	5	580	591

| Table 7.27 | | The Results for the Tool Life Experiment |

Speed (rpm)	Depth (in.)	Feed (ips)	Tool Life (min)
1000	0.02	2	200
2000	0.02	2	225
1000	0.08	2	28
2000	0.08	2	35
1000	0.02	10	150
2000	0.02	10	154
1000	0.08	10	17
2000	0.08	10	21

Table 7.28 | Remote-Control Car Battery Experiment

AA battery	Connector	Battery Temperature	Time to Discharge
Low	Standard	Cold	72
High	Standard	Cold	612
Low	Gold	Cold	75
High	Gold	Cold	490
Low	Standard	Ambient	93
High	Standard	Ambient	489
Low	Gold	Ambient	94
High	Gold	Ambient	493

a. Estimate all of the main effects and interactions. Use software as appropriate.
b. Plot the estimated effects on a normal probability plot. Again, use software as appropriate.
c. Interpret your results.

7.17 Ferrer and Romero (1995) conducted an unreplicated 2^4 factorial design to determine the effects of the amount of glue (x_1), predrying temperature (x_2), tunnel temperature (x_3), and pressure (x_4) on the adhesive force obtained in an adhesive process of polyurethane sheets. Table 7.30 gives the results.

a. Estimate all of the main effects and interactions. Use software as appropriate.
b. Plot the estimated effects on a normal probability plot. Again, use software as appropriate.
c. Interpret your results.

Table 7.29 | The Responses for the Coating Thickness Experiment

(1)	0.575
a	0.585
b	0.680
ab	0.590
c	0.665
ac	0.585
bc	0.915
abc	0.785

Table 7.30	The Results for the Adhesive Experiment

x_1	x_2	x_3	x_4	y
−1	−1	−1	−1	3.80
1	−1	−1	−1	4.34
−1	1	−1	−1	3.54
1	1	−1	−1	4.59
−1	−1	1	−1	3.95
1	−1	1	−1	4.83
−1	1	1	−1	4.86
1	1	1	−1	5.28
−1	−1	−1	1	3.29
1	−1	−1	1	2.82
−1	1	−1	1	4.59
1	1	−1	1	4.68
−1	−1	1	1	2.73
1	−1	1	1	4.31
−1	1	1	1	5.16
1	1	1	1	6.06

⇒ 7.3 Half Fractions of the 2^k Factorial

Table 7.31 presents the numbers of treatment combinations required for various 2^k factorial designs. For even moderate values of k, the 2^k factorial design can require a prohibitive number of experimental runs. Often, particularly when screening important factors, engineers must construct experiments involving six or more factors. Consider an experimental situation that involves aircraft engines where each experimental unit costs $500,000 to test! Such expense is not uncommon in engineering experiments. A full factorial in even five factors probably is prohibitive. In other cases, time provides a significant barrier. In the semiconductor industry, engineers often must produce significant results in less than two weeks. In such a situation, full 2^k factorial experiments are impossible.

Fortunately, we can use *fractional factorial designs* to reduce the total number of treatment combinations while preserving the basic factorial structure of the experiment. These designs use the "main effects" principle to reduce

Table 7.31	The Numbers of Treatment Combinations Required for a Full 2^k Factorial Design for Various Numbers of Factors

k	Number of Treatment Combinations
2	4
3	8
4	16
5	32
6	64
7	128
8	256
9	512

VOICE OF EXPERIENCE

Fractional factorial designs are important for screening out unimportant factors.

the number of treatment combinations. Essentially, the main effects principle states that main effects tend to dominate two-factor interactions, two-factor interactions tend to dominate three-factor interactions, and so on. The full 2^k factorial design allows us to estimate all of the interactions. As k gets larger, more and more of these interactions probably are not important. Fractional factorial designs "sacrifice" the ability to estimate the higher-order interactions in order to reduce the number of treatment combinations.

Half Fraction of the 2^3

Recall the steel springs with cracks example where we seek to determine the influence of steel temperature (factor A), amount of carbon (factor B), and quenching oil temperature (factor C) on the percent of steel springs that exhibit no signs of cracking. Suppose we can afford only four runs. How can we carry out this experiment?

Recall the table of contrasts for the full 2^3 factorial, given in Table 7.32. Under the main effects principle, the one effect least likely to be important is the three-factor interaction. We need a design with only four treatment combinations. Consider using the four treatment combinations that have $+1$ in the $x_1 x_2 x_3$ column as our design, which the following table summarizes:

x_1	x_2	x_3
1	-1	-1
-1	1	-1
-1	-1	1
1	1	1

Table 7.32 | Table of Contrasts for a Full 2^3 Factorial Experiment

I	x_1	x_2	x_3	$x_1 x_2$	$x_1 x_3$	$x_2 x_3$	$x_1 x_2 x_3$
1	−1	−1	−1	+1	+1	+1	−1
1	1	−1	−1	−1	−1	+1	+1
1	−1	1	−1	−1	+1	−1	+1
1	1	1	−1	+1	−1	−1	−1
1	−1	−1	1	+1	−1	−1	+1
1	1	−1	1	−1	+1	−1	−1
1	−1	1	1	−1	−1	+1	−1
1	1	1	1	+1	+1	+1	+1

Table 7.33 is the resulting table of contrasts. First, note that the column for the three-factor interaction ($x_1 x_2 x_3$) is the same as the intercept (I) column. As a result, the numerator for estimating the intercept is exactly the same numerator for estimating the three-factor interaction. We thus cannot uniquely estimate either! In this situation, we say that the three-factor interaction, ABC, is *aliased* with the intercept. By choosing the three-factor interaction as the basis for cutting the full factorial in half, we have lost our ability to estimate it; hence, we have sacrificed it. Next, note that the column for x_1 is the same as the column for $x_2 x_3$. As a result, factor A is aliased with the BC interaction. Consequently, our estimate of the main effect of A is the same as the estimate for the BC interaction. If this estimate is important, we have no statistical basis for determining whether it is due to the main effect of A or to the interaction of B with C. In a similar manner, factor B is aliased with the AC interaction, and factor C is aliased with the AB interaction. As a result, all of the main effects are aliased with the two-factor interactions.

Table 7.33 | Table of Contrasts for a Half Fraction of a 2^3 Factorial Experiment

I	x_1	x_2	x_3	$x_1 x_2$	$x_1 x_3$	$x_2 x_3$	$x_1 x_2 x_3$
1	1	−1	−1	−1	−1	+1	+1
1	−1	1	−1	−1	+1	−1	+1
1	−1	−1	1	+1	−1	−1	+1
1	1	1	1	+1	+1	+1	+1

What if we use the treatment combinations with $x_1 x_2 x_3 = -1$? We can show that factor A is aliased with the negative of the BC interaction, factor B with the negative of the AC interaction, and factor C with the negative of the AB interaction. Consequently, we still cannot obtain unique estimates of these effects.

How can we get around this problem? Suppose that the engineers propose this model:

$$y_i = \beta_0 + \beta_1 x_{i1} + \beta_2 x_{i2} + \beta_3 x_{i3} + \epsilon_i,$$

which assumes that only the main effects are important. With this model and the proposed design, if an estimated effect appears significant, we conclude that the corresponding main effect and not the interaction is important. The main effects principle suggests that assuming the interactions to be unimportant is often quite reasonable. In those cases where this assumption holds, we may use the proposed design. On the other hand, if one of the interactions is truly important, we mistakenly conclude that a main effect is important when it is not.

If we use $x_1 x_2 x_3 = 1$ to select the treatment combinations, then the resulting design is called the positive half fraction of the 2^3. Figure 7.14 illustrates this design. If we use $x_1 x_2 x_3 = -1$ to select the treatment combinations, the resulting design is called the negative half fraction of the 2^3. In either case, since we used $x_1 x_2 x_3$ to select the treatment combinations, we say that ABC is the *defining interaction* for the design.

In general, we denote a half (either positive or negative) fraction of the 2^3 by 2^{3-1}, where:

- 2 indicates the number of levels for each factor.
- The exponent 3 indicates the number of factors.
- The exponent -1 indicates a half (2^{-1}) fraction.

The total number of treatment combinations is $2^{3-1} = 4$.

The Alias Structure

We can always use the table of contrasts to determine which effects are aliased with one another. A quicker method uses modulo 2 arithmetic (clock arithmetic with only two numbers: 0 and 1). In modulo 2 arithmetic, $1 + 1 = 0$.

Figure 7.14 | Illustration of a Half Fraction of a 2^3 Design

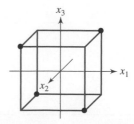

VOICE OF EXPERIENCE

For a fractional factorial experiment, there is no information for the defining interaction.

To obtain the "alias structure," we first note that because ABC is the defining interaction, it is aliased with the intercept or the identity. We thus write $I = ABC$, which is read "ABC is aliased with the intercept." We then call ABC the *defining interaction*. To determine with what A is aliased, we add A to the defining interaction, ABC, using modulo 2 arithmetic. Thus,

$$A + ABC = BC,$$

and we see that A is aliased with the BC interaction, just as before. In a similar manner, we obtain

$$B + ABC = AC$$
$$C + ABC = AB.$$

We can summarize the entire alias structure in the following table:

I	=	**ABC**
A	=	BC
B	=	AC
C	=	AB
AB	=	C
AC	=	B
BC	=	A

We often stop the number of rows after every main effect and interaction appear in the table. Using this convention, we get this alias structure:

I	=	**ABC**
A	=	BC
B	=	AC
C	=	AB

VOICE OF EXPERIENCE

Fractional factorial experiments do not uniquely estimate the terms of our model.

The alias structure in this form tells us the largest model we can estimate from the experiment. In this case, we can estimate a term associated with the intercept (I), the main effect for A, the main effect for B, and the main effect for C. We cannot estimate any of the interactions.

Using a 2^2 to Generate a 2^{3-1}

We can always generate fractional factorial designs from the table of contrasts. But again, there are quicker methods. We note that the 2^{3-1} design requires four different treatment combinations, just like the 2^2. Consider the following table of contrasts for the 2^2 factorial design:

I	x_1	x_2	$x_1 x_2$
1	−1	−1	+1
1	1	−1	−1
1	−1	1	−1
1	1	1	+1

If we can assume that the two-factor interaction is unimportant, then we can use the $x_1 x_2$ column to determine the settings for a third factor, C, whose design variable is x_3. In so doing, we make $x_3 = x_1 x_2$, which is equivalent to aliasing C with the AB interaction. The defining interaction is C + AB, which is ABC even under modulo 2 arithmetic. The resulting design is the positive half fraction of the 2^3.

This approach emphasizes the *projection* property of half fractions. If one of the factors is actually unimportant, then the half fraction becomes a full factorial in the other factors. For example, consider the positive half fraction of the 2^3. For the sake of argument, if factor C is unimportant, then the design given above becomes the following 2^2:

x_1	x_2
−1	−1
1	−1
−1	1
1	1

Half Fractions in General

Half fractions readily extend for higher numbers of factors. In each case, one uses the highest-order interaction to generate the design.

Example 7.7 | **Oil Extraction from Peanuts**

Kilgo (1988) performed an experiment to determine the effect of CO_2 pressure, CO_2 temperature, peanut moisture, CO_2 flow rate, and peanut particle size on the total yield of oil per batch of peanuts. The specific levels used are listed in Table 7.34 For economic reasons, she could afford only 16 experimental units. She generated a half fraction of the 2^5, a 2^{5-1}, design from the 2^4 full factorial design. Table 7.35 gives the 2^4 design, in design variables, plus the column for the ABCD interaction. By letting $x_5 = -x_1 x_2 x_3 x_4$, we obtain the negative half fraction with ABCDE as the defining interaction, which is what Kilgo used. Table 7.36 lists the yield results. Table 7.37 gives the alias structure. The alias structure tells us that

Table 7.34 | The Factors and Their Levels for the Oil Extraction Experiment

A Pressure (bar)	B Temp. (°C)	C Moisture (% by weight)	D Flow Rate (L/min)	E Particle Size (mm)
415	25	5	40	1.28
550	95	15	60	4.05

we can estimate a model with terms for the intercept, each main effect, and each two-factor interaction.

We can use either computer software or the table of contrasts to estimate the effects. The table of contrasts tells us how to combine the responses to form the numerators. The denominator is still one-half the number of factorial treatment combinations. In this design, we have 16 factorial treatment combinations, so our denominator is 8. Table 7.38 lists the estimated effects, and Figure 7.15 is the resulting normal probability plot. This plot indicates that the two important

Table 7.35 | The Experimental Design for the Oil Extraction Experiment

x_1	x_2	x_3	x_4	$x_1 x_2 x_3 x_4$
−1	−1	−1	−1	1
1	−1	−1	−1	−1
−1	1	−1	−1	−1
1	1	−1	−1	1
−1	−1	1	−1	−1
1	−1	1	−1	1
−1	1	1	−1	1
1	1	1	−1	−1
−1	−1	−1	1	−1
1	−1	−1	1	1
−1	1	−1	1	1
1	1	−1	1	−1
−1	−1	1	1	1
1	−1	1	1	−1
−1	1	1	1	−1
1	1	1	1	1

| Table 7.36 | | | | | The Results for the Oil Extraction Experiment | |

x_1	x_2	x_3	x_4	x_5	Response	Yield
−1	−1	−1	−1	−1	(1)	63
1	−1	−1	−1	1	ae	21
−1	1	−1	−1	1	be	36
1	1	−1	−1	−1	ab	99
−1	−1	1	−1	1	ce	24
1	−1	1	−1	−1	ac	66
−1	1	1	−1	−1	bc	71
1	1	1	−1	1	abce	54
−1	−1	−1	1	1	de	23
1	−1	−1	1	−1	ad	74
−1	1	−1	1	−1	bd	80
1	1	−1	1	1	abde	33
−1	−1	1	1	−1	cd	63
1	−1	1	1	1	acde	21
−1	1	1	1	1	bcde	44
1	1	1	1	−1	abcd	96

effects are the main effect due to the average particle size (E) and the main effect due to temperature (B) because they do not lie close to the straight line formed by the estimated effects near zero. None of the estimated interactions appear important.

Table 7.39 is the ANOVA table generated by SAS. The overall F-test, with a p-value of 0.0001, indicates that at least one effect is important. The R^2 exceeds 0.90, which indicates that our reduced model explains more than 90% of the total variability. The tests on the estimated regression coefficients, which are half the size of the corresponding effects, confirm that both temperature and particle size are important, with p-values of 0.0006 and 0.0001, respectively, but the interaction, with a p-value of 0.9544, is clearly not.

Figures 7.16 and 7.17 give the main effects plots for particle size and temperature. These plots show that the yield decreases as the particle size increases, which makes engineering sense. As the particle size increases, the available surface area decreases, and because the surface area controls the transfer of oil to the CO_2, the larger the particle size, the smaller the yield. Similarly, the more energetic the physical system, the more likely transfer from the peanuts to the CO_2. Thus, the higher the temperature, the better the yield.

Table 7.37	The Alias Structure for the Oil Extraction Experiment

I	=	ABCDE
A	=	BCDE
B	=	ACDE
C	=	ABDE
D	=	ABCE
E	=	ABCD
AB	=	CDE
AC	=	BDE
AD	=	BCE
AE	=	BCD
BC	=	ADE
BD	=	ACE
BE	=	ACD
CD	=	ABE
CE	=	ABD
DE	=	ABC

Smaller Fractions

For more than four factors ($k > 4$), we can construct even smaller fractions than the half. We can generate two-level fractional factorial designs with eight treatment combinations for up to seven factors, with 16 treatment combinations for up to 15 factors, and so on. The designs with up to eight treatment combinations alias the interaction columns from the table of contrasts for the 2^3 factorial design. The designs with up to 15 treatment combinations use the columns for a 2^4 factorial design. The next example illustrates the method.

Example 7.8	**Thickness of Paint Coatings**

Eibl, Kess, and Pukelsheim (1992) used a sequential experimentation strategy to achieve a target coating thickness of 0.8 mm for a painting process. In their first experiment, they ran a 2^{6-3} factorial design to study the impact of these factors on the thickness:

- Belt speed (factor A)
- Tube width (factor B)
- Pump pressure (factor C)

| Table 7.38 | The Estimated Effects for the Oil Extraction Experiment |

Source	Estimated Effect
Pressure (A)	7.5
Temp. (B)	19.75
Moisture (C)	1.25
Flow rate (D)	0.0
Particle size (E)	−44.5
AB	5.25
AC	1.25
AD	−4.0
AE	−7.0
BC	3.0
BD	−1.75
BE	−0.25
CD	2.25
CE	6.25
DE	−3.5

- Paint viscosity (factor D)
- Tube height (factor E)
- Heating temperature (factor F)

For proprietary reasons, the authors did not publish the actual levels for these factors.

| Figure 7.15 | The Normal Probability Plot of the Estimated Effects for the Oil Extraction Experiment |

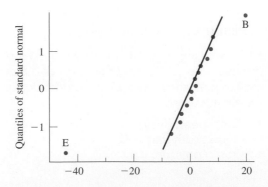

Table 7.39	The Analysis of Variance Table for the Oil Extraction Experiment

```
                         Analysis of Variance
                         Sum of          Mean
      Source       DF    Squares        Square    F Value    Prob>F
      Model         3    9481.5000      3160.50   43.0244    0.0001
      Error        12     881.5000        73.46
      C Total      15   10363.0000

           Root MSE           8.57078     R-square          0.9149
           Dep Mean          54.25000     Adj R-sq          0.8937

                         Parameter Estimates

                      Parameter    Standard    T for H0:
      Variable    DF   Estimate      Error    Parameter=0   Prob > |T|
      INTERCEPT   1     54.2500     2.142696     25.32        0.0001
      X2          1      9.8750     2.142696      4.61        0.0006
      X5          1    -22.2500     2.142696    -10.38        0.0001
      X25         1     -0.1250     2.142696     -0.06        0.9544
```

Table 7.40 lists the actual design and results. The authors generated this design from a 2^3 full factorial design by setting $x_4 = x_2 x_3$, $x_5 = x_1 x_2 x_3$, and $x_6 = x_1 x_2$. Thus, the defining interactions are BCD ($x_2 x_3 x_4$), ABCE ($x_1 x_2 x_3 x_5$), and ABF ($x_1 x_2 x_6$). This design can estimate only one interaction uniquely. In this particular case, we may ascribe its effect to the AC interaction because it is the only column from the original 2^3 not already aliased with a main effect.

Figure 7.16	The Main Effect Plot for Particle Size

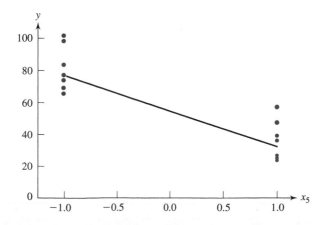

| Figure 7.17 | The Main Effect Plot for Temperature |

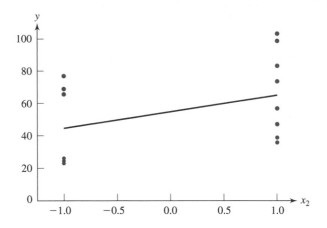

Table 7.41 gives the estimated effects, and Figure 7.18 is the resulting normal probability plot. In this case, the estimated effect associated with the belt speed (A) appears to be important. The estimated effect for tube width (B) may be important because it does not appear to fall on the line generated by the effects near zero.

Table 7.42 is the ANOVA table when we use belt speed (x_1), tube width (x_2), and their interaction ($x_1 x_2$) as our model. The overall F-test, with a p-value of 0.0261, indicates that at least one of the terms is important. The R^2 of 0.8976 indicates that the model explains approximately 90% of the total variability, which

| Table 7.40 | The Coating Thickness Experiment |

x_1	x_2	x_3	x_4	x_5	x_6	Response	y
−1	−1	−1	1	−1	1	df	1.490
1	−1	−1	1	1	−1	ade	0.835
−1	1	−1	−1	1	−1	be	1.738
1	1	−1	−1	−1	1	abf	1.130
−1	−1	1	−1	1	1	cef	1.545
1	−1	1	−1	−1	−1	ac	0.980
−1	1	1	1	−1	−1	bcd	2.175
1	1	1	1	1	1	abcdef	1.448

Table 7.41 | The Estimated Effects for the Coating Thickness Experiment

Source	Estimated Effect
A	−0.639
B	0.410
C	0.239
D	0.139
E	−0.052
F	−0.029
AC	−0.007

is fairly good. The tests on the estimated coefficients indicate that belt speed (x_1) is reasonably important, with a p-value of 0.0105, and that tube width (x_2) is probably important, with a p-value of 0.0433. The belt speed–tube width interaction is clearly not important, with a p-value of 0.8479.

Figures 7.19 and 7.20 show the main effects plots for both factors. The main effects plot for the belt speed (x_1) shows that the higher the belt speed, the thinner the coating, which makes good engineering sense. The main effects plot for the tube width shows that the thickness increases as we increase the tube width. Taken together, these plots suggest using the high level of paint speed and the low level of tube width to reduce the thickness of the coating.

Figure 7.18 | The Normal Probability Plot of the Estimated Effects for the Coating Thickness Experiment

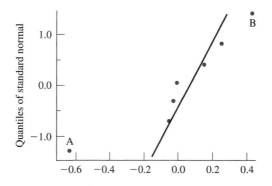

| Table 7.42 | The Analysis of Variance Table for the Coating Thickness Experiment |

Analysis of Variance

Source	DF	Sum of Squares	Mean Square	F Value	Prob>F
Model	3	1.1542664	0.384755	9.7362	0.0261
Error	4	0.1580715	0.039518		
C Total	7	1.3123379			

Root MSE	0.198791	R-square	0.8795	
Dep Mean	1.417625	Adj R-sq	0.7892	

Parameter Estimates

Variable	DF	Parameter Estimate	Standard Error	T for H0: Parameter=0	Prob > |T|
INTERCEPT	1	1.417625	0.070283	20.17	0.0001
X1	1	-0.319375	0.070283	-4.54	0.0105
X2	1	0.205125	0.070283	2.92	0.0433
X12	1	-0.014375	0.070283	-0.20	0.8479

Design Resolution

Design resolution refers to the length (number of letters) in the smallest defining or "generalized" interaction and tells the user some critical information about the alias structure. Generalized interactions are every possible modulo 2 sum of the defining interactions. For example, if we use ABCD and ABE for our defining interactions, then CDE is the resulting generalized interaction. Both the defining and the generalized interactions are aliased with the intercept. We thus must include all generalized interactions when we construct the full alias structure of a design.

For a Resolution III design, the smallest defining or generalized interaction has three letters. As a result, at least one main effect is aliased with a two-factor interaction. Example 7.8 illustrates a Resolution III design because several of the defining and generalized interactions have three letters. We typically use Resolution III designs for screening because they are the smallest designs available.

For a Resolution IV design, the smallest defining or generalized interaction has four letters. As a result, the smallest interaction aliased with a main effect is a three-factor. At least some of the two-factor interactions are aliased with each other. In the next chapter, Example 8.6 illustrates a Resolution IV design because the defining interaction is ABCD. Engineers use Resolution IV designs when they believe that some, but not all, of the two-factor interactions may be important.

Figure 7.19 | The Main Effects Plot for Belt Speed

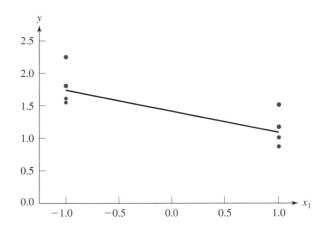

For a Resolution V design, the smallest defining or generalized interaction has five letters. The smallest interaction aliased with a main effect has four factors, and the two-factor interactions are aliased with three-factor or higher interactions. Example 7.7 illustrates a Resolution V design because the defining interaction is ABCDE. Resolution V designs are very important for process optimization.

By convention, we note the resolution of a fractional factorial design by a subscript. For example, 2_{III}^{3-1} denotes a Resolution III half fraction of a 2^3 factorial design. Similarly, 2_{IV}^{6-2} denotes a Resolution IV quarter fraction of a 2^6 factorial design.

Figure 7.20 | The Main Effects Plot for Tube Width

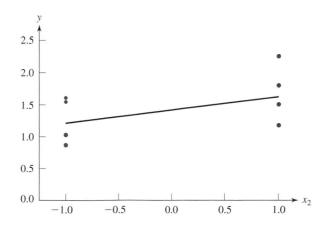

Factorial-Based Designs and Traditional Engineering Experimentation

Engineers not well trained in statistics often pursue two basic approaches: one factor at a time and "shotgun." In general, we do not recommend either of these strategies.

With one-factor-at-a-time experimentation, we run one treatment combination with every factor at a specified level, usually the low level. We then run a series of treatment combinations where we use the high level for one factor and the low level for the others. If we have k factors, then we run a series of k treatment combinations, allowing each factor to be at its high level once. The experiment thus consists of $k+1$ total treatment combinations. Figure 7.21 illustrates this experimental strategy for a three-factor situation. One-factor-at-a-time experimentation suffers from these major drawbacks:

- It cannot estimate interactions.
- It does not cover the entire experimental region.
- It produces less precise estimates of the effects than a corresponding factorial or fractional factorial design.

Efficient model estimation requires us to spread our design points evenly on the boundary of the region of interest defined by our model. Problems occur because the one-factor approach provides no information in the upper-right-hand quadrant, as Figure 7.21 illustrates.

Figure 7.22 illustrates the negative fraction of the 2^{3-1} factorial design. It is important to note that this design also cannot estimate interactions. Nonetheless, it is a much better experimental design than the one-factor-at-a-time approach illustrated in Figure 7.21. The one-factor-at-a-time approach uses only two experimental runs to estimate each factor effect: the one run where all the factors are at their low levels and the one run where the specific factor of interest is at its high level. On the other hand, the 2^{3-1} factorial design uses all four experimental

---

Figure 7.21 | An Illustration of One-Factor-at-a-Time Experimentation

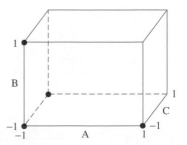

Figure 7.22 | An Illustration of Fractional Factorial Experimentation

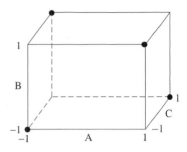

runs to estimate each factor effect: the average of the two runs at the factor's high level and the average of the two runs at the factor's low level.

Further, compare the volumes enclosed by the experimental runs in the two designs. The 2^{3-1} factorial design encloses much more of the volume of the cube. The bottom line is that the 2^{3-1} factorial design makes much better use of the resources available for the experiment than the one-factor-at-a-time approach. The benefits of using fractional factorial designs actually increase as we use more factors.

The shotgun approach randomly selects points over the region of interest. It gets its name because the pattern looks like a shotgun blast. This approach also has drawbacks:

- It tends to use many more treatment combinations and thus tends to waste resources.
- It rarely covers the experimental region well.
- It produces less precise estimates of the effects.

Some engineers pursue a shotgun approach as an expression of their creativity. In general, a better way to express creativity is in judicial selection of the appropriate factors and their ranges for the experiment.

∴ Exercises

7.18 Aceves-Mijares and colleagues (1995) consider a polysilicon deposition process commonly used in the manufacture of integrated circuits. Four factors of interest are pressure (A), temperature gradient (B), furnace temperature (C), and the location of the gas input (D). Suppose the engineers assigned to this process wish to conduct an eight-run design using the following levels:

Pressure	Temp. Gradient	Furnace Temp.	Gas Input
1 Torr	Flat	625° C	Bottom
3 Torr	Ramp	725° C	Top

   a. Construct the appropriate fractional factorial design, in the design variables, that has the highest resolution.

   b. Give this design in the natural units.

   c. Give this design in random order.

7.19 Consider the oil extraction experiment described in Example 7.10. Suppose the researcher could afford only an eight-run design.

   a. Construct the appropriate fractional factorial design, in the design variables, that has the highest resolution.

   b. Give this design in the natural units.

   c. Give this design in random order.

7.20 Consider the freeze process efficiency experiment described in Exercise **7.11**. Suppose the researcher couldn't afford the full 32-run design.

   a. Consider 16 runs. Construct the appropriate fractional factorial design, in the design variables, that has the highest resolution.

   b. Give this design in natural units.

   c. Give this design in random order.

   d. Consider 8 runs. Construct the appropriate fractional factorial design, in the design variables, that has the highest resolution.

   e. Give this design in natural units.

   f. Give this design in random order.

7.21 Consider the catapult experiment described in Exercise **7.13**. Suppose the researcher could afford only an eight-run design.

   a. Construct the appropriate fractional factorial design, in the design variables, that has the highest resolution.

   b. Give this design in natural units.

   c. Give this design in random order.

7.22 Shoemaker, Tsui, and Wu (1991) study the process that grows silicon wafers for integrated circuits. These layers need to be as uniform as possible because later processing steps form electrical devices within these layers. The smallest unit to which the engineers can apply the processing steps is a batch of wafers. The individual wafers are the observational units. In their initial experiment, the engineers seek to determine which of six possible factors truly influence the uniformity of the silicon layer. They intend to conduct follow-up experimentation using the significant factors from this initial experiment. To minimize the size of the design, they use only two levels for each factor. These are the factors and their levels:

Deposition temp.		1210	1220
Deposition time		Low	High
Argon flow rate		55%	59%
HCl etch temp.		1180	1215
HCl flow rate		10%	14%
Nozzle position		2	4

Construct a 16-run Resolution IV (2_{IV}^{6-2}) fractional factorial design in both the design variables and the natural units.

7.23 Anand, Bhadkamkar, and Moghe (1995) used a fractional factorial design to determine which of five possible factors influenced the determination of carbon in cast iron. The five factors and their levels follow:

	Morning	Afternoon
A—Testing time	Morning	Afternoon
B—KOH conc.	38	42
C—Heating time	45	75
D—Oxygen flow	Slow	Fast
E—Muffle temp.	950	1100

Table 7.43 summarizes the experimental results.

a. Identify this design.
b. Give the defining interaction or interactions.
c. Analyze the experimental results.

Table 7.43	The Carbon in Cast Iron Experiment

x_1	x_2	x_3	x_4	x_5	y
−1	−1	−1	−1	−1	3.130
−1	−1	1	1	−1	3.065
−1	1	−1	−1	1	3.105
−1	1	1	1	1	2.940
1	−1	−1	1	1	2.940
1	−1	1	−1	1	3.110
1	1	−1	1	−1	3.145
1	1	1	−1	−1	3.240

7.24   Pignatiello and Ramberg (1985) studied the impact of several factors involving the heat treatment of leaf springs. In this process, a conveyor system transports leaf spring assemblies through a high-temperature furnace. After this heat treatment, a high-pressure press induces the curvature. After the spring leaves the press, an oil quench cools it to near ambient temperature. An important quality characteristic of this process is the resulting free height of the spring. The researchers used these factors:

Factor	Low Level	High Level
High heat temp. (x_1)	1840	1880
Heating time (x_2)	23	25
Transfer time (x_3)	10	12
Hold down time (x_4)	2	3

Table 7.44 summarizes the experimental results.

a.   Identify this design.
b.   Give the defining interaction or interactions.
c.   Analyze the experimental results.

7.25   Anand, Bhadkamkar, and Moghe (1995) used a fractional factorial design to determine which of six possible factors influenced the determination of manganese in cast iron. The six factors and their levels follow:

A—Titration speed	Medium	Fast
B—Dissolution time	20	30
C—$AgNO_3$ addition	20	10
D—Persulfate addition	2	3
E—Volume $HMnO_4$	100	150
F—Sodium arsenite	0.10	0.15

Table 7.45 summarizes the actual experimental results.

a.   Identify this design.
b.   Give the defining interaction or interactions.
c.   Analyze the experimental results.

7.26   Chapman (1995) discusses an experiment to improve a photographic color slide process. For proprietary reasons, he could identify the six factors only by x_1–x_6. In addition, he transformed the response. Table 7.46 lists the experimental results in the natural units.

a.   Identify this design.
b.   Give the defining interaction or interactions.
c.   Analyze the experimental results.

| Table 7.44 | The Leaf Spring Experiment |

x_1	x_2	x_3	x_4	y
−1	−1	−1	−1	7.37
1	−1	−1	1	7.66
−1	1	−1	1	7.67
1	1	−1	−1	7.79
−1	−1	1	1	7.52
1	−1	1	−1	7.64
−1	1	1	−1	7.54
1	1	1	1	7.90

7.27 Colloidal gas aphrons (micro bubbles) are created from an anionic surfactant and can be characterized in terms of stability. The effect of pH, salt concentration, time of stirring, and temperature on the stability was studied by Jauregi, Gilmour, and Varley (1997). The data in Table 7.47 are based on their experiment.

a. Identify this design.
b. Give the defining interaction or interactions.
c. Analyze the experimental results.

7.28 Five factors in a manufacturing process for an integrated circuit were investigated with the objective of improving the process yield. The five factors were aperture setting (small and large), exposure time (percent below nominal and above nominal), develop time (30 and 45 seconds), mask dimension (small and large), and etch time (14.5 and 15.5 minutes). The data are in Table 7.48

| Table 7.45 | The Manganese in Cast Iron Experiment |

x_1	x_2	x_3	x_4	x_5	x_6	y
−1	−1	−1	−1	−1	−1	0.1745
−1	−1	−1	1	1	1	0.1340
−1	1	1	−1	−1	1	0.1630
−1	1	1	1	1	−1	0.0680
1	−1	1	−1	1	−1	0.0055
1	−1	1	1	−1	1	0.0330
1	1	−1	−1	1	1	0.0690
1	1	−1	1	−1	−1	0.0730

Table 7.46	The Photographic Color Slide Experiment

x_1	x_2	x_3	x_4	x_5	x_6	y
1.5	2.1	7	0.2	6.8	9.75	5
7	2.1	7	0.1	3.5	9.75	3
1.5	3.6	7	0.1	6.8	9.55	14
7	3.6	7	0.2	3.5	9.55	9
1.5	2.1	18	0.2	3.5	9.55	13
7	2.1	18	0.1	6.8	9.55	11
1.5	3.6	18	0.1	3.5	9.75	4
7	3.6	18	0.2	6.8	9.75	5

a. Identify this design.
b. Give the defining interaction or interactions.
c. Analyze the experimental results.

7.29 Azeredo, Da Silva and Rekab (2003) conducted an experiment to see the effect of process parameters on the force required to open an injection molded closure for medical infusion bottles. Table 7.49 summarizes the experimental results.

a. Identify this design.
b. Give the defining interaction or interactions.
c. Analyze the experimental results.
d. Can you think of a design with higher resolution? How could it be carried out in 24 runs?

Table 7.47	The Colloidal Gas Aphrons Experiment

pH	Salt Conc.	Time	Temp.	Stability
4	0.00	4	1	90
4	0.00	16	2	107
8	0.00	16	1	68
8	0.14	4	1	60
8	0.00	4	2	90
4	0.14	16	1	30
4	0.14	4	2	45
8	0.14	16	2	75

Table 7.48	The Integrated Circuit Experiment				

Aperture Setting	Exposure Time	Develop Time	Mask Dimension	Etch Time	Yield
small	−20	30	small	15.5	8
large	−20	30	small	14.5	9
small	20	30	small	14.5	34
large	20	30	small	15.5	52
small	−20	45	small	14.5	16
large	−20	45	small	15.5	22
small	20	45	small	15.5	45
large	20	45	small	14.5	60
small	−20	30	large	14.5	6
large	−20	30	large	15.5	10
small	20	30	large	15.5	30
large	20	30	large	14.5	50
small	−20	45	large	15.5	15
large	−20	45	large	14.5	21
small	20	45	large	14.5	44
large	20	45	large	15.5	63

Table 7.49	The Medical Infusion Bottle Experiment								

Injection Speed	Mold Temp.	Melt Temp.	Holding Pressure	Holding Time	Cooling Time	Ejection Speed	Force		
40	25	205	45	3	25	5	41.04	44.02	41.89
75	25	205	25	2	25	25	68.59	70.89	71.53
40	45	205	25	3	10	25	44.12	46.46	32.33
75	45	205	45	2	10	5	63.02	64.12	62.67
40	25	235	45	2	10	25	65.51	62.48	59.05
75	25	235	25	3	10	5	71.62	78.44	73.96
40	45	235	25	2	25	5	42.77	41.15	39.49
75	45	235	45	3	25	25	64.33	73.43	70.95

7.4 Case Study

Pen barrels are produced by an injection molding process. Pigment and polypropylene are mixed together, melted, and extruded into a chilled mold.

Faber-Castell ran into a problem with "warped" pen barrels when it brought a new injection molder into production. In some cases, the warp was quite significant with the barrels displaying a pronounced bow. These barrels were completely unsuitable for use. As a result, the process engineer needed to conduct an appropriate experiment whose first objective was to identify the factors that influence the warping. In Chapter 8, we discuss the design and analysis of experiments that can minimize warping.

The process engineer worked closely with the area supervisor and the operators. In a series of meetings, they devised a method for measuring the warp as a deflection in degrees. They then identified four processing factors that should

| Table 7.50 | The First Pen Barrel Warping Experiment |

x_1	x_2	x_3	x_4	Warp
0	0	0	0	9.54
−1	1	1	−1	2.06
1	−1	1	−1	2.41
−1	−1	−1	1	33.44
0	0	0	0	9.91
1	−1	−1	1	33.77
−1	1	1	1	14.84
−1	−1	−1	−1	13.52
−1	1	−1	−1	14.34
1	1	1	−1	2.76
1	1	−1	1	33.25
−1	1	−1	1	33.24
1	−1	1	1	13.21
−1	−1	1	−1	1.28
−1	−1	1	1	13.62
1	−1	−1	−1	13.90
1	1	−1	−1	13.94
1	1	1	1	14.53
0	0	0	0	10.20

Table 7.51	The Analysis of Variance Table for the First Pen Barrel Warping Experiment

Term	Effect	Coef	SE Coef	T	P
Constant		15.882	0.1125	141.20	0.000
A	0.177	0.089	0.1125	0.79	0.456
B	0.478	0.239	0.1125	2.13	0.071
C	-15.585	-7.793	0.1125	-69.28	0.000
D	15.711	7.855	0.1125	69.84	0.000
A*B	-0.181	-0.091	0.1125	-0.81	0.447
A*C	0.096	0.048	0.1125	0.43	0.682
A*D	-0.274	-0.137	0.1125	-1.22	0.263
B*C	0.443	0.221	0.1125	1.97	0.090
B*D	-0.024	-0.012	0.1125	-0.10	0.919
C*D	-3.786	-1.893	0.1125	-16.83	0.000
Ct Pt		-5.996	0.2831	-21.18	0.000

influence warping: x_1, the proportion of polypropylene in the formulation; x_2, the melt temperature in the holding chamber; x_3, the injection pressure; x_4, the temperature of the chilled water in the mold. Finally, they agreed to run a 2^4 full factorial experiment with three center runs. Table 7.50 shows the data. The specific levels used for each factor are confidential.

We see from the analysis in Table 7.51 that x_3, the injection pressure, x_4, the temperature of the chilled water in the mold, and the $x_3 x_4$ interaction are important. In addition, we see that there is serious evidence of curvature (Ct · Pt $p = 0.000$). These results strongly suggest that we conduct a follow-up experiment based on x_3 and x_4 that will allow us to model the curvature in the hopes of finding conditions that will minimize the warping, which we do in chapter 8.

7.5 Ideas for Projects

1. As a class, perform the "catapult" experiment. Many instructors use a rubber band catapult to generate experimental data. The ones we use allow you to change the tension on the rubber band in several ways. Perform the experiment to generate a prediction equation for how far the catapult throws the ball. Use the prediction equation to try to hit a target a known distance away.

2. Perform the "paper helicopter" experiment. Many instructors make paper helicopters and drop them in class. We prefer to drop them from the fourth floor of a building on campus. The goal is to design a paper helicopter that flies as long as possible.
3. Work with an instructor for a unit operations laboratory. Help a class perform an experiment to screen for important factors that affect the operation of a piece of equipment.

References

1. Aceves-Mijares, M., Murphy-Arteaga, R., Torres-Jacome, A., and Calleja-Arriaga, W. (1995). Quality assurance in polysilicon deposition using statistics. *Quality Engineering, 8, 255–262.*

2. Anand, K. N., Bhadkamkar, S. M., and Moghe, R. (1995). Wet method of chemical analysis of cast iron: Upgrading accuracy and precision through experimental design. *Quality Engineering, 22(6), 54–61.*

3. Antony, J. (2002). Training for design of experiments using a catapult. *Quality and Reliability International, 18, 29–35.*

4. Azeredo, M. B., Da Silva, S. S., and Rekab, K. (2003). Improve molded part quality. *Quality Progress, 7, 72–76.*

5. Box, G. E. P., and Bisgaard, S. (1987). The scientific context of quality improvement. *Quality Progress, 22(6), 54–61.*

6. Box, G. E. P. and Wilson, K. B. (1951). On the experimental attainment of optimum conditions. *Journal of the Royal Statistical Society, Series B, 13, 1–45.*

7. Buckner, J., Cammenga, D. J., and Weber, A. (1993). Elimination of TiN peeling during exposure to CVD tungsten deposition process using designed experiments. *Statistics in the Semiconductor Industry.* Austin, Texas: SEMATECH. Technology Transfer No. 92051125A-GEN, Vol. I, 3-45–3-71.

8. Chapman, R. E. (1995). Photochemistry multiple response co-optimization. *Quality Engineering, 8, 31–45.*

9. Eibl, S., Kess, U., and Pukelsheim, F. (1992). Achieving a target value for a manufacturing process: A case study. *Journal of Quality Technology, 24, 22–26.*

10. Ferrer, A. J., and Romero, R. (1995). A simple method to study dispersion effects from non-necessarily replicated data in industrial contexts. *Quality Engineering, 7, 747–775.*

11. Jauregi, P., Gilmour, S., and Varley, J. (1997). Characterization of colloidal gas aphrons for subsequent use for protein recovery. *The Chemical Engineering Journal, 65, 1–11.*

12. Kilgo, M. B. (1998). An application of fractional factorial experimental designs. *Quality Engineering, 1, 19–23.*

13. Kim, K. W., Kweon, S. J., Doh, Y. S., and Yi, T. (2003). Fracture toughness of polymer-modified asphalt concrete at low temperatures. *The Canadian Journal of Civil Engineering, 30, 406–413.*

14. Liu, C. H., Kan, M., and Chen, B. H. (1993). A correlation of two-phase pressure drops in screen-plate bubble column. *Canadian Journal of Chemical Engineering, 71, 460–463.*

15. Montgomery, D. C. (2004). *Design and Analysis of Experiments,* 6th ed. New York: John Wiley.

16. Myers, R. H., and Montogomery, D. C. (2002). *Response surface methodology: Process and product optimization using designed experiments.* New York: John Wiley.

17. Pignatiello, J. J., Jr., and Ramberg, J. S. (1985). Discussion of "Off-line quality control, parameter design, and the Taguchi method" by R. N. Kackar. *Journal of Quality Technology, 17,* 198–206.

18. Rao, G., and Saxena, S. C. (1993). Prediction of flue gas composistion of an incinerator based on a nonequilibrium-reaction approach. *Journal of Waste Management Association, 43,* 745–752.

19. Said, A., Hassan, E., El-Awad, A., El-Salaam, K., and El-Wahab, M. (1994). Structural changes and surface properties of $CO_xFe_{3-x}O_4$ spinels. *Journal of Chemical Technology and Biotechnology, 60,* 161–170.

20. Shina, S. G. (1991). The successful use of the Taguchi method to increase manufacturing process capability. *Quality Engineering, 3,* 333–349.

21. Shoemaker, A. C., Tsui, K., and Wu, C. F. J. (1991). Economical experimentation methods for robust design. *Technometrics, 33,* 415–427.

22. Snee, R. D. (1981). Developing blending models for gasoline and other mixtures. *Technometrics, 23,* 119–130.

23. Ul-Haq, E., White, D. A., and Adeleye, S. A. (1995). Freezing in an ultrasonic bath as a method for the decontamination of aqueous effluents. *The Chemical Engineering Journal, 57,* 53–60.

24. Wasiloff, E., and Hargritt, C. (1999). Using DOE to determine AA battery life. *Quality Progress, 3,* 67–71.

25. Yi, J., and Shrive, N. G. (2003). Behavior of hollow concrete masonary walls with one-course bond beams subjected to concentric and eccentric concentrated loading. *The Canadian Journal of Civil Engineering, 30,* 181–190.

Introduction to Response Surface Methodology

With a basic understanding of the 2^k factorial design and its fractions, we show how we can optimize a process using RSM (Sections 8.1 and 8.2). Section 8.3 outlines how to optimize a process with more than one response of interest. In Chapter 5, we learned that the process variability is at least as important as the process mean. Section 8.4 outlines appropriate methods for incorporating the process variance into our experiments and their analyses.

8.1 Sequential Philosophy of Experimentation

Engineering experiments, particularly if we seek to optimize a process or product, should proceed sequentially, where we use low-order polynomial equations as our models. Ideally, we apply what we learn from our previous experiments to plan each subsequent one.

Figure 8.1 illustrates a typical sequence of experiments for optimizing a process. We center the initial experiment at the setting of the two factors that we expect to give the best results based on whatever previous knowledge we have. For example, when we scale from bench-top research to a pilot plant operation, we start the pilot plant operation using the best settings recommended by the bench-top results. We have no reason to believe that these recommended settings really should provide the best pilot plant operation; however, we have no basis for starting anywhere else. Typically, we conduct this experiment for two purposes:

1. To screen out the unimportant factors
2. To suggest better operating conditions

Since we know that we are going to run a sequence of experiments, we should save resources for the later, more important and more expensive experiments. As a result, we tend to use the smallest two-level experiments we can, which are the Resolution III screening designs. These designs allow us only to estimate a strict

---

Figure 8.1 | An Example of a Sequential Approach to Experimentation

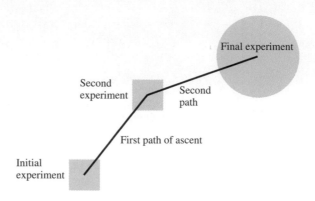

first-order model, which we hope serves as an adequate approximation to the true relationship.

Based on the results from our screening, we can estimate a model in the important factors. We can use this model to construct a *path of steepest ascent* (or *descent,* depending on whether we wish to maximize or minimize the response). We then conduct a second series of runs along this path, stopping when we begin to see a deterioration in our response.

We next conduct another two-level factorial experiment centered at the settings that have given us the best performance so far. Typically, we use a Resolution IV design at this stage. Based on these results, we can estimate another model, which we can use to suggest another path for exploration.

After the second path of exploration begins to deteriorate, we usually are quite near a true point of optimal response. In this case, we plan another experiment that will allow us to find the settings in the factors to achieve this goal. Section 8.2 outlines one common experimental strategy for optimization.

The Path of Steepest Ascent

The path of steepest ascent provides a powerful basis for process improvement. Consider a situation where we model the response by a strict first-order model with a prediction equation of the form

$$\hat{y} = b_0 + b_1 x_1 + b_2 x_2 + \cdots + b_k x_k.$$

From calculus, we know that to find either a maximum or a minimum, we must take the partial derivatives and set them equal to zero. Taking the first derivative with respect to x_j, we see that

$$\frac{\partial \hat{y}}{\partial x_j} = b_j \neq 0.$$

As a result, the maximum and minimum for the estimated response do not lie in the interior of the region of interest but on the boundary, which should not be a surprise. The simplest example of a strict first-order model is a straight line. If the slope is positive, then the maximum value occurs at $+\infty$ and the minimum at $-\infty$, and conversely if the slope is negative.

Technically, we find the path of steepest ascent by a constrained optimization technique based on Lagrangian multipliers. If we have only two factors, the path of steepest ascent is the line from the origin to the maximum response over the circle defined by

$$x_1^2 + x_2^2 = c^2 \quad \text{for any } c > 0,$$

where c is the radius of the circle. For $k \geq 3$ factors, the path of steepest ascent is the line from the origin to the maximum response over the sphere defined by

$$\sum_{j=1}^{k} x_j^2 = c^2 \quad \text{for any } c > 0.$$

The path of steepest descent searches for the minimum response. Since the path of steepest ascent represents the optimum response over spheres, we need to construct this path in the metric where spheres make the most sense. A sphere assumes that a one-unit change in one factor is equivalent to a one-unit change in any other factor. As a result, we should always construct the path of steepest ascent in terms of the design variables and then convert this line back to the natural units.

Figure 8.2 illustrates a path of steepest ascent for a two-factor experiment. We construct this path so that it runs perpendicular to the contours of constant response. In so doing, we get the greatest expected increase in our response for a one-unit change in either of the two factors.

Since the path of steepest ascent is a straight line, we define the specific settings for all of the factors when we specify the setting for any single one. Let x_1 be this "key" factor, and let x_{10} be a specific value for this factor along the desired path.

Figure 8.2 | An Illustration of the Path of Steepest Ascent

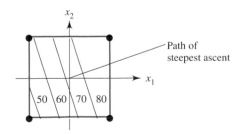

The settings for the other factors are

$$x_{j0} = \frac{b_j}{b_1} x_{10} \qquad j = 2, 3, \ldots, k.$$

This path passes through the point $(0, 0, \ldots, 0)$, which is the center, in terms of the design variables, of the region of interest.

To convert this line back to the natural units, we must formalize how to transform the natural levels to the design variables. Let c_j be the center value, in the natural units, for the jth factor. By definition, c_j is the same distance from the high and low levels for this factor. Let d_j represent this distance. Let x_{j0}^* be the specific setting for the jth factor along the path of steepest ascent; thus,

$$x_{j0}^* = c_j + d_j x_{j0}.$$

We usually pick the factor with the largest estimated coefficient in absolute value as our key factor. We construct the line by increasing this key factor by a convenient amount each time. We then run a series of experiments along this path. For a path of steepest ascent, we expect the response to increase for the first few runs along this path. At some point, we expect to encounter some curvature, and the response should begin to decrease. We stop our series of runs once we are reasonably convinced that we have passed the best setting on this path.

Example 8.1 | **Oil Extraction from Peanuts—Revisited**

Kilgo (1988) conducted an experiment to maximize the amount of oil extracted. In Example 7.7, we concluded that only temperature (x_2) and particle size (x_5) were important. Table 8.1 gives the information we need to construct the path of steepest ascent. Since x_5 has the largest coefficient in absolute value, we use it as our key factor. For a specific setting of particle size, x_{50}, along the path, the appropriate setting for temperature, x_{20}, is given by

$$x_{20} = \frac{b_2}{b_5} x_{50} = \frac{9.875}{-22.25} x_{50} = -0.44 x_{50}.$$

Table 8.1 | The Information Required to Generate the Path of Steepest Ascent for the Oil Extraction Experiment

	x_2	x_5
b_j	9.875	−22.250
c_j	60	2.665
d_j	35	1.385

Table 8.2	The Path of Steepest Ascent for the Oil Extraction Experiment

| | Design Variables | | Natural Units | | |
Run	x_{20}	x_{50}	x^*_{20}	x^*_{50}	y
1	0.444	−1.0	75.5	1.28	81
2	0.488	−1.1	77.1	1.14	84
3	0.533	−1.2	78.7	1.00	90
4	0.577	−1.3	80.2	0.86	97
5	0.622	−1.4	81.8	0.73	95
6	0.666	−1.5	83.3	0.59	92

We can convert each value of x_{20} back to the natural units by

$$x^*_{20} = c_2 + d_2 x_{20} = 60 + 35 x_{20}.$$

We can convert each value of x_{50} back to the natural units by

$$x^*_{50} = c_5 + d_5 x_{50} = 2.665 + 1.385 x_{50}.$$

Table 8.2 gives the path in both the design and the natural variables along with the observed responses. The estimated coefficient associated with the key factor, x_5, suggests that we want to make this factor as negative as possible. Thus, this path starts with $x_5 = -1$, which is the most negative value for the key factor actually used in the experiment. We see that the yield does increase initially along this path. At run 4, we observe our largest yield along the path (97). Run 5 gives a slightly smaller yield, but not so much smaller as to justify stopping the path. Run 6 gives an even smaller yield. We thus conclude that run 4 is the best setting along this path. We should center our next experiment at this point.

∷ Exercises

8.1 Construct an appropriate path of steepest descent for the uniformity experiment in Exercise **7.1**.

8.2 Construct an appropriate path of steepest ascent for the calcinations experiment in Exercise **7.2**.

8.3 Construct an appropriate path of steepest ascent for the strength of hollow masonary walls experiment in Exercise **7.3**.

8.4 Construct an appropriate path of steepest ascent for the distillation column experiment in Exercise **7.4**.

8.5 Construct an appropriate path of steepest descent for the coating thickness experiment in Exercise **7.16**.

8.6 Construct an appropriate path of steepest descent for the manganese in cast iron experiment in Exercise **7.25**.

8.7 Construct an appropriate path of steepest ascent for the photographic color slide experiment in Exercise **7.26**.

⯈ 8.2 Central Composite Designs

The Second-Order Model and Optimization

> **VOICE OF EXPERIENCE**
>
> *Second-order models are much larger than typical first-order models.*

The responses for most engineering processes display curvature in the neighborhood of their optimal settings. Strict first-order models provide no way to account for this curvature. In such a situation, we base our model on a second-order Taylor series approximation of the form

$$y_i = \beta_0 + \beta_1 x_{i1} + \beta_2 x_{i2} + \cdots + \beta_k x_{ik} + \beta_{11} x_{i1}^2 + \beta_{22} x_{i2}^2 + \cdots$$
$$+ \beta_{kk} x_{ik}^2 + \beta_{12} x_{i1} x_{i2} + \beta_{13} x_{i1} x_{i3} + \cdots + \beta_{k-1,k} x_{i,k-1} x_{ik} + \epsilon_i$$
$$= \beta_0 + \sum_{j=1}^{k} \beta_j x_{ij} + \sum_{j=1}^{k} \beta_{jj} x_{ij}^2 + \sum_{j=1}^{k-1} \sum_{j'>j}^{k} \beta_{jj'} x_{ij} x_{ij'} + \epsilon_i.$$

This model includes the $k + 1$ terms from the strict first-order model, the k pure quadratic (x_i^2) terms, and the $\binom{k}{2}$ two-factor interactions. Thus, this model contains a total of

$$(k + 1) + k + \binom{k}{2} = \frac{(k + 1)(k + 2)}{2}$$

terms, which can be quite large for even moderate choices of k.

The second-order model provides a powerful basis for selecting optimal settings for our process. From calculus, we optimize a function by taking the first partial derivatives and setting them equal to zero. We call the resulting solution the *stationary point*. There are three possibilities for the stationary point:

1. A point of maximum response
2. A point of minimum response
3. A saddle point

Figure 8.3 | A Three-Dimensional Response Surface Plot Illustrating a Stationary Point That Is a Maximum

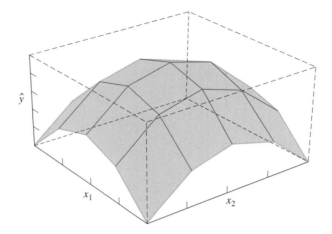

We determine the actual status of the stationary point by examining the second partial derivatives. Figures 8.3–8.5 show three-dimensional *response surface plots* to illustrate stationary points that are a maximum, a minimum, and a saddle point, respectively. Figure 8.3 shows that the response decreases as we move in any direction away from the stationary point. Similarly, Figure 8.4 shows that the response increases as we move in any direction away from the stationary point. Of most interest is Figure 8.5. With a saddle

VOICE OF EXPERIENCE

To find stationary points, we take derivatives of the second-order model with respect to each factor and set equal to zero.

Figure 8.4 | A Three-Dimensional Response Surface Plot Illustrating a Stationary Point That Is a Minimum

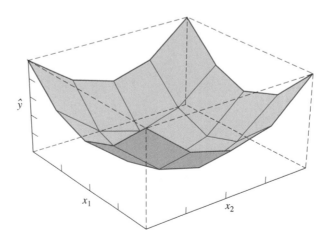

Figure 8.5 | A Three-Dimensional Response Surface Plot Illustrating a Stationary Point That Is a Saddle

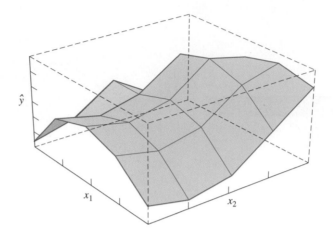

system, the response increases as we move away from the stationary point in one direction, and it decreases as we move in another direction. Figures 8.6–8.8 use *contour plots* to illustrate stationary points that are a maximum, a minimum, and a saddle point, respectively. These contour plots illustrate the same behavior as we move away from the stationary point.

Many analysts use both three-dimensional response surface and contour plots to determine optimum operating conditions for a process, particularly if they are

Figure 8.6 | A Contour Plot Illustrating a Stationary Point That Is a Maximum

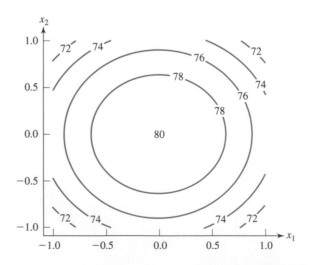

| Figure 8.7 | A Contour Plot Illustrating a Stationary Point That Is a Minimum |

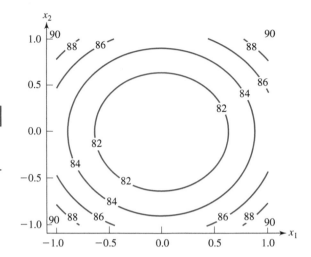

analyzing an experiment that involves only two factors. As the number of factors increases, the value of these graphical procedures rapidly decreases, and we need alternative analytical tools.

Optimizing a response over a region of interest when the prediction equation contains second-order or higher terms is a standard example of a nonlinear programming problem. Many spreadsheets have built-in routines for solving these problems—for example, the SOLVER routine in Microsoft EXCEL. The major

| Figure 8.8 | A Contour Plot Illustrating a Stationary Point That Is a Saddle |

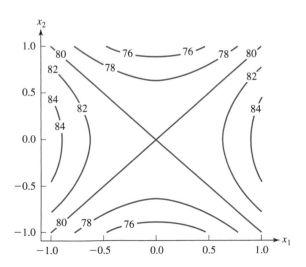

spreadsheets use good algorithms, usually based on reduced gradients. We simply need to input the appropriate prediction equations, and the spreadsheet routine finds the optimal setting. These packages should give as this optimal setting the stationary point whenever it lies in the region of interest and it is a point of optimal response. If the stationary point is not an optimum or if it lies outside the experimental region, then the point of optimal response lies on the boundary of this region. We then must supply the package with the appropriate constraints corresponding to the boundary of the experimental region. For cuboidal experimental regions, wherein each x_j must fall within the interval -1 to 1, we use the following additional constraints:

$$-1 \leq x_1 \leq 1, \ -1 \leq x_2 \leq 1, \ \ldots, \ -1 \leq x_k \leq 1.$$

For spherical experimental regions, we need the additional constraint

$$\sum_{j=1}^{k} x_j^2 \leq k.$$

With these constraints, the spreadsheet routine can find the optimal settings.

Design Construction

To estimate a second-order model, the design must satisfy these conditions:

- The design must have at least as many distinct treatment combinations as terms to estimate in the model.
- Each factor must have at least three distinct levels.

The most commonly used design to estimate the second-order model is the *central composite design (ccd)*. Figure 8.9 illustrates a ccd for two factors. Table 8.3 gives this design in design variables. The first four treatment combinations form a 2^2 factorial. The next four treatment combinations are the *axial runs* because they

---

Figure 8.9 | A Two-Factor Central Composite Design

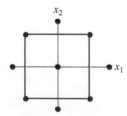

| Table 8.3 | A Two-Factor Central Composite Design |

x_1	x_2
-1	-1
1	-1
-1	1
1	1
$-\alpha$	0
α	0
0	$-\alpha$
0	α
0	0

lie on the axes defined by the design variables. The last treatment combination represents at least one center run.

The specific choice of α depends on the experimenter. These are typical choices for α:

- 1, creating a *face-centered cube* ccd
- $\sqrt{k}$, creating a *spherical* ccd
- $n_f^{0.25}$, creating a *rotatable* ccd, where n_f is the number of factorial runs used in the design

A face-centered cube has all of the treatment combinations except the center run on the surface of a cube (for two factors, on the surface of a square). A spherical ccd has all of the treatment combinations except the center run on the surface of a sphere (for two factors, on a circle). The rotatable ccd produces prediction variances that are functions only of σ^2 and the distance from the center of the design. The interested student can read about the rotatable ccd in Myers and Montgomery (2002, pp. 331–335). Often the rotatable and spherical values for α are inconvenient and, in some cases, not practical to attain. In such cases, engineers often use the closest practical values for α as they can.

Figure 8.10 illustrates a three-factor ccd. Table 8.4 gives the design in design variables. The first eight treatment combinations form a 2^3 factorial design. The next six treatment combinations are the axial runs. The last treatment combination represents the center run. In general, a ccd consists of three portions:

1. A 2^k full factorial or a Resolution V fraction of that
2. $2k$ axial runs
3. Usually, at least one center run

Figure 8.10 | A Three-Factor Central Composite Design

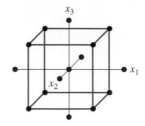

Example 8.2 | **Optimizing the Deposition Rate for a Silicon Wafer Process**

Buckner, Cammenga, and Weber (1993) ran a two-factor, spherical ccd to optimize the deposition rate for a tungsten film on a silicon wafer in terms of the process pressure and the ratio of H_2 to WF_6 in the reaction atmosphere. The ranges for the two factors are listed here:

Factor	Low Level	High Level
Pressure	4 Torr	80 Torr
H_2/WF_6	2	10

Table 8.4 | A Three-Factor Central Composite Design

x_1	x_2	x_3
-1	-1	-1
1	-1	-1
-1	1	-1
1	1	-1
-1	-1	1
1	-1	1
-1	1	1
1	1	1
$-\alpha$	0	0
α	0	0
0	$-\alpha$	0
0	α	0
0	0	$-\alpha$
0	0	α
0	0	0

	x_1	x_2	y
Table 8.5	The Tungsten Deposition Experiment		

x_1	x_2	y
−1	−1	3663
1	−1	9393
−1	1	5602
1	1	12488
−1.414	0	1984
1.414	0	12603
0	−1.414	5007
0	1.414	10310
0	0	8979
0	0	8960
0	0	8979

Since they used a rotatable ccd, the low level corresponds to a design variable of -1.414 $(-\sqrt{2})$ and the high level corresponds to a design variable of 1.414 $(\sqrt{2})$.

Let x_1 be the design variable associated with pressure, and let x_2 be the design variable associated with H_2/WF_6. Table 8.5 gives the result of this experiment. Table 8.6 gives the regression analysis of these results from the SAS statistical software package. The overall F-test, with a p-value of 0.0001, strongly suggests that at least one of the terms in our second-order model is important. The R^2 of 0.9862 indicates that our model explains almost 99% of the total variability, which is quite good. The tests on the individual coefficients suggest that all of the terms except the two-factor interaction are at least marginally important. The residual plots, which we leave as an exercise, show no problems with the data.

The analysis suggests this prediction equation:

$$\hat{y} = 8973 + 3454x_1 + 1567x_2 - 762x_1^2 - 579x_2^2.$$

The first partial derivatives of $\hat{y}$ with respect to the x's yield

$$\frac{\partial \hat{y}}{\partial x_1} = 3454 - 2(762)x_1$$

$$\frac{\partial \hat{y}}{\partial x_2} = 1567 - 2(579)x_2.$$

Table 8.6	The Regression Analysis of the Tungsten Deposition Experiment

Analysis of Variance

Source	DF	Sum of Squares	Mean Square	F Value	Prob>F
Model	5	119478764.6	23895752.92	71.279	0.0001
Error	5	1676204.3072	335240.86144		
C Total	10	121154968.91			

Root MSE	578.99988	R-square	0.9862
Dep Mean	7997.09091	Adj R-sq	0.9723
C.V.	7.24013		

Parameter Estimates

Variable	DF	Parameter Estimate	Standard Error	T for H0: Parameter=0	Prob > |T|
INTERCEP	1	8972.604062	334.28572661	26.841	0.0001
X1	1	3454.429869	204.72282801	16.874	0.0001
X2	1	1566.791836	204.72282801	7.653	0.0006
X11	1	-762.044144	243.70028813	-3.127	0.0260
X22	1	-579.489012	243.70028813	-2.378	0.0633
X12	1	289.000000	289.49994017	0.998	0.3640

Obs	Dep Var Y	Predict Value	Std Err Predict	Residual	Std Err Residual	Student Residual
1	3663.0	2898.8	457.767	764.2	354.528	2.155
2	9393.0	9229.7	457.767	163.3	354.528	0.461
3	5602.0	5454.4	457.767	147.6	354.528	0.416
4	12488.0	12941.3	457.767	-453.3	354.528	-1.279
5	1984.0	2564.4	457.712	-580.4	354.599	-1.637
6	12603.0	12333.5	457.712	269.5	354.599	0.760
7	5007.0	5598.5	457.712	-591.5	354.599	-1.668
8	10310.0	10029.4	457.712	280.6	354.599	0.791
9	8979.0	8972.6	334.286	6.3959	472.751	0.014
10	8960.0	8972.6	334.286	-12.6041	472.751	-0.027
11	8979.0	8972.6	334.286	6.3959	472.751	0.014

Figure 8.11 | The Contour Plot for the Tungsten Deposition Experiment

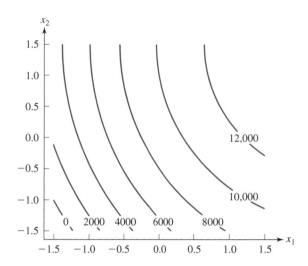

Setting these partials to zero and solving, we get $x_1 = 2.27$ and $x_2 = 1.35$. By looking at the second partial derivatives, we can show that this setting is a point of maximum response. However, this setting lies outside the experimental region, so we probably should not put much faith in it. Figure 8.11 gives the contour plot for this prediction equation. The plot suggests that we should increase both the pressure and the ratio of H_2 to WF_6 to maximize the deposition rate.

We can use the SOLVER routine in Microsoft EXCEL to find optimal conditions over the experimental region. We first must enter into the spreadsheet the prediction equation

$$\hat{y} = 8973 + 3454x_1 + 1567x_2 - 762x_1^2 - 579x_2^2.$$

Since this experiment uses a rotatable ccd, we also need to impose the constraint

$$\sum_{j=1}^{k} x_j^2 \le 2,$$

which guarantees that our solution is in the region defined by our design. The spreadsheet recommends the setting

$$x_1 = 1.253 \qquad x_2 = 0.656.$$

This setting yields a predicted deposition rate of 12,883. Naturally, we should conduct a confirmatory experiment in the neighborhood of this recommendation.

∴ *Exercises*

8.8 Consider Example **7.4**, the springs with cracks experiment.

a. Construct a face-centered cube ccd in both the design and the natural variables.

b. Construct a spherical ccd for this situation in both the design and the natural variables.

8.9 Consider Example **7.7**, the oil extraction from peanuts experiment.

a. Construct a face-centered cube ccd using all five factors in both the design and the natural variables.

b. Construct a rotatable ccd using all five factors in both the design and the natural variables.

8.10 Consider Exercise **7.3**, the strength of hollow masonary walls.

a. Construct a face-centered cube ccd with 6 center runs in both the design and natural variables.

b. Construct a spherical ccd for this situation in both the design and natural variables.

8.11 Consider Exercise **7.4**, the distillation column experiment. Construct a spherical ccd in both the design and the natural variables.

8.12 Consider Exercise **7.5**, the catapult experiment. Construct a face-centered cube ccd in both the design and the natural variables.

8.13 Consider Exercise **7.10**, the polymer filament experiment.

a. Construct a spherical ccd in both the design and the natural variables.

b. Construct a rotatable ccd in both the design and the natural variables.

8.14 Consider Exercise **7.15**, the battery life of remote-control cars.

a. Construct a face-centered cube ccd in both the design and natural variables.

b. Construct a rotatable ccd for this situation in both the design and natural variables.

8.15 Consider Exercise **7.26**, the photographic color slide experiment.

a. Construct a spherical ccd in both the design and the natural variables.

b. Construct a rotatable ccd in both the design and the natural variables.

8.16 Broderick, Lanouette, and Valade (1997) worked on optimizing a refiner operation. The objective of the study was to develop models that describe the impact of two-stage refining conditions on the handsheet properties of an ultra-high-yield bisulphite pulp in order to identify energy efficient operating strategies for optimizing pulp quality. There were four variables of interest: ESP1 (0.3 and 0.5 mm), CONS1 (9 and 15%), ESP2 (0.3 and 0.5 mm), and CONS2 (9 and 15%). Construct a face-centered cube ccd with 6 center runs in both the design and natural variables.

| Table 8.7 | The Abrasion Index Experiment |

x_1	x_2	x_3	y
−1	−1	1	102
1	−1	−1	120
−1	1	−1	117
1	1	1	198
−1	−1	−1	103
1	−1	1	132
−1	1	1	132
1	1	−1	139
−1.633	0	0	102
1.633	0	0	154
0	−1.633	0	96
0	1.633	0	163
0	0	−1.633	116
0	0	1.633	153
0	0	0	133
0	0	0	133
0	0	0	140
0	0	0	142
0	0	0	145
0	0	0	142

8.17 Derringer and Suich (1980) used a ccd to maximize an abrasion index for a tire tread compound in terms of three factors: x_1, hydrated silica level; x_2, silane coupling agent level; and x_3, sulfur level. Table 8.7 lists the actual results. Perform a thorough analysis of the results including residual plots.

8.18 Coteron, Sanchez, Martinez, and Aracil (1993) sought the settings of reaction temperature (x_1), initial amount of catalyst (x_2), and pressure (x_3) that maximize the yield of a synthetic analogue to jojoba oil. Table 8.8 gives the experimental results. Perform a thorough analysis of the results including residual plots.

8.19 Chang and Shivpuri (1994) sought the settings of furnace temperature (x_1) and die close time (x_2) that minimize the porosity of a casting. Table 8.9 lists the experimental results. Perform a thorough analysis of the results including residual plots.

Table 8.8	The Jojoba Oil Experiment

x_1	x_2	x_3	y
−1	−1	−1	17
1	−1	−1	44
−1	1	−1	19
1	1	−1	46
−1	−1	1	7
1	−1	1	55
−1	1	1	15
1	1	1	41
−1.682	0	0	8
1.682	0	0	74.5
0	−1.682	0	30
0	1.682	0	37.5
0	0	−1.682	35
0	0	1.682	30.5
0	0	0	29
0	0	0	28.5
0	0	0	30
0	0	0	27
0	0	0	28

Table 8.9	The Die Casting Experiment

x_1	x_2	y
−1	−1	21
1	−1	15
−1	1	20
1	1	15
−1	0	19
1	0	14
0	−1	19
0	1	17
0	0	17

8.20 Dumbbell shapes of high-density polyethylene were made using injection molding, cut into two pieces, and hot plate welded back together. The quality of the weld (WF) was then measured by the ratio of the yield stress of the welded bar to the mean yield stress of unwelded bars. Four control variables were involved: hot plate temperature (pt, in degrees centigrade), heating time (ht, in seconds), welding time (wt, in seconds) and pressure on the weld (wp, in bars). A central composite design was used, and the data from Buxton (1991) are given in Table 8.10. Perform a thorough analysis of the results including residual plots.

8.21 Chau (1992) studied how the fluffiness of bread was impacted by changes to six minor ingredients, labeled A, B, C, D, E, F. The response variable is specific volume. Market research has shown that, in general, people prefer light fluffy loaves, corresponding to high specific volume. A central composite design was used, and the data are in Table 8.11. Perform a thorough analysis of the results including residual plots.

8.3 Multiple Responses

In many engineering experiments, we have more than one response of interest. The settings that optimize one response may prove disastrous for at least one of the other responses. The key, then, is to find appropriate compromise operating conditions. In this section, we outline two basic approaches for jointly optimizing two or more responses:

1. The desirability function
2. Nonlinear programming approaches

Several statistical software packages include some form of the desirability function. Some spreadsheets, including EXCEL, use good reduced gradient algorithms to perform appropriate constrained optimization.

The Desirability Function

The desirability function provides an overall measure for the "goodness" of a specific setting. A large value indicates a desirable set of values for the various responses; a low value indicates an undesirable set of values. Derringer and Suich (1980) proposed an approach that determines the individual desirabilities for each response and then combines these individual desirabilities into an overall desirability. The analyst then seeks to find the settings in the factors that maximize the overall desirability.

VOICE OF EXPERIENCE

Many engineering situations involve more than one response of interest.

Table 8.10 | The Welded Bars Experiment

pt	ht	wt	wp	WF
245	20	15	2.0	0.40
245	20	15	3.0	0.77
245	20	25	2.0	0.74
245	20	25	3.0	0.67
245	40	15	2.0	0.82
245	40	15	3.0	0.84
245	40	25	2.0	0.86
245	40	25	3.0	0.89
295	20	15	2.0	0.58
295	20	15	3.0	0.77
295	20	25	2.0	0.80
295	20	25	3.0	0.82
295	40	15	2.0	0.86
295	40	15	3.0	0.83
295	40	25	2.0	0.87
295	40	25	3.0	0.82
270	30	20	1.5	*
270	30	20	3.5	0.83
270	30	10	2.5	0.66
270	30	30	2.5	0.88
270	10	20	2.5	0.81
270	50	20	2.5	0.84
220	30	20	2.5	0.81
320	30	20	2.5	0.88
270	30	20	2.5	0.82
270	30	20	2.5	0.86
270	30	20	2.5	0.80
270	30	20	2.5	0.84
270	30	20	2.5	0.86
270	30	20	2.5	0.83
270	30	20	2.5	0.81
270	30	20	2.5	0.79
270	30	20	2.5	0.80
270	30	20	2.5	0.82
270	30	20	2.5	0.86

Table 8.11 | The Fluffiness of Bread Experiment

A	B	C	D	E	F	Specific Volume
−1.0000	−1.0000	−1.0000	−1.0000	−1.0000	−1.0000	429.25
−1.0000	−1.0000	−1.0000	−1.0000	1.0000	1.0000	433.00
−1.0000	−1.0000	1.0000	1.0000	−1.0000	1.0000	454.25
−1.0000	−1.0000	1.0000	1.0000	1.0000	−1.0000	456.75
−1.0000	1.0000	−1.0000	1.0000	−1.0000	1.0000	446.75
−1.0000	1.0000	−1.0000	1.0000	1.0000	−1.0000	447.75
−1.0000	1.0000	1.0000	−1.0000	−1.0000	−1.0000	455.50
−1.0000	1.0000	1.0000	−1.0000	1.0000	1.0000	448.25
1.0000	−1.0000	−1.0000	1.0000	−1.0000	−1.0000	458.75
1.0000	−1.0000	−1.0000	1.0000	1.0000	1.0000	449.50
1.0000	−1.0000	1.0000	−1.0000	−1.0000	1.0000	463.75
1.0000	−1.0000	1.0000	−1.0000	1.0000	−1.0000	466.00
1.0000	1.0000	−1.0000	−1.0000	−1.0000	1.0000	449.50
1.0000	1.0000	−1.0000	−1.0000	1.0000	−1.0000	452.75
1.0000	1.0000	1.0000	1.0000	−1.0000	−1.0000	469.00
1.0000	1.0000	1.0000	1.0000	1.0000	1.0000	471.50
1.6644	0.0000	0.0000	0.0000	0.0000	0.0000	460.25
−1.6644	0.0000	0.0000	0.0000	0.0000	0.0000	447.50
0.0000	1.6644	0.0000	0.0000	0.0000	0.0000	463.50
0.0000	−1.6644	0.0000	0.0000	0.0000	0.0000	456.25
0.0000	0.0000	1.6644	0.0000	0.0000	0.0000	461.00
0.0000	0.0000	−1.6644	0.0000	0.0000	0.0000	454.25
0.0000	0.0000	0.0000	1.6644	0.0000	0.0000	456.25
0.0000	0.0000	0.0000	−1.6644	0.0000	0.0000	449.75
0.0000	0.0000	0.0000	0.0000	1.6644	0.0000	464.00
0.0000	0.0000	0.0000	0.0000	−1.6644	0.0000	461.00
0.0000	0.0000	0.0000	0.0000	0.0000	1.6644	456.00
0.0000	0.0000	0.0000	0.0000	0.0000	−1.6644	450.75
0.0000	0.0000	0.0000	0.0000	0.0000	0.0000	461.00

The individual desirabilities depend on which of these options we choose:

- Maximize the response of interest.
- Minimize the response of interest.
- Achieve a specific target value for the response of interest.

Derringer and Suich use a scale from 0, which represents completely undesirable, to 1, fully desirable, for their individual desirability functions.

This approach makes the most sense if we consider the target value case first. Let $\hat{y}$ be the predicted value for the response, let y_T be the specific target value for the response of interest, let y_L be the smallest possible value that has any desirability, and let y_U be the largest possible value that has any desirability. One approach defines the desirability for this response by

$$d = \begin{cases} 0 & \text{for } \hat{y} < y_L \\ \dfrac{\hat{y} - y_L}{y_T - y_L} & \text{for } y_L \leq \hat{y} \leq y_T \\ \dfrac{y_U - \hat{y}}{y_U - y_T} & \text{for } y_T < \hat{y} \leq y_U \\ 0 & \text{for } \hat{y} > y_U. \end{cases}$$

With this definition, we give a desirability of 0 to any predicted value for the response less than y_L or greater than y_U. If the predicted value is exactly at the target value, we give it a desirability of 1. The farther the predicted value is from the target, the lower desirability we give it. Derringer and Suich actually proposed the following slight modification:

$$d = \begin{cases} 0 & \text{for } \hat{y} < y_L \\ \left(\dfrac{\hat{y} - y_L}{y_T - y_L}\right)^s & \text{for } y_L \leq \hat{y} \leq y_T \\ \left(\dfrac{y_U - \hat{y}}{y_U - y_T}\right)^t & \text{for } y_T < \hat{y} \leq y_U \\ 0 & \text{for } \hat{y} > y_U. \end{cases}$$

The exponents s and t provide greater flexibility in assigning the desirability within the range of interest. As Figure 8.12 illustrates, values of the exponents less than 1 make more of the range highly desirable, and values greater than 1 concentrate the highly desirable values near the target. Some packages use only $s = t = 1$, which reduces to the approach we originally outlined.

Next, suppose we wish to maximize the response. Let y_L be the smallest desirable value for this response, and let y_U be a fully desirable value. Basically, y_U represents the point of diminishing returns. Any value larger than y_U does not yield any real additional benefit. In some cases, y_U represents a true bound for the response. For example, we cannot achieve concentrations greater than

Figure 8.12 | The Desirability Function for a Target Value

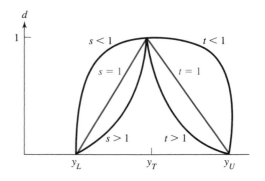

100%. In other cases, y_U is some arbitrary value larger than the largest observed response. For this situation, Derringer and Suich proposed

$$d = \begin{cases} 0 & \text{for } \hat{y} < y_L \\ \left(\dfrac{\hat{y} - y_L}{y_U - y_L} \right)^s & \text{for } y_L \leq \hat{y} \leq y_U \\ 1 & \text{for } \hat{y} > y_U. \end{cases}$$

Figure 8.13 illustrates this desirability function.

Now suppose we wish to minimize the response. Let y_U be the largest desirable value for this response, and let y_L be a fully desirable value. Basically, y_L represents the point of diminishing returns. Any value smaller than y_L does not

Figure 8.13 | The Desirability Function for Maximizing a Response

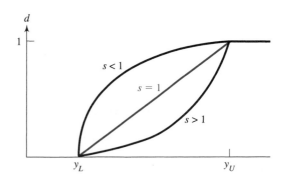

Figure 8.14 | The Desirability Function for Minimizing a Response

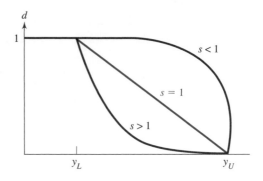

yield any real additional benefit. Derringer and Suich proposed

$$
d = \begin{cases}
1 & \text{for } \hat{y} < y_L \\
\left(\dfrac{y_U - \hat{y}}{y_U - y_L}\right)^s & \text{for } y_L \le \hat{y} \le y_U \\
0 & \text{for } \hat{y} > y_U.
\end{cases}
$$

Figure 8.14 illustrates this desirability function.

Once we have the individual desirabilities, we need to combine them in a meaningful way. We observe that if any of the individual responses is completely undesirable, then the overall desirability also should be completely undesirable. Similarly, the overall desirability should be 1 if and only if all of the individual responses are completely desirable. Suppose we have m responses of interest. Let $d_1, d_2, \ldots, d_m$ be the individual desirabilities. Derringer and Suich defined the overall desirability, D, by

$$
D = (d_1 \cdot d_2 \cdots d_m)^{1/m} = \left(\prod_{j=1}^{m} d_j\right)^{1/m}
$$

which is the geometric mean of the desirabilities.

The conditions recommended by this approach represent an explicit compromise among the various responses *based on the individual prediction equations.* We should always perform at least one confirmatory run at the recommended settings to ensure that all of the responses are satisfactory.

Example 8.3 | **The Chemical Reactor Experiment**

Myers and Montgomery (2002) outline an experiment in which the engineers seek to find the settings for reaction time (x_1), reaction temperature (x_2), and the amount of catalyst (x_3) that maximize the conversion (y_1) of a polymer and achieve a target value of 57.5 for the thermal activity (y_2). Management has set a lower bound of

Table 8.12	The Chemical Reactor Experiment				

x_1	x_2	x_3	y_1	y_2
−1	−1	−1	74	53.2
1	−1	−1	51	62.9
−1	1	−1	88	53.4
1	1	−1	70	62.6
−1	−1	1	71	57.3
1	−1	1	90	67.9
−1	1	1	66	59.8
1	1	1	97	67.8
−1.682	0	0	76	59.1
1.682	0	0	79	65.9
0	−1.682	0	85	60.0
0	1.682	0	97	60.7
0	0	−1.682	55	57.4
0	0	1.682	81	63.2
0	0	0	81	59.2
0	0	0	75	60.4
0	0	0	76	59.1
0	0	0	83	60.6
0	0	0	80	60.8
0	0	0	91	58.9

80 for the conversion. The maximum possible value is 100. Management has set a lower bound of 55 and an upper bound of 60 for the thermal activity. Table 8.12 presents the experimental results. The engineers could justify this full second-order prediction equation for conversion:

$$\hat{y}_1 = 81.09 + 1.03x_1 + 4.04x_2 + 6.20x_3 - 1.83x_1^2 + 2.94x_2^2$$
$$-5.19x_3^2 + 2.13x_1x_2 + 11.37x_1x_3 - 3.87x_2x_3.$$

The engineers proposed this prediction equation for the thermal activity:

$$\hat{y}_2 = 60.5 + 4.26x_1 + 2.23x_3.$$

The engineers used the Derringer and Such approach to propose optimal operating conditions with $s = 1$ for conversion and with $s = t = 1$ for thermal activity.

Based on the prediction equations, the package recommended a setting of

$$x_1 = -0.389 \qquad x_2 = 1.682 \qquad x_3 = -0.484.$$

This setting gives a predicted conversion of 95.21% and a predicted thermal activity of 57.50. The overall desirability for this setting is 0.8720, which is reasonably close to 1.

Nonlinear Programming Approaches

Jointly optimizing two or more responses when the prediction equations contain second-order or higher terms is a standard example of a nonlinear programming problem. Many spreadsheets have built-in routines for solving these problems—for example, the SOLVER routine in Microsoft EXCEL. The major spreadsheets use good algorithms, usually based on reduced gradients. We simply need to input the appropriate prediction equations and constraints and specify one response as the "key." The spreadsheet routine finds the optimal setting. These routines are not guaranteed to find a solution within the experimental region unless we specify some additional constraints. For cuboidal experimental regions—that is, when we use a face-centered cube ccd—each x_j must fall within the interval -1 to 1, which implies these additional constraints:

$$-1 \le x_1 \le 1, \quad -1 \le x_2 \le 1, \quad \dots, \quad -1 \le x_k \le 1.$$

For spherical experimental regions, we need the additional constraint

$$\sum_{j=1}^{k} x_j^2 \le k.$$

With these additional constraints, the spreadsheet routine may not find a feasible solution. Then we must relax one or more of the constraints to find a solution.

The nonlinear programming solutions tend to be similar to those obtained by the Derringer and Such method, but not identical. If we specify a target range for the nonlinear programming approaches, they often give solutions that fall on the boundary of that range rather than at the specified target value. If the specific target value is important, we should impose the equality constraint $y = y_T$ instead of the constraint $y_L \le y \le y_U$.

| Example 8.4 | **The Chemical Reactor Experiment—Revisited** |

We can use the SOLVER routine in Microsoft EXCEL to find optimal conditions. As before, we use this second-order prediction equation for conversion:

$$\hat{y}_1 = 81.09 + 1.03x_1 + 4.04x_2 + 6.20x_3 - 1.83x_1^2 + 2.94x_2^2$$
$$- 5.19x_3^2 + 2.13x_1x_2 + 11.37x_1x_3 - 3.87x_2x_3$$

and the following prediction equation for thermal activity:

$$\hat{y}_2 = 60.5 + 4.26x_1 + 2.23x_3.$$

Recall that we seek to maximize the conversion. We thus specify conversion, y_1, as the key response and tell the routine that we want to maximize it. Since we have a target value of 57.5 for the thermal activity, we specify the constraint

$$y_2 = 57.5.$$

Since this experiment uses a rotatable ccd, we need to impose the additional constraint

$$\sum_{j=1}^{k} x_j^2 \le 3,$$

which guarantees that our solution will be in the region defined by our design. The spreadsheet recommends the setting

$$x_1 = -0.477 \qquad x_2 = 1.608 \qquad x_3 = -0.433.$$

This setting gives a conversion of 94.00% and a thermal activity of 57.5. On the whole, this recommendation appears quite similar to the one we obtained from the Derringer and Suich approach.

Exercises

8.22 Derringer and Suich (1980) discuss an experiment that studies the effect of hydrated silica level (x_1), silane coupling agent (x_2), and sulfur (x_3) on four responses: PICO abrasion index (y_1), modulus (y_2), elongation (y_3), and hardness (y_4). Here are the desirable conditions for these responses:

$$y_1 > 120$$
$$y_2 > 1000$$
$$400 < y_3 < 600$$
$$60 < y_4 < 75.$$

Table 8.13 lists the experimental results.

a. Use the Derringer and Suich approach to generate recommended settings for this process.

b. Use a nonlinear programming package such as EXCEL SOLVER to generate recommended settings.

Table 8.13 | The Derringer and Suich Experiment

x_1	x_2	x_3	y_1	y_2	y_3	y_4
−1	−1	−1	103	490	640	62.5
1	−1	−1	120	860	410	65
−1	1	−1	117	800	570	77.5
1	1	−1	139	1090	380	70
−1	−1	1	102	900	470	67.5
1	−1	1	132	1289	270	67
−1	1	1	132	1270	410	78
1	1	1	198	2294	240	74.5
−1.633	0	0	102	770	590	76
1.633	0	0	154	1690	260	70
0	−1.633	0	96	700	520	63
0	1.633	0	163	1540	380	75
0	0	−1.633	116	2184	520	65
0	0	1.633	153	1784	290	71
0	0	0	133	1300	380	70
0	0	0	133	1300	380	68.5
0	0	0	140	1145	430	68
0	0	0	142	1090	430	68
0	0	0	145	1260	390	69
0	0	0	142	1344	390	70

8.23 Chang and Shivpuri (1994) studied the effect of furnace temperature (x_1) and die close time (x_2) on the hole diameter (y_1), the casting porosity (y_2), and the temperature difference (y_3) for a casting process. The hole diameter was measured as the difference between the measured value and the target; thus, the ideal value was 0. The engineers sought to find the settings for the factors that made the thickness 0 and minimize the porosity and the temperature difference. Table 8.14 presents the experimental results.

    a. Use the Derringer and Suich approach to generate recommended settings for this process.

    b. Use a nonlinear programming package such as EXCEL SOLVER to generate recommended settings.

8.24 Chapman (1995) discusses an experiment to improve a photographic color slide process. For proprietary reasons, he could identify the six factors only by x_1–x_6.

| Table 8.14 | The Die Casting Experiment |

x_1	x_2	y_1	y_2	y_3
−1	−1	3	21	80
1	−1	1	15	101
−1	1	4	20	87
1	1	2	15	108
−1	0	8	19	85
1	0	5	14	106
0	−1	2	19	92
0	1	3	17	96
0	0	7	17	95

He called the responses of interest RMAX (y_1), GMAX (y_2), GLD (y_3), and BLD (y_4). Each of these responses is measured as a difference from target; thus, the ideal value for each is 0. Table 8.15 gives the experimental results in the natural units.

a. Use the Derringer and Suich approach to generate recommended settings for this process.
b. Use a nonlinear programming package such as EXCEL SOLVER to generate recommended settings.

| Table 8.15 | The Photographic Color Slide Experiment |

x_1	x_2	x_3	x_4	x_5	x_6	y_1	y_2	y_3	y_4
1.5	2.1	7	0.2	6.8	9.75	4	−8	7	5
7	2.1	7	0.1	3.5	9.75	−4	−17	3	3
1.5	3.6	7	0.1	6.8	9.55	11	2	16	14
7	3.6	7	0.2	3.5	9.55	2	−9	12	9
1.5	2.1	18	0.2	3.5	9.55	3	−5	16	13
7	2.1	18	0.1	6.8	9.55	0	−12	12	11
1.5	3.6	18	0.1	3.5	9.75	−2	−13	5	4
7	3.6	18	0.2	6.8	9.75	−5	−22	5	5

Table 8.16 | The Distillation Column Experiment

x_1	x_2	x_3	y_1	y_2
210	270	50	93.9	3698
230	270	50	88.1	−589
210	300	50	92.7	3598
230	300	50	90.1	3638
210	270	80	94.4	2962
230	270	80	13.5	−5564
210	300	80	97.4	5088
230	300	80	99.9	5087
210	285	65	99.9	5424
230	285	65	98.8	5479
220	270	65	97.4	5503
220	300	65	95.2	5381
220	285	50	90.5	3665
220	285	80	71.8	−5694
220	285	65	99.9	5458
220	285	65	97.6	5445
220	285	65	99.9	5464
220	285	65	92.9	3273

8.25   A chemical engineer conducted an experiment on a distillation column to determine the effects of condenser temperature (x_1), reboil temperature (x_2), and reflux ratio (x_3) on two responses: the concentration of the product in the distillate (y_1) and the daily profit from running the column (y_2). Management has set a minimum average concentration of 94% and a minimum daily profit of $4000.00. Table 8.16 summarizes the experimental results in the natural units.

a. Use the Derringer and Suich approach to generate recommended settings for this process.

b. Use a nonlinear programming package such as EXCEL SOLVER to generate recommended settings.

8.26   Colloidal gas aphrons (micro bubbles) are created from an anionic surfactant and can be characterized in terms of stability. The effect of pH, salt concentration, time of stirring and temperature on the stability and gas hold-up was studied by

	The Colloidal Gas Experiment				
pH	**Salt Conc.**	**Time**	**Temp.**	**Stability**	**Gas-hold**
4	0.00	4	1	90	0.17
4	0.00	16	2	107	0.32
8	0.00	16	1	68	0.32
8	0.14	4	1	60	0.20
8	0.00	4	2	90	0.26
4	0.14	16	1	30	0.12
4	0.14	4	2	45	0.17
8	0.14	16	2	75	0.28

Table 8.17

Jauregi, Gilmour, and Varley (1997). The data in Table 8.17 are based on their experiment.

a. Use the Derringer and Suich approach to generate recommended settings.
b. Use a nonlinear programming package such as EXCEL SOLVER to generate recommended settings.

8.4 Experimental Designs for Quality Improvement

In classical experimental design, the term *optimize* tends to mean either to maximize or to minimize a response of interest. For example, we may wish to find the settings for polymerization temperature and molecular weight that maximize the yield of a particular polymer from a chemical reactor. In a slightly different context, we may wish to find the settings that minimize the costs associated with operating this reactor. In either case, optimize does coincide with the notion of maximize or minimize.

VOICE OF EXPERIENCE

In many engineering processes, the variation is just as important as the mean.

Engineers are beginning to realize that to optimize a process does not always mean to maximize or to minimize. Consider the manufacture of a ballpoint pen. The fit between the cap and the barrel is a primary quality characteristic. If the fit is too tight, the customer cannot remove the cap. If the fit is too loose, the cap falls off. The caps and the barrels are made by separate injection molding processes. In this case, do we really want to make the outside diameter of these pen barrels as large or as small as possible? Rather, we seek to find the conditions for the injection molding process that produce the outside diameters as close to a stated nominal as possible with minimum variability. Ideally, we

Figure 8.15 | The Japanese "Loss Function"

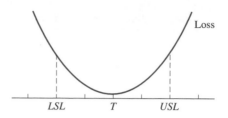

would like to find process conditions that produce the barrel to the target diameter with no variability whatsoever. In many engineering situations like this one, we must consider both the mean and the variance of the characteristic of interest.

Figure 8.15 illustrates the Japanese philosophy towards this problem. The Japanese view any part that does not achieve the target value as having some tangible loss of value, which they quantify by a quadratic "loss function." In this figure, *LSL* is the lower specification limit, *USL* is the upper specification limit, and *T* is the target value for this particular characteristic of interest. Historically in Western manufacturing, any part that falls within the specification limits is considered to have full value. The Japanese believe that a part may be within specifications and still be considered "poor," however, just not quite poor enough to be rejected. The quadratic loss function provides some basis for quantifying the loss from just being within specifications. There are two results of such a philosophy:

1. We must seek conditions that minimize "loss."
2. We must consider both the mean and the variance of the characteristic of interest.

Taguchi Robust Parameter Design

The Japanese engineer Genichi Taguchi proposed a methodology for minimizing the loss function. In this text, we concentrate on his concept of robust parameter design. Fundamental to this approach are the concepts of control and noise factors.

Control and Noise Factors   *Control factors* are those that the experimenter can readily control not only for the purposes of an experiment but also in the process. We denote the design variables associated with the control factors by *x*'s. *Noise factors* are those that the experimenter either cannot or will not control directly *in the process* although he or she may be able to control them for the purposes of an experiment. We denote the design variables associated with the noise factors by z's. Although we may be able to fix the levels of the noise factors in our experiment, they are truly random variables in the process, which provides some complexity to any formal analysis.

Example 8.5	**Cake Mix Experiment**

Box and Jones (1992) discuss an experiment involving a cake mix. The control factors are the amounts of flour, shortening, sugar, and egg powder because these four factors actually make up the cake mix itself. The manufacturer has direct control over these factors, both in the experiment and in the actual process. The noise factors are the consumer's oven temperature and the cooking time. The manufacturer routinely tests its products under extremely well-controlled conditions. It performs constant maintenance on its ovens and routinely checks the temperatures. In addition, it uses precise timers on each oven. As a result, the manufacturer can accurately control the oven temperature and the cooking time for the purposes of an experiment. However, in the process, the consumer controls the oven temperature and the cooking time. As a result, these two factors truly are random variables *in the process*.

The Crossed Array   The basic goal of parameter design is to find the settings for the control factors that are most robust to the noise factors. To achieve this end, Taguchi crosses two arrays:

1. A design for the control factors (the *inner* or *control array*)
2. A design for the noise factors (the *outer* or *noise array*)

By *cross*, we mean that each point of the inner array is replicated according to a design in the noise factors called the outer array. As a result, parameter designs often require a large total number of experimental runs. Typically, the inner and outer arrays are Resolution III in order to reduce the overall size of the experiment. Often, the inner and outer arrays do not allow the estimation of control-by-control or noise-by-noise interactions. However, the crossed array does allow us to estimate all of the interactions between the control and noise factors. Later, we shall see that the control-by-noise interactions provide the key for making a process robust to the noise factors. By robust we mean the process is as insensitive as possible to changes in the noise factors.

An important question is: Why run the experiment in the noise factors? We seek to find the settings in the control factors that are most robust to the noise factors. If we choose the high and low levels for the noise factors ± 1 properly, then they correspond to maximal noise. By running the factorial experiment in the noise factors, we are intentionally making the process as noisy as possible for each setting of the control factors. If a setting achieves our target objective for the mean of the response with acceptable variation when faced with maximum noise, we should expect it to behave well under normal operating conditions.

Table 8.18 | The Control Array for the Cake Mix Experiment

x_1	x_2	x_3	x_4
−1	−1	−1	−1
1	−1	−1	1
−1	1	−1	1
1	1	−1	−1
−1	−1	1	1
1	−1	1	−1
−1	1	1	−1
1	1	1	1

Example 8.6 | **Cake Mix Experiment—Revisited**

In this experiment, the amounts of flour (x_1), shortening (x_2), sugar (x_3), and egg powder (x_4) are the control factors. The oven temperature (z_1) and baking time (z_2) are the noise factors. Table 8.18 gives an appropriate inner or control array, which is the 2_{IV}^{4-1} fractional factorial. The following table is an appropriate outer or noise array, which is a 2^2 full factorial:

z_1	z_2
−1	−1
1	−1
−1	1
1	1

Table 8.19 is the crossed array, where we replicate each setting of the control array by the noise array. In this case, the control array consists of eight runs, and the noise array consists of four. The crossed array consists of $8 \cdot 4$ or 32 runs.

Orthogonal Arrays and Plackett–Burman Designs   The Taguchi school proposes the use of "orthogonal arrays" for their designs. In general, the designs have these properties:

- They allow the estimation of "main effects."
- They *do not allow* the estimation of any interactions.

Two examples of these orthogonal arrays are

1. Resolution III fractional factorial designs
2. Plackett–Burman (1946) designs, both two- and three-level

Table 8.19	The Full Crossed Array for the Cake Mix Experiment

x_1	x_2	x_3	x_4	z_1	z_2
−1	−1	−1	−1	−1	−1
−1	−1	−1	−1	1	−1
−1	−1	−1	−1	−1	1
−1	−1	−1	−1	1	1
1	−1	−1	1	−1	−1
1	−1	−1	1	1	−1
1	−1	−1	1	−1	1
1	−1	−1	1	1	1
−1	1	−1	1	−1	−1
−1	1	−1	1	1	−1
−1	1	−1	1	−1	1
−1	1	−1	1	1	1
1	1	−1	−1	−1	−1
1	1	−1	−1	1	−1
1	1	−1	−1	−1	1
1	1	−1	−1	1	1
−1	−1	1	1	−1	−1
−1	−1	1	1	1	−1
−1	−1	1	1	−1	1
−1	−1	1	1	1	1
1	−1	1	−1	−1	−1
1	−1	1	−1	1	−1
1	−1	1	−1	−1	1
1	−1	1	−1	1	1
−1	1	1	−1	−1	−1
−1	1	1	−1	1	−1
−1	1	1	−1	−1	1
−1	1	1	−1	1	1
1	1	1	1	−1	−1
1	1	1	1	1	−1
1	1	1	1	−1	1
1	1	1	1	1	1

The Plackett–Burman designs that use three levels for each factor allow the estimation of the linear and pure quadratic terms but do not allow estimation of the two-factor or higher interactions. We construct Plackett–Burman designs by cyclically permuting a "baseline."

The two-level Plackett–Burman designs use a multiple of four distinct design runs. An n-point Plackett–Burman design can support the estimation of up to $n - 1$ main effects. When n equals a power of 2, the Placket–Burman design is equivalent to a fractional factorial. The baselines for some common two-level Plackett–Burman designs follow:

12-point:	1	1	−1	1		1	1	−1	−1	−1	1	−1
20-point:	1	1	−1	−1		1	1	1	1	−1	1	
	−1	1	−1	−1		−1	−1	1	1	−1		
24-point:	1	1	1	1	1	−1	1	−1	1	1	−1	−1
	1	1	−1	−1		1	−1	1	−1	−1	−1	−1

We can generate these designs in either of two ways. We can use the baseline as the first column of our design matrix. To construct the second column, we move the bottom element of the first column to the top position. We then move each element of the first column down one position. Once we have generated the required number of columns, we add a row of −1's.

To construct a 12-run Plackett–Burman design for five factors, each at two levels, we start with the baseline, given in Table 8.20. The first two columns

---

Table 8.20 | The Baseline for the 12-Run Plackett–Burman Design

$$
\begin{array}{c}
1 \\
1 \\
-1 \\
1 \\
1 \\
1 \\
-1 \\
-1 \\
-1 \\
1 \\
-1
\end{array}
$$

Table 8.21	The First Two Columns in Generating the 12-Run Plackett–Burman Design

1	−1
1	1
−1	1
1	−1
1	1
1	1
−1	1
−1	−1
−1	−1
1	−1
−1	1

follow as in Table 8.21. Then Table 8.22 shows the first five columns. Finally, we add the row of −1's. The full design is shown in Table 8.23.

Another way to construct the Plackett–Burman design is to use the baseline to generate the first row. We generate the other rows by cyclically permuting the first row. Once again, the last row consists of all −1's. A k-factor design consists of any k columns from this matrix.

Table 8.22	The First Five Columns in Generating the 12-Run Plackett–Burman Design

1	−1	1	−1	−1
1	1	−1	1	−1
−1	1	1	−1	1
1	−1	1	1	−1
1	1	−1	1	1
1	1	1	−1	1
−1	1	1	1	−1
−1	−1	1	1	1
−1	−1	−1	1	1
1	−1	−1	−1	1
−1	1	−1	−1	−1

| | Table 8.23 | The Entire 12-Run Plackett–Burman Design for Five Factors |

x_1	x_2	x_3	x_4	x_5
1	−1	1	−1	−1
1	1	−1	1	−1
−1	1	1	−1	1
1	−1	1	1	−1
1	1	−1	1	1
1	1	1	−1	1
−1	1	1	1	−1
−1	−1	1	1	1
−1	−1	−1	1	1
1	−1	−1	−1	1
−1	1	−1	−1	−1
−1	−1	−1	−1	−1

We construct the three-level Plackett–Burman design in a similar manner. Two common three-level designs involve 9 and 27 points. Their baselines follow:

9-point:	−1	0	−1	−1	−1	−1	0	0	
27-point:	−1	−1	0	−1	0	1	0	0	1
	−1	0	0	0	−1	−1	1	−1	1
	0	1	1	0	−1	1	1	1	

If n is the total number of points in the three-level design, then this design can accommodate up to $(n - 1)/2$ factors.

Contributions and Drawbacks   The greatest contributions of this total approach are

- It seriously considers the variance over a region of interest.
- It provides a rationale for modeling the behavior of the noise in terms of the control factors.

The traditional RSM often buries the impact of the noise factors in ϵ_i, the random error term. Even worse, RSM traditionally assumes that the variability is constant over the entire region of interest. On the other hand, Taguchi hopes that the variance is not constant over the region of interest so that he can find suitable interactions between the control factors and the noise that minimize the variability.

The basic drawbacks to the Taguchi approach are

- It uses an unnecessarily limited number of designs that do not adequately deal with interactions.
- There are better, simpler, and more efficient analyses.
- The designs often are much larger than required because they completely cross the noise and the control factors.
- It does not go far enough to model the variance.

A proper application of RSM can overcome most of these drawbacks. Since the traditional analysis of Taguchi parameter designs has many flaws, we have elected not to summarize it in this book.

Using RSM to Find Conditions Insensitive to the Noise

Many statisticians and engineers propose the use of the "combined array," which is based on these ideas:

- Propose a single model in both the control and noise factors.
- Run a design specifically for the model proposed.

In the process,

- We can estimate some of the control-by-control interactions.
- We can allocate our experimental resources more efficiently.
- In some cases, the resulting designs are significantly smaller. (Usually, if we use a fractional factorial, the combined array is about the same size as the corresponding crossed array.)

With the combined array, we can use the basic techniques outlined in this chapter to determine operating conditions that are robust to the noise factors. The key task is to find suitable interactions between the control and noise factors. In particular, we hope to find a level for at least one control factor for which the response changes as little as possible no matter the level of at least one noise factor. The next example illustrates this point.

Example 8.7	**Leaf Springs Experiment**

Pignatiello and Ramberg (1985) studied a heat treatment process for leaf springs used in trucks. The engineers assigned to this process need to achieve a target height of 8.0 in. with minimal variability. Table 8.24 lists the four control factors and their levels. The engineers think that the primary source of variability in the process is the oil quench temperature, which they used as the single noise factor, E. This temperature ranged from a target value of 140 to a target value of 160. They ran a 2_{IV}^{4-1} fractional factorial design in the control factors. They ran each treatment combination of the 2_{IV}^{4-1} design at both the low level and the high level of the

Table 8.24	The Factors and Their Levels for the Leaf Springs Experiment

Factor	Low	High
A—Peak Heat Temp.	1840	1880
B—Heating Time	25	23
C—Transfer Time	12	10
D—Hold Time	2	3

oil quench temperature. The entire design was replicated a total of three times. Table 8.25 summarizes the design and the results. With this design, we can estimate the AB, AC, and BC interactions among the control factors and the AE, BE, CE, DE, ABE, ACE, and BCE control-by-noise interactions.

Table 8.26 gives the regression analysis from the SAS statistical software package. The residual analysis, which is not given, indicates no problems. We see that all the main effects except C, the transfer time, are important. None of the control-by-control interactions appears to be significant. The AE, heating temperature-by-oil quench temperature, and the BE, heating time-by-oil quench temperature, interactions appear to be significant. Figure 8.16 gives the AE interaction plot. The lines look almost parallel. As a result, we see about the same effect due to the changes in the oil quench temperature for both levels of the heating temperature. We cannot minimize the impact of variation in the oil quench temperature by choosing one level of heating temperature over the other. Figure 8.17 gives the BE interaction plot. In this case, the lines are definitely not parallel, which indicates that the effect of the oil quench temperature does depend on the level chosen for the heating time. This plot suggests that using the +1 level for the heating time

Table 8.25	The Leaf Springs Experiment

A	B	C	D	E = −1			E = −1		
−1	−1	−1	−1	7.78	7.78	7.81	7.50	7.25	7.12
1	−1	−1	1	8.15	8.18	7.88	7.88	7.88	7.44
−1	1	−1	1	7.50	7.56	7.50	7.50	7.56	7.50
1	1	−1	−1	7.59	7.56	7.75	7.63	7.75	7.56
−1	−1	1	1	7.94	8.00	7.88	7.32	7.44	7.44
1	−1	1	−1	7.69	8.09	8.06	7.56	7.69	7.62
−1	1	1	−1	7.56	7.62	7.44	7.18	7.18	7.25
1	1	1	1	7.56	7.81	7.69	7.81	7.50	7.59

| Table 8.26 | The ANOVA Table for the Leaf Springs Experiment |

Analysis of Variance

		Sum of	Mean		
Source	DF	Squares	Square	F Value	Prob>F
Model	15	2.43621	0.16241	9.811	0.0001
Error	32	0.52973	0.01655		
C Total	47	2.96595			

Root MSE	0.12866	R-square	0.8214	
Dep Mean	7.63604	Adj R-sq	0.7377	
C.V.	1.68494			

Parameter Estimates

Variable	DF	Parameter Estimate	Standard Error	T for H0: Parameter=0	Prob > |T|
INTERCEP	1	7.636042	0.01857090	411.183	0.0001
A	1	0.110625	0.01857090	5.957	0.0001
B	1	-0.088125	0.01857090	-4.745	0.0001
C	1	-0.014375	0.01857090	-0.774	0.4446
D	1	0.051875	0.01857090	2.793	0.0087
E	1	-0.129792	0.01857090	-6.989	0.0001
AB	1	-0.008542	0.01857090	-0.460	0.6487
AC	1	-0.009792	0.01857090	-0.527	0.6017
BC	1	-0.017708	0.01857090	-0.954	0.3475
AE	1	0.042292	0.01857090	2.277	0.0296
BE	1	0.082708	0.01857090	4.454	0.0001
CE	1	-0.026875	0.01857090	-1.447	0.1576
DE	1	0.013542	0.01857090	0.729	0.4712
ABE	1	-0.005208	0.01857090	-0.280	0.7809
ACE	1	0.020208	0.01857090	1.088	0.2846
BCE	1	-0.023542	0.01857090	-1.268	0.2141

| Figure 8.16 | The Peak Heat Temperature and Oil Quench Temperature Interaction Plot |

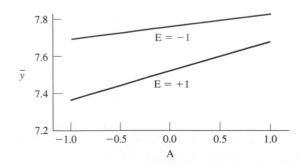

Figure 8.17 | The Heating Time and Oil Quench Temperature Interaction Plot

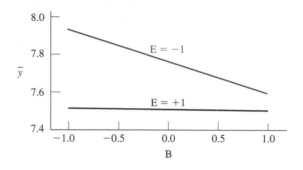

---

will minimize the effect of the oil quench temperature. Since we do not plan to control the oil quench temperature in the process, choosing the high level for the heating time will help to reduce the variability in the leaf spring heights. We can adjust the average height of these leaf springs by properly choosing the levels for the other factors.

A Dual-Response Approach

Another approach recognizes that the process mean and the process variance form a dual-response problem. As a result, we can apply the multiple-response analyses discussed in Section 8.3 to find settings that achieve a target condition for the mean and minimize the process variance. In this situation, the variance, or a suitable transformation, is the key response.

VOICE OF EXPERIENCE

A natural dual-response system is the response and the variability.

Suppose that an n-point design has been replicated such that each design point has been run a total of $m \geq 2$ times. Any reasonable method may be applied to generate the replication, including the use of an outer array. Let s_i^2 be the estimated variance at the ith design point, and let $t(s_i^2)$ be a suitable transformation of the variance. In this approach, we estimate separate models for y and $t(s_i^2)$.

Much work has been and currently is being done on modeling variance. Most authors suggest using the natural logarithm of the variance, $\log(s_i^2)$, but the theory for this transformation requires moderate to large amounts of replication ($m \geq 10$) to justify this approach. A reasonable alternative uses the standard deviation, which is the square root transformation.

Example 8.8 | **The Printing Study Experiment**

Vining and Myers (1990) analyze an experiment that originally appeared in Box and Draper (1988). This experiment studied the effect of speed (x_1), pressure (x_2),

Table 8.27	The Printing Ink Experiment

i	x_1	x_2	x_3	y_{i1}	y_{i2}	y_{i3}	$\bar{y}_{i\cdot}$	s_i
1	−1	−1	−1	34	10	28	24.0	12.5
2	0	−1	−1	115	116	130	120.3	8.4
3	1	−1	−1	192	186	263	213.7	42.8
4	−1	0	−1	82	88	88	86.0	3.7
5	0	0	−1	44	178	188	136.7	80.4
6	1	0	−1	322	350	350	340.7	16.2
7	−1	1	−1	141	110	86	112.3	27.6
8	0	1	−1	259	251	259	256.3	4.6
9	1	1	−1	290	280	245	271.7	23.6
10	−1	−1	0	81	81	81	81.0	0.0
11	0	−1	0	90	122	93	101.7	17.7
12	1	−1	0	319	376	376	357.0	32.9
13	−1	0	0	180	180	154	171.3	15.0
14	0	0	0	372	372	372	372.0	0.0
15	1	0	0	541	568	396	501.7	92.5
16	−1	1	0	288	192	312	264.0	63.5
17	0	1	0	432	336	513	427.0	88.6
18	1	1	0	713	725	754	730.7	21.1
19	−1	−1	1	364	99	199	220.7	133.8
20	0	−1	1	232	221	266	239.7	23.5
21	1	−1	1	408	415	443	422.0	18.5
22	−1	0	1	182	233	182	199.0	29.4
23	0	0	1	507	515	434	485.3	44.6
24	1	0	1	846	535	640	673.7	158.2
25	−1	1	1	236	126	168	176.7	55.5
26	0	1	1	660	440	403	501.0	138.9
27	1	1	1	878	991	1161	1010.0	142.5

and distance (x_3) on a printing machine's ability to apply coloring inks on package labels. Table 8.27 summarizes the experiment, which was a 3^3 complete factorial with three runs at each design point ($m = 3$).

Vining and Myers sought the conditions that minimize the variability while achieving a target value of 500.0 for the mean response. The fitted response surface for the response itself was

$$\hat{\mu} = 314.7 + 177x_1 + 109.4x_2 + 131.5x_3$$
$$+ 66x_1x_2 + 75.5x_1x_3 + 43.6x_2x_3 + 82.8x_1x_2x_3.$$

The fitted response surface for the standard deviation was

$$\hat{\sigma} = 48.0 + 29.2x_3.$$

Consider using the Derringer–Such desirability approach to minimize $\hat{\sigma}$ over the cube defined by the 3^3 design subject to these constraints:

- $\hat{\mu} = 500.0$ (acceptable range: 490–510).
- A maximum acceptable value for $\hat{\sigma}$ is 60.

The resulting best settings are

$$x_1 = 1.00 \qquad x_2 = 1.00 \qquad x_3 = -0.5,$$

which yield an estimated standard deviation of 33.4 and a desirability of 0.6662.

Consider using the SOLVER tool in EXCEL to minimize $\hat{\sigma}$ over the cube defined by the 3^3 design subject to the constraint

$$\hat{\mu} = 500.0.$$

The resulting best settings again are

$$x_1 = 1.00 \qquad x_2 = 1.00 \qquad x_3 = -0.50.$$

Replicating a full second-order design can be expensive. Vining and Schaub (1996) present an interesting alternative that recognizes that models for the variance tend to be of lower order than models for the mean. This approach replicates only a Resolution III first-order portion of a second-order design. Vining and Schaub note that the replicated 3^3, which uses 81 runs, contains as a subset a replicated central composite design. Table 8.28 gives the replicated factorial design used by Vining and Schaub. We use all of these data to estimate the mean model. To estimate the variance model, we compute the sample standard deviation for each of the four replicated points, as shown in this table:

x_1	x_2	x_3	s
1	−1	−1	42.8
−1	1	−1	27.6
−1	−1	1	133.8
1	1	1	142.5

Table 8.28	The Replicated Factorial Design from the Printing Ink Experiment

x_1	x_2	x_3	y
−1	−1	−1	34
1	−1	−1	192
1	−1	−1	186
1	−1	−1	263
−1	1	−1	141
−1	1	−1	110
−1	1	−1	86
1	1	−1	280
−1	−1	1	364
−1	−1	1	99
−1	−1	1	199
1	−1	1	408
−1	1	1	168
1	1	1	878
1	1	1	991
1	1	1	1161
−1	0	0	180
1	0	0	396
0	−1	0	122
0	1	0	513
0	0	−1	188
0	0	1	515

The fitted response surface for the response itself was

$$\hat{\mu} = 310.2 + 163x_1 + 102x_2 + 148x_3$$
$$+ 79.1x_1x_2 + 77.5x_1x_3 + 55.2x_2x_3 + 81.1x_1x_2x_3.$$

The fitted response surface for the standard deviation was

$$\hat{\sigma} = 86.7 + 6.0x_1 - 1.6x_2 + 51.5x_3.$$

The resulting best settings from SOLVER in EXCEL are

$$x_1 = 1.00 \qquad x_2 = 1.00 \qquad x_3 = -0.43.$$

| Table 8.29 | The Factors and Their Levels for the Plasma Etcher Experiment |

Level	A Gap (cm)	B Pressure (m Torr)	C C$_2$F$_6$ flow (sccm)	D Power (W)
Low (−1)	0.80	450	125	275
High (1)	1.20	550	200	325

⫶ Exercises

8.27 Yin and Jillie (1987) studied a nitride etch process on a single-wafer plasma etcher. The response of interest was the etch rate. These were the experimental factors:

A   Gap, the spacing between the anode and the cathode
B   Pressure in the reactor chamber
C   Flow rate of the reactant gas (C$_2$F$_6$)
D   Power applied to the cathode

| Table 8.30 | The Plasma Etcher Experiment |

Run	A Gap	B Pressure	C C$_2$F$_6$	D Power	Etch rate Å/min
1	−1	−1	−1	−1	550
2	1	−1	−1	−1	669
3	−1	1	−1	−1	604
4	1	1	−1	−1	650
5	−1	−1	1	−1	633
6	1	−1	1	−1	642
7	−1	1	1	−1	601
8	1	1	1	−1	635
9	−1	−1	−1	1	1037
10	1	−1	−1	1	749
11	−1	1	−1	1	1052
12	1	1	−1	1	868
13	−1	−1	1	1	1075
14	1	−1	1	1	860
15	−1	1	1	1	1063
16	1	1	1	1	729

	Table 8.31	The Cake Mix Experiment

x_1	x_2	x_3	z_1	z_2	y
−1	−1	−1	−1	−1	1.3
−1	−1	−1	1	1	3.1
1	−1	−1	1	−1	5.5
1	−1	−1	−1	1	3.2
−1	1	−1	1	−1	1.2
−1	1	−1	−1	1	1.5
1	1	−1	−1	−1	3.7
1	1	−1	1	1	4.2
−1	−1	1	1	−1	3.5
−1	−1	1	−1	1	2.3
1	−1	1	−1	−1	4.1
1	−1	1	1	1	6.3
−1	1	1	−1	−1	1.9
−1	1	1	1	1	2.2
1	1	1	1	−1	5.8
1	1	1	−1	1	5.5

For the purposes of this discussion, factors A (gap) and B (pressure) are the control factors. Factors C (flow rate) and D (power) are the noise factors. The factor levels used were as listed in Table 8.29. Table 8.30 summarizes the design and observed results. Perform an appropriate combined array analysis of this experiment.

8.28 Box and Jones (1992) outline a cake mix experiment in which the control factors are amount of flour (x_1), amount of shortening (x_2), and amount of egg powder (x_3). The noise factors are baking temperature (z_1) and baking time (z_2). Table 8.31 summarizes the design and its results. Perform an appropriate combined array analysis.

8.29 Schubert and colleagues (1992) conducted an experiment with a catapult to determine the effects of hook (x_1), arm length (x_2), start angle (x_3), and stop angle (x_4) on the distance that the catapult throws a ball. They threw the ball three times for each setting of the factors. Table 8.32 lists the experimental results. Find the settings for these factors that will throw the ball 100 in. with minimum variability.

8.30 An industrial experiment was performed to study the influence of several controllable factors on the mean value and the variation in the percentage of shrinkage of products made by injection molding. The controllable factors were cycle time, mold temperature, cavity thickness, holding pressure, injection speed, holding

Table 8.32	The Catapult Experiment

x_1	x_2	x_3	x_4	y		
−1	−1	−1	−1	28.0	27.1	26.2
−1	−1	1	1	46.3	43.5	46.5
−1	1	−1	1	21.9	21.0	20.1
−1	1	1	−1	52.9	53.7	52.0
1	−1	−1	1	75.0	73.1	74.3
1	−1	1	−1	127.7	126.9	128.7
1	1	−1	−1	86.2	86.5	87.0
1	1	1	1	195.0	195.9	195.7

time, and gate size. The noise factors were percentage regrind, moisture content, and ambient temperature. The data from Engel and Huele (1996) are given in Table 8.33. Perform the appropriate combined array analysis of this experiment.

8.5 Case Study

In the Chapter 7 case study, we studied an injection molding process for pen barrels. Faber-Castell had a new injection molding process that had problems with warped barrels. Faber-Castell ran an initial experiment to screen the important factors. We learned that the warping appeared to be due to the injection pressure and the temperature of the chilled water used in the mold. We also discovered significant curvature in the experimental region. This presence of curvature suggests augmenting the original 2^4 plus center runs to a central composite design. Table 8.34 gives the full second-order design using the rotatable value for α. The augmented runs are in block 2.

The analysis in Table 8.35 suggests the following operating conditions: $x_3 = 0.6527$ and $x_4 = -1.414$. Faber-Castell ran a series of confirmatory runs at these conditions, which confirmed that these conditions were satisfactory.

⊳ 8.6 Ideas for Projects

1. As a class, perform the "catapult" experiment. Many instructors use a rubber band catapult to generate experimental data. The ones we use allow you to change the tension on the rubber band in several ways. Perform the experiment to generate a prediction equation for how far the catapult

| Table 8.33 | | The Percent Shrinkage Experiment |

Cycle	Mold	Cavity	Pressure	Speed	Time	Gate	Regrind	Moisture	Temp.	Shrinkage
−1	−1	−1	−1	−1	−1	−1	−1	−1	−1	2.2
−1	−1	−1	−1	−1	−1	−1	−1	1	1	2.1
−1	−1	−1	−1	−1	−1	−1	1	−1	1	2.3
−1	−1	−1	−1	−1	−1	−1	1	1	−1	2.3
−1	−1	−1	1	1	1	1	−1	−1	−1	0.3
−1	−1	−1	1	1	1	1	−1	1	1	2.5
−1	−1	−1	1	1	1	1	1	−1	1	2.7
−1	−1	−1	1	1	1	1	1	1	−1	0.3
−1	1	1	−1	−1	1	1	−1	−1	−1	0.5
−1	1	1	−1	−1	1	1	−1	1	1	3.1
−1	1	1	−1	−1	1	1	1	−1	1	0.4
−1	1	1	−1	−1	1	1	1	1	−1	2.8
−1	1	1	1	1	−1	−1	−1	−1	−1	1.0
−1	1	1	1	1	−1	−1	−1	1	1	1.9
−1	1	1	1	1	−1	−1	1	−1	1	1.8
−1	1	1	1	1	−1	−1	1	1	−1	1.0
1	−1	1	−1	1	−1	1	−1	−1	−1	3.0
1	−1	1	−1	1	−1	1	−1	1	1	3.1
1	−1	1	−1	1	−1	1	1	−1	1	3.0
1	−1	1	−1	1	−1	1	1	1	−1	3.0
1	−1	1	1	−1	1	−1	−1	−1	−1	2.1
1	−1	1	1	−1	1	−1	−1	1	1	4.2
1	−1	1	1	−1	1	−1	1	−1	1	−1.0
1	−1	1	1	−1	1	−1	1	1	−1	3.1
1	1	−1	−1	1	1	−1	−1	−1	−1	4.0
1	1	−1	−1	1	1	−1	−1	1	1	1.9
1	1	−1	−1	1	1	−1	1	−1	1	4.6
1	1	−1	−1	1	1	−1	1	1	−1	2.2
1	1	−1	1	−1	−1	1	−1	−1	−1	1.0
1	1	−1	1	−1	−1	1	−1	1	1	1.9
1	1	−1	1	−1	−1	1	1	−1	1	1.9
1	1	−1	1	−1	−1	1	1	1	−1	1.8

Table 8.34 | The Follow-up Pen Barrel Warping Experiment

X_3	X_4	Warp	Block
0.000	0.000	9.5408	1
1.000	−1.000	2.0647	1
1.000	−1.000	2.4075	1
−1.000	1.000	33.4354	1
0.000	0.000	9.9129	1
−1.000	1.000	33.7727	1
1.000	1.000	14.8448	1
−1.000	−1.000	13.5184	1
−1.000	−1.000	14.3446	1
1.000	−1.000	2.7584	1
−1.000	1.000	33.2454	1
−1.000	1.000	33.2368	1
1.000	1.000	13.2066	1
1.000	−1.000	1.2758	1
1.000	1.000	13.6236	1
−1.000	−1.000	13.9000	1
−1.000	−1.000	13.9417	1
1.000	1.000	14.5301	1
0.000	0.000	10.2025	1
0.000	0.000	10.0495	2
−1.414	0.000	29.5770	2
0.000	0.000	9.7864	2
0.000	1.414	25.5054	2
0.000	−1.414	2.4379	2
1.414	0.000	7.1114	2
0.000	0.000	9.1131	2

throws the ball. Use the prediction equation to try to hit a target a known distance away.

2. Perform the "paper helicopter" experiment. Many instructors make paper helicopters and drop them in class. We prefer to drop them from the fourth floor of a building on campus. The goal is to design a paper helicopter that flies as long as possible.

Table 8.35	Analysis for the Follow-up Experiment

Term	Coef	SE Coef	T	P
Constant	9.76755	0.2055	47.539	0.000
Block	-0.05707	0.1161	-0.492	0.629
C	-7.82296	0.1125	-69.512	0.000
D	7.91557	0.1125	70.335	0.000
C*C	4.19545	0.1754	23.915	0.000
D*D	2.00852	0.1754	11.449	0.000
C*D	-1.89293	0.1258	-15.045	0.000

3. Download the distillation column simulator from the web page (http://stat.vt.edu/~vining). This program realistically simulates the performance of a distillation column. A typical assignment is to have students optimize the yield and the profit. For most students, this project brings together most of the concepts of experimental design and process optimization.

4. Work with an instructor for a unit operations laboratory. Help a class perform a series of experiments to optimize the operation of a piece of equipment.

5. Perform a factorial experiment of your own choosing. It is a good idea to require such a project as part of this course. Most students bake cookies or pop popcorn. Some are very creative. If you are over 21, you may want to consider brewing beer, which is a complex biochemical process.

⫶ References

1. Box, G. E. P., and Draper, N. R. (1988). *Empirical model building and response surfaces*. New York: John Wiley.
2. Box, G. E. P., Hunter, W. G., and Hunter, J. S. (1978). *Statistics for experimenters*. New York: John Wiley.
3. Box, G. E. P., and Jones, S. (1992). Designing products that are robust to the environment. *Total Quality Management, 3*, 265–282.
4. Box, G. E. P., and Wilson, K. B. (1951). On the experimental attainment of optimum conditions. *Journal of the Royal Statistical Society, Series B, 13*, 1–45.
5. Broderick, G., Lanouette, R., and Valade, J. L. (1997). Optimizing refiner operation with statistical modeling. *The Canadian Journal of Chemical Engineering, 75*, 79–87.
6. Buckner, J., Cammenga, D. J., and Weber, A. (1993). Elimination of TiN peeling during exposure to CVD tungsten deposition process using designed experiments. *Statistics in the Semiconductor Industry*. Austin, Texas: SEMATECH. Technology Transfer No. 92051125A-GEN, Vol. I, 3-45–3-71.

7. Buxton, J. R. (1991). Some comments on the use of response variable transformations in empirical modeling. *Applied Statistics, 40,* 391–400.

8. Chang, S. I., and Shivpuri, R. (1994). A multiple-objective decision-making approach for assessing simultaneous improvement in die life and casting quality in a die-casting process. *Quality Engineering, 7,* 371–383.

9. Chapman, R. E. (1995). Photochemistry multiple response co-optimization. *Quality Engineering, 8,* 31–45.

10. Chau, A. K-M. (1992). A comparison of methods of analysis for a quality improvement study. *MSc Thesis,* University of Southhampton.

11. Coteron, A., Sanchez, N., Martinez, M., and Aracil, J. (1993). Optimisation of synthesis of an analogue of jojoba oil using a fully central composite design. *Canadian Journal of Chemical Engineering, 71,* 485–488.

12. Derringer, G., and Suich, R. (1980). Simultaneous optimization of several response variables. *Journal of Quality Technology, 12,* 214–219.

13. Engel, J. and Huele, A. F. (1996). A generalized linear modeling approach to robust design. *Technometrics, 38,* 365–373.

14. Jauregi, P., Gilmour, S., and Varley, J. (1997). Characterization of colloidal gas aphrons for subsequent use for protein recovery. *The Chemical Engineering Journal, 65,* 1–11.

15. Kilgo, M. B. (1988). An application of fractional factorial experimental designs. *Quality Engineering, 1,* 19–23.

16. Montgomery, D.C. (2001). *Response surface methodology: Process and product optimization using designed experiments.* 4th ed. New York: John Wiley.

17. Myers, R. H., and Montgomery, D. C. (2002). *Response surface methodology: Process and product optimization using designed experiments.* New York: John Wiley.

18. Pignatiello, J. J., Jr., and Ramberg, J. S. (1985). Discussion of "Off-line quality control, parameter design, and the Taguchi method" by R. N. Kackar. *Journal of Quality Technology, 17,* 198–206.

19. Plackett, R. L., and Burman, J. P. (1946). The design of optimum multifactorial experiments. *Biometrika, 33,* 305–325.

20. Schubert, K., Kerber, M. W., Schmidt, S. R., and Jones, S. E. (1992). The catapult problem: Enhanced engineering modeling using experimental design. *Quality Engineering, 4,* 463–473.

21. Vining, G. G., and Myers, R. H. (1990). Combining Taguchi and response surface philosophies: A dual response approach. *Journal of Quality Technology, 22,* 15–22.

22. Vining, G. G., and Schaub, D. (1996). Experimental designs for estimating both mean and variance functions. *Journal of Quality Technology, 28,* 135–147.

23. Yin, G. Z., and Jillie, D. W. (1987). Orthogonal design for process optimization and its application in plasma etching. *Solid State Technology, 30*(5), 127–132.

9

Coda

⟫ 9.1 The Themes of This Course

This course introduced a varied arsenal of important tools for approaching engineering problems. Of course, tools alone solve nothing. They prove useful only when engineers appreciate how and when they should apply the tools. In a single semester, we cannot discuss in adequate detail all of the statistical methodologies important to engineering. At best, we can develop a sound foundation and then build from this foundation to provide a brief overview of some of the more important statistical tools and to show their relevance to modern engineering practice.

Laying Foundations

The first part of this course laid the foundation for the application of statistics to real engineering problems. These basic concepts were emphasized:

- The engineering method
- Data collection
- Graphical analysis of data
- Probabilistic modeling of data
- Formal estimation

Throughout this development, we tried to illustrate the practical application of these methods to real engineering problems.

The engineering method uses abstract models to solve real problems. In the process, statistics should serve as a handmaiden, guiding the interplay between the concrete world of the data and the theoretical world of the model. Statistics and statistical thinking should provide the bridge between these two worlds through the proper understanding of data: how to collect them and how to interpret them. Variability, the true engineering adversary, clouds our ability to make sound engineering decisions. The proper appreciation of statistics provides the basis for piercing through this cloud of uncertainty.

The bridge from theory to the actual problem requires data. Too often engineers do not appreciate the importance of statistical thinking when they collect data. These data must address specific questions. Statistics simply seeks to ensure that the data address these questions as efficiently and effectively as possible. Engineers are learning that they should consult with statisticians to develop effective methods for obtaining these data, whether they plan to pursue a purely observational study or a designed experiment. Statisticians are learning that they should not make the engineers' life more difficult; rather, they should ensure that engineers can collect the data they need within the constraints they face. This dialogue leads to better engineering and statistical practice.

Graphical displays provide a quick, intuitive way to learn from the data. At a minimum, they give us this information:

- Typical values for the data
- The variability in the data
- The nature of the parent distribution for the data
- Any outliers or other interesting features in the data

With this information, we can begin to sketch an appropriate formal analysis of the data. In some cases, displays answer completely the questions of interest. In other cases, they point out potential difficulties in the formal analyses.

Proper modeling of random behavior allows us to see, at least partially, through the cloud of variability. Probability forms the language for this modeling. With this language, we can begin to describe in a meaningful way the behavior of engineering data through distributions. Naturally, the appropriate model for engineering data depends on the specific situation. Nonetheless, a fundamental understanding both of the engineering context and the appropriate statistical consequences provides a powerful basis for describing the variability inherent in engineering problems. Often the statistical theory underlying these approaches may seem subtle, complex, abstract, and removed from the basic reality of the problem, yet this theory provides the very foundation for modeling the phenomena of interest so that the engineer may find an appropriate solution to the problem at hand.

Estimation and testing are the bridge from the abstract engineering model to the actual data. We begin to see the engineering method at its fullest. In testing, we use data to confirm the reasonableness of our presuppositions concerning the problem at hand. We use appropriate models for the random behavior of the data as the basis for these decisions. In estimation, we use these models to provide a basis for constructing a range of plausible values for the parameter of interest. This range in turn provides a basis for evaluating the quality of the estimated value.

Building on the Foundations

In the first portion of this course, we tended to use relatively simplistic scenarios to lay these foundations. In the latter portion, we were able to extend these basic concepts to more complicated and more interesting engineering situations:

- Process monitoring
- Linear modeling
- Analyzing formal experiments

Throughout this development, we showed how to use simple graphical methods to check the assumptions underlying our analyses.

Engineers, particularly in manufacturing, constantly must monitor critical processes, but the variability inherent in these data complicates this activity. Appropriate probabilistic models in conjunction with formal testing procedures provide a basis for seeing through the cloud created by the variability and for detecting unwanted changes in these processes.

The engineering method depends on models to solve problems. In most interesting situations, the response of interest depends on several other engineering characteristics. Linear models are a useful basis for approximating this relationship. Statistics and statistical thinking through regression analysis provide an efficient and effective method for estimating and testing this relationship. Thorough residual analysis allows us to determine where the relationship predicts well and where we may be able to improve it. Even more important, such an analysis can suggest improvements to the model that allow it to predict the data better.

Finally, when we have a firm idea of the nature of the relationship between the independent variables (the factors) and the response, we can use designed experiments to collect the data as efficiently as possible. By knowing which relationships are important to estimate and which are not, we can dramatically reduce the amount of data we must collect. Through a sequential approach to experimentation, we use as few resources as possible in the early phases of experimentation when we merely wish to screen out the less important factors. In the process, we save resources for later experimentation when we wish to develop a model in terms of the important factors that will allow us to optimize the engineering process.

9.2 Integrating the Themes

Engineers are discovering that they must seek continuing education in statistics, just as they must seek continuing education in their specific engineering disciplines, in order to keep current in their fields. Here are some important statistical applications for good engineering practice:

- Both linear and nonlinear modeling
- More experimental design and analyses
- Process monitoring
- Process and product reliability

In this course, we have been able to provide only a brief introduction to these topics. A solid second semester of engineering statistics would explore these areas in more detail.

Modeling engineering data, particularly when the data come from an observational study, often requires more statistical machinery. A second course in engineering statistics should first review the material presented in this course, with particular emphasis on using the computer and on residual analysis. Next, it should outline more formal methods for analyzing residuals. In the process, it should discuss in more detail the concepts of *leverage points* and *influential observations,* which often can dominate the least squares estimation of the model. We should wonder about the quality of our model when just one or two observations appear to control the relationship. When the data come from an observational study, we often require more statistical machinery for *selecting the most appropriate model.* Observational studies can produce analyses where the overall F-test is strongly significant, the R^2 is much greater than .9, and none of the individual coefficients are significant. A second course in engineering statistics should discuss the techniques available for selecting the most appropriate model in such situations. Closely related to the model selection issue is the problem of *multicollinearity,* where relationships exist among the independent variables. The relationships among the independent variables obscure their relationship with the response. Finally, although linear models work well in a variety of settings, many engineering situations require models that are *nonlinear* in the parameters. Engineering theory often suggests an underlying mechanism that involves the solution of a system of differential equations. This system often dictates a model that is inherently nonlinear. A second course should outline the statistical techniques for estimating and analyzing such models.

Many issues remain for the proper design and analysis of engineering experiments. Commonly, engineers run experiments that involve categorical factors with more than two levels, which leads to the concept of *analysis of variance* (ANOVA). Engineers also face constraints on how they run experiments, which leads to the concept of *blocking* introduced in Chapter 2. A second course in engineering statistics should outline the appropriate analysis when we run the design in blocks. In other cases, we face the problem of easy-to-change and hard-to-change factors, which leads into the concept of a *split-plot* experiment. In such an experiment, the experimental unit for the easy-to-change factor is actually an observational unit for the hard-to-change factor. For example, in Chapter 2, we outlined an example involving sewer pipe. Suppose we wish to run an experiment that uses the firing regime for these pipes as one factor and the pipe's formulation as the second factor. The experimental unit for the firing regime is the set of pipes that see a specific setting of the furnace conditions. The experimental unit for the formulation is the individual pipe. We can form the set of pipes that see a specific setting of the firing conditions so that a pipe from each formulation is present. Such an experiment makes a lot of common sense, but it also presents certain complications for the analysis. Finally, a second course should discuss in much more detail how engineers can use *experimental design for quality improvement.*

Engineers monitor processes to detect changes as quickly and as economically as possible. In this course, we introduced the basic Shewhart control chart, which is the foundation for all other monitoring procedures. A second course in engineering statistics should provide a basic overview of the Shewhart charts, including the use of runs rules. It then should go into more detail about such methods as the *cumulative sum* (CUSUM) and the *exponentially weighted moving average* (EWMA) charts, which use past data as well as the current sample to determine whether the process is in control. Another important topic is the *economic design* of control charts, which seeks an appropriate balance among the various costs and risks associated with process monitoring. In many cases, we wish to monitor several characteristics of interest for the same product or process. For example, in a paint process, we wish to monitor simultaneously such characteristics as color, viscosity, and some measure of settling. Monitoring several characteristics at the same time leads to the concept of *multivariate control charts*. Finally, products inevitably must meet certain specifications. *Process capability analysis* provides a basis for predicting how well a process can meet the specifications for its products.

An important aspect of engineering deals with process and product reliability. In this course, we have outlined only a few of the distributions often used in reliability analysis. A second course should spend much more time on *modeling process and product life times*. Estimating these models requires data on the time until failure. Waiting until a product or a process fails often requires more time than we can afford, so we use *accelerated* testing, where the process or product faces more severe conditions (usually temperature) that should cause it to fail sooner. Even under accelerated testing, we may face time constraints that require us to end the test before every item or process fails, which leads to the concept of *censoring*. Finally, many products are actually repairable after failure. Another important concept is a *renewal process*, which formally considers this fact.

⯈ **9.3** Statistics and Engineering

Statistics can never replace sound engineering theory and intuition. At their best, statistical thinking and good statistical practice complement an engineer's basic education and instincts. Statistics provides an important set of tools: nothing more, nothing less. Good engineering practice combines these tools with sound engineering theory. Properly done, statistics can reinforce, complement, and even enhance engineering theory. Properly applied, statistics and statistical thinking can unleash the creative spirit of engineers, helping them to achieve new vistas and goals. Statisticians can never replace engineers, but statistics can help engineers do their jobs better.

Appendix

Table 1	The Cumulative Distribution Function for the Standard Normal Distribution—Negative Values for Z

z	.00	.01	.02	.03	.04	.05	.06	.07	.08	.09
−3.0	.0013	.0013	.0013	.0012	.0012	.0011	.0011	.0011	.0010	.0010
−2.9	.0019	.0018	.0017	.0017	.0016	.0016	.0015	.0015	.0014	.0014
−2.8	.0026	.0025	.0024	.0023	.0023	.0022	.0021	.0021	.0020	.0019
−2.7	.0035	.0034	.0033	.0032	.0031	.0030	.0029	.0028	.0027	.0026
−2.6	.0047	.0045	.0044	.0043	.0041	.0040	.0039	.0038	.0037	.0036
−2.5	.0062	.0060	.0059	.0057	.0055	.0054	.0052	.0051	.0049	.0048
−2.4	.0082	.0080	.0078	.0075	.0073	.0071	.0069	.0068	.0066	.0064
−2.3	.0107	.0104	.0102	.0099	.0096	.0094	.0091	.0089	.0087	.0084
−2.2	.0139	.0136	.0132	.0129	.0125	.0122	.0119	.0116	.0113	.0110
−2.1	.0179	.0174	.0170	.0166	.0162	.0158	.0154	.0150	.0146	.0143
−2.0	.0228	.0222	.0217	.0212	.0207	.0202	.0197	.0192	.0188	.0183
−1.9	.0287	.0281	.0274	.0268	.0262	.0256	.0250	.0244	.0239	.0233
−1.8	.0359	.0351	.0344	.0336	.0329	.0322	.0314	.0307	.0301	.0294
−1.7	.0446	.0436	.0427	.0418	.0409	.0401	.0392	.0384	.0375	.0367
−1.6	.0548	.0537	.0526	.0516	.0505	.0495	.0485	.0475	.0465	.0455
−1.5	.0668	.0655	.0643	.0630	.0618	.0606	.0594	.0582	.0571	.0559
−1.4	.0808	.0793	.0778	.0764	.0749	.0735	.0721	.0708	.0694	.0681
−1.3	.0968	.0951	.0934	.0918	.0901	.0885	.0869	.0853	.0838	.0823
−1.2	.1151	.1131	.1112	.1093	.1075	.1056	.1038	.1020	.1003	.0985
−1.1	.1357	.1335	.1314	.1292	.1271	.1251	.1230	.1210	.1190	.1170
−1.0	.1587	.1562	.1539	.1515	.1492	.1469	.1446	.1423	.1401	.1379
−0.9	.1841	.1814	.1788	.1762	.1736	.1711	.1685	.1660	.1635	.1611
−0.8	.2119	.2090	.2061	.2033	.2005	.1977	.1949	.1922	.1894	.1867
−0.7	.2420	.2389	.2358	.2327	.2297	.2266	.2236	.2206	.2177	.2148
−0.6	.2743	.2709	.2676	.2643	.2611	.2578	.2546	.2514	.2483	.2451
−0.5	.3085	.3050	.3015	.2981	.2946	.2912	.2877	.2843	.2810	.2776

Table 1	The Cumulative Distribution Function for the Standard Normal Distribution—Negative Values for Z (*Continued*)

z	.00	.01	.02	.03	.04	.05	.06	.07	.08	.09
−0.4	.3446	.3409	.3372	.3336	.3300	.3264	.3228	.3192	.3156	.3121
−0.3	.3821	.3783	.3745	.3707	.3669	.3632	.3594	.3557	.3520	.3483
−0.2	.4207	.4168	.4129	.4090	.4052	.4013	.3974	.3936	.3897	.3859
−0.1	.4602	.4562	.4522	.4483	.4443	.4404	.4364	.4325	.4286	.4247
−0.0	.5000	.4960	.4920	.4880	.4840	.4801	.4761	.4721	.4681	.4641
0.0	.5000	.5040	.5080	.5120	.5160	.5199	.5239	.5279	.5319	.5359
0.1	.5398	.5438	.5478	.5517	.5557	.5596	.5636	.5675	.5714	.5753
0.2	.5793	.5832	.5871	.5910	.5948	.5987	.6026	.6064	.6103	.6141
0.3	.6179	.6217	.6255	.6293	.6331	.6368	.6406	.6443	.6480	.6517
0.4	.6554	.6591	.6628	.6664	.6700	.6736	.6772	.6808	.6844	.6879
0.5	.6915	.6950	.6985	.7019	.7054	.7088	.7123	.7157	.7190	.7224
0.6	.7257	.7291	.7324	.7357	.7389	.7422	.7454	.7486	.7517	.7549
0.7	.7580	.7611	.7642	.7673	.7704	.7734	.7764	.7794	.7823	.7852
0.8	.7881	.7910	.7939	.7967	.7995	.8023	.8051	.8078	.8106	.8133
0.9	.8159	.8186	.8212	.8238	.8264	.8289	.8315	.8340	.8365	.8389
1.0	.8413	.8438	.8461	.8485	.8508	.8531	.8554	.8577	.8599	.8621
1.1	.8643	.8665	.8686	.8708	.8729	.8749	.8770	.8790	.8810	.8830
1.2	.8849	.8869	.8888	.8907	.8925	.8944	.8962	.8980	.8997	.9015
1.3	.9032	.9049	.9066	.9082	.9099	.9115	.9131	.9147	.9162	.9177
1.4	.9192	.9207	.9222	.9236	.9251	.9265	.9279	.9292	.9306	.9319
1.5	.9332	.9345	.9357	.9370	.9382	.9394	.9406	.9418	.9429	.9441
1.6	.9452	.9463	.9474	.9484	.9495	.9505	.9515	.9525	.9535	.9545
1.7	.9554	.9564	.9573	.9582	.9591	.9599	.9608	.9616	.9625	.9633
1.8	.9641	.9649	.9656	.9664	.9671	.9678	.9686	.9693	.9699	.9706
1.9	.9713	.9719	.9726	.9732	.9738	.9744	.9750	.9756	.9761	.9767
2.0	.9772	.9778	.9783	.9788	.9793	.9798	.9803	.9808	.9812	.9817
2.1	.9821	.9826	.9830	.9834	.9838	.9842	.9846	.9850	.9854	.9857
2.2	.9861	.9864	.9868	.9871	.9875	.9878	.9881	.9884	.9887	.9890
2.3	.9893	.9896	.9898	.9901	.9904	.9906	.9909	.9911	.9913	.9916
2.4	.9918	.9920	.9922	.9925	.9927	.9929	.9931	.9932	.9934	.9936
2.5	.9938	.9940	.9941	.9943	.9945	.9946	.9948	.9949	.9951	.9952
2.6	.9953	.9955	.9956	.9957	.9959	.9960	.9961	.9962	.9963	.9964
2.7	.9965	.9966	.9967	.9968	.9969	.9970	.9971	.9972	.9973	.9974
2.8	.9974	.9975	.9976	.9977	.9977	.9978	.9979	.9979	.9980	.9981
2.9	.9981	.9982	.9983	.9983	.9984	.9984	.9985	.9985	.9986	.9986
3.0	.9987	.9987	.9987	.9988	.9988	.9989	.9989	.9989	.9990	.9990

Source: Tables 1–4 adapted from Tables 1, 8, 12, 18 in *Biometrika Tables for Statisticians, Vol. 1,* Third Edition, edited by E. S. Pearson and H. O. Hartley, pp. 110–114, 136–137, 146, 170–173. Copyright ©1996. Adapted by permission of the Cambridge University Press and the Biometrika Trustees.

Table 2	The Cumulative Distribution Function for the t-Distribution

df	90%	95%	97.5%	99%	99.5%	99.9%
1	3.078	6.314	12.706	31.821	63.657	318.309
2	1.886	2.920	4.303	6.965	9.925	22.327
3	1.638	2.353	3.183	4.541	5.841	10.215
4	1.533	2.132	2.777	3.747	4.604	7.173
5	1.476	2.015	2.571	3.365	4.032	5.893
6	1.440	1.943	2.447	3.143	3.708	5.208
7	1.415	1.895	2.365	2.998	3.500	4.785
8	1.397	1.860	2.306	2.897	3.355	4.501
9	1.383	1.833	2.262	2.822	3.250	4.297
10	1.372	1.812	2.228	2.764	3.169	4.144
11	1.363	1.796	2.201	2.718	3.106	4.025
12	1.356	1.782	2.179	2.681	3.055	3.930
13	1.350	1.771	2.160	2.650	3.012	3.852
14	1.345	1.761	2.145	2.625	2.977	3.787
15	1.341	1.753	2.132	2.603	2.947	3.733
16	1.337	1.746	2.120	2.584	2.921	3.686
17	1.333	1.740	2.110	2.567	2.898	3.646
18	1.330	1.734	2.101	2.552	2.879	3.611
19	1.328	1.729	2.093	2.540	2.861	3.580
20	1.325	1.725	2.086	2.528	2.845	3.552
21	1.323	1.721	2.080	2.518	2.831	3.527
22	1.321	1.717	2.074	2.508	2.819	3.505
23	1.319	1.714	2.069	2.500	2.807	3.485
24	1.318	1.711	2.064	2.492	2.797	3.467
25	1.316	1.708	2.060	2.485	2.788	3.450
26	1.315	1.706	2.056	2.479	2.779	3.435
27	1.314	1.703	2.052	2.473	2.771	3.421
28	1.313	1.701	2.048	2.467	2.763	3.408
29	1.311	1.699	2.045	2.462	2.756	3.396
30	1.310	1.697	2.042	2.457	2.750	3.385
40	1.303	1.684	2.021	2.423	2.705	3.307
50	1.299	1.676	2.009	2.403	2.678	3.262
60	1.296	1.671	2.000	2.390	2.660	3.232
80	1.292	1.664	1.990	2.374	2.639	3.195
100	1.290	1.660	1.984	2.364	2.626	3.174
200	1.286	1.653	1.972	2.345	2.601	3.132
∞	1.282	1.645	1.960	2.326	2.576	3.090

Table 3 | The Cumulative Distribution Function for the χ^2 Distribution

df	0.1%	0.135%	0.5%	1.0%	2.5%	5.0%	10.0%
1	0.000	0.000	0.000	0.000	0.001	0.004	0.016
2	0.002	0.003	0.010	0.020	0.051	0.103	0.211
3	0.024	0.030	0.072	0.115	0.216	0.352	0.584
4	0.091	0.106	0.207	0.297	0.484	0.711	1.064
5	0.210	0.238	0.412	0.554	0.831	1.145	1.610
6	0.381	0.423	0.676	0.872	1.237	1.635	2.204
7	0.598	0.656	0.989	1.239	1.690	2.167	2.833
8	0.857	0.931	1.344	1.647	2.180	2.733	3.490
9	1.152	1.241	1.735	2.088	2.700	3.325	4.168
10	1.479	1.584	2.156	2.558	3.247	3.940	4.865
11	1.834	1.954	2.603	3.053	3.816	4.575	5.578
12	2.214	2.350	3.074	3.571	4.404	5.226	6.304
13	2.617	2.768	3.565	4.107	5.009	5.892	7.042
14	3.041	3.206	4.075	4.660	5.629	6.571	7.790
15	3.483	3.662	4.601	5.229	6.262	7.261	8.547
16	3.942	4.135	5.142	5.812	6.908	7.962	9.312
17	4.416	4.624	5.697	6.408	7.564	8.672	10.085
18	4.905	5.126	6.265	7.015	8.231	9.390	10.865
19	5.407	5.641	6.844	7.633	8.907	10.117	11.651
20	5.921	6.169	7.434	8.260	9.591	10.851	12.443
21	6.447	6.707	8.034	8.897	10.283	11.591	13.240
22	6.983	7.256	8.643	9.542	10.982	12.338	14.041
23	7.529	7.814	9.260	10.196	11.689	13.091	14.848
24	8.085	8.382	9.886	10.856	12.401	13.848	15.659
25	8.649	8.959	10.520	11.524	13.120	14.611	16.473
26	9.222	9.543	11.160	12.198	13.844	15.379	17.292
27	9.803	10.135	11.808	12.879	14.573	16.151	18.114
28	10.391	10.735	12.461	13.565	15.308	16.928	18.939
29	10.986	11.341	13.121	14.256	16.047	17.708	19.768
30	11.588	11.954	13.787	14.953	16.791	18.493	20.599
40	17.916	18.385	20.707	22.164	24.433	26.509	29.051
50	24.674	25.235	27.991	29.707	32.357	34.764	37.689

| Table 3 | The Cumulative Distribution Function for the χ^2 Distribution (*Continued*) |

df	90%	95%	97.5%	99%	99.5%	99.865%	99.9%
1	2.705	3.841	5.024	6.635	7.879	10.273	10.827
2	4.605	5.991	7.378	9.210	10.597	13.215	13.815
3	6.251	7.815	9.348	11.345	12.838	15.630	16.266
4	7.779	9.488	11.143	13.277	14.860	17.800	18.467
5	9.236	11.070	12.832	15.086	16.750	19.821	20.515
6	10.645	12.591	14.449	16.812	18.547	21.739	22.458
7	12.017	14.067	16.013	18.475	20.278	23.580	24.322
8	13.361	15.507	17.534	20.090	21.955	25.361	26.124
9	14.684	16.919	19.023	21.666	23.589	27.093	27.877
10	15.987	18.307	20.483	23.209	25.188	28.785	29.588
11	17.275	19.675	21.920	24.725	26.757	30.442	31.264
12	18.549	21.026	23.337	26.217	28.299	32.069	32.909
13	19.812	22.362	24.736	27.688	29.819	33.671	34.528
14	21.064	23.685	26.119	29.141	31.319	35.250	36.123
15	22.307	24.996	27.488	30.578	32.801	36.808	37.697
16	23.542	26.296	28.845	32.000	34.267	38.347	39.252
17	24.769	27.587	30.191	33.409	35.718	39.870	40.790
18	25.989	28.869	31.526	34.805	37.156	41.377	42.312
19	27.203	30.143	32.852	36.191	38.582	42.871	43.820
20	28.412	31.410	34.170	37.566	39.997	44.351	45.315
21	29.615	32.670	35.479	38.932	41.401	45.820	46.797
22	30.813	33.924	36.781	40.289	42.796	47.278	48.268
23	32.007	35.172	38.076	41.638	44.181	48.725	49.728
24	33.196	36.415	39.364	42.980	45.558	50.163	51.178
25	34.381	37.652	40.646	44.314	46.928	51.591	52.620
26	35.563	38.885	41.923	45.642	48.290	53.011	54.052
27	36.741	40.113	43.194	46.963	49.645	54.423	55.476
28	37.916	41.337	44.461	48.278	50.993	55.828	56.892
29	39.087	42.557	45.722	49.588	52.336	57.225	58.301
30	40.256	43.773	46.979	50.892	53.672	58.615	59.703
40	51.805	55.758	59.342	63.691	66.766	72.209	73.402
50	63.167	67.505	71.420	76.154	79.490	85.374	86.661

Table 4 | Percentiles of the F-Distribution

Upper 10% point of the F-distribution

		Degrees of Freedom for Numerator											
	1	2	3	4	5	6	7	8	9	10	11	12	13
1	39.9	49.5	53.6	55.8	57.2	58.2	58.9	59.4	59.9	60.2	60.5	60.7	60.9
2	8.53	9.00	9.16	9.24	9.29	9.33	9.35	9.37	9.37	9.38	9.40	9.41	9.41
3	5.54	5.46	5.39	5.34	5.31	5.28	5.27	5.25	5.24	5.23	5.22	5.22	5.21
4	4.54	4.32	4.19	4.11	4.05	4.01	3.98	3.95	3.94	3.92	3.91	3.90	3.89
5	4.06	3.78	3.62	3.52	3.45	3.40	3.37	3.34	3.32	3.30	3.28	3.27	3.26
6	3.78	3.46	3.29	3.18	3.11	3.05	3.01	2.98	2.96	2.94	2.92	2.90	2.89
7	3.59	3.26	3.07	2.96	2.88	2.83	2.78	2.75	2.72	2.70	2.68	2.67	2.65
8	3.46	3.11	2.92	2.81	2.73	2.67	2.62	2.59	2.56	2.54	2.52	2.50	2.49
9	3.36	3.01	2.81	2.69	2.61	2.55	2.51	2.47	2.44	2.42	2.40	2.38	2.36
10	3.29	2.92	2.73	2.61	2.52	2.46	2.41	2.38	2.35	2.32	2.30	2.28	2.27
11	3.23	2.86	2.66	2.54	2.45	2.39	2.34	2.30	2.27	2.25	2.23	2.21	2.19
12	3.18	2.81	2.61	2.48	2.39	2.33	2.28	2.24	2.21	2.19	2.17	2.15	2.13
13	3.14	2.76	2.56	2.43	2.35	2.28	2.23	2.20	2.16	2.14	2.12	2.10	2.08
14	3.10	2.73	2.52	2.39	2.31	2.24	2.19	2.15	2.12	2.10	2.07	2.05	2.04
15	3.07	2.70	2.49	2.36	2.27	2.21	2.16	2.12	2.09	2.06	2.04	2.02	2.00
16	3.05	2.67	2.46	2.33	2.24	2.18	2.13	2.09	2.06	2.03	2.01	1.99	1.97
17	3.03	2.64	2.44	2.31	2.22	2.15	2.10	2.06	2.03	2.00	1.98	1.96	1.94
18	3.01	2.62	2.42	2.29	2.20	2.13	2.08	2.04	2.00	1.98	1.95	1.93	1.92
19	2.99	2.61	2.40	2.27	2.18	2.11	2.06	2.02	1.98	1.96	1.93	1.91	1.89
20	2.97	2.59	2.38	2.25	2.16	2.09	2.04	2.00	1.96	1.94	1.91	1.89	1.87
21	2.96	2.57	2.36	2.23	2.14	2.08	2.02	1.98	1.95	1.92	1.90	1.87	1.86
22	2.95	2.56	2.35	2.22	2.13	2.06	2.01	1.97	1.93	1.90	1.88	1.86	1.84
23	2.94	2.55	2.34	2.21	2.11	2.05	1.99	1.95	1.92	1.89	1.87	1.84	1.83
24	2.93	2.54	2.33	2.19	2.10	2.04	1.98	1.94	1.91	1.88	1.85	1.83	1.81
25	2.92	2.53	2.32	2.18	2.09	2.02	1.97	1.93	1.89	1.87	1.84	1.82	1.80
26	2.91	2.52	2.31	2.17	2.08	2.01	1.96	1.92	1.88	1.86	1.83	1.81	1.79
27	2.90	2.51	2.30	2.17	2.07	2.00	1.95	1.91	1.87	1.85	1.82	1.80	1.78
28	2.89	2.50	2.29	2.16	2.06	2.00	1.94	1.90	1.87	1.84	1.81	1.79	1.77
29	2.89	2.50	2.28	2.15	2.06	1.99	1.93	1.89	1.86	1.83	1.80	1.78	1.76
30	2.88	2.49	2.28	2.14	2.05	1.98	1.93	1.88	1.85	1.82	1.79	1.77	1.75
32	2.87	2.48	2.26	2.13	2.04	1.97	1.91	1.87	1.83	1.81	1.78	1.76	1.74
34	2.86	2.47	2.25	2.12	2.02	1.96	1.90	1.86	1.82	1.79	1.77	1.75	1.73
36	2.85	2.46	2.24	2.11	2.01	1.94	1.89	1.85	1.81	1.78	1.76	1.73	1.71
38	2.84	2.45	2.23	2.10	2.01	1.94	1.88	1.84	1.80	1.77	1.75	1.72	1.70
40	2.84	2.44	2.23	2.09	2.00	1.93	1.87	1.83	1.79	1.76	1.74	1.71	1.70
42	2.83	2.43	2.22	2.08	1.99	1.92	1.86	1.82	1.78	1.75	1.73	1.71	1.69
44	2.82	2.43	2.21	2.08	1.98	1.91	1.86	1.81	1.78	1.75	1.72	1.70	1.68
46	2.82	2.42	2.21	2.07	1.98	1.91	1.85	1.81	1.77	1.74	1.71	1.69	1.67
48	2.81	2.42	2.20	2.07	1.97	1.90	1.85	1.80	1.77	1.73	1.71	1.69	1.67
50	2.81	2.41	2.20	2.06	1.97	1.90	1.84	1.80	1.76	1.73	1.70	1.68	1.66
60	2.79	2.39	2.18	2.04	1.95	1.87	1.82	1.77	1.74	1.71	1.68	1.66	1.64
70	2.78	2.38	2.16	2.03	1.93	1.86	1.80	1.76	1.72	1.69	1.66	1.64	1.62
80	2.77	2.37	2.15	2.02	1.92	1.85	1.79	1.75	1.71	1.68	1.65	1.63	1.61
90	2.76	2.36	2.15	2.01	1.91	1.84	1.78	1.74	1.70	1.67	1.64	1.62	1.60
100	2.76	2.36	2.14	2.00	1.91	1.83	1.78	1.73	1.69	1.66	1.64	1.61	1.59
125	2.75	2.35	2.13	1.99	1.89	1.82	1.77	1.72	1.68	1.65	1.62	1.60	1.58
150	2.74	2.34	2.12	1.98	1.89	1.81	1.76	1.71	1.67	1.64	1.61	1.59	1.57
200	2.73	2.33	2.11	1.97	1.88	1.80	1.75	1.70	1.66	1.63	1.60	1.58	1.56
300	2.72	2.32	2.10	1.96	1.87	1.79	1.74	1.69	1.65	1.62	1.59	1.57	1.55
500	2.72	2.31	2.09	1.96	1.86	1.79	1.73	1.68	1.64	1.61	1.58	1.56	1.54
1000	2.71	2.31	2.09	1.95	1.85	1.78	1.72	1.68	1.64	1.61	1.58	1.55	1.53

Degrees of Freedom for Denominator

Table 4 | Percentiles of the *F*-Distribution (*Continued*)

Upper 10% point of the *F*-distribution

		Degrees of Freedom for Numerator													
		14	15	16	17	18	19	20	25	30	40	50	100	150	200
	1	61.1	61.2	61.3	61.5	61.6	61.7	61.7	62.1	62.3	62.5	62.7	63.0	63.1	63.2
	2	9.42	9.42	9.43	9.43	9.44	9.44	9.44	9.45	9.46	9.47	9.47	9.48	9.48	9.49
	3	5.20	5.20	5.20	5.19	5.19	5.19	5.18	5.17	5.17	5.16	5.15	5.14	5.14	5.14
	4	3.88	3.87	3.86	3.86	3.85	3.85	3.84	3.83	3.82	3.80	3.80	3.78	3.77	3.77
	5	3.25	3.24	3.23	3.22	3.22	3.21	3.21	3.19	3.17	3.16	3.15	3.13	3.12	3.12
	6	2.88	2.87	2.86	2.85	2.85	2.84	2.84	2.81	2.80	2.78	2.77	2.75	2.74	2.73
	7	2.64	2.63	2.62	2.61	2.61	2.60	2.59	2.57	2.56	2.54	2.52	2.50	2.49	2.48
	8	2.48	2.46	2.45	2.45	2.44	2.43	2.42	2.40	2.38	2.36	2.35	2.32	2.31	2.31
	9	2.35	2.34	2.33	2.32	2.31	2.30	2.30	2.27	2.25	2.23	2.22	2.19	2.18	2.17
	10	2.26	2.24	2.23	2.22	2.22	2.21	2.20	2.17	2.16	2.13	2.12	2.09	2.08	2.07
	11	2.18	2.17	2.16	2.15	2.14	2.13	2.12	2.10	2.08	2.05	2.04	2.01	1.99	1.99
	12	2.12	2.10	2.09	2.08	2.08	2.07	2.06	2.03	2.01	1.99	1.97	1.94	1.93	1.92
	13	2.07	2.05	2.04	2.03	2.02	2.01	2.01	1.98	1.96	1.93	1.92	1.88	1.87	1.86
	14	2.02	2.01	2.00	1.99	1.98	1.97	1.96	1.93	1.91	1.89	1.87	1.83	1.82	1.82
	15	1.99	1.97	1.96	1.95	1.94	1.93	1.92	1.89	1.87	1.85	1.83	1.79	1.78	1.77
	16	1.95	1.94	1.93	1.92	1.91	1.90	1.89	1.86	1.84	1.81	1.79	1.76	1.74	1.74
	17	1.93	1.91	1.90	1.89	1.88	1.87	1.86	1.83	1.81	1.78	1.76	1.73	1.71	1.71
	18	1.90	1.89	1.87	1.86	1.85	1.84	1.84	1.80	1.78	1.75	1.74	1.70	1.68	1.68
	19	1.88	1.86	1.85	1.84	1.83	1.82	1.81	1.78	1.76	1.73	1.71	1.67	1.66	1.65
	20	1.86	1.84	1.83	1.82	1.81	1.80	1.79	1.76	1.74	1.71	1.69	1.65	1.64	1.63
	21	1.84	1.83	1.81	1.80	1.79	1.78	1.78	1.74	1.72	1.69	1.67	1.63	1.62	1.61
	22	1.83	1.81	1.80	1.79	1.78	1.77	1.76	1.73	1.70	1.67	1.65	1.61	1.60	1.59
	23	1.81	1.80	1.78	1.77	1.76	1.75	1.74	1.71	1.69	1.66	1.64	1.59	1.58	1.57
	24	1.80	1.78	1.77	1.76	1.75	1.74	1.73	1.70	1.67	1.64	1.62	1.58	1.56	1.56
	25	1.79	1.77	1.76	1.75	1.74	1.73	1.72	1.68	1.66	1.63	1.61	1.56	1.55	1.54
	26	1.77	1.76	1.75	1.73	1.72	1.71	1.71	1.67	1.65	1.61	1.59	1.55	1.54	1.53
	27	1.76	1.75	1.74	1.72	1.71	1.70	1.70	1.66	1.64	1.60	1.58	1.54	1.52	1.52
	28	1.75	1.74	1.73	1.71	1.70	1.69	1.69	1.65	1.63	1.59	1.57	1.53	1.51	1.50
	29	1.75	1.73	1.72	1.71	1.69	1.68	1.68	1.64	1.62	1.58	1.56	1.52	1.50	1.49
	30	1.74	1.72	1.71	1.70	1.69	1.68	1.67	1.63	1.61	1.57	1.55	1.51	1.49	1.48
	32	1.72	1.71	1.69	1.68	1.67	1.66	1.65	1.62	1.59	1.56	1.53	1.49	1.47	1.46
	34	1.71	1.69	1.68	1.67	1.66	1.65	1.64	1.60	1.58	1.54	1.52	1.47	1.46	1.45
	36	1.70	1.68	1.67	1.66	1.65	1.64	1.63	1.59	1.56	1.53	1.51	1.46	1.44	1.43
	38	1.69	1.67	1.66	1.65	1.63	1.62	1.61	1.58	1.55	1.52	1.49	1.45	1.43	1.42
	40	1.68	1.66	1.65	1.64	1.62	1.61	1.61	1.57	1.54	1.51	1.48	1.43	1.42	1.41
	42	1.67	1.65	1.64	1.63	1.62	1.61	1.60	1.56	1.53	1.50	1.47	1.42	1.40	1.40
	44	1.66	1.65	1.63	1.62	1.61	1.60	1.59	1.55	1.52	1.49	1.46	1.41	1.39	1.39
	46	1.65	1.64	1.63	1.61	1.60	1.59	1.58	1.54	1.52	1.48	1.46	1.40	1.39	1.38
	48	1.65	1.63	1.62	1.61	1.59	1.58	1.57	1.54	1.51	1.47	1.45	1.40	1.38	1.37
	50	1.64	1.63	1.61	1.60	1.59	1.58	1.57	1.53	1.50	1.46	1.44	1.39	1.37	1.36
	60	1.62	1.60	1.59	1.58	1.56	1.55	1.54	1.50	1.48	1.44	1.41	1.36	1.34	1.33
	70	1.60	1.59	1.57	1.56	1.55	1.54	1.53	1.49	1.46	1.42	1.39	1.34	1.31	1.30
	80	1.59	1.57	1.56	1.55	1.53	1.52	1.51	1.47	1.44	1.40	1.38	1.32	1.30	1.28
	90	1.58	1.56	1.55	1.54	1.52	1.51	1.50	1.46	1.43	1.39	1.36	1.30	1.28	1.27
	100	1.57	1.56	1.54	1.53	1.52	1.50	1.49	1.45	1.42	1.38	1.35	1.29	1.27	1.26
	125	1.56	1.54	1.53	1.51	1.50	1.49	1.48	1.44	1.41	1.36	1.34	1.27	1.25	1.23
	150	1.55	1.53	1.52	1.50	1.49	1.48	1.47	1.43	1.40	1.35	1.33	1.26	1.23	1.22
	200	1.54	1.52	1.51	1.49	1.48	1.47	1.46	1.41	1.38	1.34	1.31	1.24	1.21	1.20
	300	1.53	1.51	1.49	1.48	1.47	1.46	1.45	1.40	1.37	1.32	1.29	1.22	1.19	1.18
	500	1.52	1.50	1.49	1.47	1.46	1.45	1.44	1.39	1.36	1.31	1.28	1.21	1.18	1.16
	1000	1.51	1.49	1.48	1.46	1.45	1.44	1.43	1.38	1.35	1.30	1.27	1.20	1.16	1.15

Degrees of Freedom for Denominator

Table 4 Percentiles of the *F*-Distribution (*Continued*)

Upper 5% point of the *F*-distribution

	Degrees of Freedom for Numerator												
Degrees of Freedom for Denominator	**1**	**2**	**3**	**4**	**5**	**6**	**7**	**8**	**9**	**10**	**11**	**12**	**13**
1	161	200	216	225	230	234	237	239	241	242	243	244	245
2	18.5	19.0	19.2	19.2	19.3	19.3	19.4	19.4	19.4	19.4	19.4	19.4	19.4
3	10.1	9.55	9.28	9.12	9.01	8.94	8.89	8.85	8.81	8.79	8.76	8.74	8.73
4	7.71	6.94	6.59	6.39	6.26	6.16	6.09	6.04	6.00	5.96	5.94	5.91	5.89
5	6.61	5.79	5.41	5.19	5.05	4.95	4.88	4.82	4.77	4.74	4.70	4.68	4.66
6	5.99	5.14	4.76	4.53	4.39	4.28	4.21	4.15	4.10	4.06	4.03	4.00	3.98
7	5.59	4.74	4.35	4.12	3.97	3.87	3.79	3.73	3.68	3.64	3.60	3.57	3.55
8	5.32	4.46	4.07	3.84	3.69	3.58	3.50	3.44	3.39	3.35	3.31	3.28	3.26
9	5.12	4.26	3.86	3.63	3.48	3.37	3.29	3.23	3.18	3.14	3.10	3.07	3.05
10	4.96	4.10	3.71	3.48	3.33	3.22	3.14	3.07	3.02	2.98	2.94	2.91	2.89
11	4.84	3.98	3.59	3.36	3.20	3.09	3.01	2.95	2.90	2.82	2.82	2.79	2.76
12	4.75	3.89	3.49	3.26	3.11	3.00	2.91	2.85	2.80	2.72	2.72	2.69	2.66
13	4.67	3.81	3.41	3.18	3.03	2.92	2.83	2.77	2.71	2.67	2.63	2.60	2.58
14	4.60	3.74	3.34	3.11	2.96	2.85	2.76	2.70	2.65	2.60	2.57	2.53	2.51
15	4.54	3.68	3.29	3.05	2.90	2.79	2.71	2.64	2.59	2.54	2.51	2.48	2.45
16	4.49	3.63	3.24	3.01	2.85	2.74	2.66	2.59	2.54	2.49	2.46	2.42	2.40
17	4.45	3.59	3.20	2.96	2.81	2.70	2.61	2.55	2.49	2.45	2.41	2.38	2.35
18	4.41	3.55	3.16	2.93	2.77	2.66	2.58	2.51	2.46	2.41	2.37	2.34	2.31
19	4.38	3.52	3.13	2.90	2.74	2.63	2.54	2.48	2.42	2.38	2.34	2.31	2.28
20	4.35	3.49	3.10	2.87	2.71	2.60	2.51	2.45	2.39	2.35	2.31	2.28	2.25
21	4.32	3.47	3.07	2.84	2.68	2.57	2.49	2.42	2.37	2.32	2.28	2.25	2.22
22	4.30	3.44	3.05	2.82	2.66	2.55	2.46	2.40	2.34	2.30	2.26	2.23	2.20
23	4.28	3.42	3.03	2.80	2.64	2.53	2.44	2.37	2.32	2.27	2.24	2.20	2.18
24	4.26	3.40	3.01	2.78	2.62	2.51	2.42	2.36	2.30	2.25	2.22	2.18	2.15
25	4.24	3.39	2.99	2.75	2.60	2.49	2.40	2.34	2.28	2.24	2.20	2.16	2.14
26	4.23	3.37	2.98	2.74	2.59	2.47	2.39	2.32	2.27	2.22	2.18	2.15	2.12
27	4.21	3.35	2.96	2.73	2.57	2.46	2.37	2.31	2.25	2.20	2.17	2.13	2.10
28	4.20	3.34	2.95	2.71	2.56	2.45	2.36	2.29	2.24	2.19	2.15	2.12	2.09
29	4.18	3.33	2.93	2.70	2.55	2.43	2.35	2.28	2.22	2.18	2.14	2.10	2.08
30	4.17	3.32	2.92	2.69	2.53	2.42	2.33	2.27	2.21	2.16	2.13	2.09	2.06
32	4.15	3.29	2.90	2.67	2.51	2.40	2.31	2.24	2.19	2.14	2.10	2.07	2.04
34	4.13	3.28	2.88	2.65	2.49	2.38	2.29	2.23	2.17	2.12	2.08	2.05	2.02
36	4.11	3.26	2.87	2.63	2.48	2.36	2.28	2.21	2.15	2.11	2.07	2.03	2.00
38	4.10	3.24	2.85	2.62	2.46	2.35	2.26	2.19	2.14	2.09	2.05	2.02	1.99
40	4.08	3.23	2.84	2.61	2.45	2.34	2.25	2.18	2.12	2.08	2.04	2.00	1.97
42	4.07	3.22	2.83	2.59	2.44	2.32	2.24	2.17	2.11	2.06	2.03	1.99	1.96
44	4.06	3.21	2.82	2.58	2.43	2.31	2.23	2.16	2.10	2.05	2.01	1.98	1.95
46	4.05	3.20	2.81	2.57	2.42	2.30	2.22	2.15	2.09	2.04	2.00	1.97	1.94
48	4.04	3.19	2.80	2.57	2.41	2.29	2.21	2.14	2.08	2.03	1.99	1.96	1.93
50	4.03	3.18	2.79	2.56	2.40	2.29	2.20	2.13	2.07	2.03	1.99	1.95	1.92
60	4.00	3.15	2.76	2.53	2.37	2.25	2.17	2.10	2.04	1.99	1.95	1.92	1.89
70	3.93	3.13	2.74	2.50	2.35	2.23	2.14	2.07	2.02	1.97	1.93	1.89	1.86
80	3.96	3.11	2.72	2.49	2.33	2.21	2.13	2.06	2.00	1.95	1.91	1.88	1.84
90	3.95	3.10	2.71	2.47	2.32	2.20	2.11	2.04	1.99	1.94	1.90	1.86	1.83
100	3.94	3.09	2.70	2.45	2.31	2.19	2.10	2.03	1.97	1.93	1.89	1.85	1.82
125	3.92	3.07	2.68	2.44	2.29	2.17	2.08	2.01	1.96	1.91	1.87	1.83	1.80
150	3.90	3.06	2.66	2.43	2.27	2.16	2.07	2.00	1.94	1.89	1.85	1.82	1.79
200	3.89	3.04	2.65	2.42	2.26	2.14	2.06	1.98	1.93	1.88	1.84	1.80	1.77
300	3.87	3.03	2.63	2.40	2.24	2.13	2.04	1.97	1.91	1.86	1.82	1.78	1.75
500	3.86	3.01	2.62	2.39	2.23	2.12	2.03	1.96	1.90	1.85	1.81	1.77	1.74
1000	3.85	3.00	2.61	2.38	2.22	2.11	2.02	1.95	1.89	1.84	1.80	1.76	1.73

Table 4 | Percentiles of the *F*-Distribution (*Continued*)

Upper 5% point of the *F*-distribution

	Degrees of Freedom for Numerator													
	14	15	16	17	18	19	20	25	30	40	50	100	150	200
1	245	246	246	247	247	248	248	249	250	251	252	253	253	254
2	19.4	19.4	19.4	19.4	19.4	19.4	19.4	19.5	19.5	19.5	19.5	19.5	19.5	19.5
3	8.71	8.70	8.69	8.68	8.67	8.67	8.66	8.63	8.62	8.59	8.58	8.55	8.54	8.54
4	5.87	5.86	5.84	5.83	5.82	5.81	5.80	5.77	5.75	5.72	5.70	5.66	5.65	5.65
5	4.64	4.62	4.60	4.59	4.58	4.57	4.56	4.52	4.50	4.46	4.44	4.11	4.39	4.69
6	3.96	3.94	3.92	3.91	3.90	3.88	3.87	3.83	3.81	3.77	3.75	3.71	3.70	3.69
7	3.53	3.51	3.49	3.48	3.47	3.46	3.44	3.40	3.38	3.34	3.32	3.27	3.23	3.25
8	3.24	3.22	3.20	3.19	3.17	3.16	3.15	3.11	3.08	3.04	3.02	2.97	2.93	2.95
9	3.03	3.01	2.99	2.97	2.96	2.95	2.94	2.89	2.86	2.83	2.80	2.76	2.71	2.73
10	2.86	2.85	2.83	2.81	2.80	2.79	2.77	2.73	2.70	2.66	2.64	2.59	2.54	2.56
11	2.74	2.72	2.70	2.69	2.67	2.66	2.65	2.60	2.57	2.53	2.51	2.46	2.41	2.43
12	2.64	2.62	2.60	2.58	2.57	2.56	2.54	2.50	2.47	2.43	2.40	2.35	2.3.	2.32
13	2.55	2.53	2.51	2.50	2.48	2.47	2.46	2.41	2.38	2.34	2.31	2.26	2.21	2.23
14	2.48	2.46	2.44	2.43	2.41	2.40	2.39	2.34	2.31	2.27	2.24	2.19	2.17	2.16
15	2.42	2.40	2.38	2.37	2.35	2.34	2.33	2.28	2.25	2.20	2.18	2.12	2.10	2.10
16	2.37	2.35	2.33	2.32	2.30	2.29	2.28	2.23	2.19	2.15	2.12	2.07	2.05	2.04
17	2.33	2.31	2.29	2.27	2.26	2.24	2.23	2.18	2.15	2.10	2.08	2.02	2.00	1.99
18	2.29	2.27	2.25	2.23	2.22	2.20	2.19	2.14	2.11	2.06	2.04	1.98	1.96	1.95
19	2.26	2.23	2.21	2.20	2.18	2.17	2.16	2.11	2.07	2.03	2.00	1.94	1.92	1.91
20	2.22	2.20	2.18	2.17	2.15	2.14	2.12	2.07	2.04	1.99	1.97	1.91	1.89	1.88
21	2.20	2.18	2.16	2.14	2.12	2.11	2.10	2.05	2.01	1.96	1.94	1.88	1.86	1.84
22	2.17	2.15	2.13	2.11	2.10	2.08	2.07	2.02	1.98	1.94	1.91	1.85	1.83	1.82
23	2.15	2.13	2.11	2.09	2.08	2.06	2.05	2.00	1.96	1.91	1.88	1.82	1.80	1.79
24	2.13	2.11	2.09	2.07	2.05	2.04	2.03	1.97	1.94	1.89	1.86	1.80	1.78	1.77
25	2.11	2.09	2.07	2.05	2.04	2.02	2.01	1.96	1.92	1.87	1.84	1.78	1.76	1.75
26	2.09	2.07	2.05	2.03	2.02	2.00	1.99	1.94	1.90	1.85	1.82	1.76	1.74	1.73
27	2.08	2.06	2.04	2.02	2.00	1.99	1.97	1.92	1.88	1.84	1.81	1.74	1.72	1.71
28	2.06	2.04	2.02	2.00	1.99	1.97	1.96	1.91	1.87	1.82	1.79	1.73	1.70	1.69
29	2.05	2.03	2.01	1.99	1.97	1.96	1.94	1.89	1.85	1.81	1.77	1.71	1.69	1.64
30	2.04	2.01	1.99	1.98	1.96	1.95	1.93	1.88	1.84	1.79	1.76	1.70	1.67	1.66
32	2.01	1.99	1.97	1.95	1.94	1.92	1.91	1.85	1.82	1.77	1.74	1.67	1.64	1.63
34	1.99	1.97	1.95	1.93	1.92	1.90	1.89	1.83	1.80	1.75	1.71	1.65	1.62	1.61
36	1.98	1.95	1.93	1.92	1.90	1.88	1.87	1.81	1.78	1.73	1.69	1.62	1.60	1.59
38	1.96	1.94	1.92	1.90	1.88	1.87	1.85	1.80	1.76	1.71	1.68	1.61	1.58	1.57
40	1.95	1.92	1.90	1.89	1.87	1.85	1.84	1.78	1.74	1.69	1.66	1.59	1.56	1.55
42	1.94	1.91	1.89	1.87	1.86	1.84	1.83	1.77	1.73	1.68	1.65	1.57	1.55	1.53
44	1.92	1.90	1.88	1.86	1.84	1.83	1.81	1.76	1.72	1.67	1.63	1.56	1.53	1.52
46	1.91	1.89	1.87	1.85	1.83	1.82	1.80	1.75	1.71	1.65	1.62	1.55	1.52	1.51
48	1.90	1.88	1.86	1.84	1.82	1.81	1.79	1.74	1.70	1.64	1.61	1.54	1.51	1.49
50	1.89	1.87	1.85	1.83	1.81	1.80	1.78	1.73	1.69	1.63	1.60	1.52	1.50	1.48
60	1.86	1.84	1.82	1.80	1.78	1.76	1.75	1.69	1.65	1.59	1.56	1.48	1.45	1.44
70	1.84	1.81	1.79	1.77	1.75	1.74	1.72	1.66	1.62	1.57	1.53	1.45	1.42	1.40
80	1.82	1.79	1.77	1.75	1.73	1.72	1.70	1.64	1.60	1.54	1.51	1.43	1.39	1.38
90	1.80	1.78	1.76	1.74	1.72	1.70	1.69	1.63	1.59	1.53	1.49	1.41	1.38	1.36
100	1.79	1.77	1.75	1.73	1.71	1.69	1.68	1.62	1.57	1.52	1.48	1.39	1.36	1.34
125	1.77	1.75	1.73	1.71	1.69	1.67	1.66	1.59	1.55	1.49	1.45	1.36	1.33	1.31
150	1.76	1.73	1.71	1.69	1.67	1.66	1.64	1.58	1.54	1.48	1.44	1.34	1.31	1.29
200	1.74	1.72	1.69	1.67	1.66	1.64	1.62	1.56	1.52	1.46	1.41	1.32	1.28	1.26
300	1.72	1.70	1.68	1.66	1.64	1.62	1.61	1.54	1.50	1.43	1.39	1.30	1.26	1.23
500	1.71	1.69	1.66	1.64	1.62	1.61	1.59	1.53	1.48	1.42	1.38	1.28	1.23	1.21
1000	1.70	1.68	1.65	1.63	1.61	1.60	1.58	1.52	1.47	1.41	1.36	1.26	1.22	1.19

Degrees of Freedom for Denominator

Table 4 | Percentiles of the *F*-Distribution (*Continued*)

Upper 1% point of the *F*-distribution

					Degrees of Freedom for Numerator								
	1	2	3	4	5	6	7	8	9	10	11	12	13
1	4057	5000	5403	5625	5764	5859	5928	5981	6022	6056	6083	6106	6126
2	98.5	99.0	99.2	99.2	99.3	99.3	99.4	99.4	99.4	99.4	99.4	99.4	99.4
3	34.1	30.8	29.5	28.7	28.2	27.9	27.7	27.5	27.3	27.2	27.1	27.1	27.0
4	21.2	18.0	16.7	16.0	15.5	15.2	15.0	14.8	14.7	14.5	14.5	14.4	14.3
5	16.3	13.3	12.1	11.4	11.0	10.7	10.5	10.3	10.2	10.1	9.96	9.89	9382
6	13.7	10.9	9.78	9.15	8.75	8.47	8.26	8.10	7.98	7.87	7.79	7.72	7.66
7	12.2	9.55	8.45	7.85	7.46	7.19	6.99	6.84	6.72	6.62	6.54	6.47	6.41
8	11.3	8.65	7.59	7.01	6.63	6.37	6.18	6.03	5.91	5.81	5.73	5.67	5.61
9	10.6	8.02	6.99	6042	6.06	5.80	5.61	5.47	5.35	5.26	5.18	5.11	5.05
10	10.0	7.56	6.55	5.99	5.64	5.39	5.20	5.06	4.94	4.85	4.77	4.71	4.65
11	9.65	7.21	6.22	5.67	5.32	5.07	4.89	4.74	4.63	4.54	4.46	4.40	4.34
12	9.33	6.93	5.95	5.41	5.06	4.82	4.64	4.50	4.39	4.30	4.22	4.16	4.10
13	9.07	6.70	5.74	5.21	4.86	4.62	4.44	4.30	4.19	4.10	4.02	3.96	3.91
14	8.88	6.51	5.56	5.04	4.69	4.46	4.28	4.14	4.03	3.94	3.86	3.80	3.75
15	8.68	6.36	5.42	4.89	4.56	4.32	4.14	4.00	3.89	3.80	3.73	3.67	3.61
16	8.53	6.23	5.29	4.77	4.44	4.20	4.03	3.89	3.78	3.69	3.62	3.55	3.50
17	8.40	6.11	5.19	4.67	4.10	4.10	3.93	3.79	3.68	3.59	3.52	3.46	3.40
18	8.29	6.01	5.09	4.58	4.25	4.01	3.84	3.71	3.60	3.51	3.43	3.37	3.32
19	8.18	5.93	5.01	4.50	4.17	3.94	3.77	3.63	3.52	3.43	3.36	3.30	3.24
20	8.10	5.85	4.94	4.43	4.10	3.87	3.70	3.56	3.46	3.37	3.29	3.23	3.18
21	8.02	5.78	4.87	4.37	4.04	3.81	3.64	3.51	3.40	3.31	3.24	3.17	3.12
22	7.92	5.72	4.82	4.31	3.99	3.76	3.59	3.45	3.35	3.26	3.18	3.12	3.07
23	7.88	5.66	4.76	4.26	3.94	3.71	3.54	3.41	3.30	3.21	3.14	3.07	3.02
24	7.82	5.61	4.72	4.22	3.90	3.67	3.50	3.36	3.26	3.17	3.09	3.03	2.98
25	7.77	5.57	4.68	4.18	3.85	3.36	3.46	3.32	3.22	3.13	3.06	2.99	2.94
26	7.72	5.53	4.64	4.14	3.82	3.59	3.42	3.29	3.18	3.09	3.02	2.96	2.90
27	7.68	5.49	4.60	4.11	3.78	3.56	3.39	3.26	3.15	3.06	2.99	2.93	2.87
28	7.64	5.45	4.57	4.07	3.75	3.53	3.36	3.23	3.12	3.03	2.96	2.90	2.84
29	7.60	5.42	4.54	4.04	3.73	3.50	3.33	3.20	3.09	3.00	2.93	2.87	2.81
30	7.56	5.39	4.51	4.02	3.80	3.47	3.30	3.17	3.07	2.98	2.91	2.84	2.79
32	7.50	5.34	4.46	3.97	3.65	3.43	3.26	3.13	3.02	2.93	2.86	2.80	2.74
34	7.44	5.29	4.42	3.93	3.61	3.39	3.22	3.09	2.98	2.89	2.82	2.76	2.70
36	7.42	5.25	4.38	3.89	3.57	3.35	3.18	3.05	2.95	2.86	2.79	2.72	2.67
38	7.35	5.21	4.34	3.86	3.54	3.32	3.15	3.02	2.92	2.83	2.75	2.69	2.64
40	7.31	5.18	4.31	3.83	3.51	3.29	3.12	2.99	2.89	2.80	2.73	2.66	2.61
42	7.28	5.15	4.29	3.80	3.49	3.27	3.10	2.97	2.86	2.78	2.70	2.64	2.59
44	7.25	5.12	4.26	3.78	3.47	3.24	3.08	2.95	2.84	2.75	2.68	2.62	2.56
46	7.22	5.10	4.24	3.76	3.44	3.22	3.06	2.93	2.82	2.73	2.66	2.60	2.54
48	7.19	5.08	4.22	3.74	3.43	3.20	3.04	2.91	2.80	2.71	2.64	2.58	2.53
50	7.17	5.06	4.20	3.72	3.41	3.19	3.02	2.89	2.78	2.70	2.63	2.56	2.51
60	7.08	4.98	4.13	3.65	3.34	3.12	2.95	2.82	2.72	2.63	2.56	2.50	2.44
70	7.01	4.92	4.07	3.60	3.29	3.07	2.91	2.78	2.67	2.59	2.51	2.45	2.40
80	6.96	4.88	4.04	3.56	3.26	3.04	2.87	2.74	2.64	2.55	2.48	2.42	2.36
90	6.93	4.85	4.01	3.53	3.23	3.01	2.84	2.72	2.61	2.52	2.45	2.39	2.33
100	6.90	4.82	3.98	3.51	3.21	2.99	2.82	2.69	2.59	2.50	2.43	2.37	2.31
125	6.84	4.78	3.94	3.47	3.17	2.95	2.79	2.66	2.55	2.47	2.39	2.33	2.28
150	6.81	4.75	3.91	3.45	3.14	2.92	2.76	2.63	2.53	2.44	2.37	2.31	2.25
200	6.76	4.71	3.88	3.41	3.11	2.89	2.73	2.60	2.50	2.41	2.34	2.27	2.22
300	6.72	4.68	3.85	3.38	3.08	2.86	2.70	2.57	2.47	2.38	2.31	2.24	2.19
500	6.69	4.65	3.82	3.36	3.05	2.84	2.68	2.55	2.44	2.36	2.28	2.22	2.17
1000	6.66	4.63	3.80	3.34	3.04	2.82	2.66	2.53	2.43	2.34	2.27	2.20	2.15

Degrees of Freedom for Denominator

Table 4 | Percentiles of the *F*-Distribution (*Continued*)

Upper 1% point of the *F*-distribution

						Degrees of Freedom for Numerator									
	14	**15**	**16**	**17**	**18**	**19**	**20**	**25**	**30**	**40**	**50**	**100**	**150**	**200**
1	6143	6157	6170	6181	6192	6201	6209	6240	6261	6287	6303	6334	6345	6350
2	99.4	99.4	99.4	99.4	99.4	99.4	99.4	99.5	99.5	99.5	99.5	99.5	99.5	99.5
3	26.9	26.9	26.8	26.8	2608	26.7	26.7	26.6	26.5	26.4	26.4	26.2	26.2	26.5
4	14.2	14.2	14.2	14.1	14.1	14.0	14.0	13.9	13.8	13.7	13.7	13.6	13.5	13.5
5	9.77	9.72	9.68	9.64	9.61	9.58	9.55	9.45	9.38	9.29	9.24	9.13	9.09	9.08
6	7.60	7.56	7.52	7.48	7.45	7.42	7.40	7.30	7.23	7.14	7.09	6.99	6.95	6.93
7	6.36	6.31	6.28	6.24	6.21	6.18	6.16	6.06	5.99	5.91	5.86	5.75	5.72	5.70
8	5.56	5.52	5.48	5.44	5.41	5.38	5.36	5.26	5.20	5.12	2.07	4.95	4.93	4.91
9	5.01	4.96	4.92	4.89	4.86	4.83	4.81	4.71	4.65	4.57	4.52	4.41	4.38	4.36
10	4.60	4.56	4.52	4.49	4.46	4.43	4.41	4.31	4.25	4.17	4.12	4.01	3.98	3.93
11	4.29	4.25	4.21	4.18	4.15	4.12	4.10	4.01	3.94	3.86	3.81	3.71	3.67	..66
12	4.05	4.01	3.97	3.94	3.91	3.88	3.86	3.76	3.70	3.62	3.57	3.47	3.43	3.41
13	3.86	3.82	3.78	3.75	3.72	3.69	3.66	3.57	3.51	3.43	3.38	3.27	3.24	3.22
14	3.70	3.66	3.62	3.59	3.56	3.53	3.51	3.41	3.35	3.27	3.22	3.11	3.08	3.06
15	3.56	3.52	3.49	3.45	3.42	3.40	3.37	3.28	3.21	3.13	3.08	2.98	2.94	2.92
16	3.45	3.41	3.37	3.34	3.31	3.28	3.26	3.16	3.10	3.02	0.97	2.86	2.83	2.81
17	3.35	3.31	3.27	3.24	3.21	3.19	3.16	3.07	3.00	0.92	2.87	2.76	2.73	2.71
18	3.27	3.23	3.19	3.16	3.13	3.10	3.08	2.98	2.92	2.84	2.78	2.68	2.64	2.62
19	3.19	3.15	3.12	3.08	3.05	3.03	3.00	2.91	2.84	2.76	2.71	2.60	2.57	2.55
20	3.13	3.09	3.05	3.02	2.99	2.96	2.94	2.84	2.78	2.69	2.64	2.54	2.50	2.48
21	3.07	3.03	2.99	2.96	2.93	2.90	2.88	2.79	2.72	2.64	2.58	2.48	2.44	2.42
22	3.02	2.98	2.94	2.91	2.88	2.85	2.83	2.73	2.67	2.58	2.53	2.42	2.38	2.36
23	2.97	2.93	2.89	2.86	2.83	2.80	2.78	2.69	2.62	2.54	2.48	2.37	2.34	2.32
24	2.93	2.89	2.85	2.82	2.79	2.76	2.74	2.64	2.58	2.49	2.44	2.33	2.29	2.27
25	2.89	2.85	2.81	2.78	2.75	2.72	2.70	2.60	2.54	2.45	2.40	2.29	2.25	2.23
26	2.86	2.81	2.78	2.75	2.72	2.69	2.66	2.57	2.50	2.42	2.36	2.25	2.21	2.19
27	2.82	2.78	2.75	2.71	2.68	2.66	2.63	2.54	2.47	2.38	2.33	2.22	2.18	2.16
28	2.79	2.75	2.72	2.68	2.65	2.63	2.60	2.51	2.44	2.35	2.30	2.19	2.15	2.13
29	2.77	2.73	2.69	2.66	2.63	2.60	2.57	2.48	2.41	2.33	2.27	2.16	2.12	2.10
30	2.74	2.70	2.66	2.63	2.60	2.57	2.55	2.45	2.39	2.30	2.25	2.13	2.09	2.07
32	2.70	2.65	2.62	2.58	2.55	2.53	2.50	2.41	2.34	2.25	2.20	2.08	2.04	2.02
34	2.66	2.61	2.58	2.54	2.51	2.49	2.46	2.37	2.30	2.21	2.16	2.04	2.00	1.98
36	2.62	2.58	2.54	2.51	2.48	2.45	2.43	2.33	2.26	2.18	2.12	2.00	1.96	1.94
38	2.59	2.55	2.51	2.48	2.45	2.42	2.40	2.30	2.23	2.14	2.09	1.97	1.93	1.90
40	2.56	2.52	2.48	2.45	2.42	2.39	2.37	2.27	2.20	2.11	2.06	1.94	1.90	1.87
42	2.54	2.50	2.46	2.43	2.40	2.37	2.34	2.25	2.18	2.09	2.03	1.91	1.87	1.85
44	2.52	2.47	2.44	2.40	2.37	2.35	2.32	2.22	2.15	2.07	2.01	1.89	2.84	1.82
46	2.50	2.45	2.42	2.38	2.35	2.33	2.30	2.20	2.13	2.04	1.99	1.86	1.82	1.80
48	2.48	2.44	2.40	2.37	2.33	2.34	2.28	2.18	2.12	2.02	1.97	1.84	1.80	1.78
50	2.46	2.42	2.38	2.35	2.32	2.29	2.27	2.17	2.10	2.01	1.95	1.82	1.78	1.76
60	2.39	2.35	2.31	2.28	2.25	2.22	2.20	2.10	2.03	1.94	1.88	1.75	1.70	1.68
70	2.35	2.31	2.27	2.23	2.20	2.18	2.15	2.05	1.98	1.89	1.83	1.70	1.65	1.62
80	2.31	2.27	2.23	2.20	2.17	2.14	2.12	2.01	1.94	1.85	1.79	1.65	1.61	1.58
90	2.29	2.24	2.21	2.17	2.14	2.11	2.09	1.99	1.92	1.80	1.76	1.62	1.57	1.55
100	2.27	2.22	2.19	2.15	2.12	2.09	2.07	1.97	1.89	1.80	1.74	1.60	1.55	1.52
125	2.23	2.19	2.15	2.11	2.08	2.05	2.03	1.93	1.85	1.73	1.69	1.55	1.50	1.47
150	2.20	2.16	2.12	2.09	2.06	2.03	2.00	1.90	1.83	1.73	1.66	1.52	1.46	1.43
200	2.17	2.13	2.09	2.06	2.03	2.00	1.97	1.87	1.79	1.69	1.63	1.48	1.42	1.39
300	2.14	2.10	2.06	2.03	1.99	1.97	1.94	1.84	1.76	1.66	1.59	1.44	1.38	1.35
500	2.12	2.07	2.04	2.00	1.97	1.94	1.92	1.81	1.74	1.63	1.57	1.41	1.34	1.31
1000	2.10	2.06	2.02	1.98	1.95	1.92	1.90	1.79	1.72	1.61	1.54	1.38	1.32	1.28

Degrees of Freedom for Denominator

Table 5 | Control Chart Constants

Observations in Sample, n	Chart For Averages — Factors for Control Limits			Chart for Standard Deviations — Factors for Central Line		Chart for Standard Deviations — Factors for Control Limits				Chart for Ranges — Factors for Central Line		Chart for Ranges — Factors for Control Limits				
	A	A_2	A_1	c_4	$1/c_4$	B_3	B_4	B_5	B_6	d_2	$1/d_2$	d_3	D_1	D_2	D_3	D_4
2	2.121	1.880	2.659	0.7979	1.2533	0	3.267	0	2.606	1.128	0.8862	0.853	0	3.686	0	3.267
3	1.732	1.023	1.954	0.8862	1.1284	0	2.568	0	2.276	1.693	0.5908	0.888	0	4.358	0	2.575
4	1.500	0.729	1.628	0.9213	1.0854	0	2.266	0	2.088	2.059	0.4857	0.880	0	4.698	0	2.282
5	1.342	0.577	1.427	0.9400	1.0638	0	2.089	0	1.964	2.326	0.4299	0.864	0	4.918	0	2.114
6	1.225	0.483	1.287	0.9515	1.0510	0.030	1.970	0.029	1.874	2.534	0.3946	0.848	0	5.079	0	2.004
7	1.134	0.419	1.182	0.9594	1.0424	0.118	1.882	0.113	1.806	2.704	0.3698	0.833	0.205	5.204	0.076	1.924
8	1.061	0.373	1.099	0.9650	1.0363	0.185	1.815	0.179	1.751	2.847	0.3512	0.820	0.388	5.307	0.136	1.864
9	1.000	0.337	1.032	0.9693	1.0317	0.239	1.761	0.232	1.707	2.970	0.3367	0.808	0.547	5.393	0.184	1.816
10	0.949	0.308	0.975	0.9727	1.0281	0.284	1.716	0.276	1.669	3.078	0.2149	0.797	0.686	5.469	0.223	1.777
11	0.905	0.285	0.927	0.9754	1.0253	0.321	1.679	0.313	1.637	3.173	0.3152	0.787	0.811	5.535	0.256	1.744
12	0.866	0.266	0.886	0.9776	1.0230	0.354	1.646	0.346	1.610	3.258	0.3069	0.778	0.923	5.594	0.283	1.717
13	0.832	0.249	0.850	0.9794	1.0210	0.382	1.618	0.374	1.585	3.336	0.2998	0.770	1.025	5.647	0.307	1.693
14	0.802	0.235	0.817	0.9810	1.0194	0.406	1.594	0.399	1.563	3.407	0.2935	0.763	1.118	5.696	0.328	1.672
15	0.755	0.223	0.789	0.9823	1.0180	0.428	1.572	0.421	1.544	3.472	0.2880	0.756	1.203	5.740	0.347	1.653
16	0.750	0.212	0.763	0.9835	1.0168	0.448	1.552	0.440	1.526	3.532	0.1831	0.750	1.282	5.782	0.363	1.637
17	0.728	0.203	0.739	0.9845	1.0157	0.466	1.534	0.458	1.511	3.588	0.2787	0.744	1.356	5.820	0.378	1.622
18	0.707	0.194	0.718	0.9854	1.0148	0.482	1.518	0.475	1.496	3.640	0.2747	0.739	1.424	5.856	0.391	1.609
19	0.688	0.187	0.698	0.9862	1.0140	0.497	1.503	0.490	1.483	3.689	0.2711	0.733	1.489	5.889	0.404	1.596
20	0.671	0.180	0.680	0.9869	1.0132	0.510	1.490	0.504	1.470	3.735	0.2677	0.729	1.549	5.921	0.415	1.585
21	0.655	0.173	0.663	0.9876	1.0126	0.523	1.477	0.516	1.459	3.778	0.2647	0.724	1.606	5.951	0.425	1.575
22	0.640	0.167	0.647	0.9882	1.0120	0.534	1.466	0.528	1.448	3.819	0.2618	0.720	1.660	5.979	0.435	1.565
23	0.626	0.162	0.633	0.9887	1.0114	0.545	1.455	0.539	1.438	3.858	0.2592	0.716	1.711	6.006	0.443	1.557
24	0.612	0.157	0.619	0.9892	1.0109	0.555	1.445	0.549	1.429	3.895	0.2567	0.712	1.759	6.032	0.452	1.548
25	0.600	0.135	0.606	0.9896	1.0105	0.565	1.435	0.559	1.420	3.931	0.2544	0.708	1.805	6.056	0.459	1.541

Source: From *Manual on Presentation of Data and Control Chart Analysis*, Sixth Edition, p. 91. Copyright © American Society for Testing and Materials. Reprinted with permission.

Answers to Selected Exercises

⁚ Chapter 1

1.1   (a) coating thickness   (b) viscosity (continuous)   (c) low and high
(d) with only one factor, this is the same as part c   (e) the experimental unit and
the observational unit are an individual part   (f) randomly select panels from the two levels
of viscosity

1.3   (a) plate thickness after 200 charge–discharge cycles   (b) porosity (continuous)
(c) low and high   (d) with only one factor, this is the same as part c
(e) the experimental unit and the observational unit are a plate
(f) for each ED bath, randomly select 5 plates from both low and high porosity

1.5   (a) rating   (b) method (categorical)   (c) motor and research
(d) with only one factor, this is the same as part c   (e) the experimental unit and
the observational unit are a blend   (f) randomly select a blend of a
certain octane, divide the blend in two, and randomly assign each half to a method

1.7   (a) maximum output (in fluid ounces) per hour   (b) brand of humidifier
(categorical)   (c) A and B   (d) only one factor, this is the same as part c
(e) the experimental unit and the observational unit are a humidifier
(f) assign the 16 humidifiers to the chamber in random order

1.9   (a) amount of chlorine   (b) vendor (categorical)   (c) A, B, C, D, E
(d) only one factor, this is the same as part c   (e) the experimental unit and
the observational unit are a sponge   (f) test the 20 sponges in random order

1.11   (a) pressure to separate the cap from the bottle   (b) injection speed, mold
temperature, and cooling time (all continuous)   (c) injection speed (40 and 75),
mold temperature (25 and 45), cooling time (10 and 25)
(e) (1) the experimental unit and the observational unit are a cavity (2) for each of
the 4 shots, randomly assign a treatment to a cavity
(f) (1) the experimental unit is a shot, and the observational units are the cavities
(2) assign the treatments to the shots in random order

1.13   (a) vertical component of a dynamometric reading   (b) angle of edge level
(continuous) and type of cut (categorical)   (c) angle of edge level (15 & 30),
type of cut (continuous and interrupted)   (e) the experimental unit and
observational unit are a piece of metal   (f) for each piece of metal, carry out the four cuts
in random order

1.15   (a) texture of a cake   (b) amount of flour, amount of egg powder, amount of oil,
temperature of oven (all continuous)   (c) amount of flour (low and high),
amount of egg powder (low and high), amount of oil (low and high), temperature of
oven (375 and 400)   (e) no   (f) the experimental unit and observational
unit are the oven   (g) the experimental unit and observational unit are a cake

⫶ Chapter 2

2.21 median = 80, Q1 = 78, Q3 = 81, UIF = 85.5, UOF = 90, LIF = 73.5, LOF = 69

2.23 median = 2.7, Q1 = 1.8, Q3 = 3.2, UIF = 5.3, UOF = 7.4, LIF = −.3, LOF = −2.4

2.25 median = 18.1, Q1 = 13.8, Q3 = 21.5, UIF = 33.05, UOF = 44.6, LIF = 2.25, LOF = −9.3, mild outliers (35.1, 40.3), extreme outliers (52.3, 79.2)

2.27 median = 119, Q1 = 87, Q3 = 182, UIF = 324.5, UOF = 467, LIF = −55.5, LOF = −198, extreme outlier (511)

2.29 median = 2.0018, Q1 = 2.0015, Q3 = 2.0021, UIF = 2.003, UOF = 2.0039, LIF = 2.0006, LOF = 1.9997

2.31 median = 1331, Q1 = 865, Q3 = 1727, UIF = 3020, UOF = 4313, LIF = −428, LOF = −1721, mild outlier (3830)

2.33 Supplier 1: median = 85.35, Q1 = 84.6, Q3 = 86.25, UIF = 88.725, UOF = 91.2, LIF = 82.125, LOF = 79.65
Supplier 2: median = 85.25, Q1 = 83.65, Q3 = 87.6, UIF = 93.525, UOF = 99.45, LIF = 77.725, LOF = 71.8

2.35 Period 1: median = 5.5, Q1 = 4.5, Q3 = 8, UIF = 13.25, UOF = 18.5, LIF = −.75, LOF = −6
Period 2: median = 2.5, Q1 = 1.5, Q3 = 4, UIF = 7.75, UOF = 11.5, LIF = −2.25, LOF = −6

2.37 Time Period 1: median = 33.5, Q1 = 33.25, Q3 = 33.8, UIF = 34.625, UOF = 35.45, LIF = 32.425, LOF = 31.60
Time Period 2: median = 34.6, Q1 = 33.9, Q3 = 34.75, UIF = 36.025, UOF = 37.30, LIF = 32.625, LOF = 31.35

2.39 Freshmen: median = 19, Q1 = 18, Q3 = 21, UIF = 25.5, UOF = 30, LIF = 13.5, LOF = 9
Upperclassmen: median = 23, Q1 = 21, Q3 = 23, UIF = 26, UOF = 29, mild outlier (17), LIF = 18, LOF = 15, extreme outlier (15)

⫶ Chapter 3

3.1 (a) 0.896 (b) 0.104 (c) 0.600 (d) 0.480, 0.693

3.3 (a) 0.050 (b) 0.995 (c) 3.830 (d) 0.331, 0.575

3.5 (a) 0.210 (b) 0.689 (c) 1.000, 0.735, 0.857

3.7 (a) 0.896 (b) 0.09 (c) 0.478 (d) 0.802, 0.8955

3.9 (a) 0.8171 (b) 0.1829 (c) 0.2, 0.196, 0.4427

3.11 (a) 0.000 (b) 0.446 (c) 0.25, 0.2375, 0.4873

3.13 (a) (1) 0.0007 (2) 0.5021 (3) 0.64, 0.5376, 0.7332
(b) (1) 0.2479 (2) 0.2518 (3) 1.28, 1.075, 1.037

3.15 (a) 0.3801 (b) 0.2832 (c) 1.05, 0.9765, 0.9882

3.17 (a) 0.9427 (b) 0.3366 (c) 7.6, 0.38, 0.6164 (d) 0.4, 0.38, 0.6164

3.19 (a) 0.2501 (b) 0.0352 (c) 3.0, 2.4, 1.549 (d) 12, 2.4, 1.549

3.21 (a) 0.2231 (b) 0.7769 (c) 0.0141 (d) 1.5, 1.5, 1.2247

3.23 (a) 0.4724 (b) 0.0332 (c) 0.75, 0.75, 0.866

3.25 (a) 0.1044 (b) 5, 5, 2.236

3.27 (a) 0.0821 (b) 0.7127 (c) 2.5, 2.5, 1.581

3.29 (a) 0.3679 (b) 0.5488 (c) yes (d) 1, 1, 1

3.31 (a) 0.1029 (b) 0.3 (c) 3.33, 7.78, 2.79 (d) 0.0908

3.33 (a) 0.0387 (b) 0.90 (c) 10, 90, 9.48 (d) 0.018 (e) 40, 360, 18.97

3.35 (a) 0.0364 (b) 14.28, 189.8, 13.78 (c) 0.0171 (d) 42.86, 569.38, 23.86

3.37 (a) 0.0776 (b) 0.9212 (c) 0.08, 0.076, 0.2757

3.39 (a) 0.025 (b) 0.861 (c) 20 (d) 400, 20

3.41 (a) 0.4512 (b) 0.4066

3.43 (d) (1) 0.5 (2) 0.167 (3) 15 (4) 75, 8.66

3.45 (b) 0.206 (c) $5\sqrt{\pi}/4$ (d) 1.341, 1.158

3.47 (a) 0.383 (b) 0.3085

3.49 (a) 0.1056 (b) 0.1056 (c) 0.1034 (d) $\mu < 90.76$

3.51 (a) 0.3745 (b) 0.7548

3.53 (a) 0.0062 (b) 0.8664 (c) 74.12

3.55 (a) 0.905 (b) 0.0038

3.57 (a) 0.8026 (b) large enough n for CLT

3.59 (a) 0.9812 (b) 0.2263 (c) large enough n for CLT

3.61 (a) 0.0000 (b) 0.0174 (c) 0.9826 (d) CLT; stem-leaf/normal prob. plot

3.63 (a) 35.40 (b) 0.1685 (c) 31, 36 (d) large enough n for CLT

3.65 (a) 0.9075 (b) −2.05 (c) well-behaved pop.(distribution); stem-leaf/normal prob. plot

3.67 (a) 98.7, 8.23, 2.87 (b) −1.43 (c) well-behaved pop.(distribution); stem-leaf/normal prob. plot

3.69 (a) 18.8 (b)−0.43 (c) well-behaved pop. (distribution); stem-leaf/normal prob. plot

3.71 (a) 142.65, 9644.18, 98.204 (b) −4.02 (c) well-behaved pop. (distribution); stem-leaf/normal prob. plot (d) largest:500, smallest:142.65

3.73 (a) 0.9868 (b) 0.5398

3.75 (a) 0 (b) 0.0721

3.77 (a) 0 (b) 0.8577 (c) 0.4274 (d) 0.4271

∴ Chapter 4

4.1 (a) (5.01, 6.59) (b) 48 (c) CLT; stem-leaf/normal prob. plot

4.3 (a) (1.0752, 1.1928) (b) 139 (c) CLT; stem-leaf/normal prob. plot

4.5 (a) (0.8984, 0.9166) (b) 27 (c) CLT; stem-leaf/normal prob. plot

4.7 (a) (58.75, 75.05) (b) 27 (c) CLT; stem/leaf, normal prob. plot

4.9 (a) (2.14, 2.51) (b) 7 (c) CLT; stem/leaf, normal prob. plot

4.11 (a) H_a: $\mu > 100$, $z = 0.35$, Fail to reject H_0 (b) 0.3632 (d) 0.0336 (e) 161 (f) CLT; stem-leaf/normal prob. plot

4.13 (a) H_a: $\mu \neq 6500$, $z = -0.16$, Fail to reject H_0 (b) 0.8728 (d) 0.4840 (e) 46 (f) CLT; stem-leaf/normal prob. plot

4.15 (a) H_a: $\mu > 60$, $z = 2.18$, Fail to reject H_0 (b) 0.0146 (d) 0.2005 (e) 64 (f) CLT; stem-leaf/normal prob. plot

4.17 (a) H_a: $\mu \neq 2.2$, $z = 0.32$, Fail to reject H_0 (b) 0.7490 (d) 0.5438 (e) 17 (f) CLT; stem-leaf/normal prob. plot

4.21 (a) H_a: $\mu \neq 8$, $t = -0.85$, Fail to reject H_0 (b) (7.919, 8.033) (c) (7.57, 8.38) (d) well-behaved pop.(distribution); stem-leaf/normal prob. plot

4.23 (a) H_a: $\mu \neq 4.5$, $t = -0.408$, Fail to reject H_0 (b) (4.19, 4.73) (c) (3.79, 5.13) (d) well-behaved pop.(distribution); stem-leaf/normal prob. plot

4.25 (a) H_a: $\mu \neq 200$, $t = 5.81$, Reject H_0 (b) (203.54, 209.93) (c) (188.94, 224.53) (d) well-behaved pop.(distribution); stem-leaf/normal prob. plot

4.27 (a) H_a: $\mu \neq 100$, $t = -0.6653$, Fail to reject H_0 (b) (97.98, 101.03) (c) (91.73, 107.29) (d) well-behaved pop.(distribution); stem-leaf/normal prob. plot

4.29 (a) H_a: $\mu \neq 196$, $t = 2.22$, Reject H_0 (b) (196.02, 196.21) (c) (195.83, 196.39) (d) well-behaved pop.(distribution); stem-leaf/normal prob. plot

4.31 (a) H_a: $p < 0.12$, $z = -1.56$, Fail to reject H_0 (b) (0.055, 0.121)

4.33 (a) H_a: $p > 0.10$ $z = 5.05$, Reject H_0 (b) (0.135, 0.301)

4.35 (a) H_a: $p > 0.10$, $z = 7.16$, Reject H_0 (b) (0.196, 0.324)

4.37 (a) H_a: $p < 0.07$, $z = -0.55$, Fail to reject H_0 (b) (0.03, 0.09)

4.39 (b) H_a: $\mu_H - \mu_L > 0$, $t = 0.924$, Fail to reject H_0 (c) (−0.168, 0.446) (d) two independent samples from well-behaved populations with equal variances; stem-leafs/normal prob. plots

4.41 (b) H_a: $\mu_A - \mu_V \neq 0$, $t = -3.03$, Reject H_0 (c) (−497.07, −12.93) (d) two independent samples from well-behaved populations with equal variances; stem-leafs/normal prob. plots

4.43 (b) H_a: $\mu_1 - \mu_2 \neq 0$, $t = 2.14$, Reject H_0 (c) (0.16, 3.24) (d) two independent samples from well-behaved populations with equal variances; stem-leafs/normal prob. plots

4.45 (b) H_a: $\mu_A - \mu_B \neq 0$, $t = 0.16$, Fail to reject H_0 (c) (−2294.9, 2692.9) (d) two independent samples from well-behaved pop. with equal variances; stem-leafs/normal prob. plots

4.47   (b) H_a: $\mu_A - \mu_B \neq 0$, $t = -1.38$, Fail to reject H_0 (c) $(-180.4, 30.54)$ (d) two independent samples from well-behaved pop. with equal variances; stem-leafs/normal prob. plots

4.49   (a) H_a: $\delta \neq 0$, $t = 0.11$, Fail to reject H_0 (b) $(-0.277, 0.3053)$ (c) paired data; differences sampled from well-behaved pop.; stem-leaf/normal prob. plot of differences

4.51   (a) H_a: $\delta \neq 0$, $t = -26.97$, Reject H_0 (b) $(-0.227, -0.194)$ (c) paired data; differences sampled from well-behaved pop; stem-leaf/normal prob. plot of differences

4.53   (a) H_a: $\delta > 0$, $t = 2.213$, Reject H_0 (b) $(-0.01, 1.15)$ (c) paired data; differences sampled from well-behaved pop; stem-leaf/normal prob. plot of differences

4.55   (a) H_a: $\delta \neq 0$, $t = -16.56$, Reject H_0 (b) $(-0.72, -0.52)$ (c) paired data; differences sampled from well-behaved pop; stem-leaf/normal prob. plot of differences

4.57   (a) H_a: $\sigma^2 = 60$, $\chi^2 = 4018.4$, Reject H_0 (b) $(5931.75, 18376.67)$ (c) well-behaved pop.(distribution); stem-leaf/normal prob. plot

4.59   (a) H_a: $\sigma^2 = 144$, $\chi^2 = 2.278$, Reject H_0 (b) $(8.335, 26.456)$ (c) well-behaved pop. (distribution); stem-leaf/normal prob. plot

4.61   (a) $\sigma^2 > 70$, $\chi^2 = 9$, Fail to reject H_0 (b) $(35.93, 288.99)$ (c) well-behaved pop.; stem-leaf/normal prob. plot

4.65   (a) $\sigma_1^2 \neq \sigma_2^2$, $F = 2.09$, Fail to reject H_0 (b) two independent samples from well-behaved populations; stem-leafs/normal prob. plots

4.67   (a) $\sigma_1^2 \neq \sigma_2^2$, $F = 2.94$, Reject H_0 (b) two independent samples from well-behaved populations; stem-leafs/normal prob. plots

4.69   (a) $\sigma_1^2 \neq \sigma_2^2$, $F = 2.06$, Fail to reject H_0 (b) two independent samples from well-behaved populations; stem-leafs/normal prob. plots.

4.71   (b) H_a: Median < 80, p-value $= 0.0946$, Reject H_0 (c) No   (d) Yes, natural log transformation   (e) H_a: $\mu < 4.382$, $t = -2.32$, Reject H_0

4.73   (b) $W = 143$, p-value $= 0.7075$, Fail to reject H_0

4.75   (b) $W = 122$, p-value $= 0.2123$, Fail to reject H_0

4.77   (b) $t = -2.41$, p-value $= 0.037$, Reject H_0 (c) $W = 9.5$, p-value $= 0.041$, Reject H_0

⟫ Chapter 5

5.1 $\overline{X}$-chart, LCL $= 0.91$, UCL $= 1.09$

5.3 $\overline{X}$-chart, LCL $= 88$, UCL $= 112$

5.5 $\overline{X}$-chart, LCL $= 10.366$, UCL $= 10.634$

5.7 $\overline{X}$-chart, LCL $= 92$, UCL $= 104$

5.9   (a) R-chart, LCL $= 0$, UCL $= 4.56$ (b) $\overline{X}$-chart, LCL $= 1.39$, UCL $= 3.87$

5.11   (a) R-chart, LCL $= 0$, UCL $= 15.23$ (b) $\overline{X}$-chart, LCL $= 29.91$, UCL $= 38.21$

5.13 (a) R-chart, LCL = 0, UCL = 15.44   (b) $\overline{X}$-chart, LCL = 15.15, UCL = 23.57

5.15 (a) R-chart, LCL = 0, UCL = 1.418   (b) $\overline{X}$-chart, LCL = 1.606, UCL = 2.733
       (d) R-chart, LCL = 0.208, UCL = 2.053   (e) $\overline{X}$-chart, LCL = 1.691, UCL = 2.452

5.17 (a) s^2-chart, LCL = 0.03, UCL = 4.34   (b) $\overline{X}$-chart, LCL = 1.31, UCL = 3.95

5.19 (a) s^2-chart, LCL = 0.22, UCL = 37.67   (b) $\overline{X}$-chart, LCL = 30.16, UCL = 37.96

5.21 (a) s^2-chart, LCL = 0.27, UCL = 44.61   (b) $\overline{X}$-chart, LCL = 15.11, UCL = 23.61

5.23 (a) s^2-chart, LCL = 0, UCL = 0.691   (b) $\overline{X}$-chart, LCL = 1.61, UCL = 2.73

5.25 (a) X-chart, LCL = 7.468, UCL = 8.532

5.27 (a) X-chart, LCL = −0.56, UCL = 0.56

5.29 (a) X-chart, LCL = 9.50, UCL = 27.3

5.31 np-chart, LCL = 7.42, UCL = 32.98

5.33 np-chart, LCL = 3.72, UCL = 26.42

5.35 np-chart, LCL = 0.51, UCL = 18.61

5.37 np-chart, LCL = 7.06, UCL = 32.34

5.41 c-chart, LCL = −1.46, no LCL, UCL = 12.86

5.43 c-chart, LCL = 1.25, UCL = 21.48

5.45 c-chart, LCL = 0.41, UCL = 19.19

5.47 c-chart, LCL = 0, UCL = 8.103

⋗ Chapter 6

6.4 (b) Modulus = $8.22 - 0.0602\%$ PET
       (c)

```
Predictor        Coef        Stdev      t-ratio         p
Constant       8.2206       0.3512        23.41     0.000
%PET         -0.060165     0.004355      -13.82     0.000

s = 0.3897      R-sq = 97.4%      R-sq(adj) = 96.9%

Analysis of Variance

SOURCE        DF          SS           MS         F        p
Regression     1       28.991       28.991    190.86    0.000
Error          5        0.759        0.152
Total          6       29.750
```

6.5 (a) Visc $= 1.28 - 0.00876$Temp

(b)

```
Predictor        Coef        Stdev      t-ratio         p
Constant      1.28151      0.04687       27.34      0.000
Temp        -0.0087578    0.0007284     -12.02      0.000

s = 0.04743     R-sq = 96.0%     R-sq(adj) = 95.4%

Analysis of Variance

SOURCE        DF          SS          MS          F        p
Regression    1       0.32529     0.32529     144.58    0.000
Error         6       0.01350     0.00225
Total         7       0.33879
```

6.16 The regression equation is %imp $= -13.9 + 0.100$temp $+ 0.514$conc

```
Predictor        Coef        Stdev      t-ratio         p
Constant       -13.86       30.55        -0.45      0.659
temp           0.0996       0.1965        0.51      0.622
conc           0.5139       0.4290        1.20      0.256

s = 0.6254      R-sq = 11.6%     R-sq(adj) = 0.0%

Analysis of Variance

SOURCE        DF          SS          MS          F        p
Regression    2       0.5656      0.2828       0.72     0.507
Error         11      4.3030      0.3912
Total         13      4.8686

SOURCE        DF        SEQ SS
temp          1        0.0043
conc          1        0.5613
```

6.19 The regression equation is index $= -18.3 + 0.112$iron $+ 0.398$almn $+ 1.42$ph

```
Predictor        Coef        Stdev      t-ratio         p
Constant       -18.25       20.75        -0.88      0.402
iron          0.11216      0.03083        3.64      0.005
almn           0.3983       0.1184        3.36      0.008
ph             1.421        2.663         0.53      0.607

s = 4.545       R-sq = 95.0%     R-sq(adj) = 93.3%
```

```
Analysis of Variance

SOURCE        DF           SS          MS         F         p
Regression     3        3535.8      1178.6     57.06     0.000
Error          9         185.9        20.7
Total         12        3721.7


SOURCE        DF        SEQ SS
iron           1        3070.5
almn           1         459.4
ph             1           5.9
```

⟩ Chapter 7

7.6   (a) effect of Velocity -1.9375, effect of Viscosity 3.5375, Interaction -0.3125
(b) using coded variables $+1$, -1 The regression equation is
$K = 23.4 - 0.97\text{vlcty} + 1.77\text{vscsty} - 0.16v.v$

```
Predictor        Coef       Stdev     t-ratio        p
Constant       23.356       1.405       16.63     0.000
vlcty          -0.969       1.405       -0.69     0.504
vscsty          1.769       1.405        1.26     0.232
v.v            -0.156       1.405       -0.11     0.913

s = 5.619        R-sq = 14.7%      R-sq(adj) = 0.0%

Analysis of Variance

SOURCE        DF           SS          MS         F         p
Regression     3         65.46       21.82      0.69     0.575
Error         12        378.88       31.57
Total         15        444.34


SOURCE        DF        SEQ SS
vlcty          1         15.02
vscsty         1         50.06
v.v            1          0.39
```

Index

2² factorial design, 453, 454
2³ factorial design, 29, 468–471

A

additive law of probability, 98
adjacents, 64
alias structure, 493
analysis of variance (ANOVA) table, 381, 382
assignable cause, 277
autocorrelation, 285, 286, 287
average run length, 341–344, 345

B

Bayes rule, 99
binomial distribution, 107–111, 329
    continuity correction, 165–166
    mean, 109
    normal approximation, 164–167, 221–222, 329
    probability function, 109
    standard deviation, 109
    variance, 109
boxplot, 63–65
    adjacents, 64
    computer generated, 69–72, 73–76
    designed experiment, 80–83
    lower inner fence, 64
    lower outer fence, 65
    parallel, 65–67
    upper inner fence, 64
    upper outer fence, 65
    whisker, 64

C

capability indices, 294, 295
c chart, 335–338
    lower control limit, 335–338
    upper control limit, 335–338
center runs, 478, 527

central composite design (ccd), 30, 526, 527
Central Limit Theorem, 145–146, 154, 164, 279
χ^2 distribution, 249, 250, 251, 252, 253, 313
coefficient of determination, 377, 378, 400
collinearity diagnostics, 442
collinearity, 438–440
combined array, 555
complements, 98, 100
completely randomized design, 26, 27
condition number, 442
conditional probability, 94–96
confidence interval, 178–181
    determining sample size, 181, 182, 226
    difference of two independent means, 232
    one-sided, 212
    paired data, 242
    proportions, 225, 226
    single mean, 179–180, 211, 215
    single variance, 249, 250
confirmatory experiment, 5
continuity correction, 165–166
continuous improvement, 1
contour plot, 524
control array, 549
control chart, 2, 278–280, 287, 288
    average run length, 341–344, 345
    false alarm, 280
    in control process, 277
    lower control limit, 280, 281
    model, 275
    out of control process, 277
    process mean, 275
    process variance, 275
    runs rules, 344–347

    subgroup, 278, 285–287
    upper control limit, 280, 281
control factors, 548
critical region, 191, 192–193, 210–211, 214, 223, 224, 231, 241, 251, 254
crossed array, 549
cumulative distribution function (cdf), 102, 103, 125
CUSUM chart, 347–351

D

degrees of freedom, 156, 210–211, 214, 231, 241, 251, 254, 313, 314, 372, 400
Deming, W. Edwards, 278
DeMorgan's law, 100
depth, 44, 58–62
design resolution, 502
design variable, 454
designed experiment, 11, 14, 16, 363
desirability function, 537–540
deterministic model, 9, 11
distribution
    continuous, 125
    expected values, 104, 105, 128, 129
    general, 102
    discrete, 103–105
dual response approach, 558

E

effects, 16, 454, 455–458, 470–472
empirical rule, 105, 107, 111, 131
engineering method, 1, 3
estimator, 175
    point, 175
    unbiased, 177
event, 91
EWMA, 352–354
expected value, 104, 105, 128, 129

the graphic edge

rick poynor

First published in 1993
by Booth-Clibborn Editions
12 Percy Street
London W1P 9FB

Reprinted November 1993

Printed and bound in Hong Kong
by Dai Nippon Printing Co.

Copyright ©1993 Rick Poynor

ISBN 1873968 078

Published and distributed
in the United Kingdom.
Direct mail rights in Europe:
Internos Books
12 Percy Street
London W1P 9FB
England

Distributors for the rest of the world:
Hearst Books International
1350 Avenue of the Americas
New York NY 10019
United States of America

Distributed to the book trade and art trade
in the United States and Canada by:
North Light Books,
an imprint of F&W Publications, Inc.
1507 Dana Avenue
Cincinnati, Ohio 45207
United States of America

Book design: Sean Perkins
Design assistance: Wendelin Hess
Concept photography: Richard J. Burbridge
Project photography: Gareth McCarthy
Editorial assistance: Liz Farrelly

Acknowledgements

Thanks to Why Not Associates, Vicky Wilson,
Gerard Forde, Phoa Kia Boon, Williams and
Phoa, Adrian Caddy, Imagination, Carl Mooney,
Alan Drew, Vancouver Art Gallery, Lawrence
Weiner, Warner Books

The captions and artwork in this book are based
on material supplied by the designers whose
work is included. While every effort has been
made to ensure their accuracy, Booth-Clibborn
Editions does not under any circumstances
accept any responsibility for any errors
or omissions.

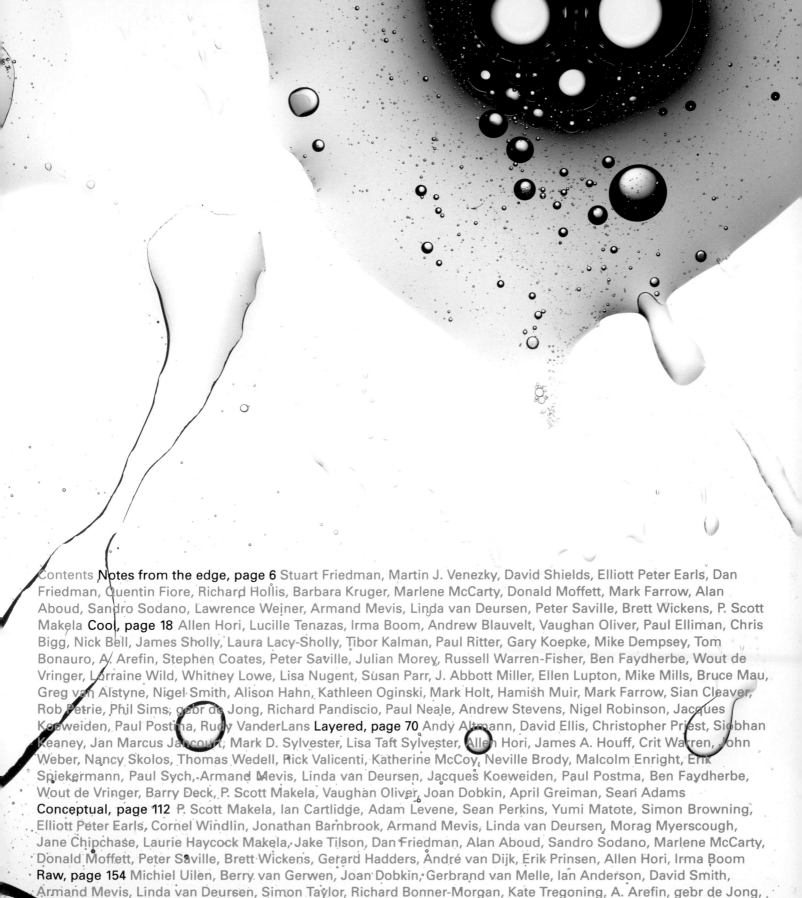

Notes from the edge This book has been conceived as a companion volume to an earlier book, *Typography Now: The Next Wave*. It is not a "part two" in the Hollywood sense in which an original idea becomes progressively thinner with each sequel or remake. Instead, it attempts to cover ground which the earlier book, by definition, could not examine. I should make it clear at the outset that this is not a book specifically about typography, or for that matter about a specific kind of typography. Hardly any of the work included is wholly typographic in content and examples of new typefaces are not shown, except where they feature as an integral part of a design. *The Graphic Edge*, as its title suggests, is a book about graphic design. It acknowledges the stylistic and expressive possibilities of the new digital typography, but argues the need at this point to go beyond the letterform as an end in itself and to reassess the ways in which word and image can interact to create graphic meaning. What this collection does share with *Typography Now* is a desire to piece together and begin to make sense of work by designers who in many cases choose to operate at the edge of the profession, or at the very least to distance themselves from the commercial mainstream. In the spirit of the work itself, this is a highly personal selection rather than the official conclusion of a jury or committee, but one that has been made with the journalistic and critical aim of showing new developments and trying to pin down that elusive phenomenon, the contemporary mood. It concentrates on work that looks of its time and which in the long view stands a good chance of coming to define the visual landscape of its time. In other words, it explores the characteristics that make the graphic design of the early 1990s different from what has gone before. These editorial aims mean that much else that is interesting or worthwhile, but less overtly concerned with formal and structural experiment, has been excluded. Other categories of graphic design are better dealt with at the proper length elsewhere.

The designs collected here, with a handful of exceptions, come from Great Britain, the United States and the Netherlands. Almost all of them were executed in the 1990s. They are divided into four thematic categories – "Cool", "Layered", "Conceptual" and "Raw" – intended to exemplify recent trends. As with any attempt at classification, the categories are not always mutually exclusive. There are many designs that could be included in only one category, but there are others that would fit reasonably smoothly into two or sometimes three categories. In such cases, classification has been determined by the characteristics that seem the most salient, and by the design's relationship to other pieces less ambiguously placed in the same category. It would have been possible to continue to break down categories until the book was divided into ten or more sections. Within the "Raw" section, for instance, I might have separated hand-drawn letterforms from machine generated typeforms. My purpose, however, was not to achieve a detailed taxonomy of graphic styles so much as a broad sequence of representative moods and themes. The editing and arrangement of designs within a given category should help to

reveal connections, confluences and contrasts of style and approach without the need for explicit division and labelling.

The category titles are intended to be self-explanatory, but before going on to a more general discussion I will offer some brief definitions. In the work included in the "Cool" category, compositional frameworks tend to be open and uncluttered, with white space showing through; pictorial elements are relatively simple and bold. Sans serif typefaces are preferred, though not obligatory, and much of the work shows clear signs of descent from earlier Swiss models. In "Layered", on the other hand, type and imagery are woven into compositions of much greater intricacy and congestion. In the most contorted and baroque examples, the design is saturated with visual and verbal information, leaving the viewer to decide what is most important and where to begin. The choice of typefaces is more varied and there is frequent use of the quirkier computer fonts. Stylistically many of the designs in "Conceptual" bear a resemblance to those in "Cool", but the work has a more personal and at times enigmatic quality; it is the kind of design most likely to be accused of

Photographic studies showing how
words influence our reading of an image
Project set by Dan Friedman
Designer: Stuart Friedman
Yale University
USA, 1992

"There are two kinds of light – the glow that illuminates,
and the glare that obscures." *James Thurber*

89

masquerading as art. There are signs of influence
from the art world in the use of type on image and
some of the projects have in fact been created for
artists or exhibitions. Sometimes this leads to poetic
mystification, sometimes it results in work of unusual
penetration and rigour. The designs in "Raw" are
equally personal, but the effect is less cerebral and the
level of formal idiosyncrasy much higher. This is the
least "designerly" work: spontaneity is preferred to
detachment, rough edges to professional finish,
emotional expression to suffocating good taste.

   Typography Now was criticised for showing
such a high proportion of work from what might be
described as the cultural arena, and the same thing
could doubtless be said of this collection. The reason
for this emphasis is simple: if you go looking for
experimentation, this is where you tend to find it.
The implication of the criticism is that a design
approach is somehow only valid if it presupposes,
and addresses itself to, the widest possible audience.
Despite the proliferating evidence to the contrary, the
idea of "universality" in graphic design dies hard.
A distinction is also drawn, sometimes explicitly,
between the supposedly esoteric domain of culture –

inhabited by whom? culture vultures? – and the
so-called "real" world inhabited by the rest of us.
This is a very peculiar and divisive view of culture and
one that bears little relation to what happens in
reality. All kinds of people watch films, visit the theatre,
read books and magazines, listen to records, and
go to exhibitions and concerts. These are not small
audiences, though it would hardly matter if they were.
What objection could there be in principle to forms of
design that treat these audiences as visually aware
and offer them graphic messages as sophisticated as
the cultural forms the messages promote? Perhaps
this work has implications for other areas of graphic
design. Perhaps it can function as a kind of "research
and development". I believe it often does. But even
if most of it proved to have no wider application,
this would not negate its effectiveness for its intended
audience, or detract from its interest as a form of
cultural expression.

   Much the same arguments can be applied to
the even more contentious area of "design about
design", the self-referential promotional projects and
student exercises for which the primary audience will
inevitably be other designers. *The Graphic Edge*, like

How can I show you devotion?
Type study
Designer: Martin J. Venezky
Cranbrook Academy of Art
USA, 1992

Silence
Poster
Designer: David Shields
Cranbrook Academy of Art
USA, 1992

Keep Him Away with the Left Jab!
Poster
Designer/illustrator: Elliott Peter Earls
Cranbrook Academy of Art
USA, 1991

Typography Now, includes a number of examples of this kind of work, from established professionals and students at Yale University and Cranbrook Academy of Art in the US and the Royal College of Art in London. This is not because the ideas contained within these designs can necessarily be translated directly into corporate identities and annual reports – this is far too limiting as a criterion for validity – but because such projects offer valuable safety zones in which designers can legitimately push at the boundaries of the discipline, and because the ideas (generally in a more diffuse form) may eventually come to nourish their commercial work and that of other professionals.

What this collection points up is that even at the experimental end of graphic design there are now two divergent currents: broadly speaking, the complex and the clean. Among the reactions to the kind of work collected in *Typography Now* were those who felt we had reached a point of maximum typographic density and distortion which left only one place to go: back to simplicity. "The limits may have been stretched so far that reevaluation is the logical next move," the critic Steven Heller wrote in an editorial in the *American Institute of Graphic Arts Journal* (vol. 10 no. 1, 1992).

In an issue of *Emigre* – the house magazine of experimental typographers – titled "Starting from Zero" (no. 19, 1991), Rudy VanderLans posed much the same question. Pointing to celebrated renunciations of complexity, from Jan Tschichold's classical conversion to Neville Brody's personal rediscovery of Helvetica, he asked: "Does all experimentation in graphic design lead to the simplification of graphic design? Are the graphic designers who concern themselves with complex solutions merely slow learners who try out the wildest schemes only to come to one conclusion, that less is more?"

Heller was inclined to doubt that simplicity was making a comeback, while the community of international type iconoclasts represented by *Emigre* shows no sign to date of experiencing a crisis of conscience and switching to centred type. And yet, as this book seeks to show, throughout the period in which the more flamboyant experimentalists gained so much publicity, there have been other equally committed attempts to make a contemporary design less reliant on sensational surface effect and more closely wedded to content. Of central importance here, but still little seen or discussed by graphic designers,

The Medium is the Massage
Book spreads
Designer: Quentin Fiore
Random House
USA, 1967

Ways of Seeing
Book cover
Designer: Richard Hollis
Penguin Books
Great Britain, 1972

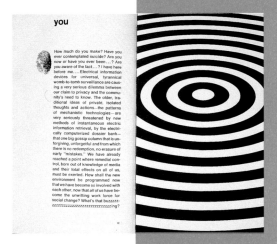

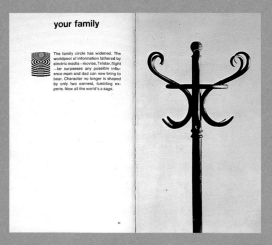

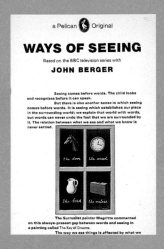

10 11

are two catalogues designed by Dan Friedman for the Deste Foundation for Contemporary Art in Athens.

Friedman, once identified with the post-modern American New Wave, is now a "Radical Modernist" who believes it is the narrow mannerisms of corporate modernism which are ethically and aesthetically bankrupt, not the transforming social and cultural vision of the movement itself. His designs for the catalogues *Artificial Nature* (1990) and *Post Human* (1992) belong to a small but visionary group of publications created since the 1960s which have sought to question and extend the word-dependent format of the printed book. The catalogues' most obvious precursors are the *Commentaires* of Chris Marker (1961 and 1967), collections of his film scripts; the designs of Quentin Fiore, in particular for the Marshall McLuhan reader, *The Medium is the Massage* (1967) – see *Eye* no. 8 vol. 2, 1993; and Richard Hollis' designs for John Berger's *Ways of Seeing* (1972) and Colin MacCabe's *Godard: images, sounds, politics.* (1980). In these books, the stream of picture material does not merely illustrate the words, but becomes an active force in the development of the argument, flowing into the text and out again and

sometimes carrying key ideas without any need of literary reinforcement.

Artificial Nature and *Post Human* work in broadly similar ways. A central essay, written by the exhibitions' curator, Jeffrey Deitch, is bracketed, front and back, by full-page pictures showing work by the artists together with brilliantly researched sequences of lurid photolibrary images germane to the shows' themes. In *Artificial Nature*, the reader is given three ways of accessing Deitch's deadpan argument that nature as a pure, untampered with, elemental phenomenon no longer exists; that we now inhabit an artificially constructed, "post-natural" world of bio-technology, beauty treatments, open-heart surgery, scented air-fresheners and virtual reality headsets. It is possible to read the large-print essay straight through as an ordinary linear text; to scan the keywords highlighted in black bars and key phrases picked out in red brush script; or to browse through the pictures, which are captioned with provocative words or slogan-like phrases drawn from the main essay: "Genuine nature may now be more artificial than natural", "Could it happen that the next generation will be our last generation of real humans?"

Artificial Nature
Exhibition poster (reverse side)
Designer: Dan Friedman
Client: Deste Foundation
for Contemporary Art, Athens
USA, 1990

Love for Sale
Book cover
Designer: Barbara Kruger
Abrams
USA, 1990

Knock. Knock.
Postcard
Designer: Donald Moffett
Co-production by Donald Moffett and
Jersey City Museum
USA, 1992

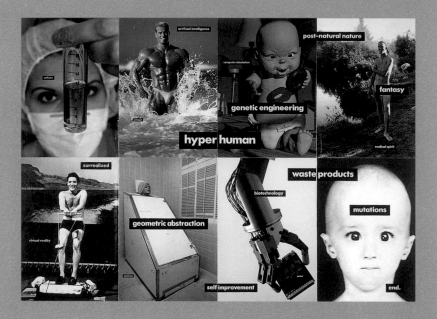

12 13

What makes these catalogues so fresh to eyes jaded by a surfeit of unnecessary styling is that they use only as much "design" as they need. It is not that they look undesigned – both books have a strong image and presence as objects. But this is precisely because Friedman has the confidence to allow the revelatory picture material to speak for itself. The simple Futura captions in *Artificial Nature* are fast-reading conductors for the catalogues' bolts of humour and satirical shocks; the pared-down graphic language is well suited to situations where the editorial message must be communicated with some urgency. It is an approach already familiar from the politically motivated, gender-based work of American artist Barbara Kruger. Indeed Kruger's career – she started as a magazine designer and picture editor at Condé Nast – underlines the complex chain of influence operating in this area, as the graphic strategies she learned as a professional designer and applied with such force as a media artist feed back into graphic design and become part of its vernacular. The Kruger format – a single telling word or phrase stamped across an image – can now be seen everywhere, from the activist graphics and commercial design projects of Donald Moffett and

Marlene McCarty, alias Bureau, in New York, to the record sleeve designs of Mark Farrow and HIV/AIDS awareness cards of Alan Aboud and Sandro Sodano in London.

For designers working at the cutting edge of graphic design, the contemporary art scene remains one of the most valuable testbeds for new ideas. And it is hardly surprising that art based on an examination of photographic, printed and televisual media should in turn prove so seductive and suggestive for designers. In the 1960s, Pop Art's transformation of printed sources from commercial art and popular culture triggered a wave of pop graphics. In the 1990s, in much the same way, the new graphic design throws out echoes of artists such as Edward Ruscha, Lawrence Weiner, John Baldessari and Richard Prince. The Dutch design duo Armand Mevis and Linda van Deursen, the English company Cartlidge Levene, and Elliott Peter Earls, a postgraduate student at the Cranbrook Academy of Art, create designs whose structural geometry and strange dislocated moods show many points of similarity with art of this kind. Peter Saville re-appropriates the Marlboro Man advertising icon,

Heart
by Pet Shop Boys
Record sleeve
Designer: Mark Farrow
Design company: 3a
Photographer: Eric Watson
Client: EMI Records
Great Britain, 1988

Biennale des Friedens
Poster
Artist: Lawrence Weiner
USA, 1985

Self-promotional advertisement
Designers: Simon Browning,
Sean Perkins, Yumi Matote
Design company: Cartlidge Levene
Great Britain, 1991

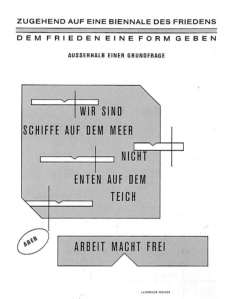

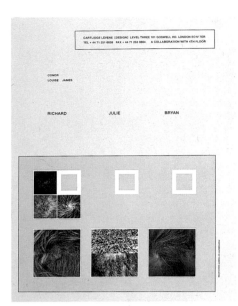

first appropriated by Richard Prince, for the English rock group New Order and designs a series of gloomy press ads for the fashion designer Yohji Yamamoto whose startling disjunctions between client and image were inspired by the art magazine advertising of the Bruno Bischofberger Gallery in Zurich. The gallery used banal Swiss scenes of a rifle range, pig slaughtering and traditional embroidery to represent artists like Andy Warhol and Francesco Clemente. Saville captures the anti-fashion mood of late 1991 with a drooping flower, a downbeat factory scene and a melting hot rod.

Such strategies assume a high degree of visual literacy on the part of the viewer. These particular designs are not visually complex, but they are conceptually subtle. It is not necessary to identify the references to understand this work; it is necessary to accept a more tangential approach to communication than graphic design orthodoxy has traditionally allowed. It is a recurrent criticism of contemporary graphics that it contains too much "self-expression", that designers are putting their own creative interests before the needs of their clients, and that the client's message is being muddled, misunderstood or lost in the process. The trouble with these criticisms is that they indiscriminately lump the good with the bad, as though one feeble play, or a thousand feeble plays, made every play feeble. This is because they start from an untenable premise – that absolute, selfless neutrality is both desirable and possible for all kinds of graphic design – which leaves no room for manoeuvre. Real examples are rarely named or discussed, making it hard to respond to these assertions, nor is "self-expression", in this implicitly pejorative sense, defined with any precision. It is hard to imagine what a poster, book, calendar, or brochure design entirely devoid of "self" would be like, unless it was generated entirely by machine, so presumably the real issue (notwithstanding the rhetoric) is a matter of degree. There comes a point for these critics, whatever the context, at which there is just too much subjective self and not enough objective message in a design. At this point they cry "art".

The problem, I believe, lies in the way we define our terms. We take two categories, "art" and "design", define them as two quite different processes (despite the fact that they clearly have a great deal in common) and insist that visual communication falls

Regret
by New Order
Record sleeve
Art director: Peter Saville
Designer: Brett Wickens
Design company: Pentagram
Photographer: photolibrary
Client: CentreDate/London Records
Great Britain, 1993

Hover Hover
Billboard
Designers: Armand Mevis, Linda van Deursen
Design company: Mevis & van Deursen
Client: Stedelijk Museum, Amsterdam
The Netherlands, 1991

neatly into one category or the other, with no blurring or overlap allowed. It is not clear why we feel that the integrity of these categories has to be defended so dogmatically, except that designers have a vested professional interest in design, the art world has a similar interest in preserving the untouchability of art, and the education of designers and artists, by splitting the two groups, tends to reflect and reinforce these polarities. Historically, of course, before the two fields became so thoroughly professionalised the distinctions were less rigidly drawn. There are plenty of twentieth-century examples of individuals who have operated effectively in both camps without troubling about their job descriptions, but a historical digression is not necessary here. My point is rather more pragmatic and rooted in the present: whether the more conservative critics of the profession like it or not, there is now a large body of "design", including many of the examples in this book, which implicitly, by its very existence, challenges these definitions. Denouncing it will not make it go away or solve the problem. So why not accept that the communication arts are evolving, that there is good work in this area as well as mediocre and derivative work (as in any other branch

of design), and enlarge our restricting and no longer viable definitions instead?

The situation is if anything about to become even more complex. Graphic designers have not been especially quick to grasp the implications of digital multimedia, and the earliest examples of the genre have badly lacked their imaginative input. This will slowly change as designers come to terms with the technology in their own studios and begin to think beyond the paper applications of the tools. Multimedia by its very nature is about the collapse of traditional boundaries and the flowing together of categories into new forms of communication, about which we can as yet only speculate. The designers likely to make the biggest impact in this area will be those who can cope with the challenge of conceptual uncertainty.

If Dan Friedman's work represents one pole of the new graphic design, then Cranbrook Academy of Art graduate P. Scott Makela stands at the other. Friedman has expressed doubts about the divisive effects of digital culture. While his designs imply many of the structural and associational freedoms of multimedia, they do not suggest some new, differently textured form of information experience; they belong

Redefining Display
Project for *Design Quarterly*
Magazine cover (front and back)
Designer: P. Scott Makela with Alex Tylevich
Magazine designer: Laurie Haycock Makela
Client: Walker Art Center, Minneapolis
USA, Winter 1993

14 15

to the 500-year tradition of the printed book. Makela, a self-styled cyberpunk who uses the boast "100% digital" as his calling card, is leaving paper behind. He is equally at home with the technologies of design, music and video and is increasingly merging all three. Scanning one of his clotted and centreless digital designs is like floating across an electronic firestorm, wondering where it would be safe to touch down. The effect, he notes, is entirely deliberate. "When I see a lot of white space in a design, and when something is really 'arranged', the design becomes contained by the borders . . . like pieces on a plate. I have a need to move off the plate. What I am trying to do is grab a chunk of experience and have that bleed off all edges" (*Emigre* no. 23, 1992). Printed projects such as Makela's prospectus for The Minneapolis College of Art and Design and visual essay for the publication *Rethinking Design* are perhaps best interpreted as two-dimensional electronic paintings of multimedia environments for which we do not yet have the technology. Stunning as they are, it is debatable whether or not they fully succeed as paper-bound communication simply because – as with other digital designs of this complexity – the frozen page and its

static frame work against images that feel as though their natural condition is fluid, reactive and temporal.

Makela himself seems to acknowledge the transitional nature of such designs. In a remarkable speculative project for the journal *Design Quarterly* (no. 158, Winter 1993), modelled on his own studio experiences of switching between software programs, E-mail, electronic bulletin boards, CD player and satellite TV, he visualises the electronic office screens of a priest, a plumber, a motorcycle mechanic and a multimedia artist. The priest listens to a man in a teleconfessional and sifts through religious texts, while the plumber files invoices, deals with an angry client and dreams of a fishing trip. Each of these fictional characters can arrange the on-screen electronic information exactly as they would the papers, gadgets and personal belongings on a desk. For anyone who is sceptical about the aesthetic possibilities of multimedia, given the rather meagre evidence so far, Makela's project will have resonance. Here, with a detail and conviction few other graphic designers working in this field have achieved, we see a glimpse of what the dawning fusion of art, design and technology may hold.
Rick Poynor

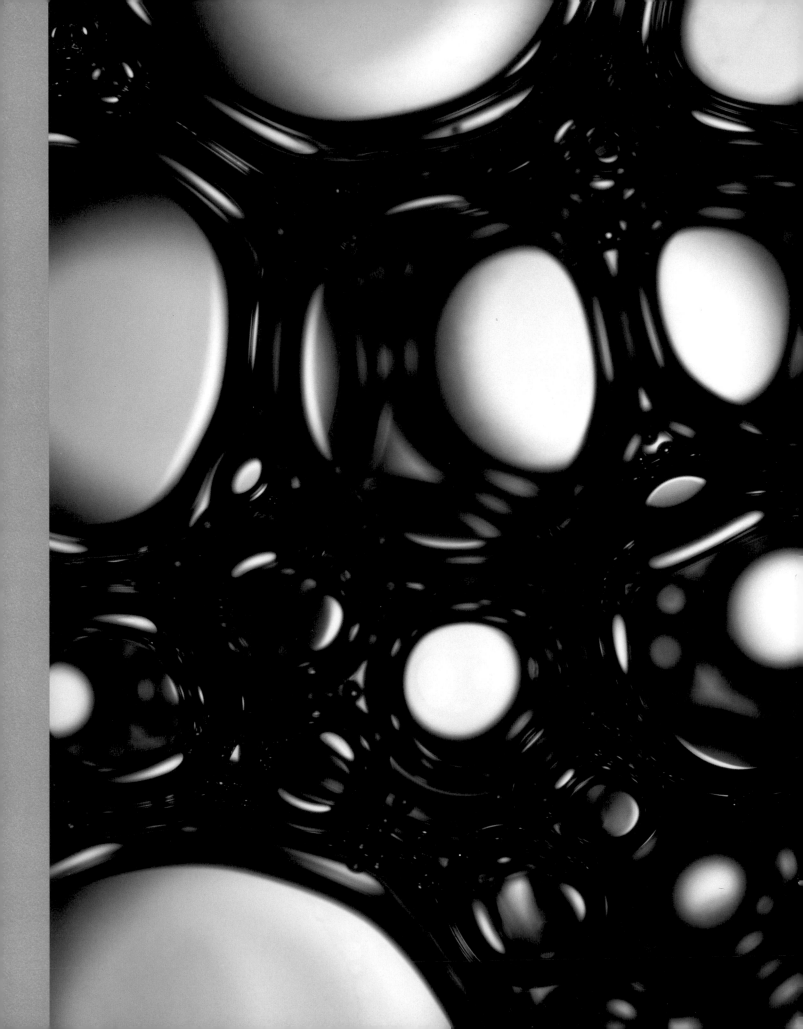

cool

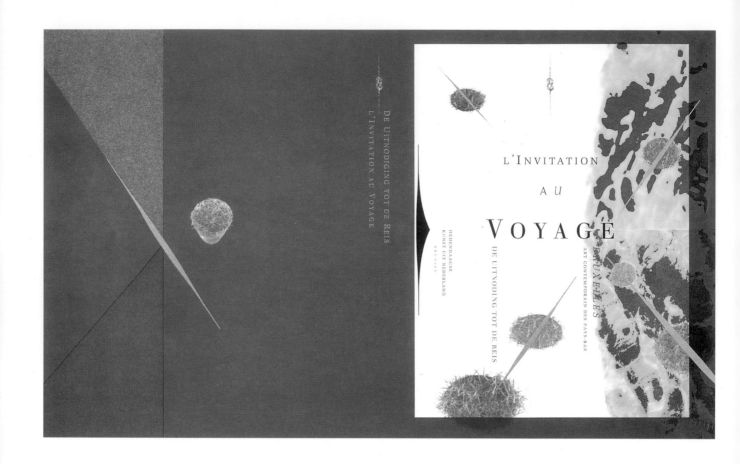

L'Invitation au Voyage
Book cover
Designer: Allen Hori
Design company: Hard Werken Design
Client: Rijksdienst Beeldende Kunst,
The Hague
The Netherlands, 1991

Adopt a Book
Poster
Designer: Lucille Tenazas
Design company: Tenazas Design
Photographer: Peter de Lory
Client: California College of Arts and Crafts
USA, 1992

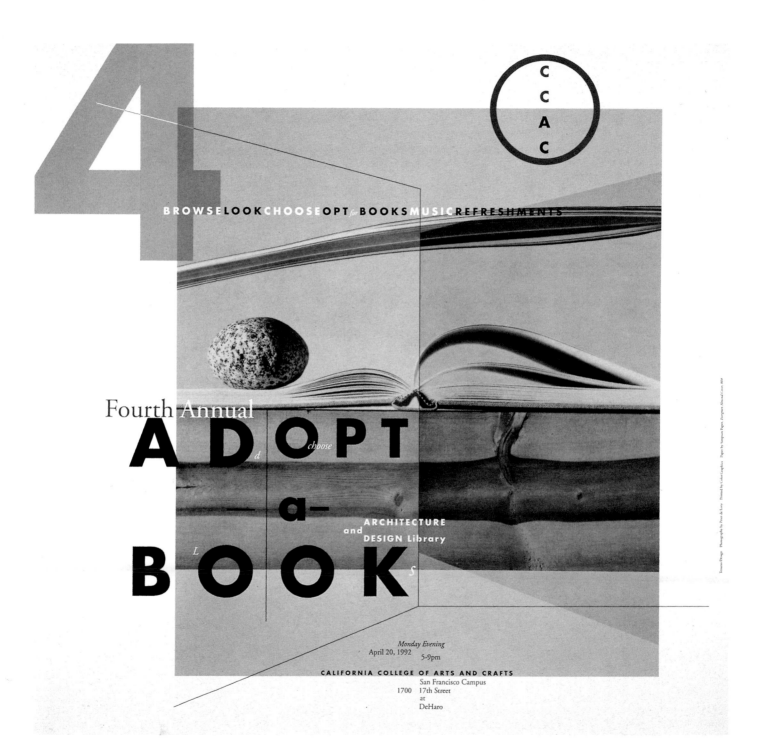

4

BROWSELOOKCHOOSEOPT for **BOOKSMUSICREFRESHMENTS**

Fourth Annual
ADOPT d choose
a–
L
BOOK s

ARCHITECTURE
and
DESIGN Library

CCAC

Monday Evening
April 20, 1992 5–9pm

CALIFORNIA COLLEGE OF ARTS AND CRAFTS
San Francisco Campus
1700 17th Street
at
DeHaro

MAGGIO 92

SETTIMANA 22

L 25   M 26   M 27   G 28   V 29   S 30   D 31

AIR

sikkens

FLAPS

Tactics of Posture
Poster
Designer: Andrew Blauvelt
Client: Washington University Gallery of Art
USA, 1991

Anatomy Calendar 93
April
Designer: Vaughan Oliver
Design company: v23
Photographer: Jim Friedman
Client: 4AD
Great Britain, 1992

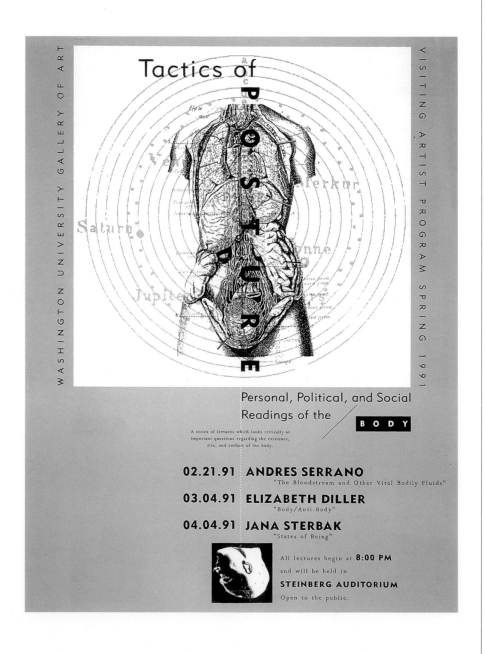

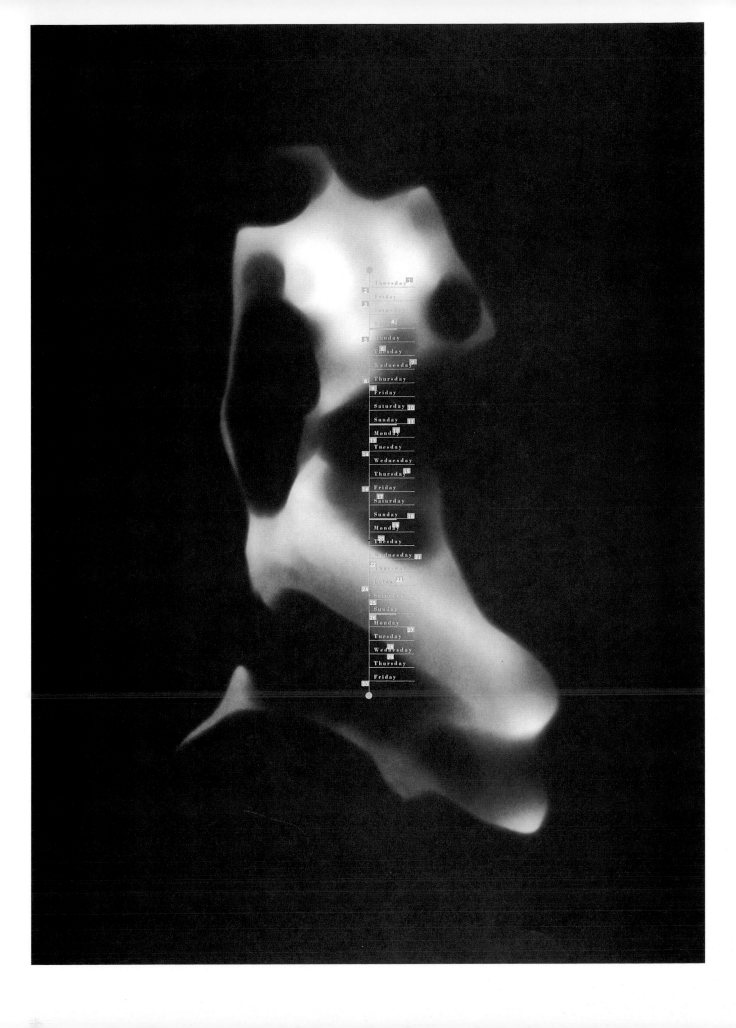

Systems Manual
Brochure for hi-fi consultant
Designer: Paul Elliman
Photographer: Moira Bogue
Client: The Cornflake Shop
Great Britain, 1991

Multi-room
Advertisement for hi-fi consultant
Designer: Paul Elliman
Photographer: Moira Bogue
Client: The Cornflake Shop
Great Britain, 1991

Cobalt Blue
Album cover
Designers: Vaughan Oliver, Chris Bigg
Design company: v23
Illustrator: Russell Mills
Photographer: Simon Larbalestier
Modelmaker: Pirate
Client: 4AD
Great Britain, 1992

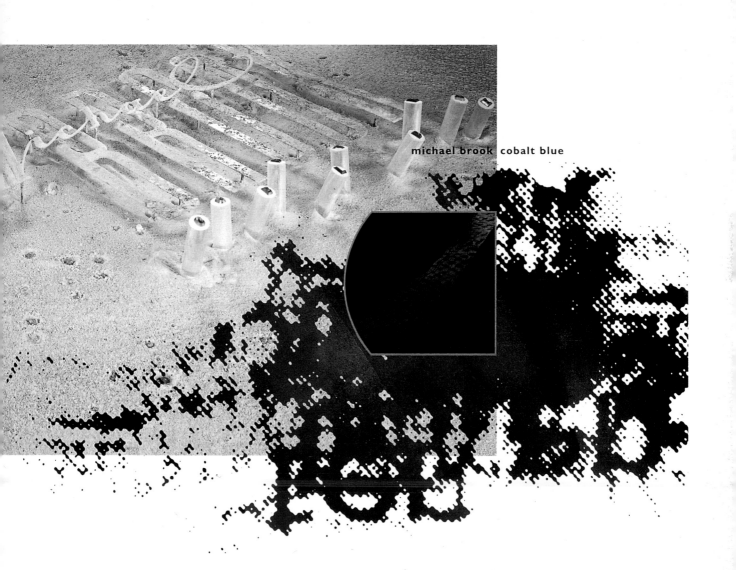

michael brook cobalt blue

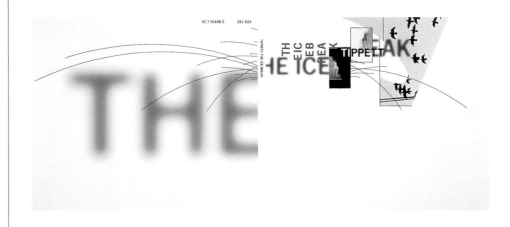

VC 7 91448-2   261 624

THE ICE BREAK
TIPPETT

typegners: a biographical directory
by Ron Eason & Sarah Rookledge, edited by Phill Baines & Gordon Rookledge

5

SAREMA PRESS

Over a period of thirty years **Ron Eason** has worked on the production of books and periodicals with leading British printers, including The Westerham Press and the former Gavin Martin Ltd. Married with two children, he lives in the Surrey countryside. Born in 1962, **Sarah Rookledge** graduating from Warwick University where she studied film and literature after working for Salisbury Publishing Services Ltd, she entered book publishing with a move to Sarema Press (Publishers) Ltd. She now works as a short-story writer and freelance journalist and lives in London. **Phil Baines** (born 1958), has worked as a freelance graphic designer since graduating from the Royal College of Art in 1987, combining this with teaching part-time at Central Saint Martin's College of Art & Design in London. **Gordon Rookledge**, 57, is the founder (1983), and Managing Director of art and graphic book publishers Sarema Press (Publishers) Ltd. He is on the committee of the Wynken de Worde Society; married with three children he lives in both London and Brighton.

'The Handbook of Type Designers differs from other type books by putting the typefaces into context and providing fascinating information about the people behind the typefaces.'

Jon Barnbrook

As a one-off/only pre-pub-
lication offer this book is available at a discounted price

Rookledge's
International **hand-
book of type
desi-**
gners:

offer expires 30-9-91

The Ice Break
CD packaging
Designer: Nick Bell
Art director: Jeremy Hall
Photographer: Nick Bell
Client: Virgin Classics
Great Britain, 1991

Rookledge's International
Handbook of Type Designers
Leaflet
Designer: Nick Bell
Client: Sarema Press
Great Britain, 1991

Typography as Discourse
Poster (reverse side)
Designers: Andrew Blauvelt, James Sholly,
Laura Lacy-Sholly
Photographer: Andrew Blauvelt
Client: Art Directors Club of Indiana
USA, 1991

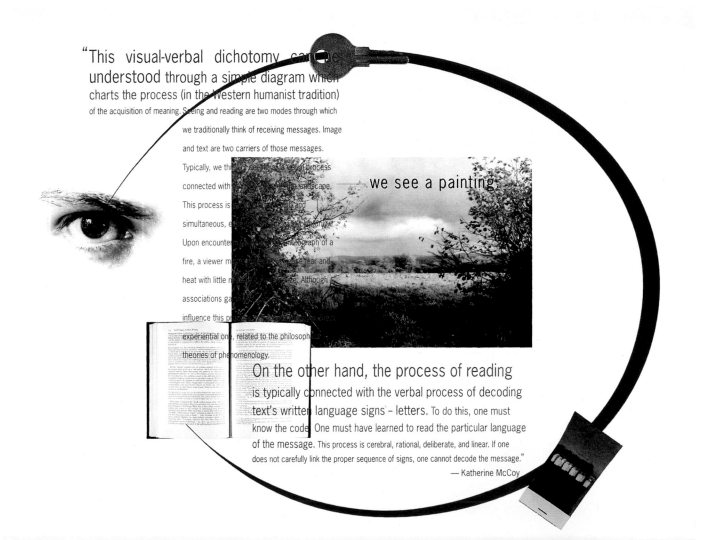

"This visual-verbal dichotomy can be understood through a simple diagram which charts the process (in the Western humanist tradition) of the acquisition of meaning. Seeing and reading are two modes through which we traditionally think of receiving messages. Image and text are two carriers of those messages.

Typically, we think of seeing as a verbal process connected with ... landscape, we see a painting. This process is ... and simultaneous, ... voluntarily. Upon encountering ... photograph of a fire, a viewer might ... fear and heat with little ... Although ... associations ga... influence this p... experiential one, related to the philosoph... theories of phenomenology.

On the other hand, the process of reading is typically connected with the verbal process of decoding text's written language signs – letters. To do this, one must know the code. One must have learned to read the particular language of the message. This process is cerebral, rational, deliberate, and linear. If one does not carefully link the proper sequence of signs, one cannot decode the message."
— Katherine McCoy

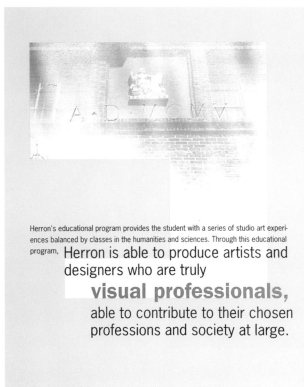

Herron's educational program provides the student with a series of studio art experiences balanced by classes in the humanities and sciences. Through this educational program, Herron is able to produce artists and designers who are truly

visual professionals,

able to contribute to their chosen professions and society at large.

Students at the Herron School of Art are surrounded by the vibrancy and intensity of a visual arts environment. The School's 400 students work closely with the full-time faculty, with a student teacher ratio of approximately 11:1.

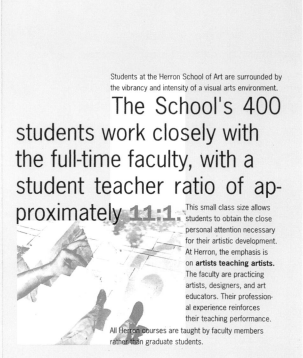

This small class size allows students to obtain the close personal attention necessary for their artistic development. At Herron, the emphasis is on **artists teaching artists.** The faculty are practicing artists, designers, and art educators. Their professional experience reinforces their teaching performance. All Herron courses are taught by faculty members rather than graduate students.

The combination of an excellent faculty, well-designed programs of instruction, and outstanding equipment and facilities provides an ideal educational setting for the preparation of

a career in the visual arts.

Graduates of the School's programs can be found in professional positions throughout the United States and in various parts of the world.

Many visual artists can be found working independently while others take careers in arts administration with galleries, agencies, and museums.

Designers can be found employed in various positions within advertising agencies, design offices, corporations, and not-for-profit institutions.

Graduates of the program in art education also are certified to teach art in the public schools of Indiana.

Herron Viewbook
Brochure spreads
Designer: Andrew Blauvelt
Photographer: Andrew Blauvelt
Client: Herron School of Art
USA, 1991

Podriga Pipeline Appeal
Poster
Designer: Nick Bell
Photographers: Robbie Mahoney,
Colin Edwards
Client: Romanian Project (UK)
Great Britain, 1992

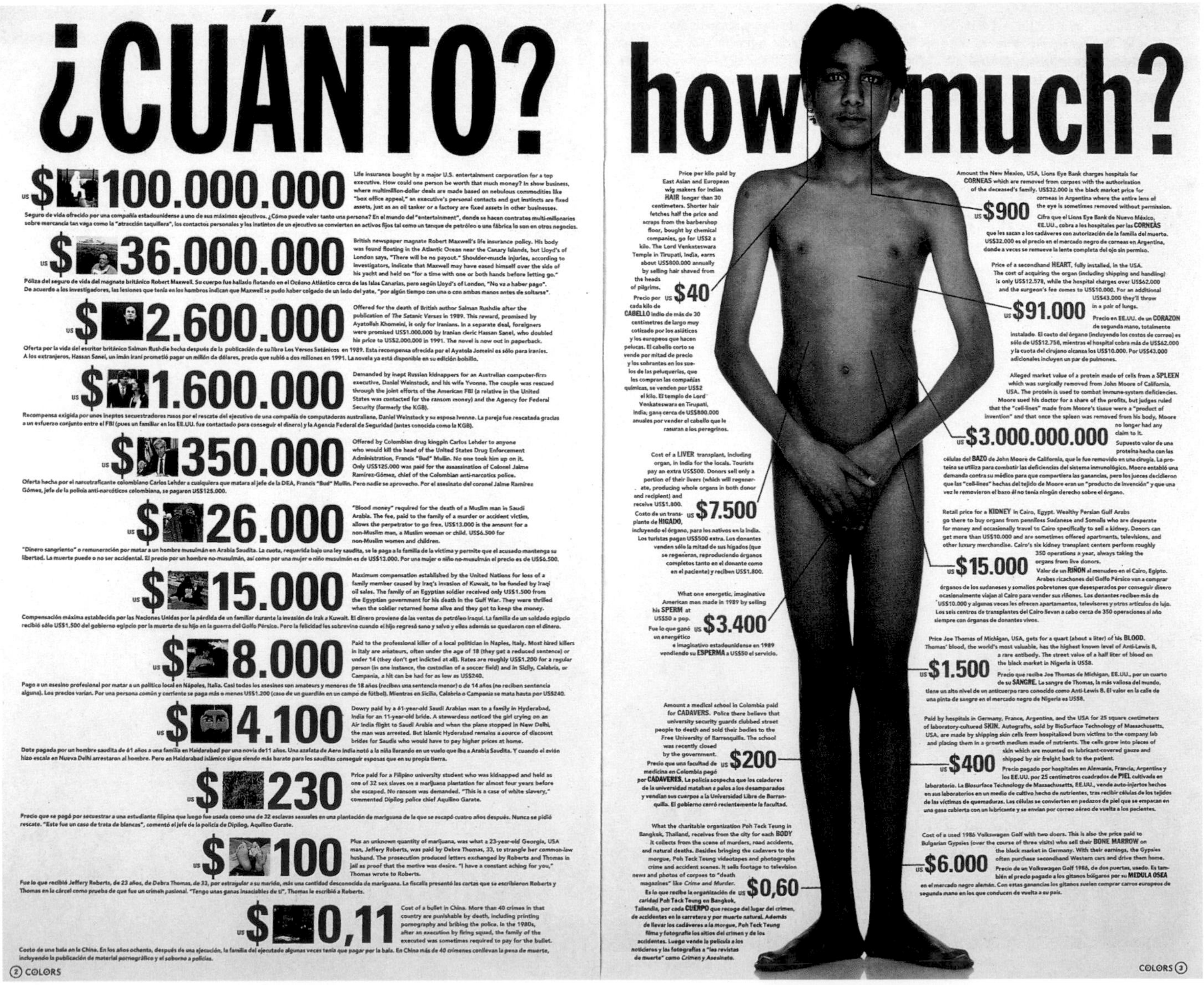

¿CUÁNTO?

US$ 100.000.000

Life insurance bought by a major U.S. entertainment corporation for a top executive. How could one person be worth that much money? In show business, where multimillion-dollar deals are made based on nebulous commodities like "box office appeal," an executive's personal contacts and gut instincts are fixed assets, just as an oil tanker or a factory are fixed assets in other businesses.

Seguro de vida ofrecido por una compañía estadounidense a uno de sus máximos ejecutivos. ¿Cómo puede valer tanto una persona? En el mundo del "entertainment", donde se hacen contratos multi-millonarios sobre mercancía tan vaga como la "atracción taquillera", los contactos personales y los instintos de un ejecutivo se convierten en activos fijos tal como un tanque de petróleo o una fábrica lo son en otros negocios.

US$ 36.000.000

British newspaper magnate Robert Maxwell's life insurance policy. His body was found floating in the Atlantic Ocean near the Canary Islands, but Lloyd's of London says, "There will be no payout." Shoulder-muscle injuries, according to investigators, indicate that Maxwell may have eased himself over the side of his yacht and held on "for a time with one or both hands before letting go."

Póliza del seguro de vida del magnate británico Robert Maxwell. Su cuerpo fue hallado flotando en el Océano Atlántico cerca de las Islas Canarias, pero según Lloyd's of London, "No va a haber pago". De acuerdo a los investigadores, las lesiones que tenía en los hombros indican que Maxwell se pudo haber colgado de un lado del yate, "por algún tiempo con una o con ambas manos antes de soltarse".

US$ 2.600.000

Offered for the death of British author Salman Rushdie after the publication of *The Satanic Verses* in 1989. This *reward*, promised by Ayatollah Khomeini, is only for Iranians. In a separate deal, foreigners were promised US$1.000.000 by Iranian cleric Hassan Sanei, who doubled his price to US$2.000.000 in 1991. The novel is now out in paperback.

Oferta por la vida del escritor británico Salman Rushdie hecha después de la publicación de su libro Los Versos Satánicos en 1989. Esta recompensa ofrecida por el Ayatola Jomeini es sólo para iraníes. A los extranjeros, Hassan Sanei, un imán iraní prometió pagar un millón de dólares, precio que subió a dos millones en 1991. La novela ya está disponible en su edición bolsillo.

US$ 1.600.000

Demanded by inept Russian kidnappers for an Australian computer-firm executive, Daniel Weinstock, and his wife Yvonne. The couple was rescued through the joint efforts of the American FBI (a relative in the United States was contacted for the ransom money) and the Agency for Federal Security (formerly the KGB).

Recompensa exigida por unos ineptos secuestradores rusos por el rescate del ejecutivo de una compañía de computadoras australiana, Daniel Weinstock y su esposa Ivonne. La pareja fue rescatada gracias a un esfuerzo conjunto entre el FBI (pues un familiar en los EE.UU. fue contactado para conseguir el dinero) y la Agencia Federal de Seguridad (antes conocida como la KGB).

US$ 350.000

Offered by Colombian drug kingpin Carlos Lehder to anyone who would kill the head of the United States Drug Enforcement Administration, Francis "Bud" Mullin. No one took him up on it. Only US$125.000 was paid for the assassination of Colonel Jaime Ramirez-Gómez, chief of the Colombian anti-narcotics police.

Oferta hecha por el narcotraficante colombiano Carlos Lehder a cualquiera que matara al jefe de la DEA, Francis "Bud" Mullin. Pero nadie se aprovecho. Por el asesinato del coronel Jaime Ramirez Gómez, jefe de la policía anti-narcóticos colombiana, se pagaron US$125.000.

US$ 26.000

"Blood money" required for the death of a Muslim man in Saudi Arabia. The fee, paid to the family of a murder or accident victim, allows the perpetrator to go free. US$13.000 is the amount for a non-Muslim man, a Muslim woman or child. US$6.500 for non-Muslim women and children.

"Dinero sangriento" o remuneración por matar a un hombre musulmán en Arabia Saudita. La cuota, requerida bajo una ley saudita, se le paga a la familia de la víctima y permite que el acusado mantenga su libertad. La muerte puede o no ser accidental. El precio por un hombre no-musulmán, así como por una mujer o niño musulmán es de US$13.000. Por una mujer o niño no-musulmán el precio es de US$6.500.

US$ 15.000

Maximum compensation established by the United Nations for loss of a family member caused by Iraq's invasion of Kuwait, to be funded by Iraqi oil sales. The family of an Egyptian soldier received only US$1.500 from the Egyptian government for his death in the Gulf War. They were thrilled when the soldier returned home alive and they got to keep the money.

Compensación máxima establecida por las Naciones Unidas por la pérdida de un familiar durante la invasión de Irak a Kuwait. El dinero proviene de las ventas de petróleo iraquí. La familia de un soldado egipcio recibió sólo US$1.500 del gobierno egipcio por la muerte de su hijo en la guerra del Golfo Pérsico. Pero la felicidad les sobrevino cuando el hijo regresó sano y salvo y ellos además se quedaron con el dinero.

US$ 8.000

Paid to the professional killer of a local politician in Naples, Italy. Most hired killers in Italy are amateurs, often under the age of 18 (they get a reduced sentence) or under 14 (they don't get indicted at all). Rates are roughly US$1.200 for a regular person (in one instance, the custodian of a soccer field) and in Sicily, Calabria, or Campania, a hit can be had for as low as US$240.

Pago a un asesino profesional por matar a un político local en Nápoles, Italia. Casi todos los asesinos son amateurs y menores de 18 años (reciben una sentencia menor) o de 14 años (no reciben sentencia alguna). Los precios varían. Por una persona común y corriente se paga más o menos US$1.200 (caso de un guardián en un campo de fútbol). Mientras en Sicilia, Calabria o Campania se mata hasta por US$240.

US$ 4.100

Dowry paid by a 61-year-old Saudi Arabian man to a family in Hyderabad, India for an 11-year-old bride. A stewardess noticed the girl crying on an Air India flight to Saudi Arabia and when the plane stopped in New Delhi, the man was arrested. But Islamic Hyderabad remains a source of discount brides for Saudis who would have to pay higher prices at home.

Dote pagada por un hombre saudita de 61 años a una familia en Haidarabad por una novia de11 años. Una azafata de Aero India notó a la niña llorando en un vuelo que iba a Arabia Saudita. Y cuando el avión hizo escala en Nueva Delhi arrestaron al hombre. Pero en Haidarabad islámico sigue siendo más barato para los sauditas conseguir esposas que en su propia tierra.

US$ 230

Price paid for a Filipino university student who was kidnapped and held as one of 32 sex slaves on a marijuana plantation for almost four years before she escaped. No ransom was demanded. "This is a case of white slavery," commented Dipilog police chief Aquilino Garate.

Precio que se pagó por secuestrar a una estudiante filipina que luego fue usada como una de 32 esclavas sexuales en una plantación de mariguana de la que se escapó cuatro años después. Nunca se pidió rescate. "Este fue un caso de trata de blancas", comentó el jefe de la policía de Dipilog, Aquilino Garate.

US$ 100

Plus an unknown quantity of marijuana, was what a 23-year-old Georgia, USA man, Jeffery Roberts, was paid by Debra Thomas, 33, to strangle her common-law husband. The prosecution produced letters exchanged by Roberts and Thomas in jail as proof that the motive was desire. "I have a constant aching for you," Thomas wrote to Roberts.

Fue lo que recibió Jeffery Roberts, de 23 años, de Debra Thomas, de 33, por estragular a su marido, más una cantidad desconocida de mariguana. La fiscalía presentó las cartas que se escribieron Roberts y Thomas en la cárcel como prueba de que fue un crimeh pasional. "Tengo unas ganas insaciables de ti", Thomas le escribió a Roberts.

US$ 0,11

Cost of a bullet in China. More than 40 crimes in that country are punishable by death, including printing pornography and bribing the police. In the 1980s, after an execution by firing squad, the family of the executed was sometimes required to pay for the bullet.

Costo de una bala en la China. En los años ochenta, después de una ejecución, la familia del ejecutado algunas veces tenía que pagar por la bala. En China más de 40 crímenes conllevan la pena de muerte, incluyendo la publicación de material pornográfico y el soborno a policías.

2 COLORS

how much?

US$40

Price per kilo paid by East Asian and European wig makers for Indian **HAIR** longer than 30 centimeters. Shorter hair fetches half the price and scraps from the barbershop floor, bought by chemical companies, go for US$2 a kilo. The Lord Venkateswara Temple in Tirupati, India, earns about US$800.000 annually by selling hair shaved from the heads of pilgrims.

Precio por cada kilo de **CABELLO** indio de más de 30 centimetros de largo muy cotizado por los asiáticos y los europeos que hacen pelucas. El cabello corto se vende por mitad de precio y los sobrantes en los suelos de las peluquerías, que los compran las compañías químicas, se venden por US$2 el kilo. El templo de Lord Venkateswara en Tirupati, India, gana cerca de US$800.000 anuales por vender el cabello que le rasuran a los peregrinos.

US$900

Amount the New Mexico, USA, Lions Eye Bank charges hospitals for **CORNEAS** which are removed from corpses with the authorization of the deceased's family. US$32.000 is the black market price for corneas in Argentina where the entire lens of the eye is sometimes removed without permission.

Cifra que el Lions Eye Bank de Nuevo México, EE.UU., cobra a los hospitales por las **CORNEAS** que les sacan a los cadáveres con autorización de la familia del muerto. US$32.000 es el precio en el mercado negro de corneas en Argentina, donde a veces se remueve la lente completa del ojo sin permiso.

US$91.000

Price of a secondhand **HEART**, fully installed, in the USA. The cost of acquiring the organ (including shipping and handling) is only US$12.578, while the hospital charges over US$62.000 and the surgeon's fee comes to US$10.000. For an additional US$43.000 they'll throw in a pair of lungs.

Precio en EE.UU. de un **CORAZON** de segunda mano, totalmente instalado. El costo del órgano (incluyendo los costos de correo) es sólo de US$12.758, mientras el hospital cobra más de US$62.000 y la cuota del cirujano alcanza los US$10.000. Por US$43.000 adicionales incluyen un par de pulmones.

US$3.000.000.000

Alleged market value of a protein made of cells from a **SPLEEN** which was surgically removed from John Moore of California, USA. The protein is used to combat immune-system deficiencies. Moore sued his doctor for a share of the profits, but judges ruled that the "cell-lines" made from Moore's tissue were a "product of invention" and that once the spleen was removed from his body, Moore no longer had any claim to it.

Supuesto valor de una proteína hecha con las células del **BAZO** de John Moore de California, que le fue removido en una cirugía. La proteína se utiliza para combatir las deficiencias del sistema inmunológico. Moore entabló una demanda contra su médico para que compartiera las ganancias, pero los jueces decidieron que las "cell-lines" hechas del tejido de Moore eran un "producto de invención" y que una vez le removieron el bazo él no tenía ningún derecho sobre el órgano.

US$7.500

Cost of a **LIVER** transplant, including organ, in India for the locals. Tourists pay an extra US$500. Donors sell only a portion of their livers (which will regenerate, producing whole organs in both donor and recipient) and receive US$1.800.

Costo de un transplante de **HIGADO**, incluyendo el órgano, para los nativos en la India. Los turistas pagan US$500 extra. Los donantes venden sólo la mitad de sus hígados (que se regeneran, reproduciendo órganos completos tanto en el donante como en el paciente) y reciben US$1.800.

US$15.000

Retail price for a **KIDNEY** in Cairo, Egypt. Wealthy Persian Gulf Arabs go there to buy organs from penniless Sudanese and Somalis who are desperate for money and occasionally travel to Cairo specifically to sell a kidney. Donors can get more than US$10.000 and are sometimes offered apartments, televisions, and other luxury merchandise. Cairo's six kidney transplant centers perform roughly 350 operations a year, always taking the organs from live donors.

Valor de un **RIÑON** al menudeo en el Cairo, Egipto. Arabes ricachones del Golfo Pérsico van a comprar órganos de los sudaneses y somalíes pobretones que desesperados por conseguir dinero ocasionalmente viajan al Cairo para vender sus riñones. Los donantes reciben más de US$10.000 y algunas veces les ofrecen apartamentos, televisores y otros artículos de lujo. Los seis centros de transplantes del Cairo llevan a cabo cerca de 350 operaciones al año siempre con órganos de donantes vivos.

US$3.400

What one energetic, imaginative American man made in 1989 by selling his **SPERM** at US$50 a pop.

Fue lo que ganó un energético e imaginativo estadounidense en 1989 vendiendo su **ESPERMA** a US$50 el servicio.

US$1.500

Price Joe Thomas of Michigan, USA, gets for a quart (about a liter) of his **BLOOD**. Thomas' blood, the world's most valuable, has the highest known level of Anti-Lewis B, a rare antibody. The street value of a half liter of blood on the black market in Nigeria is US$8.

Precio que recibe Joe Thomas de Michigan, EE.UU., por un cuarto de su **SANGRE**. La sangre de Thomas, la más valiosa del mundo, tiene un alto nivel de un anticuerpo raro conocido como Anti-Lewis B. El valor en la calle de una pinta de sangre en el mercado negro de Nigeria es US$8.

US$200

Amount a medical school in Colombia paid for **CADAVERS**. Police there believe that university security guards clubbed street people to death and sold their bodies to the Free University of Barranquilla. The school was recently closed by the government.

Precio que una facultad de medicina en Colombia pagó por **CADAVERES**. La policía sospecha que los celadores de la universidad mataban a palos a los desamparados y vendían sus cuerpos a la Universidad Libre de Barranquilla. El gobierno cerró recientemente la facultad.

US$400

Paid by hospitals in Germany, France, Argentina, and the USA for 25 square centimeters of laboratory-cultured **SKIN**. Autografts, sold by BioSurface Technology of Massachusetts, USA, are made by shipping skin cells from hospitalized burn victims to the company lab and placing them in a growth medium made of nutrients. The cells grow into pieces of skin which are mounted on lubricant-covered gauze and shipped by air freight back to the patient.

Precio pagado por hospitales en Alemania, Francia, Argentina y los EE.UU. por 25 centímetros cuadrados de **PIEL** cultivada en laboratorio. La Biosurface Technology de Massachusetts, EE.UU., vende auto-injertos hechos en sus laboratorios en un medio de cultivo hecho de nutrientes, tras recibir células de los tejidos de las víctimas de quemaduras. Las células se convierten en pedazos de piel que se empacan en una gasa cubierta con un lubricante y se envían por correo aéreo de vuelta a los pacientes.

US$0,60

What the charitable organization Poh Teck Teung in Bangkok, Thailand, receives from the city for each **BODY** it collects from the scene of murders, road accidents, and natural deaths. Besides bringing the cadavers to the morgue, Poh Teck Teung videotapes and photographs crime and accident scenes. It sells footage to television news and photos of corpses to "death magazines" like *Crime and Murder*.

Es lo que recibe la organización de caridad Poh Teck Teung en Bangkok, Tailandia, por cada **CUERPO** que recoge del lugar del crimen, de accidentes en la carretera y por muerte natural. Además de llevar los cadáveres a la morgue, Poh Teck Teung filma y fotografía los sitios del crimen y de los accidentes. Luego vende la película a los noticieros y las fotografías a "las revistas de muerte" como *Crimen y Asesinato*.

US$6.000

Cost of a used 1986 Volkswagen Golf with two doors. This is also the price paid to Bulgarian Gypsies (over the course of three visits) who sell their **BONE MARROW** on the black market in Germany. With their earnings, the Gypsies often purchase secondhand Western cars and drive them home.

Precio de un Volkswagen Golf 1986, de dos puertas, usado. Es también el precio pagado a los gitanos búlgaros por su **MEDULA OSEA** en el mercado negro alemán. Con estas ganancias los gitanos suelen comprar carros europeos de segunda mano en los que conducen de vuelta a su país.

COLORS 3

Colors, issue 3
Magazine spread
Designers: Tibor Kalman, Paul Ritter
Design company: M&Co
Photographer: Oliviero Toscani
Client: Benetton
USA, Fall/Winter 1992

Colors, issue 2
Magazine spread
Designers: Tibor Kalman, Gary Koepke
Design company: M&Co
Client: Benetton
USA, Spring/Summer 1992

religion* religione*

Handy pictorial answers to major theological questions.
Risposte pratiche illustrate a questioni teologiche fondamentali.

We have a lot in common: two ears, two eyes,
a silly nose and roughly twenty square
feet (and seven pounds) of skin...

SO, what's the difference?

Abbiamo parecchio in comune: due orecchi, due
occhi, un naso buffo e circa due metri
quadri (e tre chili) di pelle...

e allora, qual'è la differenza?

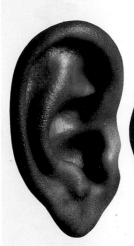

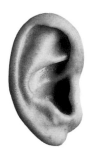

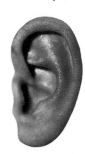

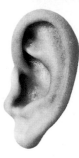

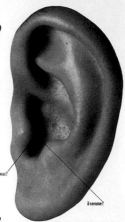

brown skin?
la pelle marrone?

black skin?
la pelle nera?

white skin?
la pelle bianca?

orange skin?
la pelle arancione?

very white skin?
la pelle bianchissima?

asian skin?
la pelle orientale?

wax?   il cerume?

So, what's the difference? It depends on who you are and where you live.

E allora? Qual'è la differenza? Dipende da chi sei e da dove vivi.

Colors, issue 4, "Race"
Magazine spreads
Designers: Tibor Kalman, Paul Ritter
Design company: M&Co
Photographer: Oliviero Toscani
Retouchers: Site One
Client: Benetton
USA, Spring/Summer 1993

what if..?
e se..?

Queen Elizabeth
Regina Elisabetta

We've told you
about **skin color,** **wax**
ear consistency
and **nose** size. Now,
what do you **really**
want to know about

Ti abbiamo parlato
del colore della **pelle**, della consistenza del
cerume e delle **naso**. Ma cosa vorresti
sapere **davvero** sulle razze diverse dalla tua **?**

people of other races? Ne eravamo sicuri.

We thought so

Power

LCO*10*

London
Chamber
Orchestra
Tchaikovsky
Vivaldi
Albinoni
Mozart
Glass
Britten
Bicât
Elgar

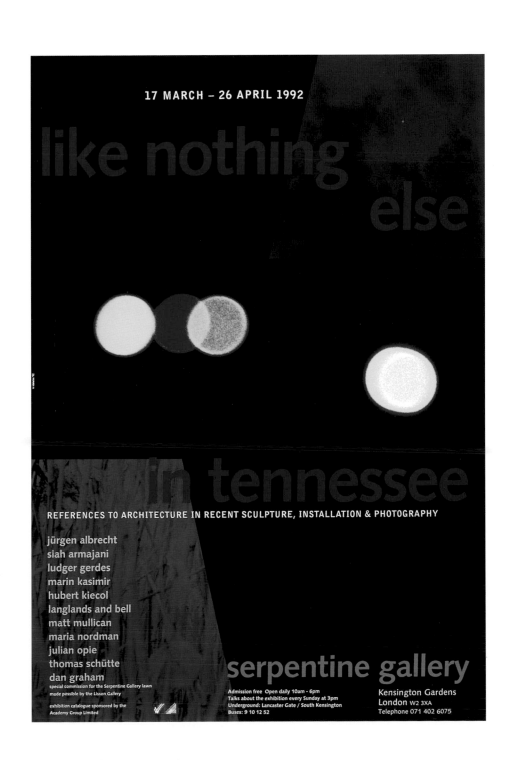

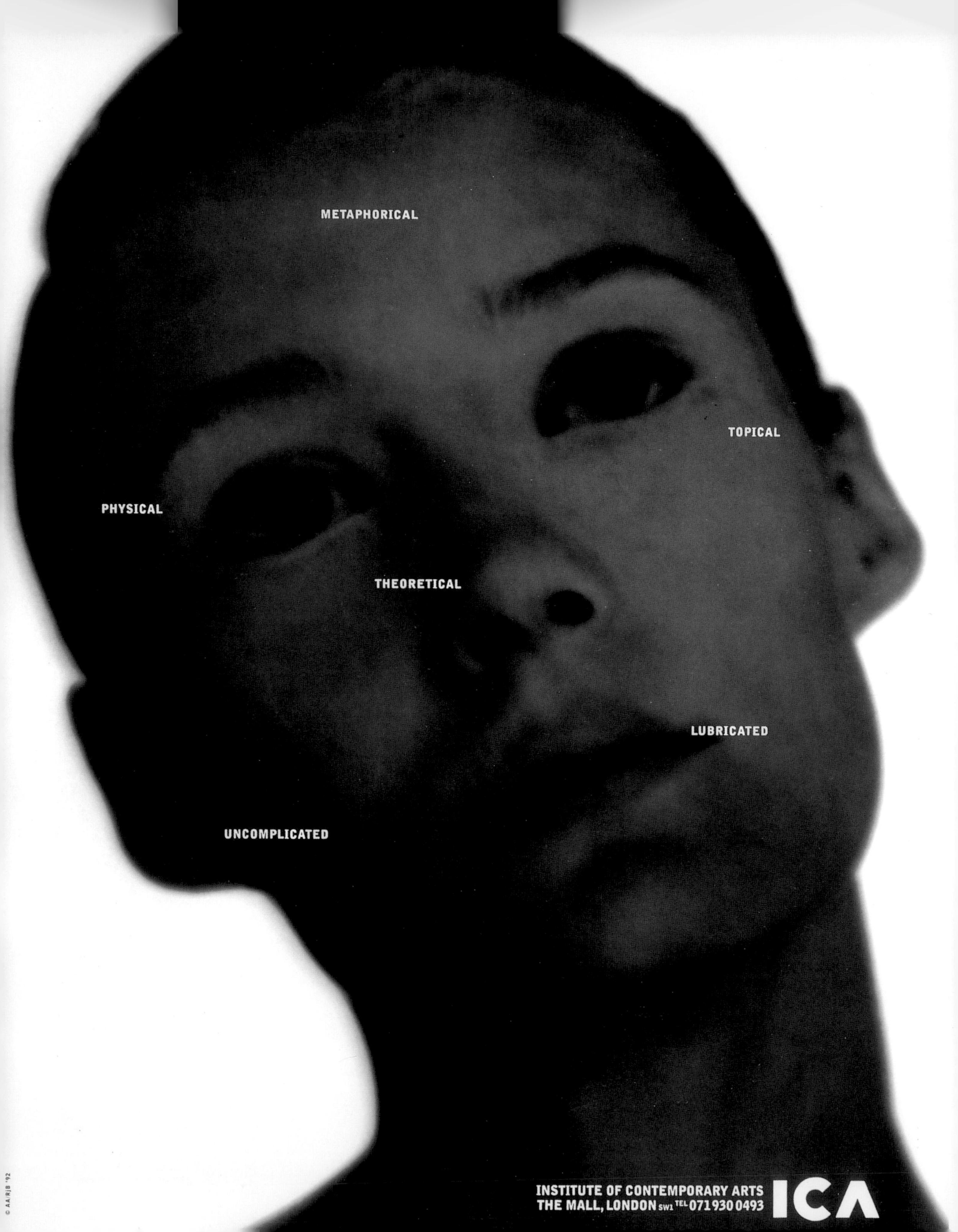

METAPHORICAL

TOPICAL

PHYSICAL

THEORETICAL

LUBRICATED

UNCOMPLICATED

INSTITUTE OF CONTEMPORARY ARTS
THE MALL, LONDON SW1 TEL 071 930 0493 **ICA**

© AA|R|B '92

Press advertisement
Designer: A. Arefin
Photographer: Richard J. Burbridge
Client: Institute of Contemporary Arts,
London
Great Britain, 1992

Halluçienda
Club poster
Designers: Peter Saville, Julian Morey
Design company: Peter Saville Associates
Photographer: Anthony Burgess,
Science Photo Library
Client: Fac 51
Great Britain, 1989

lux

Europæ

Edinburgh

Outdoor light
installations by over
forty European
artists across the
City of Edinburgh,
Scotland

For further information | fax: +44 31 557 2500

Principal funders:
Edinburgh District Council
Lothian Regional Council
Lothian and Edinburgh Enterprise Limited
European Arts Festival Fund
Scottish Arts Council
The Henry Moore Foundation
Commission of the European Communities.
Kaleidoscope Fund
Foreign & Commonwealth Office
Visiting Arts

Lux Europæ
Advertisement for exhibition
Designers: A. Arefin, Stephen Coates
Design company: Arefin Coates
Photographers: Nitin Vadukul,
Heidi Kosanuik
Client: Lux Europæ, Edinburgh
Great Britain, 1992

The Winter's Tale
Theatre programme
Designer: Russell Warren-Fisher
Photographer: Gareth McCarthy
Client: Theatre de Complicite
Great Britain, 1991

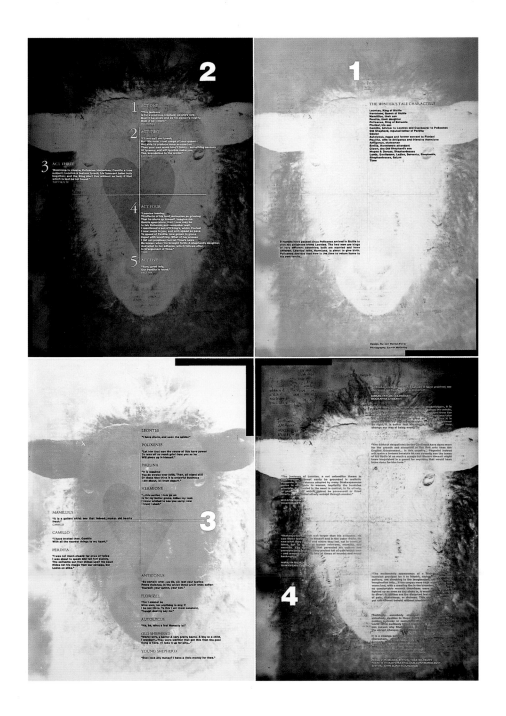

Call for Vision
Poster announcing San Francisco
waterfront competition (reverse side)
Designers: Lucille Tenazas, Todd Foreman
Design company: Tenazas Design
Photographer: Simon Yuen
Client: Center for Critical Architecture
USA, 1992

De Stad
(The City)
Architectural poster
Designers: Ben Faydherbe, Wout de Vringer
Design company: Faydherbe/De Vringer
Client: CBK-Artoteek, Dordrecht
The Netherlands, 1991

A New Architecture
Poster
Designers: Lorraine Wild, Whitney Lowe,
Lisa Nugent, Susan Parr
Design company: ReVerb/Los Angeles
Client: School of Architecture,
University of California at San Diego
USA, 1991

The Discourse of the Studio
Poster
Designers: J. Abbott Miller, Ellen Lupton
Design company: Design Writing Research
Printer: John Nichols
Client: American Collegiate Schools
of Architecture
USA, 1990

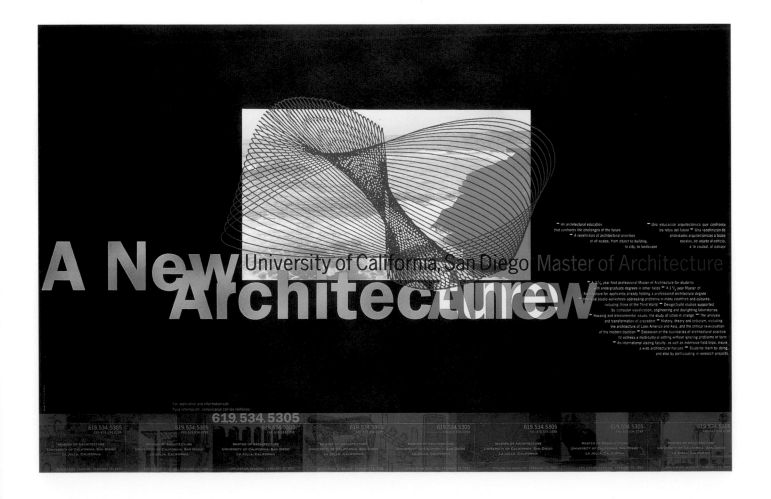

The Discourse of the Studio

CONFERENCE

1990 ACSA
Northeast Regional Meeting
School of Architecture
Princeton University

Friday, October 26 through
Saturday, October 27, 1990

A two-day conference devoted
to a discussion of the history,
current status of, and future
directions for the architectural
studio as both an educational
model and a determinant of
the language and practice of
architectural theory, criticism,
and design. Plenary speakers:

DAVID VAN ZANTEN
ROSEMARIE BLETTER
JUDITH WOLIN

EXHIBITION

The School of Architecture,
Princeton University,
is sponsoring an exhibition,
which will run concurrently
with the 1990 ACSA Northeast
Regional Meeting, of student
projects produced at schools in
the northeastern ACSA region.
The exhibition will consist of
projects from the 1989-90
academic year, representing
the most unique, successful,
and creative work produced
by students at each school.

The exhibition will run from
October 22 – November 9.

Submissions for papers and
panel discussions are due by
September 15, 1990.
Address inquiries to:
Alan Plattus
School of Architecture
Yale University
New Haven, CT 06520
tel: 203-432-2288
fax: 203-432-7175

To register, contact
Cynthia Nelson
School of Architecture
Princeton University
Princeton, NJ 08544
tel: 609-258-5018
fax: 609-258-4740

Spring 1992 Lecture Series
Architectural poster
Designer: Lorraine Wild
Design company: ReVerb/Los Angeles
Client: Southern California
Institute of Architecture
USA, 1992

the abc's of ▲■●

the bauhaus
and
design
theory

The Herb Lubalin Study Center of Design and Typography The Cooper Union for the Advancement of Science and Art

concepts   concepts

donkey — horse — mule

▲■●   house — horse — 'hourse   ▲ □ ○   ▲■●

△ □ ○   sounds   sounds

Figure 1
Visual signs

Figure 2
Kandinsky's ideal correspondence
between colors and shapes

Figure 3
Analytical drawing exercise,
Hanns Beckmann, 1929

Figure 4
Saussure: language takes shape
between two shapeless masses

Figure 5
The grid of verbal language:
vertical and horizontal relationships

Figure 6
The grid of visual language:
vertical and horizontal relationships

Figure 7
Visual signs

1 The term **translation** appears in Kandinsky's Bauhaus textbook *Point and Line to Plane*, where it refers to the act of drawing correspondences between graphic, linear marks and a range of non-graphic experiences, such as color, music, spiritual intuition, and visual perception: *"every phenomenon of the external and of the inner world can be given a linear expression— a kind of translation"* (68). Kandinsky hoped that one day all modes of expression would be **translated** through this visual script, their elements charted on one vast *"synthetic table"* or *"elementary dictionary."* ▲■● is a central example of **translation**. The series ▲■● represents Kandinsky's attempt to prove a universal correlation between color and geometry; it has become one of the most famous icons of the Bauhaus. Kandinsky conceived of these colors and shapes as a series of oppositions: yellow and blue represent the extremes of hot/cold, light/dark, and active/passive, while red is the intermediary between them. The triangle, square, and circle are graphic equivalents of the same polarities. While few designers today would accept the universal validity of the equation ▲■●, the model of visual "language" as a grammar of perceptual oppositions remains the basis of numerous textbooks of basic design.

2 Kandinsky's series ▲■● sets forth geometry as a *script* whose meaning or "content" is the primary colors, each shape serving as a graphic container surrounding a field of hue. In 1923 Kandinsky circulated a questionnaire at the Bauhaus, which asked each participant to match intuitively △, □, and ○ with the three primary colors. Labelled *"psychological test,"* the survey attempted to validate scientifically the equation ▲■●. An elementary sentence written in the language of vision, ▲■● inspired numerous objects and projects at the Bauhaus around the time of Kandinsky's questionnaire; it came to symbolize the possibility of a visual "language" that would communicate directly to the mechanics of the eye and brain, operating independently of cultural and linguistic conventions.

3 The term **translation** also appears in reference to one of Kandinsky's drawing exercises, in which students represent a still-life arrangement with a linear diagram: the image is *"completely translated into energy tensions... the over-all scheme made visible by dashed lines"* (Wingler 146). Kandinsky conceived of pictorial composition as a system of "forces"; any mark or color has a relation to such geometric or psychological oppositions as vertical/ horizontal, straight/curved, warm/cold, and active/passive. Through **translation** Kandinsky aimed to express this pattern of forces with a graphic code—the series ▲■● thus embodies the theory of visual "language" as a system of perceptual oppositions. A drawing problem similar to Kandinsky's linear object studies is assigned in many basic design courses today, in which students represent an object in pure black and white values. Often called **graphic translations**, these drawings combine the apparent objectivity of a photograph with the clarity of a letterform.

4 The term **translation** is also used in geometry, where it refers to the uniform movement of a figure in a single direction. In discussions of language, **translation** refers to the act of exchanging symbols from one system with symbols from another. What correspondences—and differences—might one draw between Kandinsky's "language" of vision and verbal language? How might one **translate** the visual sign ▲■● into the realm of the linguistic? According to the theory of the verbal sign proposed by the linguist Ferdinand de Saussure at the turn of the twentieth century, language consists of two distinct yet inseparable planes: sounds and concepts, or *signifiers* and *signifieds*. In order for the chaotic, undifferentiated mass of potential sounds to become the phonic material of language, it must be articulated into distinct, repeatable units; likewise, the plane of thought must be broken down into distinct concepts before it can be linked to material sounds. The realm of "thought" does not consist of ready-made, autonomous ideas existing independently of ready-made sounds—both planes are in themselves formless before they are cut up in relation to each other by the grid of language.

5 Saussure diagrams the **grid** of language as a series of **vertical** and **horizontal** relations. The relationship between sound and concept, or signifier and signified, is vertical: the sound "horse" is linked to the concept of a horse. Horizontally, each sign is linked to all the other signs against which it is defined: the word "horse" is opposed phonically to *house*, *hose*, and *hourse*; it is opposed conceptually to "donkey," "cow," and "mule." The link between signifier and signified is not an inherent quality of the sign, but is rather a function of the overall system. A sign is thus not an autonomous, self-contained vessel of meaning, but only has *value* in relation to other signs. Kandinsky's ▲■● is analogous in some ways to a system of linguistic signs. The series represents *vertical* links between the planes of form and color; *horizontally*, each plane is structured by the oppositions hot/cold, light/dark, and active/ passive. Similarly, Kandinsky's **translation** drawing exercise is an attempt to find a graphic equivalent for a pattern of perceptual, geometrical, and spiritual oppositions, a linear network which interprets the objects of experience.

6 The central difference between the verbal sign and the ideal of the visual sign symbolized by ▲■● is the **arbitrariness** of the link between form and concept, signifier and signified, in the verbal sign. Saussure argued that language is fundamentally *social*, depending for its survival on a shared cultural agreement; in contrast, the series ▲■● symbolized the search for a language based in *natural* laws of perception. Yet the series ▲■● itself bears cultural associations. Its kinship to children's toys carries the promise of generation, while its geometry and spectral purity allies the truth of intuition with that of science. When the forms and colors of ▲■● appear in design today, they function as transient *signs*, carrying such diverse meanings as "art," "the basics" and "modernism"; they are bound to cultural meaning by the act of quotation.

The ABC's of ▲■●:
The Bauhaus and Design Theory
Exhibition catalogue cover and spreads
Designers: Ellen Lupton,
J. Abbott Miller, Mike Mills

Design company: Design Writing Research
Photographer: Joanne Savio
Client: Herb Lubalin Study Center, Cooper Union, New York
USA, 1991

Appendix: The Gender of the Universal

Mike Mills

Western culture has constructed a dichotomy between what it labels as the "objective" quality of masculinity and the "subjective" quality of femininity. This mutually exclusive definition of gender has its analogue in Bayer's attempt to create a purely objective design which attempts to exclude, and deny, the presence of subjectivity. This appendix relates Sigmund Freud's theory of family power relations to Bayer's design methodology to show how these cultural beliefs about the meaning of masculine and feminine are developed, and to reveal that Bayer and Freud located reason and progress in the discipline and order personified by the father figure.

Mother: Freud believed that the mother "engulfs" the growing identity of the child with her nurturance. The child's total dependence on the mother in the first years of life renders him/her unable to differentiate the boundary between "self" and "mother." The child is unable to distinguish between its interior world and the exterior world. Freud argued that the child, especially the male child, must "repudiate" the mother in order to become autonomous and self-governing: "her nurturance threatens to re-engulf him with its reminder of helplessness and dependency; it must be corrected by his assertion of difference and superiority."[1] Freud labeled the mother as regressive: the child who remains bound to her is locked in a self-involved and narcissistic world.

History: Just as Freud believed the progress of the child depends on rejecting the mother, Bayer believed progress in design could only be achieved by rejecting the oppressive and "maternal" history of European culture. Modern design reacted to the self-aggrandizing, overly ornamented design of the nineteenth century: a tradition which Bayer believed to be deceitful, and—like the child who has not individuated from its mother—narcissistic. The customs of the past had to be rationalized. The need to differentiate from the mother, which Freud believed is integral to the male ego, is played out in the modernist rejection of and differentiation from the past: the child (modern design) rejects the mother (European history) and identifies with the father (American industry).

Father: Freud reinforces the common cultural belief which labels the mother as subjective and the father as objective. Freud believed that the father personifies objectivity because he brings external, social rules into the private, symbiotic relationship of the mother and child. The father grants the child a "way into the world" by asserting boundaries between the child and mother, and by inserting social norms the child must follow. He embodies an authority which Freud understood to be rational and progressive, and which the child both fears and admires. The child internalizes this "law of the father" in the form of the "super-ego": this is the agency which governs the ego, forcing the child to renounce the desire to stay united to the mother and enabling him/her to become self-governing.[2]

Progress: The demands of function work like a super-ego in the design of Bayer's Universal type. Just as the child becomes mature and responsible by internalizing the law of the father, Bayer believed that letterforms become socially responsible and progressive when the design internalizes the demands of function. Freud believed the child could only individuate from the mother by becoming objective, and similarly, Bayer believed typographic progress depended on finding objective, trans-cultural laws, which would guide the design out of the maternal arms of tradition and into the rational world of timeless laws. The dichotomy between mother and father in Freudian theory is echoed in Bayer's dichotomy between the "objective" law of progress and the "subjective" customs of the past.

1 Jessica Benjamin, *The Bonds of Love* (New York: Pantheon, 1988).

2 See Sigmund Freud, *Civilization and its Discontents* (New York: W. W. Norton, 1961).

Figure 1

Feminine: In 1896 social critic Gustav Le Bon reflected a popular prejudice when he stated that the modern crowd is irrational, volatile, and, "like women, it goes at once to extremes."[1] Similarly, the growth of machine-made commodities and such forms of mass entertainment as pulp fiction, advertising, and movies were labelled by defenders of "high" culture as an inauthentic, materialistic, and "*feminine*" threat to both traditional and modern forms of elite culture. "The swamp of big city life"... "The spreading ooze of massification," threatened to "engulf" the masculine high culture.[2] The boy's anxiety of being overwhelmed by the mother is re-enacted in high culture's fear of losing itself in the "false dreams" mass culture.

Figure 4

Subjectivity: The dichotomous thinking practiced by Bayer and Freud conflated dependence with irrationality. To be dependent on or influenced by the fluctuating and "whimsical" tastes of popular culture was to weaken the boundaries which make one autonomous and to contaminate the purity of detached reason. The over-abundance of fluctuating values of popular culture were treated as a threat to the stable, timeless authority of high culture and rational design. Design geared toward the undisciplined *appetite* of the masses rather than the refined *taste* of high culture was considered swamped in a materialisitic dream world reflecting the subjectivity and narcissism of the child who remains bound to its mother.

Masculine: While Universal type was designed for the masses, its appeal to (masculine) function over (feminine) form represents a correction of mass culture rather than an affirmation of it. The armature on which Universal was built functions like Freud's "rational" father: it represents a regulator that disciplines the letterform. The armature enabled the design to be based on objective rules that were exterior to the designer's personality and supposedly detached from the subjective and "feminine" realm of popular culture. It is this detachment which Freud believed the father embodies, and which Western culture aligns with masculinity. While this detachment is understood as neutral, it requires the *active* repudiation of subjectivity.

Figure 2

Objectivity: The ego boundaries which the child builds in the Oedipal stage make him/her autonomous from the mother. These boundaries are echoed in Bayer's dichotomy between popular culture and "functional" design, between a (regressive) history and a (progressive) future, and between (feminine) style and (masculine) rejection of style. The "neutral" objectivity which Bayer pursued can be reinterpreted as a reassertion of stable masculine ego boundaries and standards of "good" design in a world of rapidly changing values. Freud and Bayer invested belief in a science that is "premised on a radical dichotomy between subject and object and where all other experiences are accorded secondary 'feminine status.'"[3]

Figure 3

1 Gustave Le Bon, *The Crowd* (New York: Penguin, 1981) 50.
2 Andreas Huyssen, *After the Great Divide*, 52.
3 Evelyn Fox Keller, *Reflections on Gender and Science* (New Haven: Yale University Press, 1985) 87.

Figure 1 From Corbusier, *Towards a New Architecture*, 1931.
Figure 2 From Herbert Bayer, "towards a universal type," 1939-40
Figure 3 Self-portrait of Herbert Bayer's hand
Figure 4 From Herbert Bayer, "towards a universal type," 1939-40

Beyond ▲■●: Fractal Geometry

Alan Wolf

"Clouds are not spheres, mountains are not cones, coastlines are not circles... The number of distinct scales of length of natural patterns is for all practical purposes infinite. The existence of these patterns challenges us to study those forms that Euclid leaves aside as being 'formless,' to investigate the morphology of the 'amorphous'... Scientists will (I am sure) be surprised and delighted to find that not a few shapes they had to call *grainy, hydralike, in between, pimply, pocky, seaweedy, strange, tangled, tortuous, wiggly, wispy, wrinkled*, and the like, can henceforth be approached in rigorous and vigorous quantitative fashion."
Benoit Mandelbrot

Consider the apparently simple task of measuring the boundary of a fern. (We consider that the fern lies within a plane rather than having thickness, and thus is bounded by a curve.) Our measuring technique is to conform a piece of string to the perimeter of the fern as closely as possible. We then straighten the string and measure it's length with a ruler. As we use thinner and more flexible string, we are able to conform to smaller features of the perimeter, and our value for its length gradually increases.

We expect our values to converge to the "correct" length, but instead they seem to be growing to infinity! We could continue until our "string" reproduces the locations of individual atoms on the perimeter, but our interest here is in features that can be perceived by the senses or measured with macroscopic devices, so we will retain the fiction that our measurement process could have continued forever. We conclude that either the *perimeter* of the leaf is infinitely long, or that the question of its length is not well posed.

Our measurement technique suggests reasonable variants of the question, such as, "How long is a fern's perimeter when features no smaller than x are measured?" Since our eyes and other tools always have limited resolution, another reasonable question is "How long is a fern's perimeter when it is measured by x tool?" On a computer screen, for example, the size limitation is usually a screen "pixel." Thus a computer representation of a fern (such as the one shown here) is bounded by a finite number of pixels.

Now consider a cloud. The interior of a cloud is a "solid" three-dimensional object: it has length, width, and height. In this sense the interior of a cloud is like the interior of a balloon. The balloon's surface is (in our idealized discussion) perfectly smooth, and is thus an ordinary two-dimensional surface with a finite area. The cloud's surface, however, is wrinkly from its grossest features down to its microscopic structure. The cloud's *surface area* tends to infinity as the resolution of measurement increases.

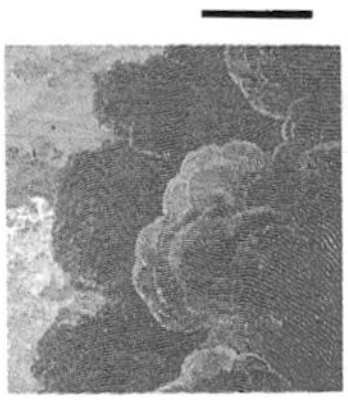

Mathematicians have quantified the wrinkliness of objects by modifying the standard definition of an object's dimension: in the new definition, points remain zero-dimensional, curves one-dimensional, and so on, but the *fractal dimension* of a cloud skin (from actual experimental measurements) has a value of about 2.3. Computer analysis of a photograph can provide such estimates. The value 2.3 suggests that a basically two-dimensional object is so wrinkled, it tends to act like a three-dimensional object.

This last remark is easily misunderstood: it is not the obviously correct statement that a crumpled two-dimensional piece of paper *occupies* a three-dimensional region of space (with some length, width, and height). What the statement *does* mean is that if the paper is wrinkled on all length scales, it tends to *densely fill* space like a solid three-dimensional object does: careful examination of the paper will not show that its "true" structure is two-dimensional, because the wrinkles never disappear at any level of scrutiny.

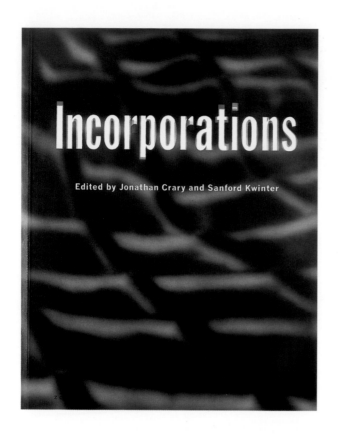

Incorporations

Edited by Jonathan Crary and Sanford Kwinter

ZONE Editors: Jonathan Crary, Michel Feher, Sanford Kwinter, Ramona Naddaff.
Editors of this volume: Jonathan Crary and Sanford Kwinter
Managing Editor: Elizabeth Felicella
Design: Bruce Mau
Design Schema: Sanford Kwinter and Bruce Mau
Production: Ed Cleary, Alison Hahn, Filip Mau, Kathleen Oginski,
Nigel Smith, Gregory Van Alstyne.

Image Research Coordinator: Elizabeth Felicella
with Naomi Jackson, Suzanne Jackson, Carole Naggar, Keith Seward,
Steve Snyder, Kristen Vallow, Timothy Wright.
Manuscript Editor: Ted Byfield
Editorial Coordinators: Ronnie Childress and Meighan Gale
with Suzanne Jackson, Linda Kelly, Rachel Robbins, Keith Seward,
Kristen Vallow, Thad Ziolkowski.
Translations: Randall Cherry, Mark Cohen, Robert Hurley, Martin Joughin,
Donald Leslie, Brian Massumi, Delphine Bechtel, Ted Byfield.
Fundraising and Advertising: Céline Cazalis de Fabel
Initiating Projects Editor: Kerri Kwinter
Special Thanks: Brian Boigon, Lois Burke, Anne Dixon, Ed Epstein, Jason Greenberg,
Virginia Heckert, Barbara Hoffman, Christian Hubert, Alex Kessler, Gus Kiley, Andy
Levine, Kerry McCarthy et AP/World Wide, Melissa Mathis, Jeffrey Meikle, Erica
Meinhardt, Anne Mensior of CLAM, Guillaume Paris, Gilles Peress, Guy Poulos, Jackie
Raynal, Wenzelin Scott, Carina Snyder, Susan Spiegel, Tim Sternberg, Alyamobisi
Williams, Thomas Jean Yen, Aperture Magazine, Avery Library, DRT Artificio,
Museum of Modern Art Photography Department, Photofest, Telescope Magazine,
University of California at Riverside Photography Museum.

We gratefully acknowledge the generous support of the National Endowment for the
Arts, Tom Cugliani Gallery, Lorence Monk Gallery, Nahon Gallery, Marian Goodman
Gallery, Vincent Wapler, John Weber Gallery, Salvatore Ala Gallery, Leo Castelli
Gallery, Vrej Baghoomian, Sidney Janis.

© 1992 Urzone, Inc.
ZONE 611 Broadway, Suite 608.
New York, NY 10012.

Printed in Canada by Bradbury Tamblyn & Boone Ltd.
Distributed by The MIT Press.
ISBN: 0-942299-29-9 (paper) 0-942299-30-2 (cloth)
Library of Congress Catalog Card Number: 88-051419.

Zone 6: Incorporations
Book cover and spreads
Designers: Bruce Mau, Greg van Alstyne,
Nigel Smith, Alison Hahn, Kathleen Oginski
Design company: Bruce Mau Design
Client: Zone Books, New York
Canada, 1992

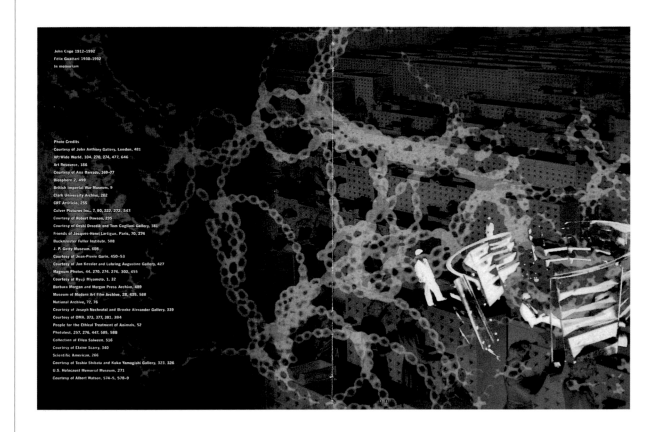

ZANDERS

				1	2
3	4	5	6	7	8	9
10	11	12	13	14	15	16
17	18	19	20	21	22	23
24	25	26	27	28		

Die Erde, vor ungefähr 4.600 Millionen Jahren geschaffen, brauchte 700 Millionen Jahre, um auf eine Temperatur unter der von kochendem Wasser abzukühlen. Erst dann begannen die Regenfälle, die die warmen, nährstoffreichen Seen bildeten, aus deren fruchtbarem Milieu schließlich das erste Leben hervorging. Die erste Spezies, die Blaualge, erschien vor ca. 800 Millionen Jahren. Etwa 465 Millionen Jahre später beherrschten die Dinosaurier die Erde. Die ersten menschenähnlichen Wesen tauchten erst vor nur 3 Millionen Jahren auf.

Created around 4,600 million years ago, the Earth took 700 million years to cool to a temperature below that of boiling water. Only then did the rains start, forming the warm nutrient-rich lakes that ultimately provided the fertile environment in which life began. The first species, the blue-green algae, appeared around 800 million years ago. Some 465 million years later the dinosaurs dominated the land. The earliest humanoids appeared a more 3 million years ago.

La terre, créée il y a près de 4.600 millions d'années, eut besoin de 700 millions d'années pour refroidir à une température inférieure à celle de l'eau bouillante. Ce n'est qu'à cette époque qu'il commence à pleuvoir, et que furent formés les lacs riches en nutriants, base du début de la vie grâce à un environnement fertile. Les premières espèces, les algues bleues, apparurent il y a 800 millions d'années. Les dinosaures dominaient le monde quelque 465 millions d'années plus tard. Il n'y a que près de 3 millions d'années qu'apparurent les premiers humanoïdes.

...förmige Galaxie, die Milch... ...rund 100 Milliarden... ...sich ungefähr... befindet sich an... der etwa... vom... für eine... braucht,... ...Kernschmelz-kraftwerk liefert die Sonne der Erde jede Sekunde mehr als 100 Millionen Milliarden Joules Energie.

Our disk-shaped galaxy, the Milky Way, contains around 100 billion stars. Formed some 5 billion years ago, the Sun is situated on the inner edge of a spiral arm approximately 300 million billion kilometres from the centre. In taking 225 million years to complete a revolution, equivalent to one cosmic year, the Sun in its lifetime has circled the centre of the Milky Way just 25 times. A great power-house of fusion reactions, the Sun provides the Earth with over 100 million billion joules of energy every second.

Près de 100 milliards d'étoiles forment la Voie Lactée, notre galaxie en forme de disque. Le soleil, formé il y a 5 milliards d'ans, est situé sur la partie intérieure d'un bras en spirale, à environ 300 millions de milliards de kilomètres du centre. Pendant son existence, le soleil n'a fait que 25 fois le tour de la Voie Lactée, puisqu'il a besoin de 225 millions d'années, soit une année cosmique, pour réaliser une révolution complète. Le soleil, générateur puissant de réactions de fusion, fournit chaque seconde plus de 100 millions de milliards de Joules d'énergie à notre terre.

Time. Machine. Future
1992 calendar: February, August
Designers: Mark Holt, Hamish Muir
Design company: 8vo
Client: Zanders
Great Britain, 1991

Museum poster series
Designers: Mark Holt, Hamish Muir
Design company: 8vo
Client: Museum Boymans-van Beuningen,
Rotterdam
Great Britain, 1990-91

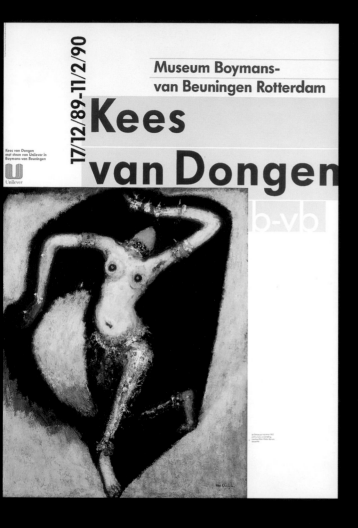

Paviljoen Van Beuningen-de Vriese

nieuw

**voor pre-industriële
gebruiksvoorwerpen,
kunstnijverheid en
industriële vormgeving**

Museum Boymans-
van Beuningen Rotterdam

Bassheads

2045896

12R 6303

Is there anybody out there?

A. Side	B. Side	Additional
Is there anybody out there? (extended).	Non verbal communication (extended).	Designed by 3a.
Written by Desry / Murphy / Imre / Various.	Written by Murphy / Desry.	
Produced by Bassheads / S. Imre.	Produced by Bassheads.	
Mixed by Desa & Ralphy.		

a DECONSTRUCTION record.

℗ 1991 the copyright in this sound recording is owned by EMI Records Ltd. © 1991 EMI Records Ltd.

a DECONSTRUCTION record.
132 Liverpool Road, London N1 1 LA.

Is there anybody out there?
Record sleeve (front and back)
Design director: Mark Farrow
Designers: Mark Farrow, Sian Cleaver,
Rob Petrie, Phil Sims
Design company: 3a
Client: Deconstruction Records
Great Britain, 1991

Mercatorplein
Poster
Design company: gebr de Jong
Client: Stadsdeel De Baarsjes
The Netherlands, 1992

K-klass
MPeople
Felix

DJ's | **Tickets**

Haçienda. Tuesday September 15th.

Mike Pickering, Daniele Davoli, Danny Rampling and Alistair Cooke.

Piccadilly Box Office, HMV, Haçienda. £7.

another steamy night out with DECONSTRUCTION.

Concert posters
Design director: Mark Farrow
Designers: Mark Farrow, Rob Petrie, Phil Sims
Design company: Farrow
Client: Deconstruction Records
Great Britain, 1992

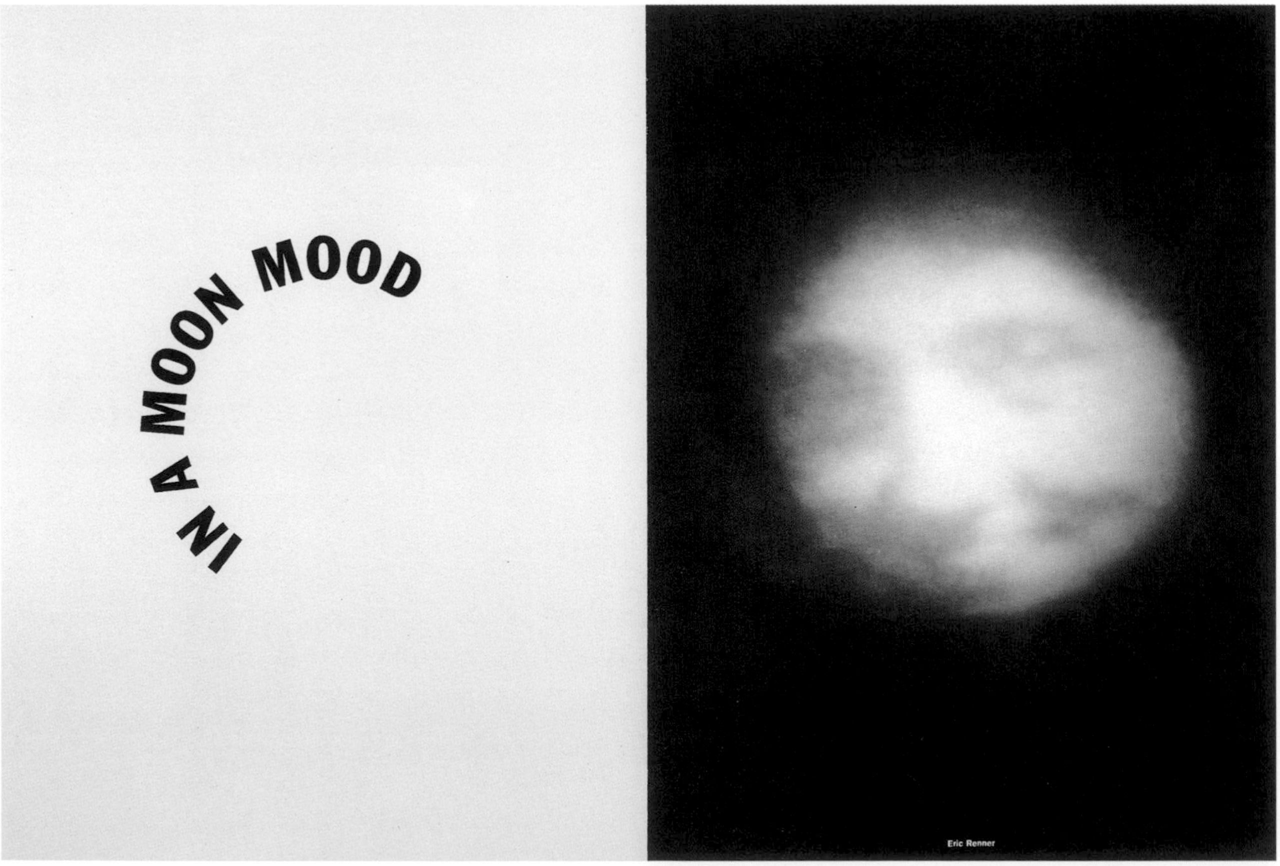

IN A MOON MOOD

Eric Renner

Interview
Magazine spread
Designers: Tibor Kalman, Richard Pandiscio
Design company: M&Co
Photographer: Eric Renner
Client: *Interview*
USA, January 1991

Self-promotional poster
Designers: Paul Neale, Andrew Stevens,
Nigel Robinson
Design company: Graphic Thought Facility
Self-published
Great Britain, 1992

RSC Fringe Festival '91
The Last Days of Don Juan
Theatre posters
Designer: Paul Neale, Andrew Stevens,
Nigel Robinson
Design company: Graphic Thought Facility
Client: Royal Shakespeare Company
Great Britain, 1991

Acts of Faith
Theatre poster
Designer: A. Arefin
Photographer: Anthony Oliver
Client: Institute of Contemporary Arts, London
Great Britain, 1992

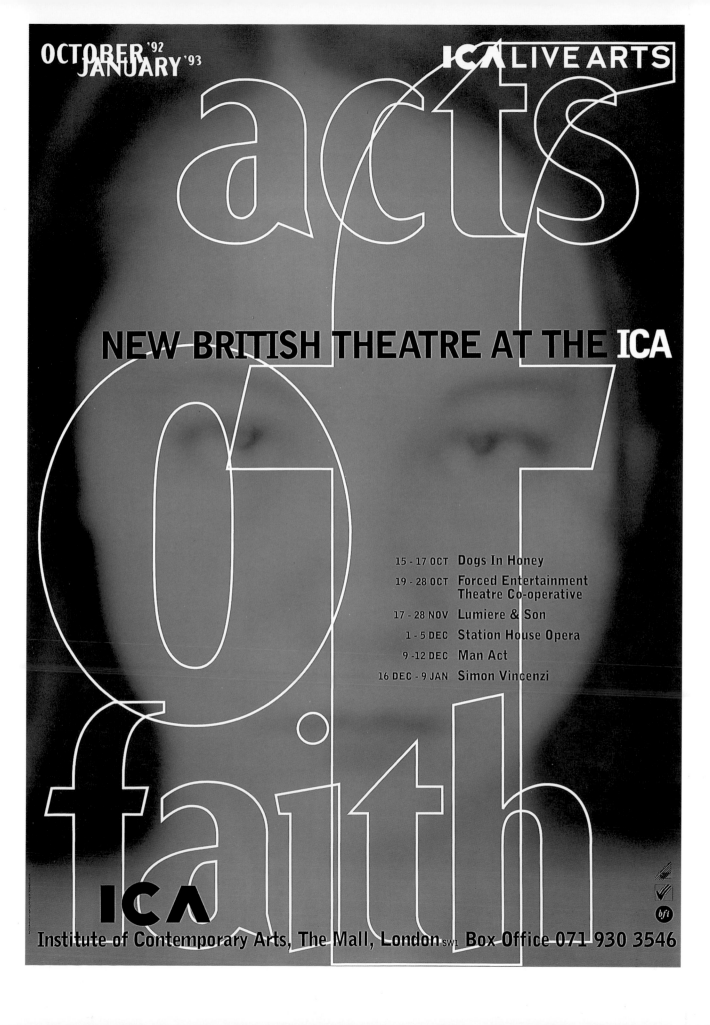

Economische Structuur 1993
Calendar: January, May
Designers: Jacques Koeweiden, Paul Postma
Design company: Koeweiden Postma
Photographer: YANi
Client: Ministry of Economic Affairs
The Netherlands, 1992

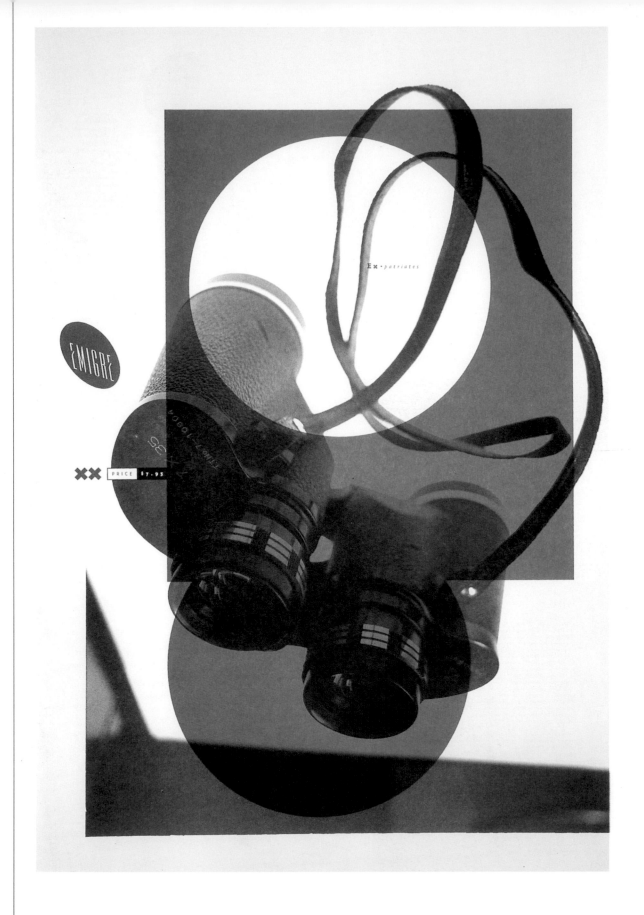

Emigre
Magazine covers
Designer: Rudy VanderLans
Design company: Emigre Graphics
Self-published
USA, 1991, 1992

layered

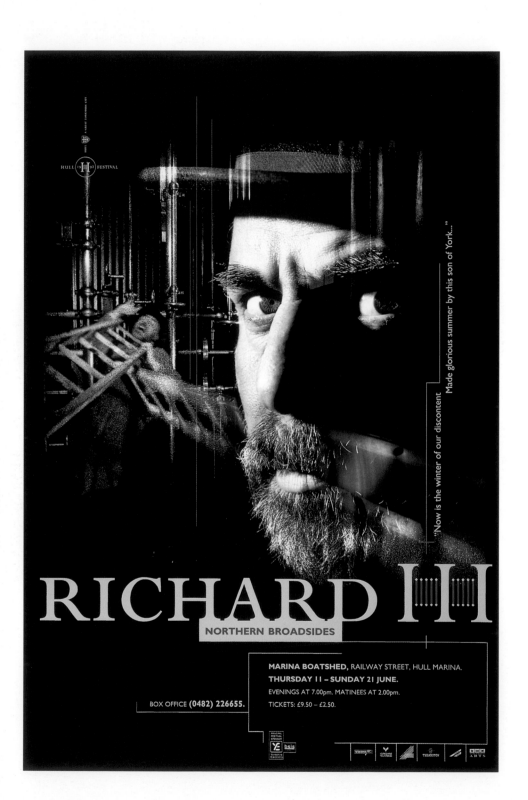

Richard III
Civil War Events
Festival posters
Designers: Andy Altmann, David Ellis,
Christopher Priest
Design company: Why Not Associates
Photographer: Nick Georghiou
Client: Hull City Council
Great Britain, 1992

HULL IS JUSTLY PROUD OF ITS HISTORIC REFUSAL
TO ALLOW CHARLES I TO ENTER THE WALLS,
IN APRIL 1642. A SIGNIFICANT TURNING POINT
IN ROYAL POPULARITY, IT JUSTIFIES THE CLAIM:
HULL – BIRTHPLACE OF
THE CIVIL WAR 1642 – 1992

CIVIL WAR

HULL 1992 FESTIVAL   HULL – A GREAT YORKSHIRE CITY

events

april 11 - may 31	**Arms and Armour Exhibition**	town docks museum
april 15	**Cromwell Film**	Ferens Art Gallery
april 23	**Witchfinder General Film**	hull film theatre
april 17 - 23	**Hull Flower and Church Arts Festival**	holy trinity church
april 23	**Civic Service with Lord Tonypandy**	holy trinity church
april 25	**Beverley Gate Refusal Re-enactment**	beverley gate
april 25	**The Governors Reception Party**	the maltings
april - december	**Hull in the 17th Century**	wilberforce house
june 15 - 19	**Hull 1642: Living History Events**	schools
june 16 - 27	**Hull City Play**	city hall
june 20 - 21	**Re-enactment of Civil War Battle**	city site
july 11 - 12	**Wilberforce House: July 1642**	wilberforce house

PRINCIPAL
FESTIVAL
SPONSOR

YE bsis

Yorkshire
Electricity

organized in association with
the English Civil War Society

BBC
ARTS

photographer jeremy hall (royal armouries), neil holmes
design: why not associates

most events are free
for further details telephone
Hull Festival hotline on 0482 223344

the inven**tive**spirit

UK PRESIDENCY
July - December 1992

NEW DESIGN FROM **BRITAIN** | DES NOUVELLES CONCEPTIONS DE **GRANDE–BRETAGNE**
| NIEUW DESIGN UIT **GROOT–BRITTANNIE**

AUTOWORLD BRUSSELS
PARC DU CINQUANTENAIRE 11 - JUBELPARK 11,
1040 BRUXELLES - BRUSSEL.
13.10.92. - 06.11.92. 10.00am - 5.00pm.

A Writing Table
Poster
Designer/illustrator: Mark D. Sylvester
Design company: Point
Client: Abigail J. Shachat
USA, 1992

30 years Amnesty International
Poster
Designer: Jan Marcus Jancourt
Client: Amnesty International/Wabnitz
Editions
USA, 1991

APRIL

APRIL IN

7:00PM

APRIL GREIMAN
APRIL 10, 1991

DESALLE AUDITORIUM, CRANBROOK ACADEMY OF ART

THEY
WILL COME
SPONSORED BY AIGA DETROIT   & CHAMPION INTERNATIONAL
THIS EVENT IS PART OF THE SERIES "CHAMPIONS OF DESIGN" WHICH IS MADE POSSIBLE IN PART BY A GRANT FROM CHAMPION INTERNATIONAL.

April in April
Lecture poster
Designer: James A. Houff
Photographer: Paul Price
Client: American Institute of
Graphic Arts, Detroit
USA, 1991

Promotional poster
Designer: James A. Houff
Client: Typocraft Printers
USA, 1990

GUIDE 2

Ohio
MEDIA EQUIPMENT

Ohio Arts Council

0 17 03 22

Access
sites

Guide to Ohio Media Equipment
Access Sites
Brochure cover
Designer/photographer: Crit Warren
Design company: Schmeltz + Warren
Client: Ohio Arts Council
USA, 1991

Towards College in Ohio 91-92
Brochure cover
Designer/photographer: Crit Warren
Design company: Schmeltz + Warren
Client: The Ohio College Association
USA, 1990

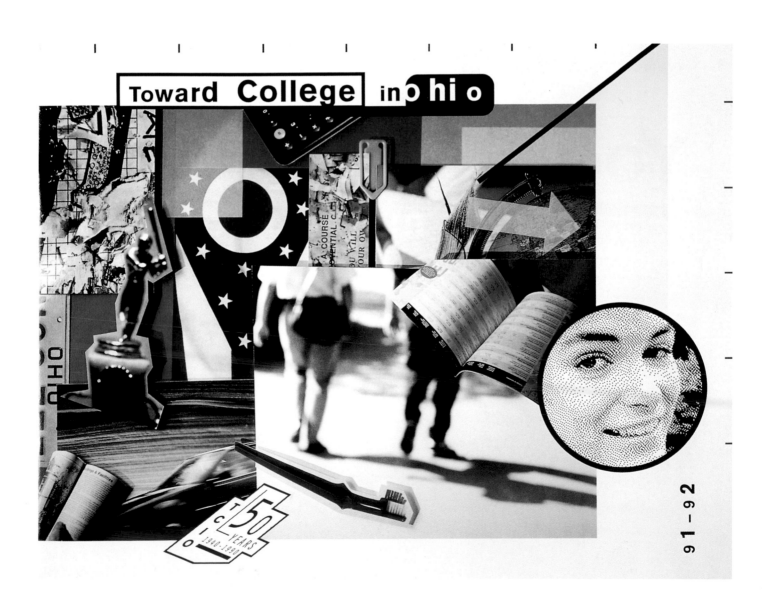

Guide to Ohio Media Equipment
Access Sites
Brochure cover
Designer/photographer: Crit Warren
Design company: Schmeltz + Warren
Client: Ohio Arts Council
USA, 1991

Towards College in Ohio 91-92
Brochure cover
Designer/photographer: Crit Warren
Design company: Schmeltz + Warren
Client: The Ohio College Association
USA, 1990

4 the Record
Posters/advertisements
Designer: John Weber
Design company: design: Weber
Client: For the Record
USA, 1992

Man ist was Man sieht
(Man is what man sees)
Personal project
Design: John Weber
Design company: design: Weber
Client: *Instant* magazine
USA, 1992

Man ist
was
sieht

was Man

TYPE 1

Balance

design:Weber;John.l

72

ACT NOW!

Fill in the coupon and start your
membership in this remarkable
book club.

1 2 3 4

5 6 7

membership requires advance deposit of $10.00.

$1.59

BONUS

OPEN HERE

16201-B
PC-31077-

FERRINGTON Guitars

1931 | 15½ inches

Spruce top, maple body and neck, ebony fretboard

Chris Difford is a guitarist, singer, and songwriter with the British band Squeeze. Working in collaboration with fellow band member Glenn Tilbrook, Difford helped develop the distinctively humane sound and colorful working-class imagery that characterize Squeeze's music.

Chris saw some of the guitars I'd made for Elvis Costello when his band Squeeze was opening for Elvis during a show, and eventually we got together to collaborate on the designs for two guitars. This small guitar was made to look like an acoustic version of a Rickenbacker-style electric like the kind John Lennon used to play.

It was a little bit tricky to do, because the curves in the points were so sharp that I had to use solid blocks of wood to make the front.
You can't tell what I did when you look at it, though, because I made it appear as if I had actually bent the wood to fit those really sharp angles.

Chris told me to make the inside of the guitar look as if it had just rolled off a factory assembly line, so I like very grungy stenciled letters. He said what I was painting it in the end that it was, in fact, a true American twist for it. All in all, the very different, you'd expect to see on an "American Bandstand." It's just right for Chris.

84 85

Ferrington Guitars
Book cover and spreads
Designers: Nancy Skolos, Thomas Wedell
Design company: Skolos/Wedell
Photographer: Thomas Wedell
Client: Callaway Editions
USA, 1992

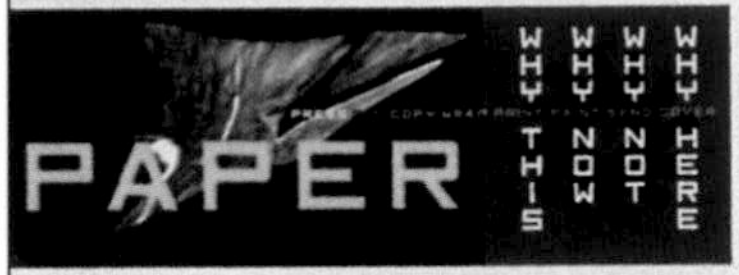

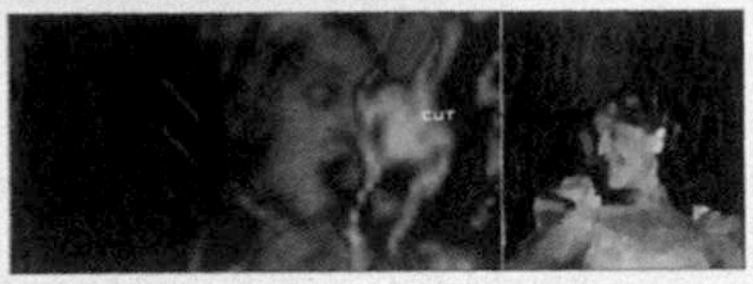

neville brody For designers and readers of the London-based magazines Arena or The Face, Neville Brody needs no introduction. The original bad-boy typographer, the visual interpreter of London's early Punk Rock scene, the champion of a visionary personal style — Brody has risen to prominence at a meteoric pace, revolutionizing the look of editorial design, advertising and typography along the way. Brody has also shouldered the burden of rapid success. Those at the top of the design profession are rampantly plagiarized and then criticized for their association with a watered-down trend. Introspective but not bitter, Brody has steadfastly moved on. When reached by phone at 8:00pm on a Sunday night, he was finishing up a project in his studio, rushing to catch a late flight to Bonn, and flat out sick with the flu. This not withstanding, he couldn't have sounded more high-spirited, as if nothing could slow him down. Are you always going at 100 miles per hour? No. . . but I do work all the time. If you're not obsessed, I believe, you're not going to get any really good work out of yourself. Are you obsessed by your work? Yes, completely. Not obsessive, mind you, but completely immersed. Recently, I read something in Emigré magazine that I wanted to ask you about. There is a line about you having taken "safe refuge in the International Style." What did you think when you read that? I wanted to reply to that, particularly since, two pages later, there is a comment that says something to the effect that "Neville Brody should have packed it in when he was still successful. . ." What can you say? Concerning the "safe refuge" remark, I want to go on record as saying that it's complete ___ ___ ___ ___. Modernism isn't a "safe refuge." My work for Arena, for example, was a modernist statement because it was a necessary reaction to everything that had gone before. There had been such a desperate search for the new that anything different became valid, even if it was completely meaningless. So the contrast between The Face and Arena was an obvious counter-reaction? Exactly. I wasn't saying, "Well, we've had our fun, now let's return to modernism." That wasn't my idea at all. Instead, I was saying: "Let's get back to basics and really think about what we're doing, think about the power of simplicity. With Arena, I wanted to show that we've all become obsessed with style for style's sake. Instead, we should return to basics and start thinking again. I also wanted to show that it was possible to use modernism as an emotive way of communicating. It sounds as though for a while there you became pretty disenchanted with the design scene. At the end of the 1980s, design had become a commodity of its own, which I've never agreed with. When design stops making people think, it becomes redundant. Design is supposed to be a way of translating ideas. It does not function well when it stands alone. Wasn't your switch to modernism actually a radical idea for its time? Yes, but like most radical ideas, it was misinterpreted. Regardless of what others may think, I do not believe in a return to modernism for its own sake or as a stylistic statement — only as a way to help us think about what we should be doing and as a way to reexamine where our responsibilities lie. For me it served as a way to ask: "What is the role of design in society?" Do you think that people are misinterpreting your backlash by calling it a retreat? Yes, completely. But as a designer, you have to keep changing. Regardless of all this criticism, you sound extremely busy. Yes, thankfully, we have a lot of clients at the moment. Yet, strangely, only one is British and none are American. Really? Yes. Most of our work is in Germany, Japan, France, Holland and Spain. I can't explain why, it just worked out that way. It's bizarre, really. What explains the shift away from Britain and America? By the nature of our work, we tend to gravitate toward societies with more forward-thinking people — people who are more willing to take risks. There's also the question of money. To give some idea, our one British client — a shelf company — elected to pay us with shelves rather than cash. What's your rate? Seventeen shelves per day. The way we see it, you can never have enough shelves.

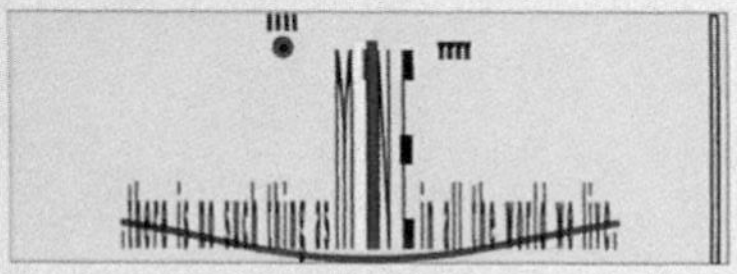

rickvalicenti

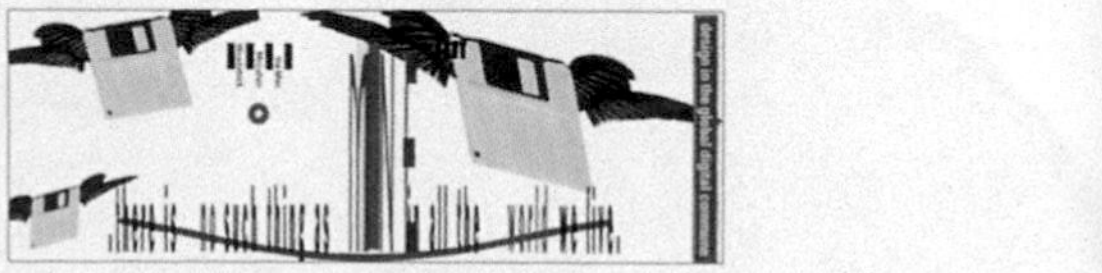

The word Thirst signifies both the name and the impassioned creative drive behind a small studio founded in Chicago by Rick Valicenti in 1988. Champions of a deeply personal approach to graphic design, Valicenti and his collaborator Michael Giammanco have provoked the design establishment with artistic, cheeky experiments that have quickly proven to be a legitimate contribution to design's discourse. Before long, gestures that were used expressively (or ironically) in the Thirst vocabulary made their way into a rash of portfolios, where their witticisms have been largely misinterpreted. Thirst, however, wasn't standing still. When we caught up with Valicenti at O'Hare International airport, he was keen to discuss cake decoration, Attitude Design, and the inspiration behind the Exquisite Corpse experiment you have before you. How has Thirst forged its distinctive, audacious style in the middle of heartland America? The only way I can answer that is to say we're just being ourselves. I'm just being me. And that's not said out of arrogance or false pride — it's just all I know. Not being trained as a graphic designer, I've had to learn how to absolutely trust my instincts. What's your impression about the state of graphic design today? A lot of the #@*% I see out there is so jive ... Well, don't print that. Let's just say a lot of design is derivative, and at this point a lot of it has become derivative of itself, so it's a self-perpetuating spiral. I've become so bored with attitude. I'd rather be on the sideline calling a spade a spade. But you're not on the sideline! At this year's AIGA conference you were at the top of the roster ... Yes, but I'm only servicing a small group of design-driven clients. So in that sense I'm not in a position where I'm supposed to appeal to the mass-market. There's a great deal of freedom in remaining small and running a lean organization. Do the Thirst imitators drag you down? They are of little concern except when they have a negative impact on my business development efforts. In any case if someone copies my typographic gestures, they may not be adopting the reasons for employing those gestures. So how do you stay ahead of the pack? The only way is through provocation. By revealing levels of thinking that are much more in-depth than anything anyone had expected. People are looking for the cake decoration and they aren't going to find it. As I see it, the trap for a designer gets set when he or she starts doing work to fulfill someone else's expectations. Then the designer become static. Thank you very much! That's the story exactly. The primary challenge is to fulfill only your own expectations; for me that often includes poking some fun — even at myself. What was your inspiration for this Exquisite Corpse project? What I told Michael, my partner, was that this project was like tribes sending smoke signals out to each other. We're saying, hey, we're having some fun over here, what's going on over there? And the part that's so exciting to me is to think that designers who respect one another can relinquish their own work to one of their peers, and then embrace someone else's work. What was the smoke signal you sent out? It's a very simple thing, partly a self-portrait in type, created from a line from Buddha: THERE IS NO SUCH THING AS MINE IN ALL THE WORLD To me this quote represented what we are doing, because here I was creating a design on a disk that I'm passing along to Katherine, who will make her mark, and then send it on to Neville, who will transform it again. I wanted to make a statement that what I was sending out was not mine, and the next person should feel free to do with it as they please. To me it's an exhilarating and unnerving way to exchange a message. All egos are checked at the door on this project? Absolutely. Everyone has said nothing is theirs. Helvetica or Franklin Gothic don't belong to any one designer; the use of warm red and black doesn't belong to any one dogma; and the use of arrows doesn't belong to the vernacular. It's all out there for us to use as communication. That is the spirit of this whole project. What do you hope will come of this Exquisite Corpse experiment? My feeling is that if seven designers can work together on this level, then there's no reason why those seven can't join up with seven others and so on. My most optimistic goal is for the world of electronic communication to be made a bit smaller, and for design collaboration to become a bit wider as a result of this global network.

erikspiekermann

In the heart of Berlin, adjacent to what used to be the symbolic and actual Wall of separation between East and West, Erik Spiekermann's twelve-year-old firm MetaDesign holds the distinction of being the largest graphic design office in Germany. With eighteen employees, the firm services a range of public and private institutions throughout the re-united country; occasionally, an overseas client in need of "a token Kraut," to use Spiekermann's term, comes knocking at MetaDesign's door. Yet despite his cultural jabs, the designer happily embraces Teutonic stereotypes. His greatest professional joy, he proclaims, comes of creating order where once there was none. Spiekermann's discovery of a Corbusian Golden Mean within the dimensions of this very page, for example, serves as a telling indication of his quest for exactitude. The obvious question for a designer — or anyone — living in Berlin right now is, What's going on? What is happening as your city undergoes such enormous changes. The problem I have in responding is that I'm too close to the situation. Even though I'm acutely aware that this is a historic moment, and that nothing like this has ever happened before, it's not easy to describe. It's not every day that a political system that has been in existence for forty years suddenly disappears. Yes, and all of the sudden I can walk 100 yards from the door of my office and I'm in East Berlin. It's odd, because {over there} everything looks different, feels different, smells different, and the people are all different, but geographically there is no difference any more. It's quite difficult for the human imagination to comprehend. Once we were two islands, and now there is no physical distance between us. And yet you are part of the same culture, you all speak the same language. That makes it even worse. We're two countries divided by a common tongue. If {the East Germans} spoke a different language, then it would be obvious and it would be easier to assimilate these kinds of cultural differences. Has the unity of Berlin resulted in any new commissions for you? One of our projects at the moment is the signage system for a major subway station in Berlin, half of which was closed for forty years while the city was divided. This is a great project because it's a public service that is truly needed, which is the type of work that I view as ideal. I would feel slightly superfluous if all we did were paper promotions. You once were quoted as saying that you weren't Neville Brody and you weren't changing the face of typography. What did you mean by that? I meant I'm not really interested in visuals. I'm not really interested in what things look like. I like solving problems. I'm a structuralist — I like to get the structure of a problem worked out, and then hand it to someone else who will then visualize it. Could you describe this process? I give people the ingredients. I say this is the grid, these are the typefaces, and here are the hierarchies. But I don't do the layout. I give people the box of tricks and they go and do the visuals. That's unusual. In my experience I find that graphic designers work much more visually than verbally. My style is reductionist simply because I haven't got the generosity, or the verve, you might say, to arrange elements across a spread so that they look natural, which is something I very much admire in Neville, but I can't do. So you see, I'm not an artist. Then it's beginning to makes sense why you were so interested in the formal components of this assignment. I have to tell you that you were the only one involved in this project who took one look at the dimensions of this page and discovered a Golden Mean based on Le Corbusier's system of proportional measurements! I always break down the spatial relationships of a page, and in this case I found that the measurement from one side to the other is the square root of eight. Don't ask me why I found that! I guess its my preoccupation with proportion, grids and the division of pages, which I find quite magical and cabalistic. So your concern is with structures that are invisible. Well somebody has to be concerned with those things. It's the grammar of design — it exists whether you recognize it or not. And yet there is something about the rules of grammar that most designers prefer to avoid. Maybe it's my German brain, but I like to know about these rules and apply them. In my opinion it's the only way you can break them, which I also enjoy doing.

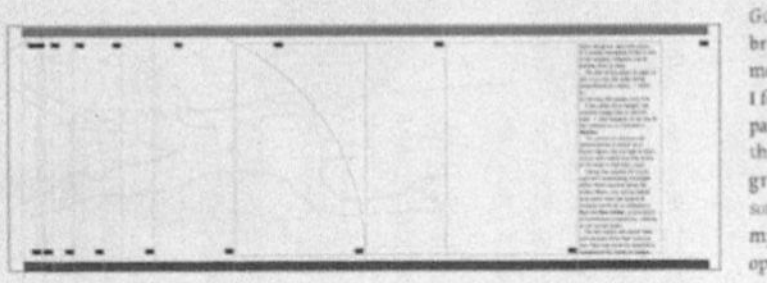

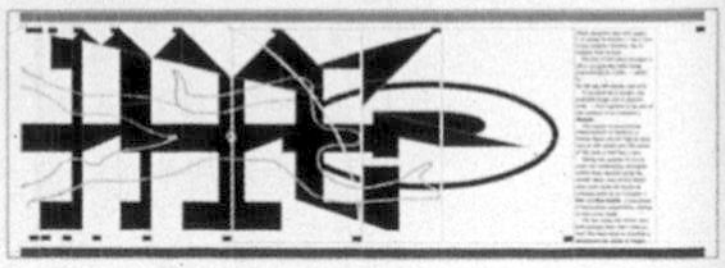

Electronic Exquisite Corpse
Brochure spreads showing three-way design collaborations
Art director: Rick Valicenti
Design company: Thirst
Designers: Neville Brody > Malcolm Enright > Katherine McCoy
Rick Valicenti > Katherine McCoy > Neville Brody
Erik Spiekermann > Paul Sych > Linda van Deursen, Armand Mevis
Client: Gilbert Paper
USA, 1991

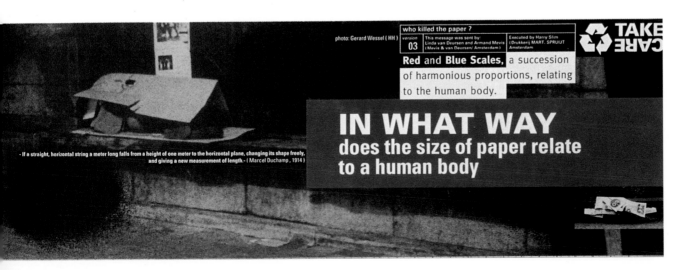

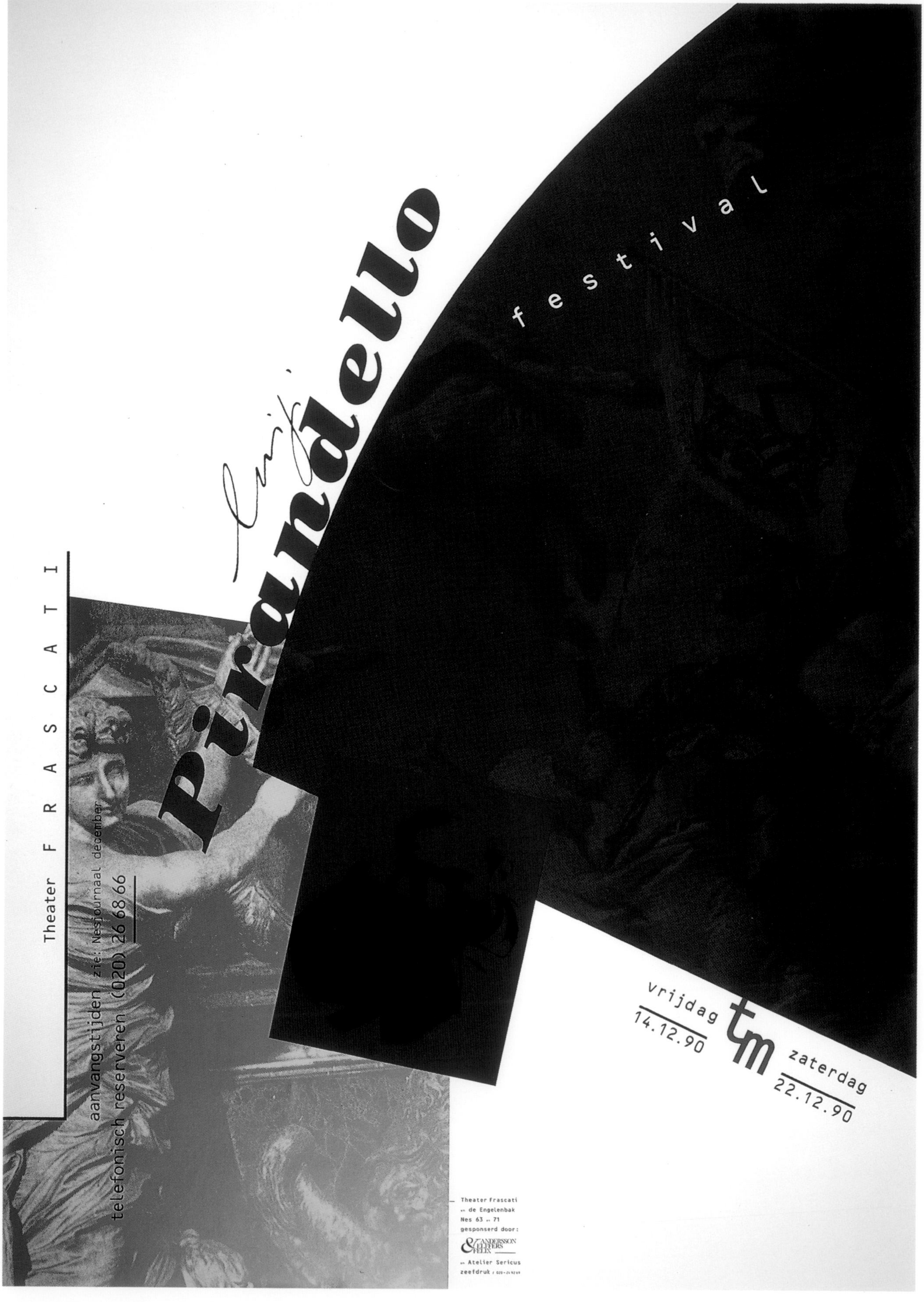
Pirandello
festival
Theater FRASCATI
aanvangstijden zie: Nesjournaal december
telefonisch reserveren (020) 26 68 66
vrijdag
14.12.90
tm
zaterdag
22.12.90
Theater Frascati
en de Engelenbak
Nes 63 en 71
gesponserd door:
Andersson Elffers Felix
en Atelier Sericus
zeefdruk

Luigi Pirandello Festival
Poster
Designers: Jacques Koeweiden, Paul Postma
Client: Theater Frascati, Amsterdam
The Netherlands, 1990

Trap 5
Dance poster
Designers: Jacques Koeweiden, Paul Postma
Design company: Koeweiden Postma
Photographer: YANi
Client: Dans Produktie, Amsterdam
The Netherlands, 1991

De Haagse Zomer
(The Hague Summer)
Festival posters
Designers: Ben Faydherbe, Wout de Vringer
Design company: Faydherbe/De Vringer
Photographer: Jordi Bover (1989)
Client: Het Gebeuren
The Netherlands, 1989, 1990, 1991

Flirting with the Edge
Poster/mailer for a conference
on creativity
Designer/type designer/
photographer: Barry Deck
Client: American Center for Design
USA, 1992

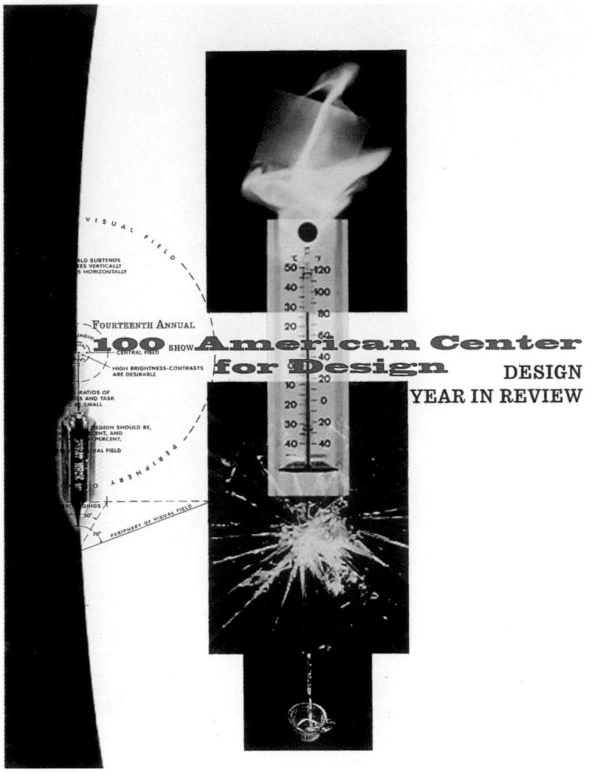
VISUAL FIELD
FOURTEENTH ANNUAL
100 SHOW
American Center for Design
DESIGN
YEAR IN REVIEW
HIGH BRIGHTNESS-CONTRASTS ARE DESIRABLE
CENTRAL FIELD
PERIPHERY
PERIPHERY OF VISUAL FIELD

JUROR'S ESSAY : **Lorraine Wild** teaches graphic design and design history at the California Institute of the Arts, where she is Director of Visual Communications. She lectures on design history, criticism and theory frequently and contributed a major essay to the book Graphic Design: A Visual Language History. She has been published widely, including ID Magazine, Print, Emigre, and the AIGA Journal. Her practice focuses on the design of architecture, fine art and design books for major museums and publishers, including the Museum of Contemporary Art in Los Angeles, Rizzoli International, and the MIT Press.

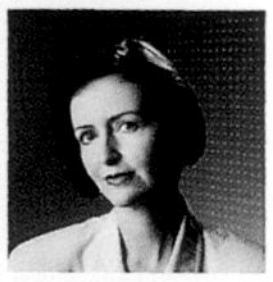

During the 1980s, the design profession underwent what might be called Competition Proliferation. Several new juried competitions were added to the old stand-bys like the New York Art Director's Club, the AIGA Book and Communication Graphics Shows, and the ACD 100 Show; perhaps this is a case where less actually is more. One of the by-products of this proliferation has been the development of a degree of cynicism on the part of many designers about the use/value of the numerous contests, old and new, and their many conventions: the high entry and hanging fees, the fuzzy criteria, the confusion of printing quality with design advancement, the weirdly provincial practice of allowing jurors to enter work, the postage-stamp sized reproductions in the annuals.

So it was with some trepidation that I agreed to serve on the jury of this year's ACD 100 Show, though I also felt that it was an honor to be asked, since this particular competition always seemed like one of the respectable ones (and tougher ones to place in, given the limit on the number of works chosen). The invitation was made more enticing by the agreement between the ACD and the show's chairperson,

A **Round Table** DISCUSSION
WITH LORRAINE WILD, RICK VERMEULEN, AND BRUCE MAU
MODERATED BY KATHERINE MCCOY

BM **We saw quite a lot of good work**, it seems to me. In the first cut we selected maybe two hundred pieces. But then on the second look, when trying to find pieces that were consistent throughout, I read more closely and realized that something was missing from much of the work, that it didn't have the quality that I thought it had.

LW One aspect of a piece would be brought to the right level; but then, when you looked more closely, you would find the underpinning of the piece to be wobbly. This of course is the pitfall of most graphic design: the exterior is often very seductive, but it doesn't get you through a second look.

BM I also realized how rarefied a climate I work in, and how so much work in graphic design is not very demanding at all. There is so little work that really reflects the kind of culture that I am interested in, corporate or otherwise.

KM **Do you mean that most of the client messages** brought to designers are not really the kind of material you want to deal with?

BM Yes, because I don't think in terms of client messages.

RV **But you work for the client, don't you?**

BM Yes, but I don't think in those terms. If I take on a project, I take it on in a collaborative sense. I don't think in terms of target audiences or market segments or the kind of terminology that seems to be present in the work that we're seeing.

RV **But doesn't that sound more like art?** I mean, there is a difference.

BM When I went to visit this man in Italy to help him do some work, he asked me "Can you help? Can you tell me how to do this the way that you do it?" I said "I'll tell you how to do this if you tell me how to do art." His response was "There is no difference between what you do and what I do."

KM **This collaborative idea** might not necessarily be dysfunctional. Collaboration could serve the client's communication objectives perfectly well.

BM I think it has to. You must be delivering the product. In marketing terminology, you must be satisfying that client.

Katherine McCoy, to eliminate the convention of jurying by consensus. I respect the two other jurors immensely, and looked forward to constructing a variation of the juried competition based on this experimental approach. It was going to be fun.

Not long before the jurying, I ran into Michael Beirut in New York, who told me that he had been kidding Kathy McCoy about inviting a jury that "promised to run the gamut from A to B!" Michael's thigh-slapper reminded me of the importance of trying to maintain some sort of objectivity (or at least to avoid acting out my own worst clichés) which of course was at odds with the whole curatorial direction that had been determined for this year's show. I began to suspect that our intention to "fix the format" would prove to be more complicated than we could possibly predict.

No amount of advanced planning quite prepares you for that first glimpse of a football-field sized room full of tables piled with print. I think we jurors quickly recognized that our experiment was about to be tested by two elements that had not been figured into the curatorial model: time (like all juries, we had only seconds to look at each piece, and to analyze whether it fit into our curatorial categories, or not), and selection (unlike curators, who slowly choose works for an exhibition from a theoretically infinite field, we had to choose from a field pre-determined by those who had sent in their work and paid their entry fees, whether they had actually read the published criteria or not).

All of this really did affect the jurying process. As I was inspecting each entry I realized that the work that I saw that was the most interesting (to me) was not necessarily in line with my announced criteria. For instance, this was not a *big* year for "history" (there was less quotation, appropriation, parody, or blatant rip-off). There was a lot of work that appeared to be informed by a knowledge of what has come before, but the issue of "history" just did not seem to jump out of the selections the way we thought it might, given that it was specifically noted in my juror's statement. But a juror's statement, of course, cannot really affect the character of the entries, since most of the entries are created long before the criteria are published. And, by and large, the work that was sent in was not radically different from previous years, reflecting the (vaguely disappointing) state-of-the-art, where the same sorts of projects elicit the same sorts

8

of solutions, over and over again. Perhaps the published criteria did influence a small percentage of the pieces that entrants chose to submit, but I also think that we jurors adjusted our criteria during the process to work with the field of entries that faced us. The result of this friction between the jury's articulated standards and the actual work produced by the entrants is this very interesting hybrid of a show, something like a curated exhibition, but not really.

A note on the jurying process, at the risk of irritating my esteemed colleagues: it didn't quite work out as planned. Despite our different statements, we shared many opinions about the work. Though there were few pieces that all three of us agreed upon, there were lots and lots that two of us liked (in all combinations), but since we had agreed that we would maintain separate selections, we were forced into a sort of flea market, "I-saw-it-first" bartering which, in the end, created artificially larger distinctions between our three sets of choices. Designers whose pieces actually got two votes might be interested to know that; but that would have taken an additional, unforeseen clerical step in the jurying process that was already more complicated by the experiment. I explain this in hope that some future jury continues to tinker with the experiment (and to acknowledge how complicated it is to alter the competition formula).

The issue of time allotted for judging was really crucial to this show because all three of us jurors put such a high value on sense and appropriateness. (I know, from Michael Beirut's witticism, that because all three of us are known for book design and cultural institutional work, that it would be easy to accuse us of impractical artiness. But like myself, Mau and Vermeulen are pragmatists who suspiciously look first for the reason behind everything.) So we felt compelled to read into each entry as best we could. Many good-looking pieces were rejected; some because they were contentless, but more often because, upon closer inspection, content and form were at odds with each other.

LW **Some of what you're saying makes sense** about that ideal of collaboration between the designer and the client. It reminds me of advice that Paul Rand gave students at Yale: *Never talk to anybody below the level of president.* That's where we started in design: a designer would know the entrepreneur president of a company, and they would get together and work as peers (or at least this is the way it's always been described). That's how the seminal work in corporate communications was created. I think it was Tibor Kalman who pointed out that these days an assistant vice president of public relations is working with a representative of a design firm who is not necessarily the person actually at the board producing the work. It makes you wonder what went wrong when you see a good corporate annual report or corporate communications piece. What allowed something good to come out of the current system? Every successful firm reaches a point where they are faced with a choice between growth and control, whether they know it or not. They grow larger, yet seem to ignore the issue of whether they are still producing work as good as when they were smaller, the kind of work which initiated their growth.

KM **Then, occasionally, you do see a firm that just stops** and says, no, wait. And they downscale. I have seen several firms do that in recent years.

LW But you do sacrifice a certain economic success.

KM **Do you always? Is there a model** for this by which economic success isn't sacrificed?

RV I don't think there is.

BM When a firm grows, the quality must be dispersed, because you have to get people who can more or less copy your work to work in a similar manner. Then the quality of their work must be reduced somewhat because it's not a singular gesture, the sort that creates very powerful statements. This is what I'm dealing with in my own situation right now: how to do projects of a bigger scale.

RV If you work on your own, you do everything on the board from start to finish. If you have assistants, you have to think through a project before it actually starts.

KM **That might change your design process** because important thinking often happens while you are in the midst of the execution.

RV What we do a lot is a sort of combination: think beforehand, have the stuff done by someone else, and then finish it or jump in at some point.

KM Rick, you said earlier that you feel the designer should be held responsible for the content of the piece. We saw quite a few pieces that dealt with environmental issues and, in a few cases, social issues. What did you think about that work, and why did you reject some of it?

RV **It's very much a fashion.** A lot of firms get into environmental issues and recycled paper, and all that stuff to look good – to be nice, to do what

American Center for Design FOURTEENTH ANNUAL 100 SHOW

Graphic design is at a fascinating moment just now, where influences flow from the fringes into mainstream practice and back in the wake of postmodernism. But the old Achilles heel of graphic design still lingers: style-mongering that is inappropriate, or which just doesn't work, and its depressing corollary, good opportunities for design that go begging. (I don't feel that a banal message necessarily dooms a project to a banal form. I was as impressed by the designers who figured out how to make something visually compelling out of a beer-and-boxing promotion as I was by the designers who produced a visually compelling booklet on human rights activism. In both cases I suspect the designers took an aggressive role in the conceptual development of the works that they produced.) It seems clear that the only way graphic designers are going to continue to justify their existence is by applying the conceptual processes of design with as much intelligence, imagination and grace to all sorts of messages that we, as the audience, will continue to seek out (or endure). But the old issue of appropriateness (and now the newer one of environmental ethics) persists, no matter who the clients are.

In my teaching, I encounter students who think that competitions are so compromised that they should just be abolished. I don't agree with this because despite all of their limitations, they serve as vehicles for the development of an identity for graphic design. After going through this process, I think that the most troublesome aspect of competitions is the narrow range of projects entered, which is still probably influenced more by entry and hanging fees than it is by strategies for jurying. But I was more than happy to contribute to this effort to move the conventions of the competition, and while I don't think we achieved our ideal, I think that it is an improvement: at least this annual will produce a record of how and why each project was selected.

9

Fourteenth Annual 100 Show
American Center for Design
Book cover, text and divider pages
Designer/illustrator: Mark D. Sylvester
Design company: Point/Cranbrook
Academy of Art
Client: American Center for Design
USA, 1991

PATH OF ROTATION

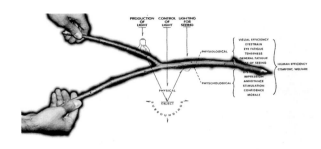

PRODUCTION OF LIGHT · CONTROL OF LIGHT · LIGHTING FOR SEEING

Juror
ROUNDTABLE DISCUSSION
Essays

Lorriane **Wild** SELECTIONS

EASIEST SEEING

THE INVISIBLE WORLD

Bruce **Mau** SELECTIONS

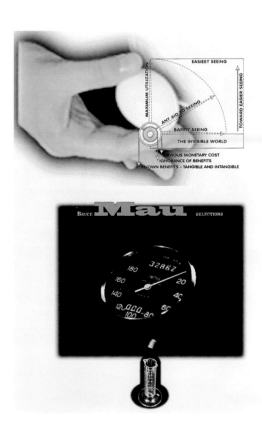

THE MINNEAPOLIS COLLEGE of ART + DESIGN

The Minneapolis College of
Art + Design 1991-1993
Prospectus cover
Designer: P. Scott Makela
Design company: P. Scott Makela Words +
Pictures for Business + Culture
Client: Minneapolis College of Art and Design
USA, 1991

Do Nothing
Visual essay from the catalogue
Rethinking Design
Designer: P. Scott Makela
Catalogue designer: Michael Bierut,
Pentagram
Client: Mohawk Paper
USA, 1992

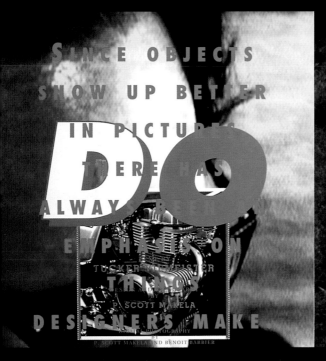

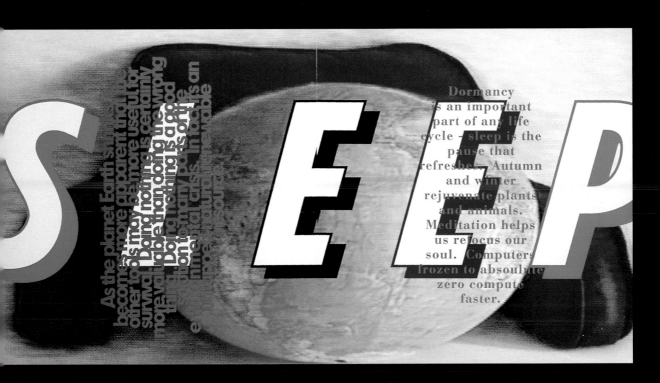

Tokyo Salamander:
American Dream Diary 1989
Spreads from an artist's book
Designer: Vaughan Oliver
Design company: v23
Artist: Shinro Ohtake
Client: Shinro Ohtake/Kyoto Shoin
Great Britain, 1993

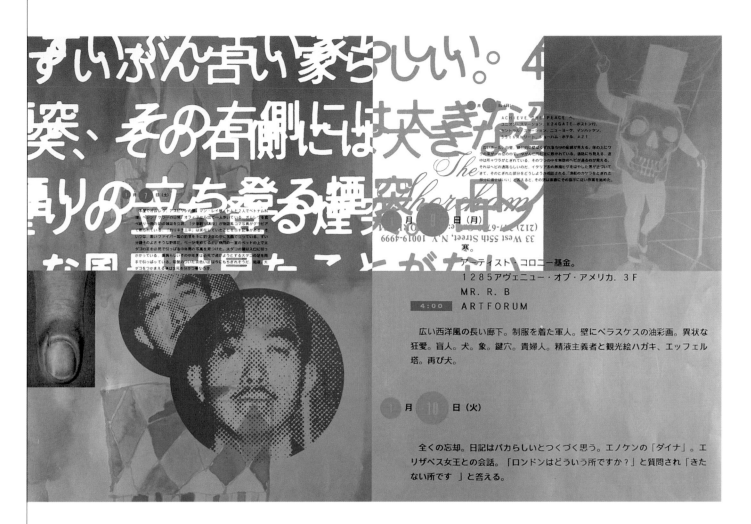

Lyceum Fellowship
Competition posters
Designers: Nancy Skolos, Thomas Wedell
Design company: Skolos/Wedell
Photographer: Thomas Wedell
Client: Lyceum Fellowship Committee
USA, 1992, 1993

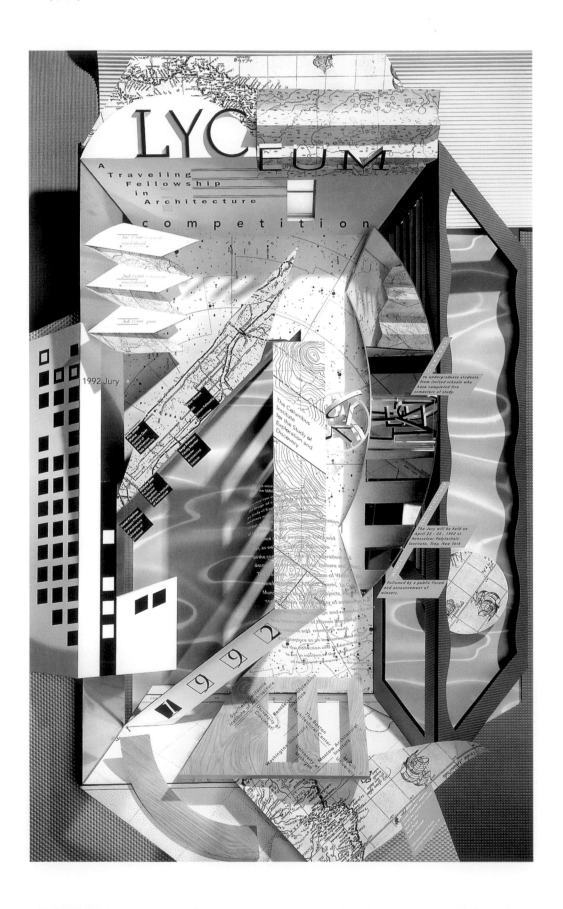

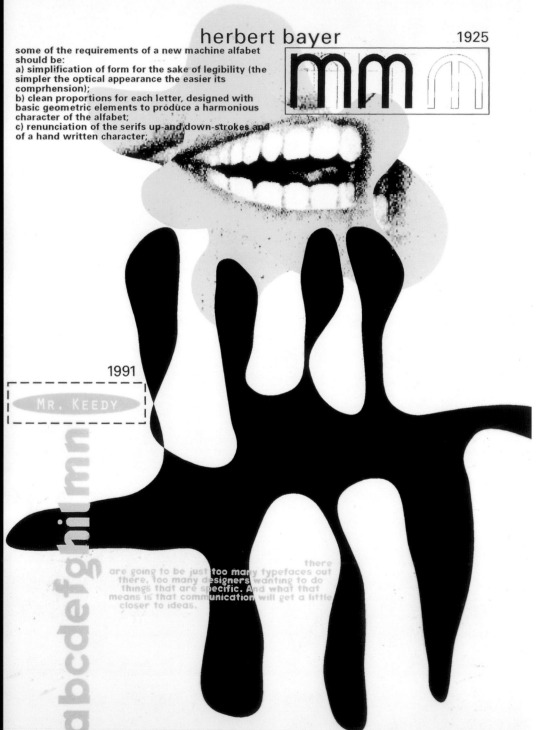

herbert bayer   1925

some of the requirements of a new machine alfabet
should be:
a) simplification of form for the sake of legibility (the
simpler the optical appearance the easier its
comprhension);
b) clean proportions for each letter, designed with
basic geometric elements to produce a harmonious
character of the alfabet;
c) renunciation of the serifs up-and-down-strokes and
of a hand written character;

1991

MR. KEEDY

abcdefghilmn

there
are going to be just too many typefaces out
there, too many designers wanting to do
things that are specific. And what that
means is that communication will get a little
closer to ideas.

Rio Festival
Poster
Designer: Neville Brody
Design company: Neville Brody Studio
Client: Design Rio, Brazil
Great Britain, 1992

Fuse
Promotional posters for interactive
typeface magazine
Designer: Neville Brody
Design company: Neville Brody Studio
Client: FontShop International
Great Britain, 1992, 1993

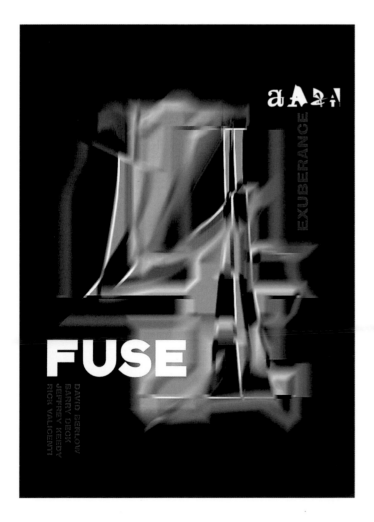

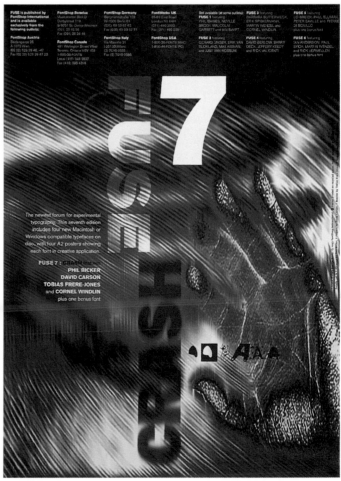

Rio Festival
Poster
Designer: Neville Brody
Design company: Neville Brody Studio
Client: Design Rio, Brazil
Great Britain, 1992

Fuse
Promotional posters for interactive
typeface magazine
Designer: Neville Brody
Design company: Neville Brody Studio
Client: FontShop International
Great Britain, 1992, 1993

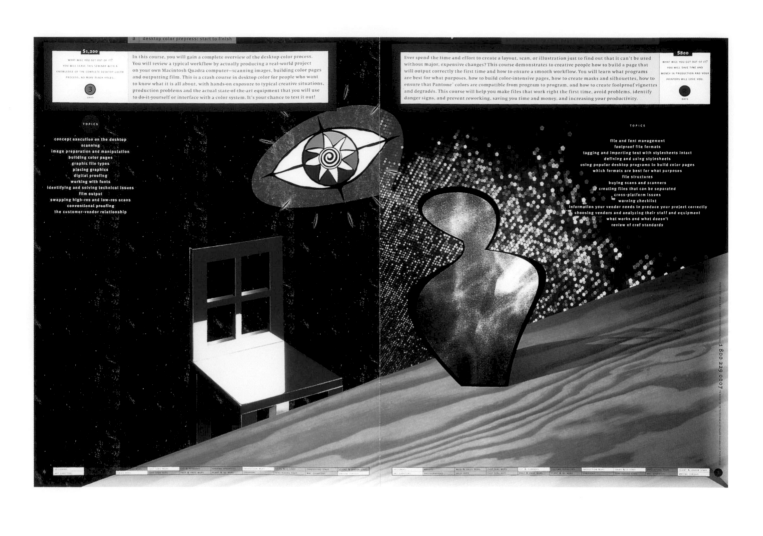

desktop color prepress: start to finish

$1,200

WHAT WILL YOU GET OUT OF IT? YOU WILL LEAVE THIS SEMINAR WITH A KNOWLEDGE OF THE COMPLETE DESKTOP COLOR PROCESS. NO MORE BLACK HOLES.

3 DAYS

In this course, you will gain a complete overview of the desktop color process. You will review a typical workflow by actually producing a real-world project on your own Macintosh Quadra computer—scanning images, building color pages and outputting film. This is a crash course in desktop color for people who want to know what it is all about, with hands-on exposure to typical creative situations, production problems and the actual state-of-the-art equipment that you will use to do-it-yourself or interface with a color system. It's your chance to test it out!

TOPICS

- concept execution on the desktop
- scanning
- image preparation and manipulation
- building color pages
- graphic file types
- placing graphics
- digital proofing
- working with fonts
- identifying and solving technical issues
- film output
- swapping high-res and low-res scans
- conventional proofing
- the customer-vendor relationship

Ever spend the time and effort to create a layout, scan, or illustration just to find out that it can't be used without major, expensive changes? This course demonstrates to creative people how to build a page that will output correctly the first time and how to ensure a smooth workflow. You will learn what programs are best for what purposes, how to build color-intensive pages, how to create masks and silhouettes, how to ensure that Pantone® colors are compatible from program to program, and how to create foolproof vignettes and degradés. This course will help you make files that work right the first time, avoid problems, identify danger signs, and prevent reworking, saving you time and money, and increasing your productivity.

$800

WHAT WILL YOU GET OUT OF IT? YOU WILL SAVE TIME AND MONEY IN PRODUCTION AND YOUR PRINTERS WILL LOVE YOU.

DAYS

TOPICS

- file and font management
- foolproof file formats
- tagging and importing text with stylesheets intact
- defining and using stylesheets
- using popular desktop programs to build color pages
- which formats are best for what purposes
- file structures
- buying scans and scanners
- creating files that can be separated
- cross-platform issues
- warning checklist
- information your vendor needs to produce your project correctly
- choosing vendors and analyzing their staff and equipment
- what works and what doesn't
- review of cref standards

1 800 229 0007

Add a Little Magic
Brochure spreads
Designer: Rick Valicenti
Typographers: Rick Valicenti, Richard Weaver, Tracy Taylor
Design company: Thirst
Photographers: Corinne Pfister, Michael Pappas
Digital imaging: Tony Klassen, Mark Rattin
Client: The Color Center
USA, 1992

g | desktop color for the production manager and customer service rep

$1,200

WHAT WILL YOU GET OUT OF IT?
YOU WILL GAIN
CONTROL OF THE DESKTOP
COLOR PROCESS.

3 DAYS

Desktop publishing has the reputation of being unmanageable. It does not have to be. This course will teach you how to manage your desktop projects without the usual desktop headaches. You will learn what desk-top publishing and desktop color are all about and how to control the process effectively—without surprises. The emphasis in this course will be on how to analyze projects to determine the scheduling, human resources, and equipment requirements needed to get a job done.

TOPICS

introduction to using the Macintosh computer
how to open and check files
questions to ask
warning signs
identifying responsibilities
scheduling strategies
understanding files, applications, and scans
coordinating project elements on the desktop
the desktop color process and its limitations
overview of desktop workflows
what works and what doesn't
common production problems to watch for
checklist: what you need from your customers for each job
checklist: how long different tasks take to accomplish
building safety into the job
traffic desktop jobs to avoid wasting time and resources
strategies for estimating desktop projects

black and white images: quality reproduction on the desktop | h

$800

WHAT WILL YOU GET OUT OF IT?
YOU WILL BE ABLE TO PRODUCE HIGH-QUALITY
BLACK AND WHITE HALFTONES AND SPECIAL
EFFECTS ON THE DESKTOP.

2 DAYS

When the project calls for one-color images, desktop halftones can be just as good or better than conventional halftones. The creative potential of desktop-generated duotones, tritones, quadtones and colorization makes black and white imagery all that more attractive. From both a creative and a production perspective, the desktop will have a big impact on black and white imagery.

TOPICS

pre-scan tips
scanning prints and transparencies for quality on desktop scanners
calibrating scanners and output devices
running tests at remote locations
resolutions for line art and halftones
understanding highlight and shadow dots
working with grayscale
maintaining sharpness and contrast
desktop retouching
posterizing images
creating duotones, tritones and quadtones
colorizing black and white images

1 800 229 0007

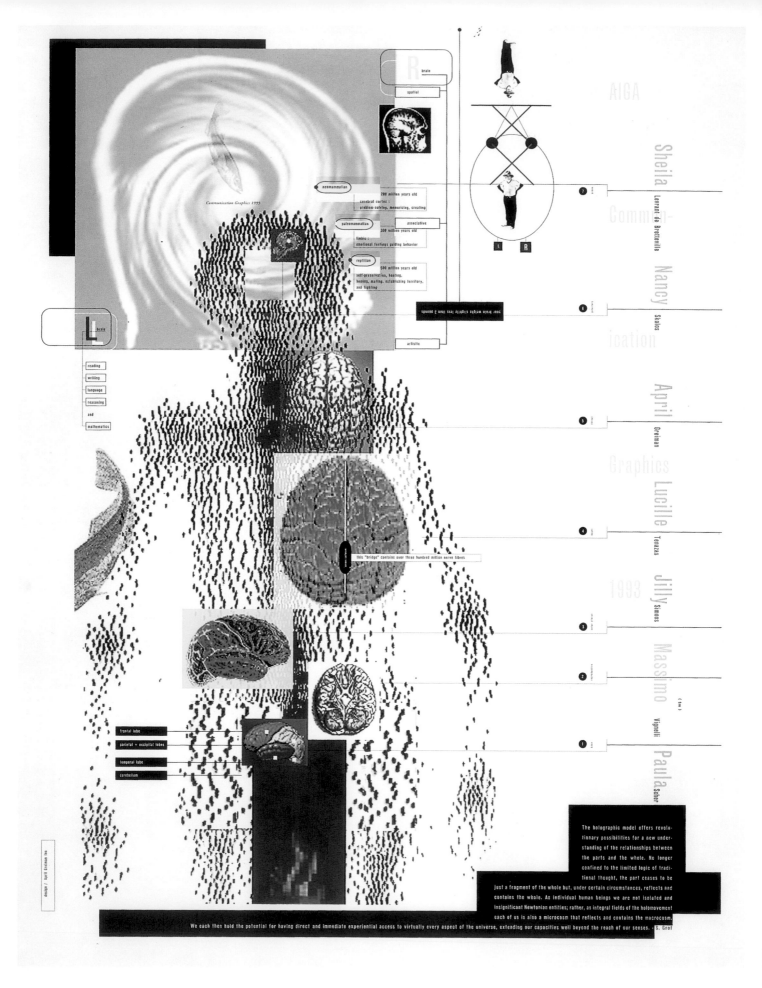

AIGA Communication Graphics 1993
Competition poster
Designer: April Greiman
Associate designer: Sean Adams
Design company: April Greiman Inc.
Client: American Institute of Graphic Arts
USA, 1992

Live-Able Benign Architecture
Poster
Designer/illustrator: Mark D. Sylvester
Design company: Point
Self-published, Cranbrook Academy of Art
USA, 1991

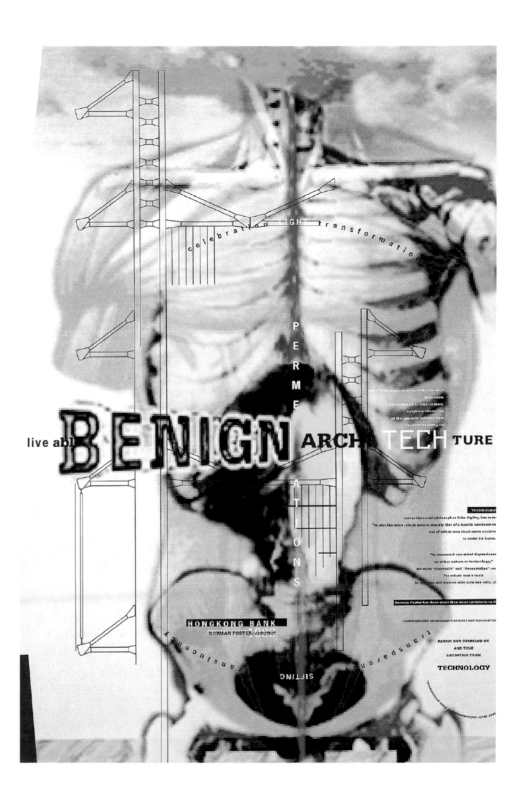

conceptual

Select Your Network
Poster
Designer: P. Scott Makela
Self-published, Cranbrook Academy of Art
USA, 1990

Self-promotional advertisement
Designers: Ian Cartlidge, Adam Levene,
Sean Perkins, Yumi Matote, Simon Browning
Design company: Cartlidge Levene
Photographer: Peter Robathan
Great Britain, 1991

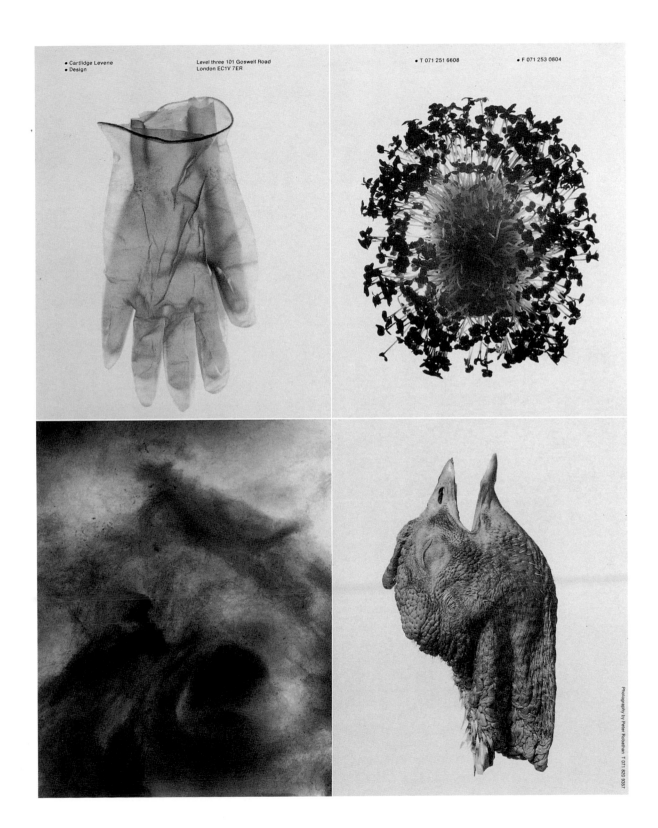

Select Your Network
Poster
Designer: P. Scott Makela
Self-published, Cranbrook Academy of Art
USA, 1990

Self-promotional advertisement
Designers: Ian Cartlidge, Adam Levene,
Sean Perkins, Yumi Matote, Simon Browning
Design company: Cartlidge Levene
Photographer: Peter Robathan
Great Britain, 1991

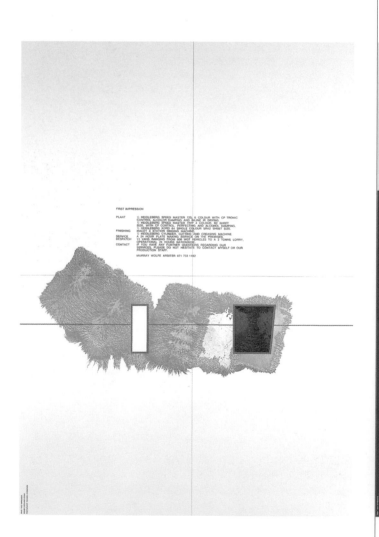

Promotional posters
Designer: Simon Browning
Design company: Cartlidge Levene
Photographer: Richard J. Burbridge
Client: First Impression Printers
Great Britain, 1991

Poster for a Geometrician
Designer: Elliott Peter Earls
Self-published, Cranbrook Academy of Art
USA, 1991

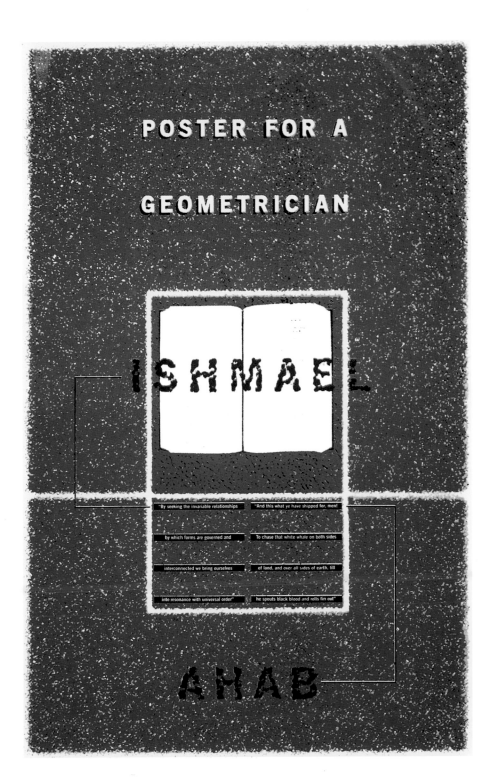

The Creative Community:
Working in Europe
Conference poster
Designers: Simon Browning, Yumi Matote
Design company: Cartlidge Levene
Photographer: Tomoko Yoneda
Client: Design Museum, London
Great Britain, 1992

Conference mailer
Designer: Cornel Windlin
Photographer: Istvan Balogh
Client: Museum für Gestaltung, Zurich
Great Britain, 1993

BILDER

MASCHINEN

STRATEGIEN

RÄTSEL

Edition Museum für Gestaltung
Eine Koproduktion mit dem Verlag Stroemfeld/Roter Stern,
Basel und Frankfurt am Main

Bitte frankieren

ZEITREISE
BILDER / MASCHINEN / STRATEGIEN / RÄTSEL

Georg Christoph Tholen, Michael Scholl, Martin Heller (Hg.)
416 Seiten, Leinen gebunden, 24.5 x 16 cm
Rund 200 Abbildungen (s/w), ISBN 3-87877-409-5
Preis Fr. 48.– zuzüglich Versandkosten

Mit Beiträgen von Cuno Affolter, Hans-Dieter Bahr, Peter Berz, Norbert
Bolz, Mathis Brauchbar, Bazon Brock, Wolfgang Coy, Friedrich Cramer,
Iris Därmann, Susanne Dudda, Peter Gendolla, Frank Haase, Martin
Heller, Jochen Hörisch, Burkhard Hofmann, I.M.A.G.E. Servaz Jörgensen,
Ralf Lukas, Werner Oeder, Hans Ulrich Reck, Sigrid Schade, Christian
W. Scheidemann, Regula Schindler, Manfred Schneider, Michael Scholl,
Georg Seeßlen, Barbara Sichtermann, Bernhard Siegert, Martin Stingelin,
Peter Suter, Georg Christoph Tholen, Elisabeth Weber, Peter Weidkuhn.

Senden an:

Ich bestelle   Exemplar(e)

Name, Vorname

Strasse

PLZ/Ort   Datum

Unterschrift

MUSEUM FÜR GESTALTUNG ZÜRICH
VERLAG
POSTFACH
CH-8031 ZÜRICH

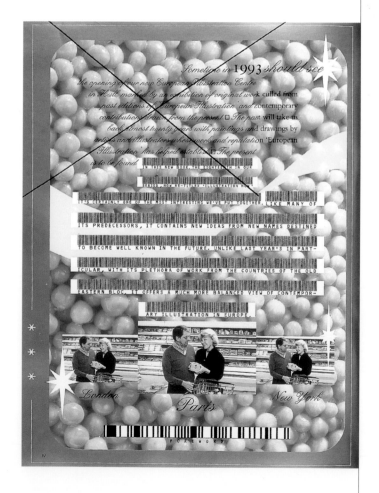

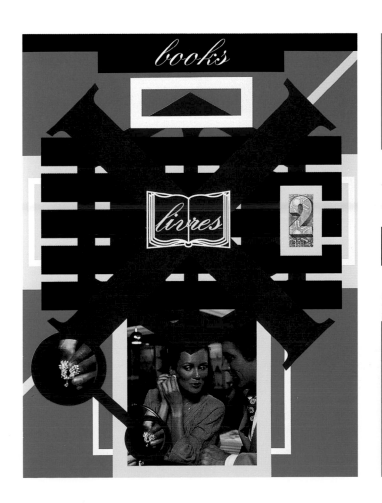

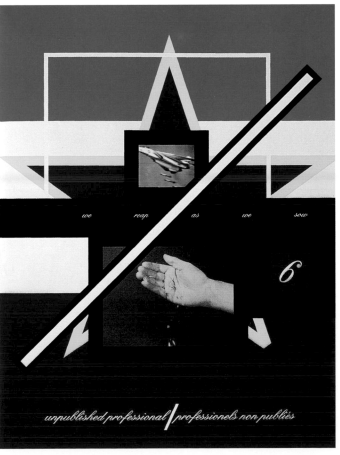

NEW YORK CITY

CARTLIDGE LEVENE 071 251 6608 RICHARD J BURBRIDGE 071 739 5786 PRECISE LITHO 071 253 568

New York City
Self-promotional advertisement
Designers: Simon Browning,
Sean Perkins, Yumi Matote
Design company: Cartlidge Levene
Photographer: Richard J. Burbridge
Great Britain, 1992

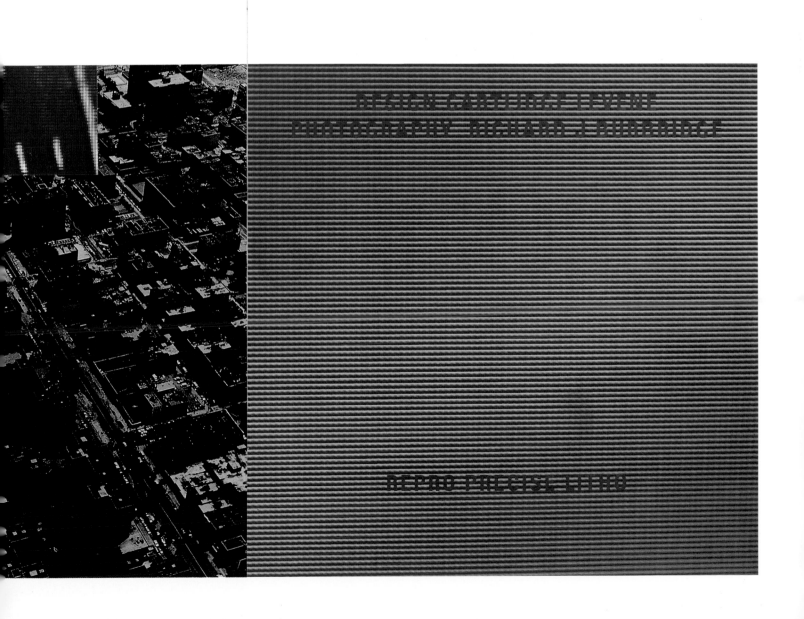

Gerald van der Kaap – Hover Hover: A Manual
Exhibition catalogue cover and spreads
Designers: Armand Mevis, Linda van Deursen
Design company: Mevis & van Deursen
Client: Stedelijk Museum, Amsterdam
The Netherlands, 1991

GERALD VAN DER KAAP

hover
hover

A MANUAL
STEDELIJK MUSEUM
AMSTERDAM

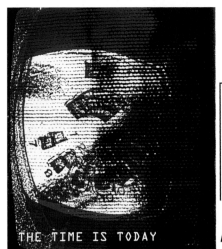

W.A.L. Beeren

Gerald Van Der Kaap (geboren 1959 Enschede) is voor het Stedelijk Museum geen onbekende. Hij nam reeds deel aan verschillende groepstentoonstellingen zoals 'Kunst voor Televisie' (1987), dat in samenwerking met de VPRO gerealiseerd werd, 'Een grote activiteit - Nieuwe generaties in de Nederlandse kunst' (1987/1988), 'Roots and Turns' (1988). In deze drie tentoonstellingen werd steeds een ander aspect van Van Der Kaap benadrukt: in de eerste, van een kunstenaar die een clip maakt voor de televisie; in de tweede, van een kunstenaar die wordt opgenomen in de geschiedenis van de Nederlandse fotografie; en in de derde, van een veelbelovende jonge Nederlandse kunstenaar die een plaats krijgt tussen verschillende vormen van beeldende kunst. Het is algemeen bekend dat hij bemoeienissen heeft met cultbands (o.a. Save the Robots) en dat hij tijdschriften uitgeeft (het fotografisch prachtblad Zien, negen nummers, 1980-1986 en het tijdschrift Blind, twee nummers, 1989-1990). Kortom een veelzijdig kunstenaar, die in het medium dat hem toch het meest eigen is - de fotografie, een hele ontwikkeling heeft doorgemaakt. Van geënsceneerde fotografie, naar zwart-witbeelden, naar met de computer gegenereerde beelden. Ik ben dan ook verheugd Gerald Van Der Kaap te kunnen uitnodigen voor een eenmanstentoonstelling -zijn eerste- in het Stedelijk. Het project 'Hover Hover' is speciaal voor het Stedelijk gerealiseerd. Het bestaat uit een tentoonstelling en een catalogus. De twee zijn complementair: De werken in de tentoonstelling zijn alleen in een andere gedaante in de catalogus terug te vinden. De imaginaire werken in de catalogus daarentegen zijn onvindbaar in de tentoonstelling. Het project 'Hover Hover' is mede mogelijk gemaakt door de productie-faciliteiten die het fotografische laboratorium Souverein bv hiedt, waarvoor ik hartelijk dank zeg.

preFACE

welcome to the hover house

Interview by Dorine Mignot

① *however-many-days:* Van Der Kaap recently described the journey at 'The New Concept' symposium in Graz, Austria; he declared that the journey would be 'an ironic Odyssey, situated in the world's financial centers, recording them in photographic images, as a monument, as highlight, just as every culture, including our own, deserves. A grand and compelling monument, and at the same time a caricature of itself. The subject calls for an image that is at once specific and general: specific in the sense that everything can be related to the manifestation of culture, commerce and power, and their revelation (these we can take with us in the form of photographs of bank buildings, brothels, monuments, means of transport, alleyways, works of art, women of all nations...); general in the striving towards the one imaginary city linked by satellites.'

The photographs to be taken will be published in book form: 'A book of photographs, in the classic sense, but placed in a different tradition, a 'carnivalesque' tradition. It will be a reversal of the genre of photojournalism. The form will be artificially borrowed from documentary photography, without the laws of motivation and involvement associated with it. The story as a pornographic poem of power'. Regarding method, Van Der Kaap wrote:

> The polish of an ad photographer. The malice of a paparazzo.

② *a theory of the carnivalesque:* Van Der Kaap got the idea after reading an article entitled 'On the carnivalesque': 'The carnivalesque represents the transposition into culture of the spirit of carnival. The carnivalesque principle abolishes hierarchies, levels social classes and artistic genres. In carnival, all that is marginalized and excluded – the mad, the scandalous, the aleatory – takes over the centre in a liberating explosion of 'otherness'. It is a symbolic defeat over death, over all that is sacred, over all that oppresses and restricts.'

reflex ; ?	! ; generate
consumer activism	so what
sky-dive ; fall (g)	hover hover (low-g/zero-g)
DON'T	**DO**
weefgetouw	zweefgetouw
the 80's	the 90's
60's (zero)	(digital zero) 10's
i = information E.T.	$\sqrt{-1} = i$ (as in happY) E.V. (en ville)
true	**yes, negative**
deconstructing lies...............	constructing lies........ (using carnivalesque strategies
mirrors	windows (0.1 portholes)
(...)	the adulterated new
everyday sign systems transformed into new meaning	new sign systems. from zero.

Architect's promotional brochure/poster
Designers: Morag Myerscough,
Jane Chipchase
Design company: Myerscough Chipchase
Client: Iain MacDonald, Arquitectura
Great Britain, 1992

Total Hoverty
Exhibition flyer/poster
Designers: Armand Mevis, Linda van Deursen
Design company: Mevis & van Deursen
Client: Gerald van der Kaap
The Netherlands, 1992

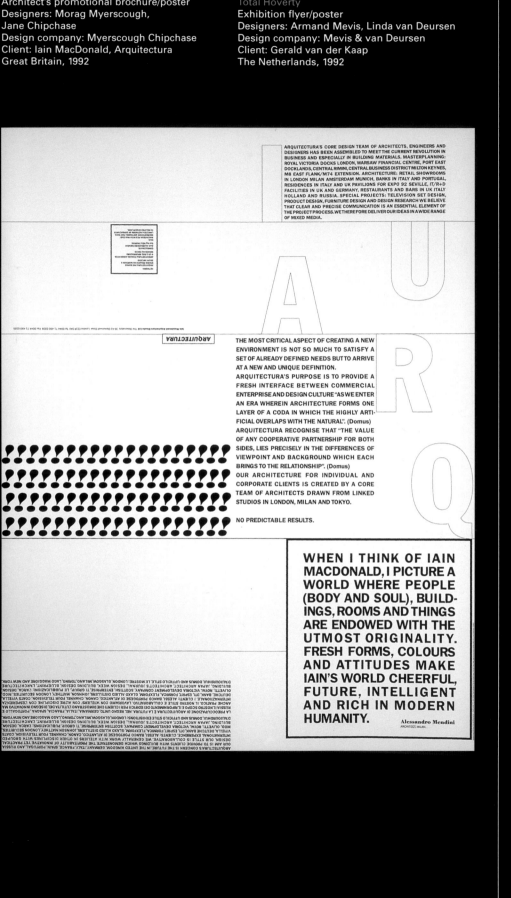

3D/4E (EXHIBITION MODEL)

Total Hoverty is based on the **BRAIN 5.1** information system that controls the installation environment. There are an almost unlimited variety of exhibitions possible through the selection and output of **BRAIN 5.1**.
A newly developed interface,
the **CHILL TERMINAL 3.2**, will provide instant mystical peak-experience.

TOTAL HOVERTY

BRAIN 5.1 (TEST) PHOTOSOFT (35MM SLIDES) > SCREENS (CAVE)

WHAT

BRAIN 5.1 (WORDS) VIDEOSOFT (VHS) > SCREENS (CAVE)

E 1 (INITIATION)
I am the natural speaker
I say
E 2 (HELL)
Old mine
I call you
I'm thinking now
2 x 2 = 2
Tokyokyotokyoto
E 3 (HOVERPOINT)
The natural speaker
I say
The piper plays
I follow him
Soft speaks a wooden sound
E 4 (HEAVEN)
I say
Forever forester
I say
I slide through lowlands
I dive into the depth
Forever I ecstasy
it has to this way
Wei mm I mm am wet
I am a vein that flows
Love keeps the pace
Words are parked like cars

E1 (INITIATION). YO. HELLO. HOW ARE YOU. GOODBYE. I GO. I WANT MY CLOTHES WASHED. I WANT THIS MENDED. A KNIFE. I CUT THE BOOK. ETERNAL DIAMOND. GO... **E2 (HELL)**...HELL. I WANT ANOTHER TABLE. PLEASE, I WANT A FACE MASSAGE. THIS IS NOT FRESH. (I NEED A FIX CAUSE I'M GOING DOWN. THE FALSE CHARACTER THAT FOLLOWS YOU AROUND.*) I'LL PRAY FOR RAIN, FAIR WEATHER, FOR HELL ... **E 3 (HOVERPOINT)**....OF COURSE. I UNDERSTAND. I KNOW EXACTLY WHAT YOU MEAN. ROOT OUT. PULL DOWN. DESTROY. LOOK AT THE SKY. GO WITH THE FLOW. WALK ON. YOUR LIFE. EARN IT. SPEND IT. SAVE IT. SPILL IT. WHERE IT CAN DO SOME GOOD. O.K. YES, NEGATIVE.... **E 4 (HEAVEN)** ...HEY YOU. I SAW YOU. BODY MYSTIC. THANKS. HERE SHE COMES. SIR. SIRE. SYRINGE. SYRINX. SISTER. SECRET NIGHT. DIE INTO ME. YOU. YO... T'

PATTERN: E1 (INITIATION) • E2 (HELL) • E3 (HOVERPOINT) • E4 (HEAVEN) • REPEAT

Heaven and hell are one word

BRAIN 5.1 (TOTAL HOVERTY) AUDIOSOFT (DAT) HEADSETS

*Our method
is techno.
Our aim
is total hoverty.*

BRAIN 5.1 (SIRENS) AUDIOSOFT (DAT) > P.A. SYSTEM

BRAIN 5.1 (BODY SONIC) AUDIOSOFT (DAT) > BODY SONIC OUTPUT

DON'T

THINK&

BRAIN 5.1 (TYPE ZERO) VIDEOSOFT (VHS) > CHILL TERMINAL

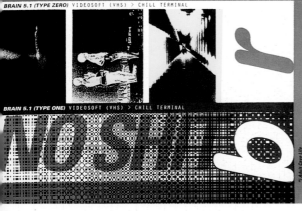

BRAIN 5.1 (TYPE ONE) VIDEOSOFT (VHS) > CHILL TERMINAL

*) Adult Cave Life

CHILLL TERMINAL 3.2
Interface for Brain 5.1
No other drugs required.

FOR ADULT CAVE LIFE.
All senses operative.
Destination: total hoverty.

PPPP Symposium
Poster
Designers: Armand Mevis, Linda van Deursen
Design company: Mevis & van Deursen
Client: Jan van Eyck Academy, Maastricht
The Netherlands, 1992

In the Spirit of Fluxus
Exhibition catalogue cover and spreads
Designers: Laurie Haycock Makela,
Mark Nelson
Client: Walker Art Center, Minneapolis
USA, 1992

HOLIDAY

above
Slowly, steadily

the bright sky

deepened

Action!

Break Break Break

Break time.

COCKTAIL LOUNGE

Mr. Emerson
sits in his leather chair.
TV on; Cheese Nachos
News time, open.
last hours' report
retreated
repeated.

Strange Compelled towards horrific
compulsion gruesome items

Make more coffee..

Back in chair,
tastes sour —
down the sink, rancid milk.

Sick,
anxious
feeling
What's the worry?
No one knows his location?
Do they?
Do they?

Little wedge
of doubt
gets a clout.
Make the first move.
Out.

nauseous.

He leaves... the apar... building... a block.

oper. for breakfast

Only 2 hours to kill

Feels nauseous.

friends downtown

Cold crisp pavement under feet slippery on old shoes.

Mooch streets,
bookshops,
maybe a bar,
yes, that's the ticket.

Holiday... across pavement.

...in through...
...ing lights reveal
...ing wallpaper.

...horseshoe bar
...orders a Rolling Rock,
no go,
...Coors instead.
Anything cold.

Sidling past the jukebox
he manoeuvres through
a maze of tables and empty
chairs and finds an
empty booth.

He slides in along
a red vinyl seat
and leans elbow
on the shiny table top
in someone else's spilt
drink.

He feels marginally better
sitting in the Holiday under the
smeared image of a Moose.

Walk across to the Holiday
take in a few beers.
That'll straighten him out.

Just the ticket.

Just the ticket

Light is provided from a red plastic lampshade.

Time for drink 2.
To the bar.

"The new American poetry as typified by the SF Renaissance (which means Ginsberg, me, Rexroth, Ferlinghetti, McClure, Corso, Gary Snyder, Philip Lamantia, Philip Whalen, I guess) is a kind of new-old Zen Lunacy poetry, writing whatever comes into your head as it comes, poetry returned to its origin, in the bardic child, truly ORAL as Ferling said, instead of gray faced Academic quibbling. Poetry & prose had for long time fallen into the false hands of the false. These new pure poets confess forth for the sheer joy of confession. They are CHILDREN. They are also childlike graybeard Homers singing in the street. They SING, they SWING. It is diametrically opposed to the Eliot shot, who so dismally advises his dreary negative rules like the objective correlative, etc. which is just a formula for more dead prose. Objective and subjective both, the poetry of a new Holy Lunacy like that of ancient times (Li Po, Hanshan, Tom O Bedlam, Kit Smart, Blake) yet it also has that mental discipline typified by the haiku (Basho, Buson), that is, the discipline of pointing out things directly, purely, concretely, no abstractions or explanations, wham wham the true blue song of man.

Jack Kerouac – The Origins of Joy in Poetry

'The word of course is one of the most powerful instruments of control... Now if you start cutting these up and rearranging them you are breaking down the control system'

"Burroughs maintains that the only way to counter recording is by counterrecording and scrambling the playback. Playback techniques that are used by others (personally and politically) to control you can be appropriated by you. This is through 'counterrecording' and playback,' a procedure of repetition, manipulation and purposeful distortion used as a tool of analysis and aggression."

Artificial Nature
Exhibition catalogue cover and spreads
Designer: Dan Friedman
Editors: Jeffrey Deitch, Dan Friedman
Client: Deste Foundation
for Contemporary Art, Athens
USA, 1990

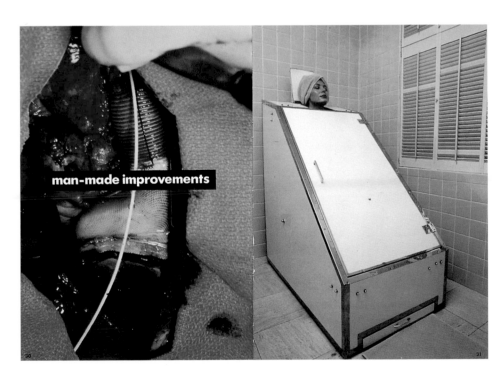

spiritualizing it

infusing it with emotion

42

43

more and more something that we are re-creating ourselves. From the greenhouse effect that could change our climate, to the green revolution that has spawned supercharged strains of grains and vegetables, nature is being reconstituted. For today's average city dweller, nature is more often than not something to be experienced on television, or on a Club Med vacation. There are still many people who earn their living from the land and the sea, and for whom nature is very real. For the rest of us, however, nature is an increasingly artificial experience.

Art and nature have been inextricably linked ever since the cave paintings of Lascaux. *Nature has traditionally been the ultimate inspiration and challenge for the artist who, depending on his or her orientation, sought to imitate it, improve upon it, or interpret it.* Our conception of nature has often been derived, in fact, from the vision of artists. Our sense of what nature is has been shaped and reshaped by artists of successive generations, such as Leone Battista Alberti, Claude Lorrain, J.M.W. Turner, and the Impressionists. Not only have artists studied nature for centuries, they have also looked to it for the revelation of basic truths. An immersion in nature and its forms gave artists, scientists, and philosophers a deeper picture of reality.

An artist can still pack a portable easel and hike into the mountains to sketch a small section of wilderness, but a truly contemporary artist might be better advised to seek truth in nature in a strip mine or in the visitors' center of a game preserve. To immerse oneself in nature today is to uncover layers of chaotic exploitation and man-made "improvements." *Genuine nature may now be more artificial than natural.*

From the genetic reconstitution of human beings, to the growth of a kind of worldwide suburbia, we are slowly but steadily replacing what we knew to be nature with a new kind of artificial nature. It is not just that an average person can easily go through years of life breathing air-conditioned air, playing football on Astroturf, and the like, but that a new model of reality is replacing our old sense of the natural order. The jungle ride at Disney World may in fact be more real to most people than the real jungle in the Amazon. Artificially colored and genetically enhanced oranges are perceived to look and taste

Artificial Nature
Exhibition catalogue spreads
Designer: Dan Friedman
Editors: Jeffrey Deitch, Dan Friedman
Client: Deste Foundation
for Contemporary Art, Athens
USA, 1990

Post

HUMAN

Post Human
Exhibition catalogue cover and spreads
Designer: Dan Friedman
Picture editors: Jeffrey Deitch, Dan Friedman
Client: Deste Foundation
for Contemporary Art, Athens
USA, 1992

The extraordinary self-transformation

of Ivana Trump is an example of this shuffling of

reality and

fantasy

into a reassembled fictional personality.

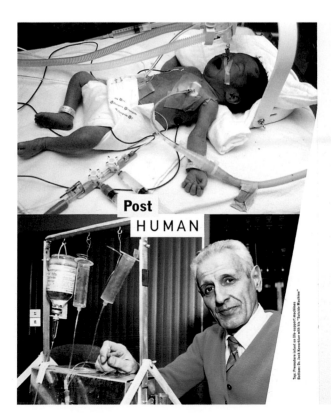

Top: Premature infant on life-support machines
Bottom: Dr. Jack Kevorkian with his "Suicide Machine"

Post
HUMAN

On most people's beauty scale, Stacey Stetler would be a 10. A blond, blue-eyed, 5-foot-11 New York model, she has confidently sashayed down the runway for Yves Saint Laurent in Paris and has graced the covers of fashion magazines. But until recently, when Ms. Stetler looked in the mirror she saw less perfection and more flaws. "I was flat-chested," Ms. Stetler said. "You couldn't tell if I was coming or going. My back protruded almost as much as my front.". . .Ms. Stetler enhanced her boyish figure by having breast implants. She is not alone.
The New York Times, 6 February 1992, front page

Stories about breast implants, crash diets, and mood drugs have moved from the health and beauty page to the front page. The public has been galvanized by explosive testimony about sexual harassment and by the sensational rape trials of public figures. Questions about the new boundaries of appropriate interpersonal behavior are attracting unprecedented interest. There is a growing sense that we should take control over our bodies and our social circumstances rather than just accepting what we inherited.

Social and scientific trends are converging to shape a new conception of the self, a new construction of what it means to be a human being. The matter-of-fact acceptance of one's "natural" looks and one's "natural" personality is being replaced by a growing sense that it is normal to reinvent oneself. The Freudian model of the "psychological person" is dissolving into a new model that encourages individuals to dispense with the anguished analysis of how subconscious childhood experiences molded their behavior. There is a new sense that one can simply construct the new self that one wants, freed from the constraints of one's past and one's inherited genetic code.

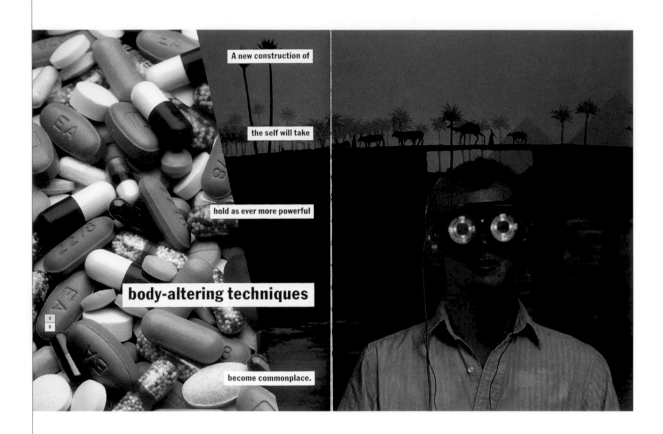

A new construction of

the self will take

hold as ever more powerful

body-altering techniques

become commonplace.

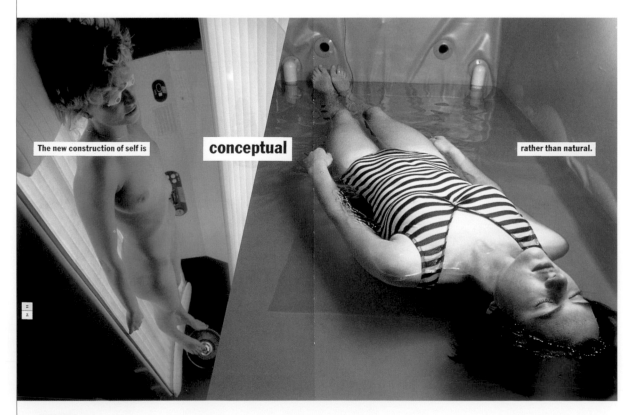

The new construction of self is **conceptual** rather than natural.

Post Human
Exhibition catalogue spreads
Designer: Dan Friedman
Picture editors: Jeffrey Deitch, Dan Friedman
Client: Deste Foundation
for Contemporary Art, Athens
USA, 1992

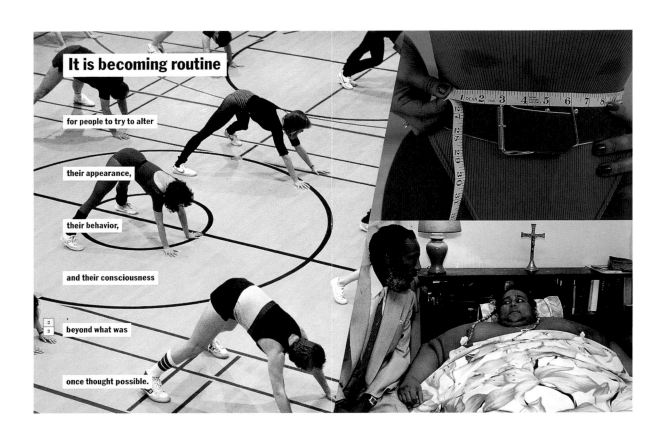

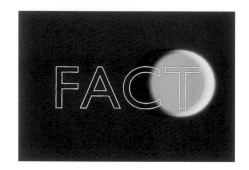

HIV/AIDS awareness cards
Designers: Alan Aboud, Sandro Sodano
Design company: Aboud-Sodano
Photographer: Sandro Sodano
Client: The Terrence Higgins Trust
Great Britain, 1992

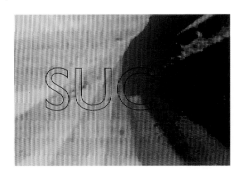

INSPIRAL CARPETS

THIS IS HOW IT FEELS: THE SINGLE

"WHEN IT COMES TO POP PROWESS THEY REMAIN IN A **TOTALLY** *DIFFERENT* **STRATOSPHERE**"

SOUNDS

Elektra

Music press advertisements
Designers: Marlene McCarty, Donald Moffett
Design company: Bureau
Photographer: photolibrary
Client: Elektra
USA, 1991

Yohji Yamamoto Autumn/Winter 91/92
Press advertising campaign
Art director: Peter Saville
Designers: Stephen Wolstenholme,
Brett Wickens
Design company: Pentagram
Photographer: photolibrary
Client: Yohji Yamamoto
Great Britain, 1991

Republic
Album cover
Art director: Peter Saville
Designer: Brett Wickens
Design company: Pentagram
Photographer: photolibrary
Client: CentreDate/London Records
Great Britain, 1993

A NEWORDER RELEASE

Republic©

Het Loie Book
(The Oil Book)
Spreads from a book
about the uses and politics of oil
Designers: Gerard Hadders,
André van Dijk, Erik Prinsen
Design company: Hard Werken Design
Client: KVGO Printers Association
The Netherlands, 1991

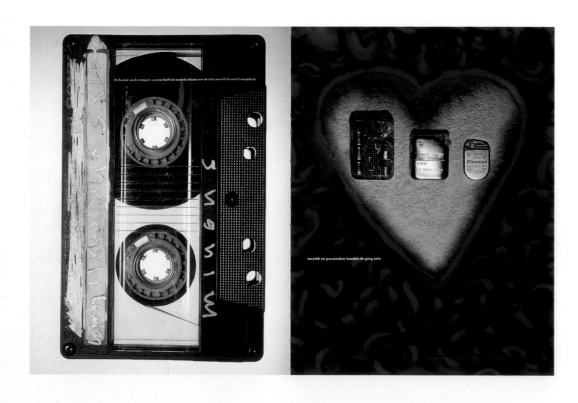

Het binnenste van de aarde is heet

Heet water borrelt of spuit uit de grond maar is nog lang geen geiser in Spijkenisse

Water stroomt altijd naar beneden, maar een glas is nog geen stuwmeer.

in water schuilt kracht

De onderwereld brandstof begint een nieuw leven
als grondstof voor produkten en energiebehoeften.

het zwarte goud gloeit

ruimte reiziger

De maan reist nu tot de verste herinnering waar aardolie de maan ooit heeft gebracht.

In de medische wetenschap vindt aardolie alledaagse toepassingen: een pilletje, een zalfje.

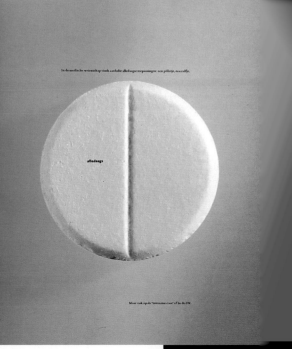

alledaags

Maar ook op de 'intensieve-care' of in de OK.

Equilibrium Vessel:
The Phases of Water/The Faces of Man
Project for *Subjective Reasoning* magazine
Designers: Gerard Hadders, Allen Hori
Creative direction: Willem Kars
Design company: Hard Werken Design
Commissioning editors: William Drenttel, Paula Scher
Client: Champion International Corporation
USA, 1993

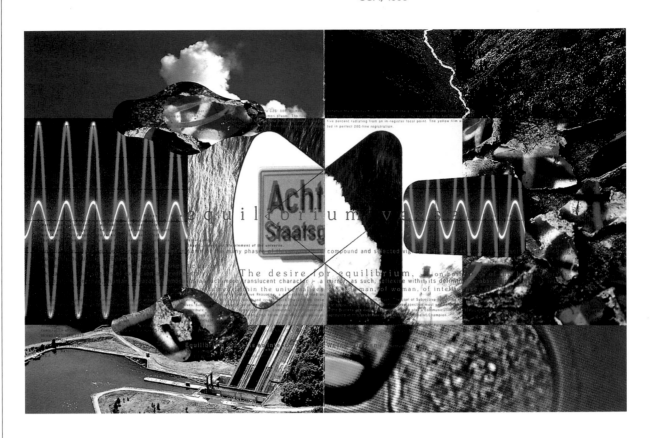

52° 5' N 5° 8' E 10.30 21031941 N° 835

52°5′ N 5°8′ E 10.30 21031941 N° 835
Birthday book cover and spreads
Designer: Irma Boom
Photographer: Paul Fentener van Vlissingen
Client: Paul Fentener van Vlissingen
The Netherlands, 1991

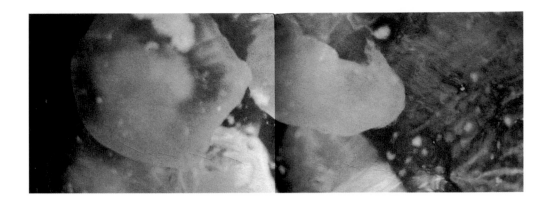

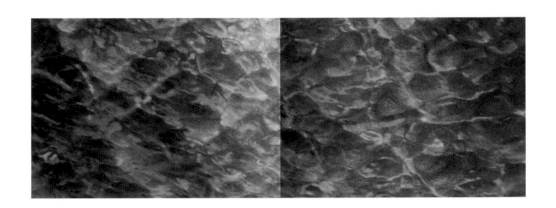

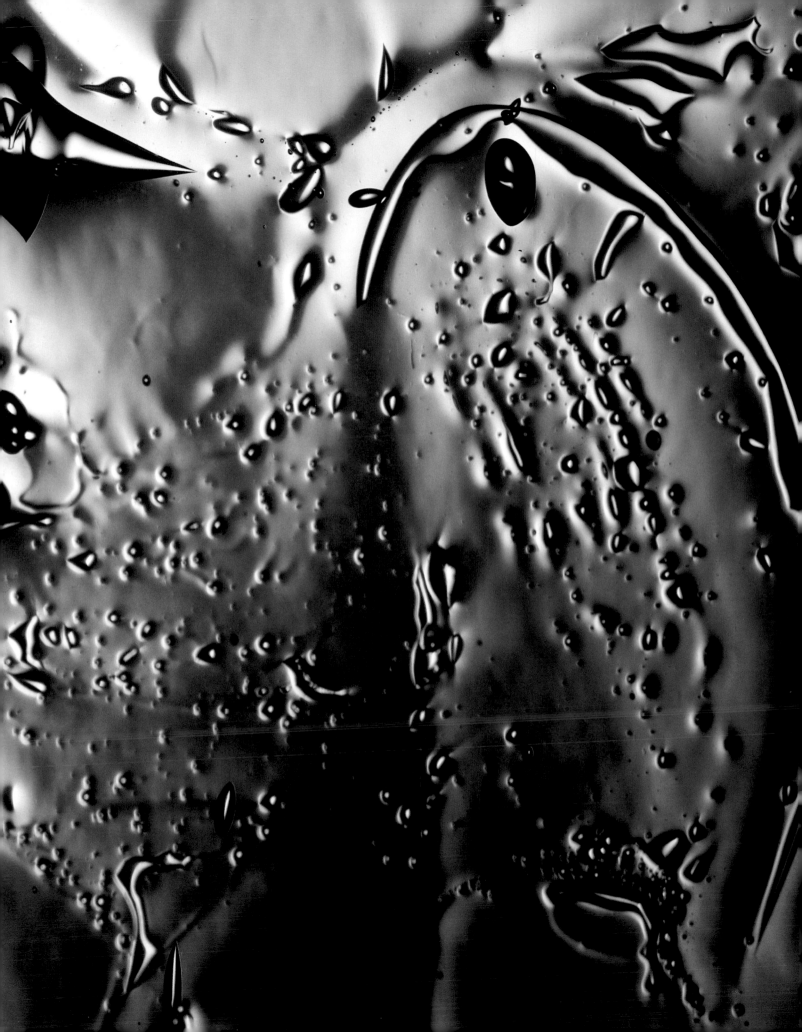

Concert posters
Designer: Michiel Uilen
Design company: Michiel Uilen Ontwerp
Client: de Salon, Groningen
The Netherlands, 1990-92

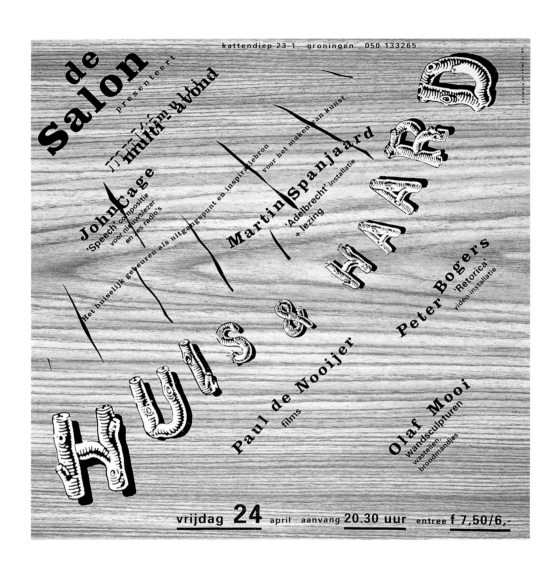

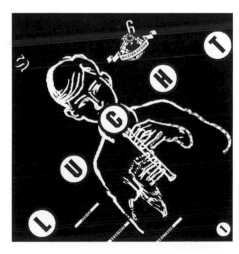

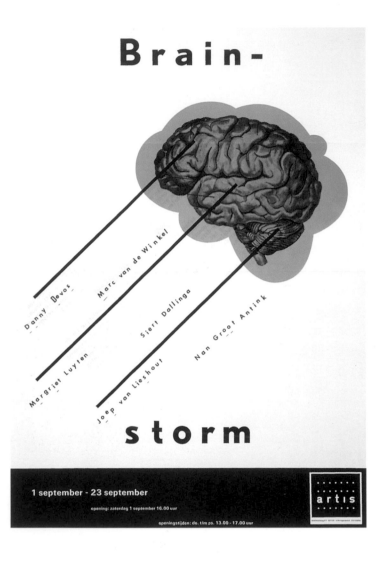

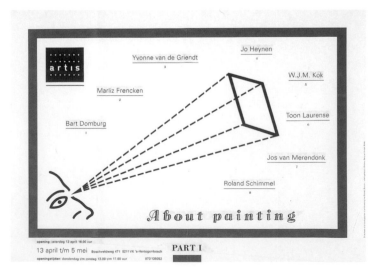

(4)

Ellis Belier
Lidy Jacobs
Teuny Tukker
Stefan Axel Valdimarsson
Dieneke Vos

met werken van

en leerlingen L.A.S. Hugo de Vries Rotterdam

15 maart t/m 22 april
opening zondag 15 maart 15.00 - 18.00 uur
Wij heten U van harte welkom
de bloemenwinkel van Geer Pouls
Nieuwe Binnenweg 182
3015 BH Rotterdam
010 - 436 59 35
openingstijden:
dinsdag t/m vrijdag 11.00 - 18.00 uur
zaterdag 11.00 - 15.00 uur

Brutto gusto

brutto gusto presenteert:

de bloemenwinkel van Geer Pouls
Nieuwe Binnenweg 182
3015 BH Rotterdam
010 436 59 35
openingstijden:
dinsdag t/m vrijdag 11.00 - 18.00 uur
zaterdag 11.00 - 15.00 uur

Leo van Veldhuizen   Paul van den Hout

HET

Lidy Jacobs   Ben Oostrum

Han Rameckers

BEHAGEN

Jenna Tas   Paula Schaper
Ellen van Eldik   Charlotte Kuiper

Christie van der Haak

Willem Speekenbrink   Vincent Rijnbende

Een romantische kersttentoonstelling.
24 november t/m 24 december
Opening: zondag 24 november 15.00 - 18.00 uur

desiGN: JoAN DOBkin

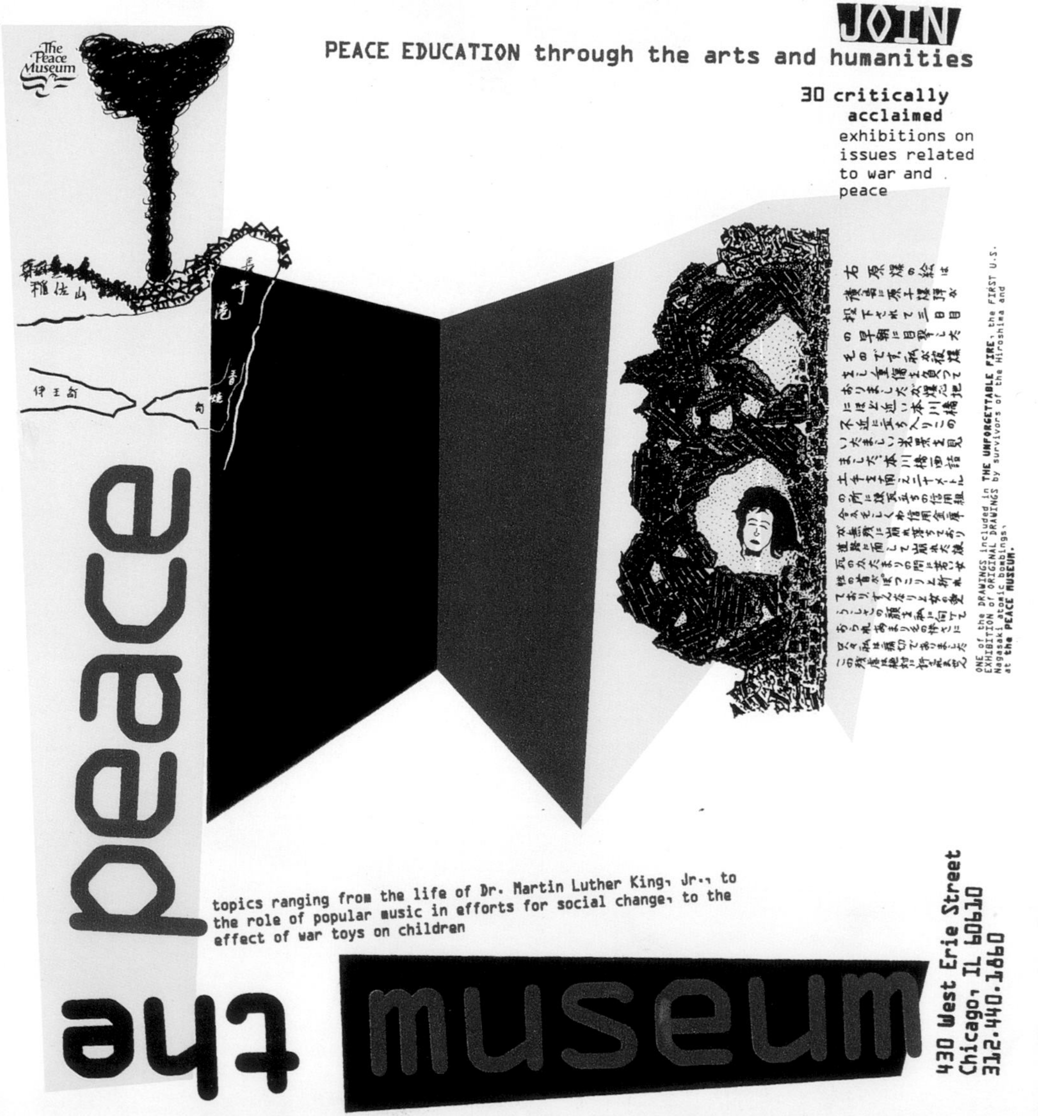

Heads Up
Concert poster
Designers: Gerbrand van Melle
Design company: AAP Designers
Client: Tivoli, Utrecht
The Netherlands, 1991

Heads Up

Tetsuo: The Iron Man
Film poster
Designer: A. Arefin
Client: Institute of Contemporary Arts,
London
Great Britain, 1991

Swoon
Film poster
Designers: Marlene McCarty, Donald Moffe
Design company: Bureau
Client: FineLine Features
USA, 1992

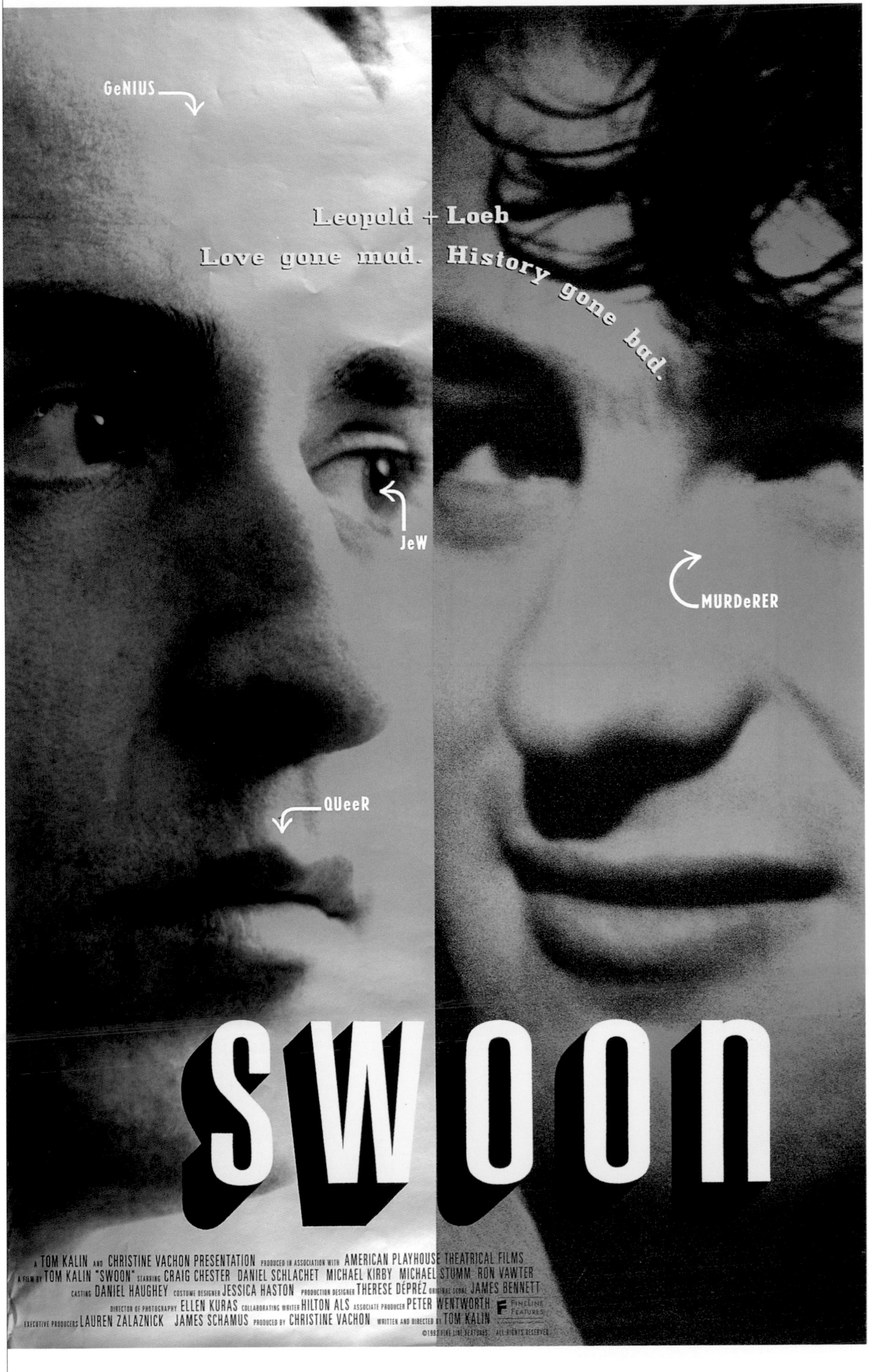
GeNIUS
Leopold + Loeb
Love gone mad. History gone bad.
JeW
MURDeRER
QUeeR
SWOON
A TOM KALIN AND CHRISTINE VACHON PRESENTATION PRODUCED IN ASSOCIATION WITH AMERICAN PLAYHOUSE THEATRICAL FILMS
A FILM BY TOM KALIN "SWOON" STARRING CRAIG CHESTER DANIEL SCHLACHET MICHAEL KIRBY MICHAEL STUMM RON VAWTER
CASTING DANIEL HAUGHEY COSTUME DESIGNER JESSICA HASTON PRODUCTION DESIGNER THERESE DÉPRÉZ ORIGINAL SCORE JAMES BENNETT
DIRECTOR OF PHOTOGRAPHY ELLEN KURAS COLLABORATING WRITER HILTON ALS ASSOCIATE PRODUCER PETER WENTWORTH
EXECUTIVE PRODUCERS LAUREN ZALAZNICK JAMES SCHAMUS PRODUCED BY CHRISTINE VACHON WRITTEN AND DIRECTED BY TOM KALIN
FINELINE FEATURES
©1992 FINE LINE FEATURES. ALL RIGHTS RESERVED

Theatre group poster
Design company: gebr de Jong
Photographers: gebr de Jong, Thijs Wolzak
Client: Waardenberg & de Jong
The Netherlands, 1991

Dutch Design/Industrial Design
Covers for a series of design books
Designers: Jacques Koeweiden, Paul Postma
Design company: Koeweiden Postma
Photographer: YANi
Client: BIS, Publishers
The Netherlands, 1992

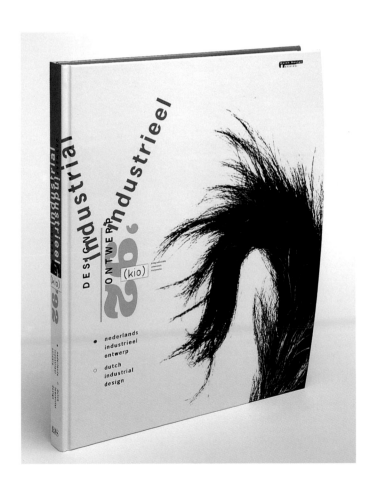

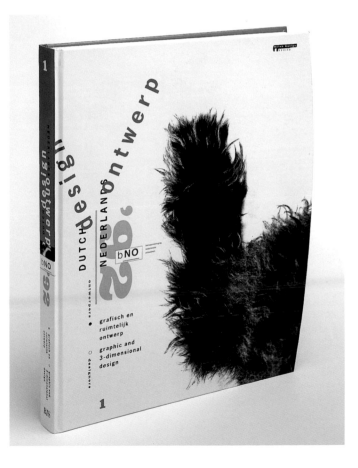

Jool Hul
Theatre group poster
Design company: gebr de Jong
Photographers: gebr de Jong,
Dominique Pelletey
Client: Waardenberg & de Jong
The Netherlands, 1989

Star
Poster for an audiovisual display
Designer: Rana Salam
Self-published, Royal College of Art
Great Britain, 1992

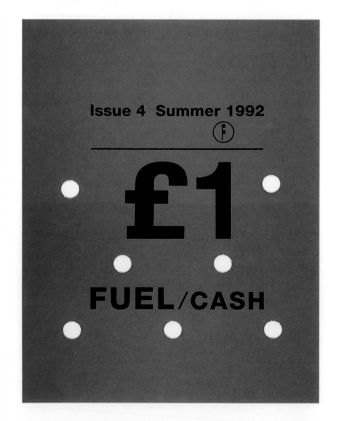

Fuel, issue 1, "Girl"
Magazine cover and spread
Designers: Peter Miles, Damon Murray,
Stephen Sorrell, Nick Oates
Design company: Fuel
Self-published, Royal College of Art
Great Britain, Spring 1991

Fuel, issue 4, "Cash"
Magazine cover and spread
Designers: Peter Miles, Damon Murray,
Stephen Sorrell
Design company: Fuel
Self-published, Royal College of Art
Great Britain, Summer 1992

MEAT

'The worst is not, so long as we can say 'This is the worst!'...'
Mass-murderer Jeffrey Dahmer took such delight in his killing, such pride. He
used to keep people's heads in his fridge. Once, he showed them to an intended
victim. 'Aren't they beautiful' he said, as if he had sculpted and cast them himself.

There have been bigger killers, possibly even madder ones, than Dahmer, but
nothing beats that image for bone-chilling horror: human heads on the shelves of
a fridge, the same shelves on which the rest of America keeps its two pint cartons
of milk and Tropicana orange juice, its jars of pesto and Jell-o moulds.

Surely there could be nothing worse. There was nothing worse, until, that is, a
new horror story came out of Argentina this year. Directors and doctors at a
mental health institution near Buenos Aires were arrested on suspicion that
patients had been killed for spare parts. Evidence accumulated from exhumed
bodies suggested that eyes and other organs had been removed and blood
drained, and that these were then sold for cash. The body of a sixteen-year-old
boy was found at the bottom of a well with his eyes missing. He was supposed to
have 'escaped' from the institution, but he was totally paralysed and could not
even feed himself.

The workings of a lunatic's mind are terrible to think about. But sometimes the
workings of the free market manage to exceed them.

Richard Preston

MART

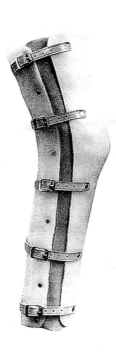

Fuel, issue 2, "Hype"
Magazine cover and spreads
Designers: Peter Miles, Damon Murray,
Stephen Sorrell, Nick Oates
Design company: Fuel
Self-published, Royal College of Art
Great Britain, Autumn 1991

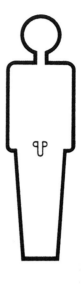

FUEL

"Yesh, lowered it, got the full kit on, Recaros, split rims, full spec, fully loaded - well sweet. Only got to sort the heads and we will be talking WELL rapid. The kitchen? Kitchen's done, sorted, so's the whole place now, top spec stuff. Miele Novotronic Hydromatic out the back and a Zanussi Jetsystem Turbodry, and the toaster's got a micro chip in it. Braun electronic sensor, but who gives a shit about toast eh, do me a favour! Electrolux? Stroll on! The only one they DO is the 1200 watt Superboost and that's for some sort of dwarf. What you want is a Hoover Turbomaster - wicked black finish, drop a body kit on it you could practically DRIVE it. Debenhams credit all the way, sweet stuff. Went up there last week, she got herself a Clairol Turbo but I really rated the Braun - the

SPEC

Silencio 1600 - oof! The dog's COCK. It's what Gary Lineker uses, a mate of mine told me. Switch that and the Clairol on together it's like you're in some serious sort of wind tunnel. Picked a portable system too - ALMOST got the Philips with the Turbo Bass Generator, but it had to be that JVC, you know, the PC-X500 - no, NOT the 300 Bazza you PLONKER - Hyper-Bass, Live Effex, choice or WHAT. Next up it's got to be one of their CDs, check the spec on this: 'XL-E44, PEM DD Converter - Pulse Edge Modulation Differential Linearity Errorless D/A'. Sweet as a NUT. Oof! Who's guffed then? I reckon when I pump it's turbocharged - a real twenty-four valve, gold-plated special - I was touching cloth for a minute there. So anyway, I s'pose a shag's out of the question?"

Richard Preston
Assistant Editor, Harpers and Queen

WARNING
THIS APPLIANCE
MUST BE EARTHED

The Cornflake Shop
37 Windmill Street London W1P 1HH
071 631 0472 Fax 071 436 7165

Earth Warning
Promotional poster
Designer: Paul Elliman
Client: The Cornflake Shop
Great Britain, 1992

Richard II
Theatre poster
Designer: Paul Elliman
Photographer: Donald Cooper
Client: Royal Shakespeare Company
Great Britain, 1991

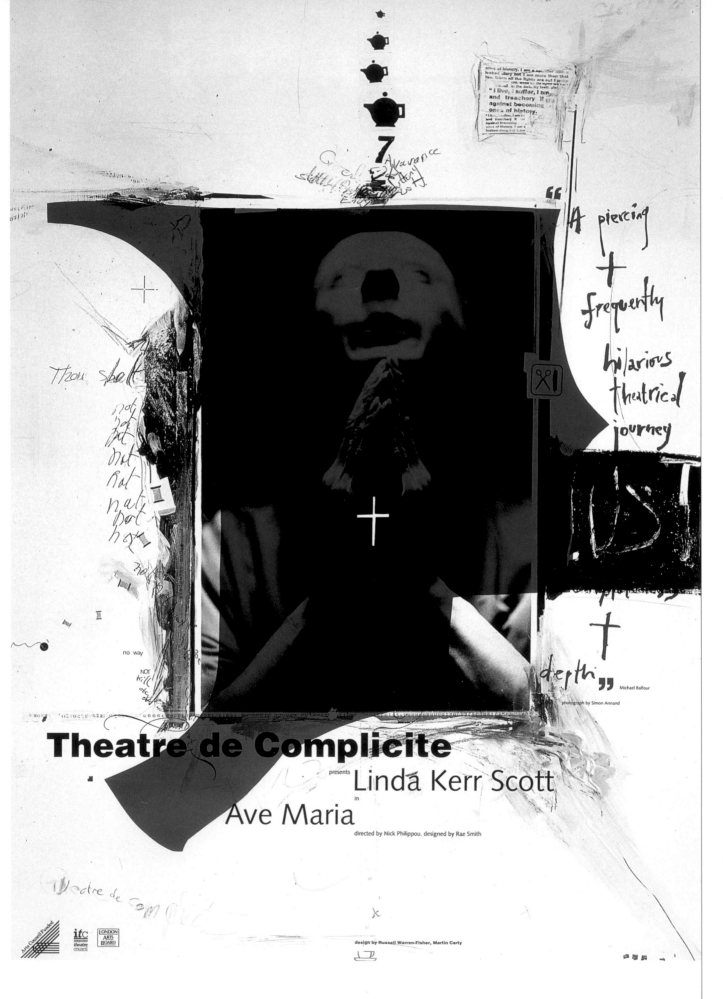

Theatre de Complicite

presents Linda Kerr Scott

in Ave Maria

directed by Nick Philippou. designed by Rae Smith

"A piercing † frequently hilarious theatrical journey ... † depth" Michael Balfour

photograph by Simon Annand

design by Russell Warren-Fisher, Martin Carty

Ave Maria
Theatre poster
Designers: Russell Warren-Fisher, Martin Carty
Photographers: Simon Annand,
Gareth McCarthy
Illustrator: Martin Carty
Client: Theatre de Complicite
Great Britain, 1992

Ray Charles: My World
CD cover
Designer: P. Scott Makela
Design company: P. Scott Makela Words +
Pictures for Business + Culture
Client: Warner Brothers Records
USA, 1992

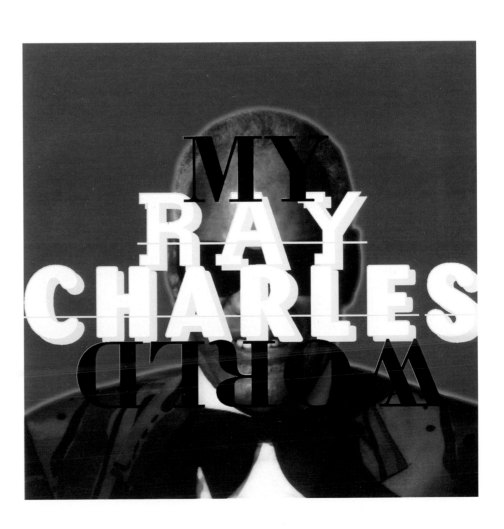

music+style
(the bible of)

PREMIERE ISSUE

rAY GUn

henry rollins
sonic youth
house of love
inspiral carpets
john wesley
harding
the lemonheads
luna
opus 111
the prodigy
mary's danish
mission uk
too much joy
david j
matt mahurin

one november 1992
$3.50 u.s $3.95 canada

11

0 38808 81260 0

Ray Gun, issue 1
Magazine cover and spreads
Designer: David Carson
Client: Ray Gun Publishing
USA, 1992

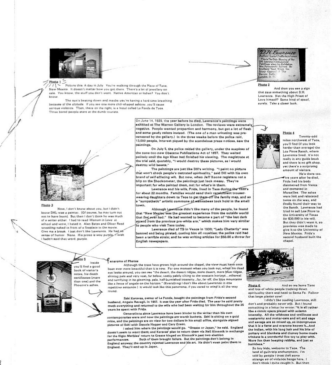

Ray Gun, issue 3
Magazine spread
Designer: David Carson
Photographer: Anthony Artiago
Client: Ray Gun Publishing
USA, 1992

Ray Gun, issue 4
Magazine spread
Designer: David Carson
Illustrator: Amy Guip
Client: Ray Gun Publishing
USA, 1993

Ray Gun, issue 2
Magazine spread
Designer: David Carson
Client: Ray Gun Publishing
USA, 1992

"What magazine are you from again?"

miXXeD s
s a g
e s

.."What are you thinking of doing?"
"committing suicide."

Love on the Airwaves

BY KAREN STILLMAN and AMY J. COHEN

in soviet
recording

BY LAUREN AGNELLI

dis-union

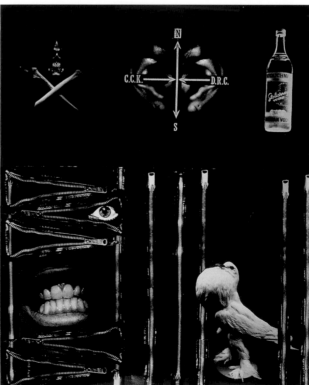

Light
Reading

THE MAN ON THE

MOON

by
Peter
Care

1:01:01.18

01:00:20.14

Last fall R.E.M. began filming *Man on the Moon*, one of the most cinematically complicated videos it or any other band has attempted. What follows is an inside look at how a music video is made, from the director's own notes.

The side of my head

hurts

from thinking in the rain

CRASH

My Head Hurts
Poster
Designer/photographer: Elliott Peter Earls
Self-published, Cranbrook Academy of Art
USA, 1992

Exhibition poster
Designer: Anne Burdick
Photographer: Julie Becker
Client: Julie Becker
USA, 1991

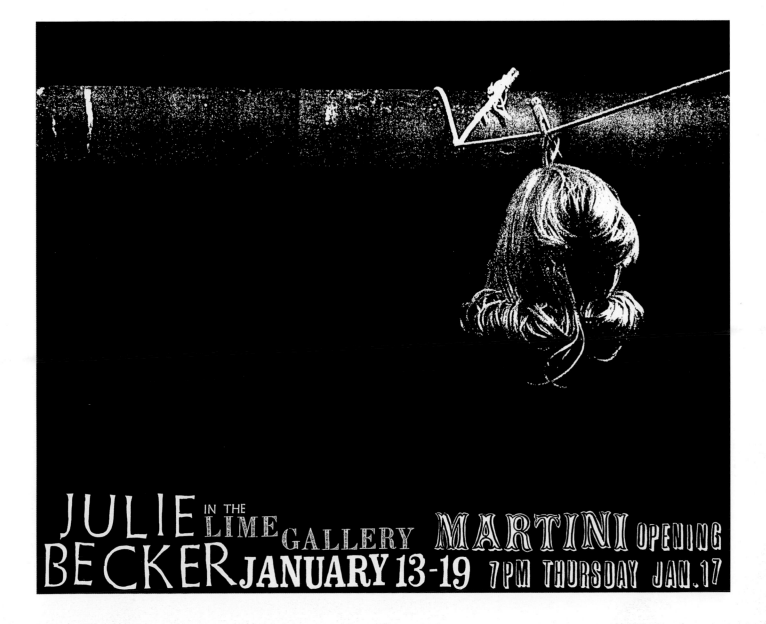

JULIE IN THE LIME GALLERY MARTINI OPENING
BECKER JANUARY 13-19 7PM THURSDAY JAN.17

Ramona 55
Record sleeve
Designer: Graham Wood
Design company: Tomato
Client: Electric Wonderland Records
Great Britain, 1992

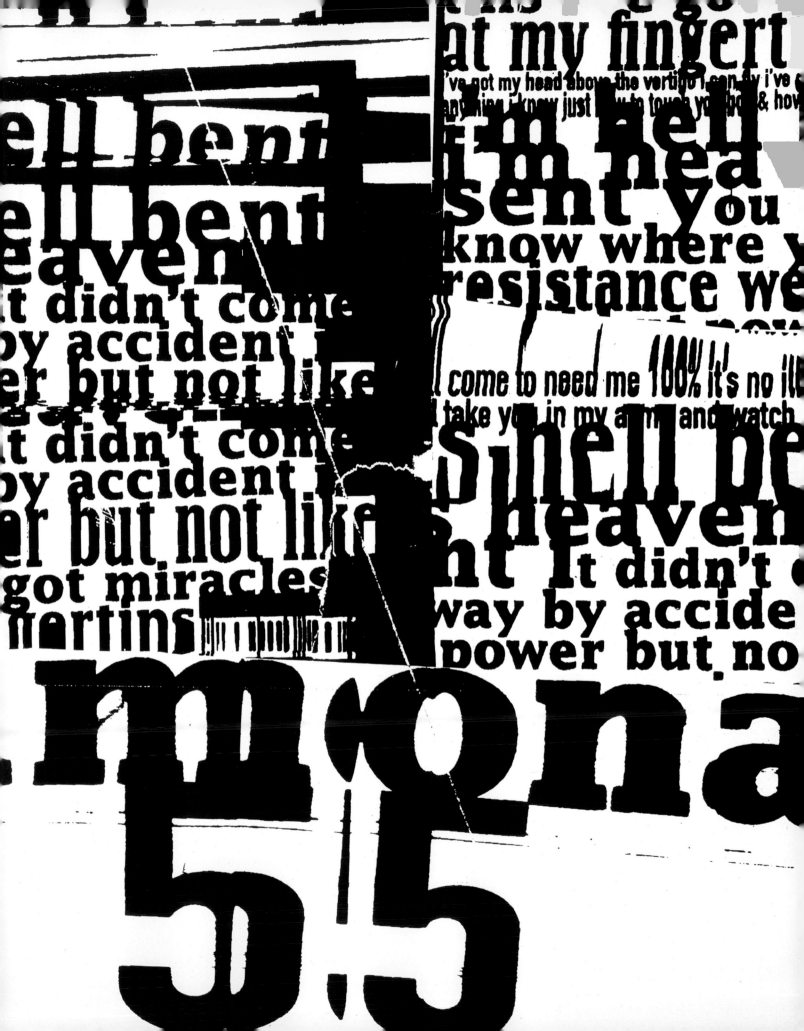

My name is Dita.
I'll be your mistress tonight.
I'll be your loved one, darling.
Turn out the light.
I'll be your sorceress,
your heart's magician.
I'm not a witch.
I'm a love technician.
I'll be your guiding light
in your darkest hour.
I'm gonna change your life.
I'm like a poison flower.
Give it up.
Do as I say.
Give it up and let me have my way.
I'll give you love.
I'll hit you like a truck.
I'll give you love......

Sex by Madonna
Book spreads
Designer: Fabien Baron
Design company: Baron & Baron
Photographer: Steven Meisel
Client: Warner Books
USA, 1992

Sex by Madonna
Book spreads
Designer: Fabien Baron
Design company: Baron & Baron
Photographer: Steven Meisel
Client: Warner Books
USA, 1992

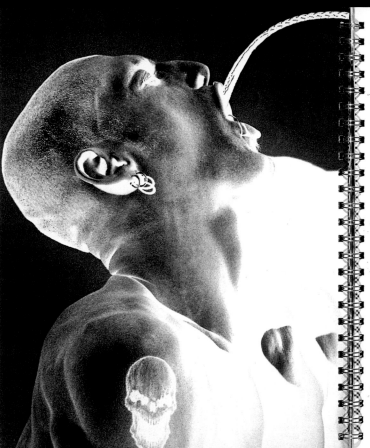

I don't think you know
what pain is. I don't
think you've gone that way.
I could bring you so much
pleasure
I'll come to you when you say.
I'm not gonna hurt you.
Just close
your eyes

I𝕤 FASHION dead

Pam Hogg, Anthony Price
Joe Caseley Harford
Bodymap John Galliano
ALLY CAPELLINO
Nicholas Georgiou
No moths in our Wardrobe

A series of talks in which top British Designers discuss the relevance and significance of fashion in the late '80s • Tickets £2.60 (plus ICA Daypass £1.00) • Institute of Contemporary Arts, The Mall, London SW1Y 5AH Box Office 01-930 3647.

ICA

present times

From chaos and danger... opportunity.

Helen Storey Autumn/Winter 1991 Collection

Is Fashion Dead?
Leaflet
Designers: Moira Bogue, Tim Hopgood
Design company: Bogue & Hopgood
Client: Institute of Contemporary Arts, London
Great Britain, 1989

Present Times
Fashion show programme
Designer: Moira Bogue
Design company: Bogue & Hopgood
Client: Helen Storey
Great Britain, 1991

Darling
Book jacket
Designer/photographer: Chip Kidd
Client: Alfred A. Knopf
USA, 1992

Is Fashion Dead?
Leaflet
Designers: Moira Bogue, Tim Hopgood
Design company: Bogue & Hopgood

Present Times
Fashion show programme
Designer: Moira Bogue
Design company: Bogue & Hopgood

Darling
Book jacket
Designer/photographer: Chip Kidd
Client: Alfred A. Knopf
USA, 1992

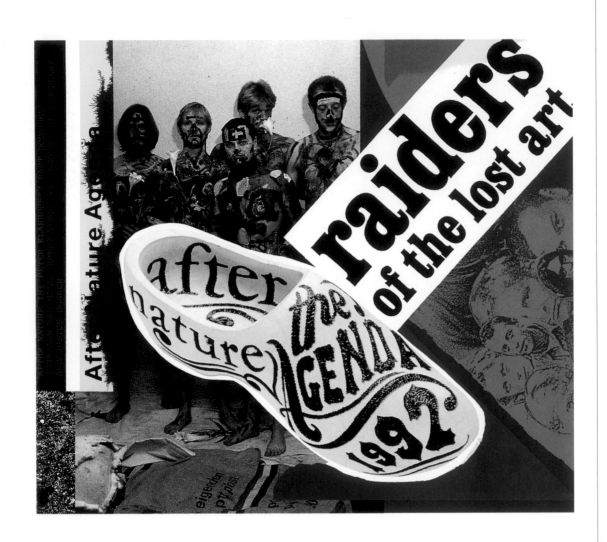

Raiders of the Lost Art
1992 diary cover
Designers: Armand Mevis, Linda van Deursen
Design company: Mevis & van Deursen
Client: After Nature
The Netherlands, 1991

NowTime, issue 2
Magazine cover
Designers: Somi Kim, Lisa Nugent, Whitney
Lowe, Susan Parr, Lorraine Wild, Caryn Aono
Design company: ReVerb/Los Angeles
Client: *NowTime*/A.R.T. Press
USA, 1992

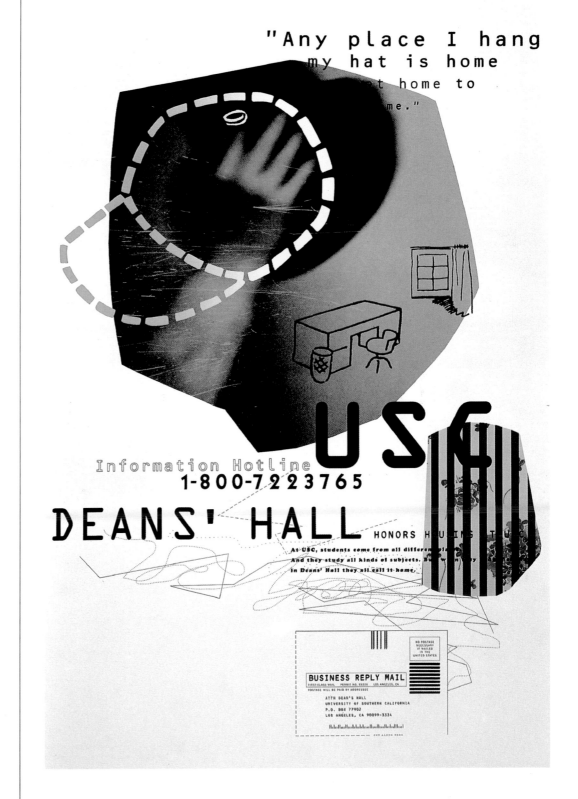

CalArts Dance
Poster
Designer: Caryn Aono
Design company: CalArts Public Affairs Office
Photographer: Steven A. Gunther
Client: CalArts
USA, 1991

T-Zone
Architectural catalogue cover
Designers: Andy Altmann, David Ellis,
Christopher Priest
Design company: Why Not Associates
Client: Design Analysis International
Great Britain, 1991

CalArts

d a n c e

faculty:

CristyneLawson, Dean
LarryAttaway, Associate Dean
RebeccaBobele, Assistant Dean
LaurenceBlake
ClareDuncan
JacquesHeim
HirokoHojo
DavidKroth
RobertOchs
MichaelRoberts
KurtWeinheimer
TinaYuan

a u d i t i o n dates:

Phoenix Jan 21
San Francisco Feb 1
Portland Feb 3
Houston Feb 7
Atlanta Feb 8
Minneapolis Feb 12
Baltimore Feb 14
Chicago Feb 15
New York City Feb 17
Washington DC Feb 20
Los Angeles Jan 18 at CalArts
Los Angeles Feb 22 at CalArts
Los Angeles Mar 7 at CalArts

Admission
to
CalArts
is based
on
talent.

Applicants
to the
performing
arts
programs
must
audition.

California
Institute
of the Arts
offers Bachelor of
Fine Arts and
Master of Fine Arts
degrees, as well as
certificate and
advanced
certificate
programs, in its
schools of Art,
Dance, Film/Video,
Music, and Theatre.
The Institute's
Division of
Critical Studies
provides the
academic courses
required of
students in the BFA
degree program.

Contact:
Office of Admissions
CalArts
24700 McBean Parkway
Valencia, CA 91355

Anyone
interested in
applying for   toll free:
any kind of
financial   1-800-292-ARTS
assistance   (in CA)
should contact
the Office of   1-800-545-ARTS
Financial Aid
well before   (other states)
the March 1,
deadline each
year.

financial Aid
toll free:
1-800-443-0480

California Institute of the Arts
is accredited by the
Accrediting Commission for
Senior Colleges
and Universities of the Western
Association of
Schools and Colleges. The
Institute's School of Dance
is an accredited institutional
member of
the National Association of
Schools of Dance.

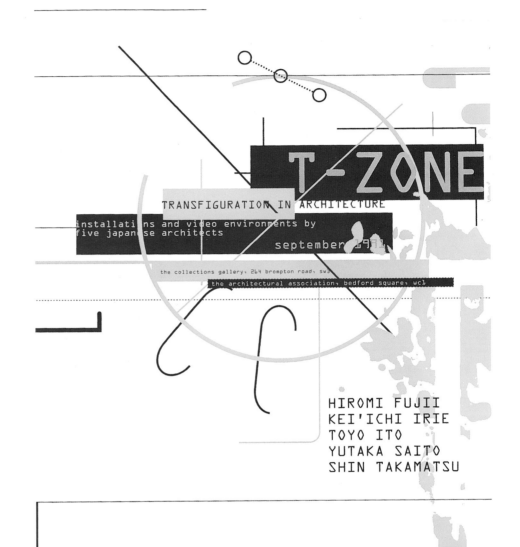

T-ZONE

TRANSFIGURATION IN ARCHITECTURE

installations and video environments by
five japanese architects

september 1991

the collections gallery, 264 brompton road, sw3

the architectural association, bedford square, wc1

HIROMI FUJII
KEI'ICHI IRIE
TOYO ITO
YUTAKA SAITO
SHIN TAKAMATSU

CalArts'
17th WoRld mUSic feStiVal '91

In MemOry Of TArAnAth RAO

NORtH INDiAn MuSic StuDents & Al u M n i   fri Apr26 3pm MAiN Gallery
NOrth InDIAn MusIc with TaLa V a d Y a   8pm MoDul ar THeatre

StudEnt JAVAnese GaMel An   Sat Apr27 5pm MaIN gallery
AFRiCAn MusIc & DaNce   8pm mODular theatRe

ClAssical JaPANese MUSiC   Sun Apr28 4pm ThEaTRe II
LATIN-AmeRICAn MusiC   8pm moDulaR thEatre

ClAssical JApANesE DaNCE   fri May3 5pm tHEAtRe II
SOUtH InDIAn MusIc & DaNce   8pm moDUlar ThEAtre

StUdeNt AFRiCAn MuSIc & DaNCe   5pm mAIn gallery
BAlIneSe MuSIC & dAnce with KecaK (monkey ch a n t)Sat May48pm moDular theaTRE
New PACiFiC   11pm MAiN GalleRy

MIdDLE-EAsTem musiC & d a n c E sun May5 4pm MaIn GalleRY
JaVAnese MuSic & d a N C e   8pm moDUlar theatrE

8 pm PERFORMances: $ 6 general, $ 3 students/seniors. OTHeR PERFormances: FREE.
6 pm DI Nners: $ 6 /person.
INFORMation: (818) 3 6 2 - 2 3 1 5 or (805) 2 5 3 - 7 8 0 0.
CALiFORnia Institute of the ARTS 24700 MCBean PARkway, ValenCIA, C A Lifornia

CalArts' 17th World
Music Festival '91
Poster
Designer: Caryn Aono
Design company: CalArts Public Affairs Office
Client: CalArts
USA, 1991

LAX: The Los Angeles Exhibition 1992
Catalogue cover and map insert
Designers: Somi Kim, Whitney Lowe,
Lisa Nugent
Design company: ReVerb/Los Angeles
Photographer: Andrew Bush
Map illustrator: Edward Fella
Client: Municipal Art Gallery, Los Angeles
USA, 1992

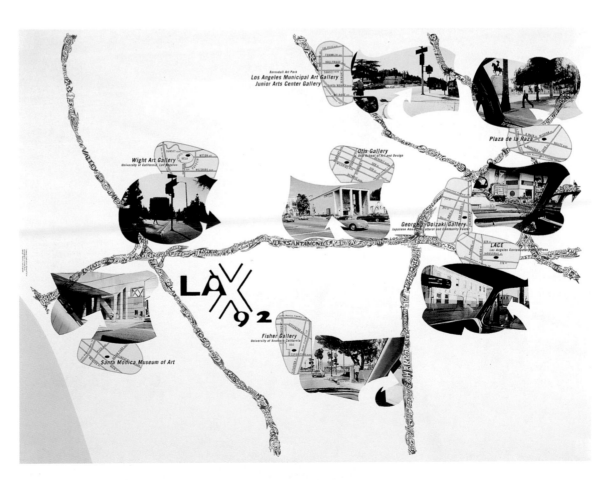

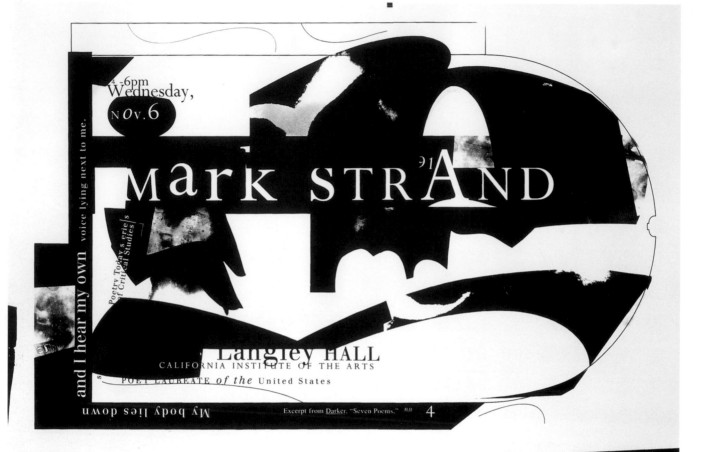

4-6pm
Wednesday,
Nov.6

voice lying next to me.

mark STRAND '91

and I hear my own

Poetry Today's erie/s
of Critical Studies

Langley HALL
CALIFORNIA INSTITUTE OF THE ARTS

POET LAUREATE *of the* United States

My body lies down

Excerpt from Darker. "Seven Poems." 4

Event posters
Designer: Gail Swanlund
Design company: CalArts Public Affairs Office
Client: CalArts
USA, 1991, 1992

The ART School IS FINALLY GIVING A Party to WELCOME Tom Lawson

on Sunday, DECEMBER 6 *snow date: the following Sunday, dec. 13TH 1992

Please Come and bring A guest DRESS WARMLY! live Music between 2 AND 3 p.m. (early supper~5 pm)

at Nancy MITCHNICK's Studio near GORMAN Compound, A LARGE SURROUNDED BY A CYCLONE fence Surrounded by Mountains on highway 138, ½ MILE PAST N2 ROAD (4 MI. OFF I-5)

Party invitation
Designer: Edward Fella
Client: CalArts
USA, 1992

Lecture announcement
Designer: Edward Fella
Client: CalArts
USA, 1992

index